Lecture Notes in Computer Science 12722

More information about this subseries at http://www.springer.com/series/7412

Bartłomiej W. Papież · Mohammad Yaqub ·
Jianbo Jiao · Ana I. L. Namburete ·
J. Alison Noble (Eds.)

Medical Image Understanding and Analysis

25th Annual Conference, MIUA 2021
Oxford, United Kingdom, July 12–14, 2021
Proceedings

Springer

Editors
Bartłomiej W. Papież (iD)
University of Oxford
Oxford, UK

Jianbo Jiao (iD)
University of Oxford
Oxford, UK

J. Alison Noble (iD)
University of Oxford
Oxford, UK

Mohammad Yaqub (iD)
Mohamed bin Zayed University
of Artificial Intelligence
Abu Dhabi, United Arab Emirates

University of Oxford
Oxford, UK

Ana I. L. Namburete (iD)
University of Oxford
Oxford, UK

ISSN 0302-9743   ISSN 1611-3349 (electronic)
Lecture Notes in Computer Science
ISBN 978-3-030-80431-2   ISBN 978-3-030-80432-9 (eBook)
https://doi.org/10.1007/978-3-030-80432-9

LNCS Sublibrary: SL6 – Image Processing, Computer Vision, Pattern Recognition, and Graphics

This Springer imprint is published by the registered company Springer Nature Switzerland AG
The registered company address is: Gewerbestrasse 11, 6330 Cham, Switzerland

Preface

We are very pleased to present the proceedings of the 25th Conference on Medical Image Understanding and Analysis (MIUA 2021), a UK-based international conference for the communication of image processing and analysis research and its application to biomedical imaging and biomedicine. The conference was supposed to be held at St Anne's College, Oxford, UK, during July 12–14; however, due to the COVID-19 pandemic, the conference was again organized as a fully virtual event. The virtual meeting was held on the same dates as originally planned, and featured presentations from the authors of all accepted papers. The conference also featured two workshops: medical image analysis using the MATLAB Deep Learning tools, organized jointly with MathWorks, and optimising deep learning with PyTorch and TensorFlow, organized with the Nvidia Corporation.

This year's edition was organized at the University of Oxford by academic members from the Medical Sciences Division: Nuffield Department of Clinical Neurosciences (https://www.ndcn.ox.ac.uk/), The Big Data Institute (https://www.bdi.ox.ac.uk/), and the Mathematical Physical and Life Sciences Division: Institute of Biomedical Engineering (http://www.ibme.ox.ac.uk/), representing Oxford's core strategic partners in medical imaging research. One of the organizers is from the Mohamed bin Zayed University of Artificial Intelligence, Abu Dhabi, UAE.

The conference was organized with sponsorship received from MathWorks (https://uk.mathworks.com/), Brainomix (https://www.brainomix.com/), the Journal of Imaging (https://www.mdpi.com/journal/jimaging), Oxford University Innovation (https://innovation.ox.ac.uk/), nVidia (https://www.nvidia.com), the Institute of Engineering and Technology (https://www.theiet.org/), and Novo Nordisk (https://www.novonordisk.co.uk/). The conference proceedings were published in partnership with Springer (https://www.springer.com).

The diverse range of topics covered in these proceedings reflect the growth in development and application of biomedical imaging. The conference proceedings feature the most recent work in the fields of: (i) image segmentation, (ii) image registration and reconstruction, (iii) biomarker detection, (iv) classification, (v) radiomics, predictive models, and quantitative imaging, (vi) biomedical simulation and modelling, and (vii) image enhancement, quality assessment, and data privacy.

Despite the COVID-19 pandemic, this year's edition of MIUA received a large number of high-quality submissions making the review process particularly competitive. In total, 77 submissions were submitted to the Conference Management Toolkit (CMT), and after an initial quality check, the papers were sent out for the peer-review process completed by the Scientific Review Committee consisting of 88 reviewers. To keep the quality of the reviews consistent with the previous editions of MIUA, the majority of the reviewers was selected from (i) a pool of previous MIUA conference reviewers and (ii) authors and co-authors papers presented at past MIUA conferences.

Each of the submissions was reviewed in a double-blind manner by at least three members of the Scientific Review Committee. Based on their recommendations, a ranking was created and the best 40 papers (52%) were accepted for presentation at the conference. Furthermore, the papers included in the proceedings were revised by the authors following feedback received from the reviewers.

Submissions were received from authors at 129 different institutes from 20 countries across 4 continents, including the UK (44), India (16), Germany (13), France (10), the USA (7), Portugal (5), Taiwan (4), and a few others. Papers were accepted from a total of 214 authors, with an average of 5 co-authors per paper.

We thank all members of the MIUA 2021 Organizing, Steering, and Scientific Review Committees. In particular, we wish to thank all who contributed greatly to the success of MIUA 2021: the authors for submitting their work, the reviewers for insightful comments improving the quality of the proceedings, the sponsors for financial support, and all participants in this year's virtual MIUA conference.

We also thank our invited keynote speakers Prof. Mihaela van der Schaar, Prof. Ben Glocker, and Prof. Aris Papageorghiou for sharing their success, knowledge, and experiences.

July 2021

Bartłomiej W. Papież
Mohammad Yaqub
Jianbo Jiao
Ana I. L. Namburete
J. Alison Noble

Organization

Program Committee Chairs

Bartłomiej W. Papież	University of Oxford, UK
Mohammad Yaqub	Mohamed bin Zayed University of Artificial Intelligence, United Arab Emirates, and University of Oxford, UK
Jianbo Jiao	University of Oxford, UK
Ana I. L. Namburete	University of Oxford, UK
J. Alison Noble	University of Oxford, UK

Steering Committee

Ke Chen	University of Liverpool, UK
Víctor González-Castro	University of León, Spain
Tryphon Lambrou	University of Lincoln, UK
Sasan Mahmoodi	University of Southampton, UK
Stephen McKenna	University of Dundee, UK
Mark Nixon	University of Southampton, UK
Nasir Rajpoot	University of Warwick, UK
Constantino Carlos Reyes-Aldasoro	City University of London, UK
Maria del C. Valdes-Hernandez	University of Edinburgh, UK
Bryan M. Williams	Lancaster University, UK
Xianghua Xie	Swansea University, UK
Xujiong Ye	University of Lincoln, UK
Yalin Zheng	University of Liverpool, UK
Reyer Zwiggelaar	Aberystwyth University, UK
Bartłomiej W. Papież	University of Oxford, UK
Ana I. L. Namburete	University of Oxford, UK
Mohammad Yaqub	Mohamed bin Zayed University of Artificial Intelligence, United Arab Emirates, and University of Oxford, UK
J. Alison Noble	University of Oxford, UK

Scientific Review Committee

Sharib Ali	University of Oxford, UK
Omar Al-Kadi	University of Jordan, Jordan
Ibrahim Almakky	Mohamed Bin Zayed University of Artificial Intelligence, United Arab Emirates
Nantheera Anantrasirichai	University of Bristol, UK

Paul Armitage — University of Sheffield, UK
John Ashburner — University College London, UK
Angelica I. Aviles-Rivero — University of Cambridge, UK
Abhirup Banerjee — University of Oxford, UK
Neslihan Bayramoglu — University of Oulu, Finland
Miguel Bernabeu — University of Edinburgh, UK
Michael Brady — University of Oxford, UK
Joshua T. Bridge — University of Liverpool, UK
Jacob Carse — University of Dundee, UK
Sevim Cengiz — Mohamed Bin Zayed University of Artificial Intelligence, United Arab Emirates

Tapabrata Chakraborty — University of Oxford, UK
Michael Chappell — University of Nottingham, UK
Ke Chen — University of Liverpool, UK
Xu Chen — University of Liverpool, UK
Zezhi Chen — Intelligent Ultrasound, UK
Wing Keung Cheung — University College London, UK
John Chiverton — University of Portsmouth, UK
Matthew J. Clarkson — University College London, UK
Timothy Cootes — University of Manchester, UK
Chengliang Dai — Imperial College London, UK
Mariia Dmitrieva — University of Oxford, UK
Qi Dou — The Chinese University of Hong Kong, Hong Kong

Bjoern Eiben — University College London, UK
Mohamed Elawady — University of Stirling, UK
Hui Fang — Loughborough University, UK
Martin Fergie — University of Manchester, UK
Alastair Gale — Loughborough University, UK
Dongxu Gao — University of Liverpool, UK
Alberto Gomez — King's College London, UK
Mara Graziani — HES-SO Valais-Wallis, Switzerland
Eyjolfur Gudmundsson — University College London, UK
Aymeric Histace — ENSEA Cergy-Pontoise, France
Yipeng Hu — University College London, UK
Mark Jenkinson — University of Oxford, UK
Bishesh Khanal — NAAMII, Nepal
Andrew King — King's College London, UK
Tryphon Lambrou — University of Lincoln, UK
Toni Lassila — University of Leeds, UK
Jonas Latz — University of Cambridge, UK
Lok Hin Lee — University of Oxford, UK
Cristian Linte — Rochester Institute of Technology, USA
Tianrui Liu — Imperial College London, UK
Dwarikanath Mahapatra — Inception Institute of Artificial Intelligence, United Arab Emirates

Contents

Image Registration, and Reconstruction

Image Segmentation

Generative Models, Biomedical Simulation and Modelling

Classification

Biomarker Detection

Exploring the Correlation Between Deep Learned and Clinical Features in Melanoma Detection

Tamal Chowdhury[1,2], Angad R.S. Bajwa[1,3], Tapabrata Chakraborti[4(✉)], Jens Rittscher[4], and Umapada Pal[1]

[1] CVPR Unit, Indian Statistical Institute, Kolkata, India
[2] National Institute of Technology (NIT), Durgapur, India
[3] National Institute of Technology (NIT), Tiruchirappalli, India
[4] IBME/BDI, Dept. of Engineering Science, University of Oxford, Oxford, UK
tapabrata.chakraborty@eng.ox.ac.uk

Abstract. Despite the recent success of deep learning methods in automated medical image analysis tasks, their acceptance in the medical community is still questionable due to the lack of explainability in their decision-making process. The highly opaque feature learning process of deep models makes it difficult to rationalize their behavior and exploit the potential bottlenecks. Hence it is crucial to verify whether these deep features correlate with the clinical features, and whether their decision-making process can be backed by conventional medical knowledge. In this work, we attempt to bridge this gap by closely examining how the raw pixel-based neural architectures associate with the clinical feature based learning algorithms at both the decision level as well as feature level. We have adopted skin lesion classification as the test case and present the insight obtained in this pilot study. Three broad kinds of raw pixel-based learning algorithms based on convolution, spatial self-attention and attention as activation were analyzed and compared with the ABCD skin lesion clinical features based learning algorithms, with qualitative and quantitative interpretations.

Keywords: Explainable artificial intelligence · Melanoma classification · Digital dermatoscopy · Attention mechanisms · Deep machine learning

1 Introduction

Among the several variants of skin lesion diseases, melanoma is the condition that puts patients' lives at risk because of its highest mortality rate, extensive class variations and complex early stage diagnosis and treatment protocol. Early detection of this cancer is linked to improved overall survival and patient health.

T. Chakraborti is funded by EPSRC SeeBiByte and UKRI DART programmes.
T. Chowdhury and A. R. S. Bajwa—First authors with equal contributions.

Manual identification and distinction of melanoma and its variants can be a challenging task and demands proper skill, expertise and experience of trained professionals. Dermatologists consider a standard set of features (popularly known as the ABCDE features) that takes into consideration the size, border irregularity, colour variation for distinguishing malignant and benign tumours. With proper segmentation boundaries these features can be extracted from images and used as inputs to machine learning algorithms for classification purposes. Also, with the recent advancements of deep learning, Convolutional neural networks (CNNs) [1] are able to differentiate the discriminative features using raw image pixels only [26]. But the decision making process of these complex networks can be opaque. Several approaches have been proposed to identify the image regions that a CNN focuses on during its decision-making process [28–30]. Van Molle et al. [3] tried to visualize the CNN learned features at the last layers and identified where these networks look for discriminative features. Young et al. [4] did a similar work towards the interpretability of these networks using GradCAM [5] and kernel SHAP [6] to show how unpredictable these models can be regarding feature selection even when displaying similar performance measures. Both these works demonstrated how pixel-based models can be misguided towards image saliency and focus on undesirable regions like skin hairs, scale marks etc. Also, attention guided CNNs were used [7,8] to solve the issue of feature localization. Though these works provide a comprehensive insight to *where* these CNNs look for unique elements in an image they are not sufficient to unveil *what* exactly these models look for and more importantly, if there is any kind of correlation with their extracted sets of features and those sought by dermatologists (the *why* question). As the consequences of a false negative can be quite severe for such diagnostic problems, it is of utmost importance to determine if the rules learned by these deep neural networks for decision making in such potential life-threatening scenarios can be backed by medical science. In this paper we have tried to address this issue by experimenting with both handcrafted ABCD features and raw pixel based features learned by a deep learning models, along with exploring if there is any correlation present between them.

2   Dataset and Methodology

2.1   Dataset: Description and Pre-processing

HAM10000 dataset [9] a benchmark data set for skin lesion classification, is used in this study. The dataset contains a total of 10015 dermoscopic images of dimensions $3 \times 450 \times 600$ distributed over 7 classes namely: melanoma (Mel, 1113 samples), melanocytic nevi (NV, 6705 samples), basal cell carcinoma (BCC, 514 samples), actinic keratosis and intraepithelial carcinoma (AKIEC, 327 samples), benign keratosis (BKL, 1099 samples), dermatofibroma (DF, 115 samples) and vascular lesions (VASC, 142 samples).

Pre-processing steps are carried out to remove the artifacts. First, the images are center cropped to extract the main lesion region and separate out the natural skin area, scale marks and shadows present due to the imaging apparatus [27]. Further, to remove the body hair and remaining scale marks, a local adaptive thresholding method is used where the threshold value of a pixel is determined by the range of intensities in its local neighbourhood. Finally, the images were enhanced using CLAHE [10] technique, and scaled using maximum pixel value. The entire dataset is divided in a 80 : 10 : 10 ratio as the training, validation and test set, respectively.

2.2 Deep Architectures

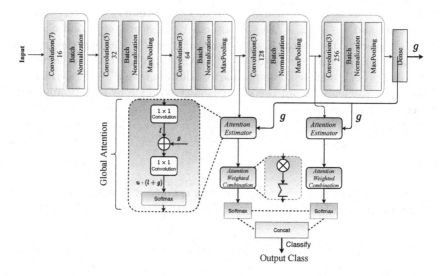

Fig. 1. An overview of the baseline CNN architecture along with the global attention modules attached to the last two convolutional blocks.

Baseline CNN: First, we designed a simple convolutional neural network with 5 convolutional blocks that servers as the baseline for other deep learning models used in this paper. Each convolutional block further consists of a convolutional layer followed by ReLU activation, max pooling (except for the first block) and batch normalization layers. Dropout layers with a dropout probability of 0.2 were used after the convolutional layers of the last two blocks to reduce overfitting. The convolutional blocks are then followed by global average pooling (GAP) [17] suitable for fine-grain classification problems and a softmax based classification layer. We used convolutional kernels with spatial extent $7, 5, 3, 3, 3$ for consecutive convolutional blocks with $16, 32, 64, 128$ and 256 feature maps, respectively.

CNN with Global Attention: Considering the importance for the network to focus on clinically relevant features, we further test the network by adding global attention modules, proposed by Jetley et al. [18] on top of the last two convolutional blocks of our baseline CNN model. The resulting network is presented in Fig. 1. which is end-to-end trainable. This method exploits the universality between local and global feature descriptors to highlight important features of an input.

First a *compatibility* score (c_i^s) is calculated using the local feature vector l_i^s and the global feature vector g as:

$$c_i^s = u(l_i^s + g) \tag{1}$$

Where, l_i^s represents the i_{th} feature map of s_{th} convolutional layer. Here $i \in \{1, 2, ldotsn\}$ and $s \in \{1, 2, \ldots S\}$ (n = number of feature maps and S = total number of layers in the network). u is the weight vector learning the universal feature sets for the relevant task. 1×1 convolutions are used to change the dimensionality of l^s to make it compatible for addition with g. Next, the attention weights a are calculated from the compatibility scores c by simply applying a softmax function function as:

$$a_i^s = \frac{exp(c_i^s)}{\sum\limits_{k=1}^{n} exp(c_k^s)} \tag{2}$$

These two operations sum up as the *attention estimator*. The final output of the attention mechanism for each block s is then calculated as:

$$g_a^s = \sum_{i=1}^{n} a_i^s \cdot l_i^s \tag{3}$$

Two such g_a^ss are concatenated as shown and a dense layer is added on top of it, to make the final prediction.

Spatial Self-attention Model: Inspired by the enormous success achieved by the transformer networks [19] in the field of natural language processing (NLP), Ramachandran et al. [20] proposed a classification framework in spatial domain entirely based on self-attention. The paper showed state-of-the-art performance on multiple popular image datasets questioning the need for convolution in vision tasks. Like convolution, the fundamental goal of self-attention also is to capture the spatial dependencies of a pixel with its neighbourhood. It does so by calculating a similarity score between a query pixel and a set of key pixels in the neighbourhood with some spatial context.

Here, we have modified the self-attention from the original work. In case of self-attention, the local neighbourhood $\mathcal{N}_k$ is denoted as the *memory block*. In contrast to the global attention modules, here the attention is calculated over a local region, which makes it flexible for using at any network depth without causing excessive computational burden. Two different matrices, query (q_{ij}) and keys (k_k), are calculated from x_{ij} and $x_{ab} \in \mathcal{N}_k$, respectively by means of linear transformations as shown in Fig. 2. Here, $q_{ij} = Qx_{ij}$ and $k_k = Kx_{ab}$ where $Q, K \in \mathbb{R}^{d_{out} \times d_{in}}$ are the query and key matrices respectively and formulated as the model parameters. Intuitively, the query represents the information (pixel) *to be matched* with a look up table containing the addresses and numerical values of a set of information represented by the keys.

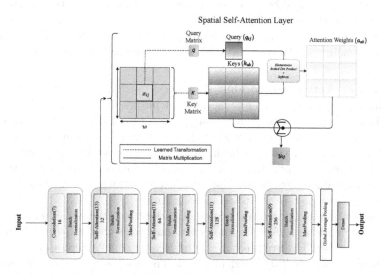

Fig. 2. An overview of the proposed self-attention model. Query (q_{ij}) and Keys (k_k) are calculated from x_{ij} and its neighbourhood x_{ab} by their linear transformations using Q and K matrices respectively.

In the original work [20], proposing the self-attention layer in spatial domain, a separate value matrix V is taken to calculate the values v which is a linear projection of the original information. Technically, in our case, keys are essentially the same thing, with the keys containing extra positional information that has been added explicitly. So, we've discarded v entirely and used k only for calculating the attention weights as well as representing the original information; that reduces the total number of model parameters. In practice, the input is divided into several parts along the depth (feature maps) and multiple convolution kernels. Multiple such query-key matrix pairs known as *heads* are used to learn distinct features from an input. Unlike [20], the single headed *normalized* attention scores in the neighbourhood $\mathcal{N}_k$ are calculated as the scaled dot product of queries and keys. Further, while calculating the attention scores, positional information is injected into the keys in the form of relative positional embedding as mentioned in [20].

$$a_{ab} = softmax_{ab} \left(\frac{q_{ij}^T \cdot k_{ab} + q_{ij}^T \cdot r_{a-i,b-j}}{\sqrt{h \times w}} \right) \tag{4}$$

Where $r_{a-i,b-j}$ is obtained by concatenating row and column offset embeddings r_{a-i} and r_{b-j} respectively, with $a-i$ and $b-j$ being row and column offsets of each element $ab \in \mathcal{N}_k$ from input x_{ij}. The attention weighted output y_{ij}^{att} corresponding to pixel x_{ij} is calculated as:

$$y_{ij}^{att} = \sum_{a,b \in \mathcal{N}_k} a_{ab} \cdot k_{ab} \tag{5}$$

Here, the same query and key matrices are used to calculate the attention outputs for each (i,j) of the input x.

Then we designed our model on the same structural backbone as our baseline CNN by replacing all the convolution layer with our self-attention layers.

Attention as Activation Model: Activation functions and attention mechanisms are typically treated as having different purposes and have evolved differently. However upon comparison, it can be seen that both the attention mechanism and the concept of activation functions give rise to non-linear adaptive gating functions [24]. To exploit both the locality of activation functions and the contextual aggregation of attention mechanisms, we use a local channel attention module, which aggregates point-wise cross-channel feature contextual information followed by sign-wise attention mechanism [24].

Our activation function resorts to point-wise convolutions [17] to realize local attention, which is a perfect fit since they map cross-channel correlations in a point-wise manner. The architecture of the local channel attention based attention activation unit is illustrated in Fig. 3. The goal is to enable the network to selectively and element-wisely activate and refine the features according to the point-wise cross-channel correlations. To reduce parameters, the attention weight $L(X) \in R^{C \times H \times W}$ is computed via a bottleneck structure.

Input(X) is first passed through a 2-D convolutional layer into a point-wise convolution of kernel size $\frac{C}{r} \times C \times 1 \times 1$ followed by batch normalization. The parameter r is the channel reduction ratio. This output is passed through a rectified linear unit (ReLU) activation function. The output of the ReLU is input to another point-wise convolution of kernel size $C \times \frac{C}{r} \times 1 \times 1$ followed by batch normalization (BN in Fig. 3). Finally, to obtain the attention weight L(X), the output is passed into a sigmoid function. It is to be noted that $L(X)$ has the same shape as the input feature maps and can thus be used to activate and highlight the subtle details in a local manner, spatially and across channels. The activated feature map X' is obtained via an element-wise multiplication with $L(X)$:

$$X' = L(X) \otimes X \tag{6}$$

In element-wise sign-attention[23], positive and negative elements receive different amounts of attention. We can represent the output from the activation function ($\mathcal{L}$) with parameters α and X'.

$$\mathcal{L}(x_i, \alpha, X') = \begin{cases} C(\alpha)R(x_i), x_i < 0 \\ X'R(x_i), x_i \geq 0 \end{cases} \tag{7}$$

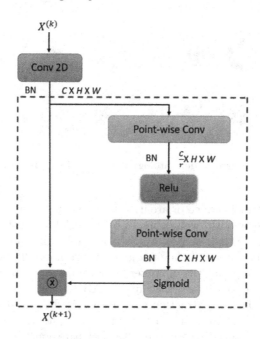

Fig. 3. Attention activation unit

Where α is a learnable parameter and $C(\cdot)$ clamps the input variable between $[0.01, 0.99]$. X' is the above calculated activated feature map. $R(X)$ is the output from standard rectified linear unit.

$$R(x_i) = \begin{cases} 0, x_i < 0 \\ x_i, x_i \geq 0 \end{cases} \tag{8}$$

This combination amplifies positive elements and suppresses negative ones. Thus, the activation function learns an element-wise residue for the activated elements with respect to ReLU which is an identity transformation, which helps mitigate gradient vanishing. We design the model based on our baseline CNN with only three blocks but with the above attentional activation function in place of ReLU.

2.3   ABCD Clinical Features and Classification

Dermatologists consider certain clinical features during the classification of malignant or benign skin lesions. A popular example is the **ABCDE** feature set [2]. In this approach, **A**symmetry, **B**order irregularity, **C**olor variation, **D**iameter and **E**volving or changing of a lesion region are taken into consideration for determining its malignancy (Ref. Fig. 4.). **A**symmetry – Melanoma is often asymmetrical, which means the shape isn't uniform. Non-cancerous moles are typically uniform and symmetrical in shape. **B**order irregularity – Melanoma often has

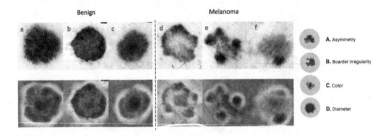

Fig. 4. ABCD features used in dermatology diagnosis of skin lesions in dermatology.

borders that aren't well defined or are irregular in shape, whereas non-cancerous moles usually have smooth, well-defined borders. Color variation – Melanoma lesions are often more than one color or shade. Moles that are benign are typically one color. Diameter – Melanoma growths are normally larger than 6mm in diameter, which is about the diameter of a standard pencil. Since we do not have time series data, we extracted the first 4 (ABCD) features for each image in our dataset. Before feature extraction, an unsupervised segmentation framework is designed based on OTSU's [11] thresholding, morphological operations, and contour detection to separate out the main lesion region from the skin. From these segmented regions, the above-mentioned set of features were extracted using several transformations and elementary mathematical functions [12,13]. Random Forest (RF) [14] and Support Vector Machines (SVM) [15] are used for the final classification with grid search [16] to find the optimal set of hyperparameters.

3   Experiments and Results

In this section, we present the experimental results, both quantitative and qualitative. First, in Table 1, we present the numerical results of the methods described in the preceding section. As evaluation metrics, we have used accuracy, AUC-ROC, precision, recall, and F1 score. Equalization sampling of minority classes was performed to tackle the problem of imbalanced dataset. All the deep learning models were trained to minimize the categorical crossentropy loss and the parameters were updated using ADAM optimizer [22].

First, we trained several traditional machine learning algorithms such as random forest and SVM, based on the ABCD features extracted as mentioned in Sect. 2.3. Grid search is used to choose the optimal set of hyperparameters and as shown in Table 1 a random forest model with 200 trees showed the best classification performance and its results are used for further comparison with the pixel-based models.

Next, multiple raw pixel-based deep learning models, as mentioned in Sect. 2.2, were trained and evaluated for the purpose of comparing and analyzing their performance with the ABCD feature based classification method, as well as to search for any feature correlation.

Table 1. Performance of different models with ABCD features and deep learned features

Method		Accuray	AUC-ROC	Precision	Recall	F1 Score
Handcrafted feature based classification	Random forest on ABCD features	75.6	76.4	75	72	73
	SVM on ABCD features	74.4	75.6	74	71	72
Raw pixel feature based classification	Baseline CNN (from scratch)	78.3	69.4	72	67	69
	CNN with global attention	82.7	75.8	78	76	77
	Self-attention based model	74.2	68.1	71	67	69
	Attention as activation based model	71.4	68.8	68	66	67

Table 2. Performance measure of the variants of spatial self-attention layer

Variations of the spatial self-attention layer	Accuracy	AUC-ROC	Precision	Recall	F1 score
Original ($k \neq v$) (Unscaled dot product) [20]	74.2	68.1	71	67	69
Proposed ($k = v$) (Unscaled dot product)	74.5	67.4	71	69	70

3.1 Quantitative Results

Table 1 shows that even with suboptimal segmentation maps ABCD features have a high discriminating power of malignancy detection and classification. Further, use of finer lesion segmentation maps obtained by a manual or supervised approach can boost the classification performance of learning algorithms utilizing these sets of features. The overall performances of the deep models are also presented.

CNNs with global attention modules showed better results compared to the baseline CNN architecture that can be explained by the improved localization and feature selection capabilities of attention modules, whereas the self-attention based model performs similar to baseline CNN. Attention as activation based model outperforms CNNs of the same size. Self attention based models face the problem that using self-attention in the initial layers of a convolutional network yields worse results compared to using the convolution stem. This problem is overcome by Attention as activation based model and is the most cost effective solution as our activation units are responsible only for activating and refining the features extracted by convolution.

Table 2 shows that the performance of our proposed variation of the spatial self-attention model is not affected when we consider keys (k) and values (v)

Table 3. Comparing alignment of deep models with ABCD features plus Random Forest

Method	Both correct (%)	ABCD features superior to raw pixel features (%)	Raw pixel features superior to ABCD features (%)	Both wrong (%)
Baseline CNN	70.2	5.4	8.1	16.3
CNN with attention	71.9	3.7	10.8	13.6
Self-attention based model	68.0	7.6	6.2	18.2
Attention as activation based model	66.5	8.9	5.5	19.1

as identical metrics. This design of the spatial self-attention layer offers similar performance at lesser parameter settings and lower computational cost.

In Fig. 5, we present the confusion matrix of stand-alone self-attention and attention as activation models on the test dataset. Both models perform well on tumor types melanoma (mel), melanocytic nevi (nv), basal cell carcinoma (bcc), actinic keratosis (akiec), intraepithelial carcinoma and benign keratosis (bkl). However, occasionally the models confuse melanocytic nevi (nv) as benign keratosis (bkl) and vascular lesions (vasc) as melanocytic nevi lesions (nv).

3.2 Alignment Between ABCD Features and Deep Learned Features

To justify the decision level correlation between deep learned features and the ABCD features, the predictions on the test dataset were analyzed using four major criteria as presented in Table 3. We find relatively higher values in the first and last columns, where both the two broad classes of algorithms either succeed or fail, clearly indicating a correlation between their sought out features. Though this is not sufficient to establish direct feature correspondence, the results point towards some clinical relevance of deep models at a decision level.

We calculate the ABCD features from the attention maps of our self attention model and the ground truth segmentation maps. We use Random Forest and Support Vector Machine models on this data. The results are presented in Table 4. These results point towards the high correspondence in the ABCD features obtained by ground truth segmentation maps (clinical features) and the

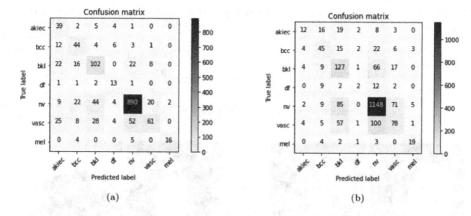

Fig. 5. Confusion matrices for (a) Self-attention, and (b) Attention as activation. Both models confuse melanocytic nevi (nv) as benign keratosis (bkl) and vascular lesions (vasc) as melanocytic nevi lesions (nv).

attention maps of self attention based model (deep learned features). We also calculate the dice score [25] to compare the similarity between the ground truth segmentation maps and the deep learning model attention maps. The average dice score calculated over all the images as presented in Table 5. These positive results help us to closely examine how the raw pixel-based neural architectures associate with the clinical feature based learning algorithms at the feature level and indicate the similarity between model predicted and ground truth lesion regions. In a few failure cases, the dice score calculated was low. We present two such examples in Fig. 6.

Table 4. Performance measure of ABCD features learned from ground truth segmentation maps and self-attention based model

Method	Accuracy	F1 score	Recall	Precision
Ground truth segmentation maps	70	60	70	62
Attention map of self-attention based model	67	54	67	45

3.3 Qualitative Results

Next, we have visually explored whether there is any direct alignment between the deep learned features and ABCD features by analyzing their global feature descriptors and segmentation maps, respectively, for a random set of test images.

Table 5. Dice score between the ground truth segmentation maps and deep architectures

Method	Dice score
Baseline CNN	79.5
Self-attention based model	73.9
Attention as activation based model	74.6

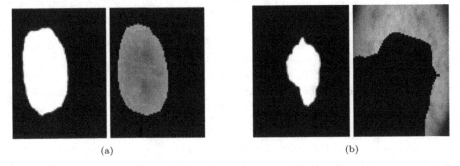

(a)   (b)

Fig. 6. (a) shows an example of correct segmentation with high dice score of 0.96. (b) is an example of incorrect output with low dice score 0.01. For each pair, we present the ground truth on the left and the model output on the right.

CAM [21] is used for visualizing the global feature descriptors for the deep classification models. From the visual results presented in Fig. 7, it is clear that the ability to precisely localize the lesion region is the most crucial quality that a model should possess. For most of the cases, whenever the attention heat maps have a satisfactory overlapping with the correct segmentation map (rows 3, 5, 6, 7) the results are correct, and whenever they differ significantly (row 2) the results are incorrect. The third column of the figure shows the activation maps of the baseline CNN to be very sparse that indicates poor localization capability, leading to many incorrect predictions. The localization capability of the attention-based models (columns 4,5 and 6) are much better than the baseline CNN that accounts for better classification results. These attention-based models have helped to pinpoint the lesion areas in the image and better addressed the fine-grain nature of the problem. Visually the localization power of the spatial self-attention and attention as activation models are quite accurate, however, in many cases, they tend to focus on the boundary regions of the image or have poor overlapping with the lesion area, which leads to incorrect predictions and suboptimal results. A good dice score suggests a descent alignment of model activations with some of the clinical features such as Asymmetry and Border irregularity, reflecting with their accuracy.

| Original Image | Segmentation Maps | Baseline CNN | CNN with Attention | Spatial self-attention | Attention as Activation |

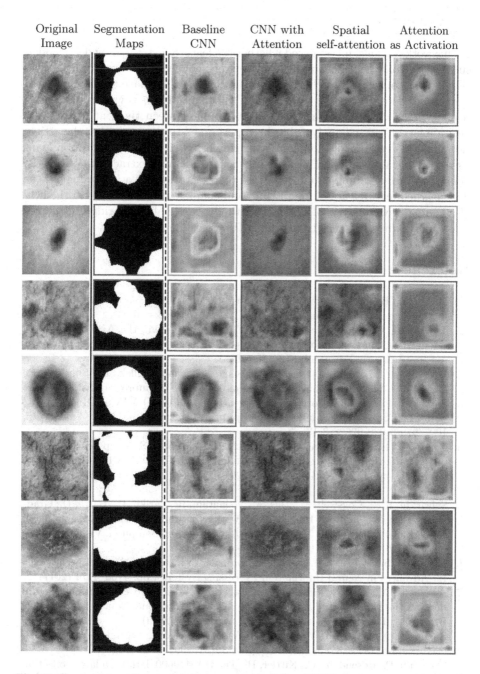

Fig. 7. Comparison of segmentation maps used for ABCD feature extraction and important regions according to deep learning models for a random set of test images. A red box around a segmentation/attention map represents incorrect prediction whereas a green box denotes correct prediction. (Color figure online)

4   Conclusion

In this work, we have investigated whether the features extracted by deep models such as convolutional networks, self-attention models and attention as activation models correlate with clinically relevant features. We have taken automated skin cancer detection as the test case and the quantitative, as well as qualitative results, point towards an underlying correlation between them at feature and decision level. A visual analysis has been performed to check whether the activation maps of deep models do possess any similarity with the segmentation maps used for clinical feature (ABCD features for skin lesion) extraction. Where the clinical features are unique and concrete representations of a lesion region, the deep learned features are more abstract and compound. However, with the help of a comparative analysis of different methods we are able to bridge the gap of trustability, when it comes to justifying their output.

References

1. LeCun, Y., Bengio, Y.: Convolutional networks for images, speech, and time series. In: The Handbook of Brain Theory and Neural Networks (1995)
2. Jensen, D., Elewski, B.E.: The ABCDEF rule: combining the ABCDE rule and the ugly duckling sign in an effort to improve patient self-screening examinations. J. Clin. Aesthetic Dermatol. **8**(2), 15 (2015)
3. Van Molle, P., De Strooper, M., Verbelen, T., Vankeirsbilck, B., Simoens, P., Dhoedt, B.: Visualizing convolutional neural networks to improve decision support for skin lesion classification. In: Stoyanov, D., et al. (eds.) MLCN/DLF/IMIMIC -2018. LNCS, vol. 11038, pp. 115–123. Springer, Cham (2018). https://doi.org/10.1007/978-3-030-02628-8_13
4. Young, K., Booth, G., Simpson, B., Dutton, R., Shrapnel, S.: Deep neural network or dermatologist? In: Suzuki, K., et al. (eds.) ML-CDS/IMIMIC -2019. LNCS, vol. 11797, pp. 48–55. Springer, Cham (2019). https://doi.org/10.1007/978-3-030-33850-3_6
5. Selvaraju, R.R., Cogswell, M., Das, A., Vedantam, R., Parikh, D., Batra, D.: Gradcam: visual explanations from deep networks via gradient-based localization. In: Proceedings of the IEEE International Conference on Computer Vision (ICCV), pp. 618–626 (2017)
6. Lundberg, S.M., Lee, S.-I.: A unified approach to interpreting model predictions. In: Advances in Neural Information Processing Systems, pp. 4765–4774 (2017)
7. Aggarwal, A., Das, N., Sreedevi, I.: Attention-guided deep convolutional neural networks for skin cancer classification. In: IEEE International Conference on Image Processing Theory, Tools and Applications, pp. 1–6 (2019)
8. Zhang, J., Xie, Y., Xia, Y., Shen, C.: Attention residual learning for skin lesion classification. IEEE Trans. Med. Imaging **38**(9), 2092–2103 (2019)
9. Tschandl, P., Rosendahl, C., Kittler, H.: The HAM10000 dataset, a large collection of multi-source dermatoscopic images of common pigmented skin lesions. Sci. Data **5**(1), 1–9 (2018)
10. Pizer, S.M., et al.: Adaptive histogram equalization and its variations. Comput. Vis. Graphics Image Process. **39**(3), 355–368 (1987)

11. Otsu, N.: A threshold selection method from gray-level histograms. IEEE Trans. Syst. Man Cybern. **9**(1), 62–66 (1979)
12. Zaqout, I.: Diagnosis of skin lesions based on dermoscopic images using image processing techniques. Pattern Recognition-Selected Methods and Applications Intech Open (2019)
13. Amaliah, B., Fatichah, C., Widyanto, M.R.: ABCD feature extraction of image dermatoscopic based on morphology analysis for melanoma skin cancer diagnosis. Jurnal Ilmu Komputer dan Informasi **3**(2), 82–90 (2010)
14. Breiman, L.: Random forests. Mach. Learn. **45**(1), 5–32 (2001)
15. Cortes, C., Vapnik, V.: Support-vector networks. Mach. Learn. **20**(3), 273–297 (1995)
16. Bergstra, J., Bardenet, R., Bengio, Y., Kégl, B.: Algorithms for hyper-parameter optimization. Advances in Neural Information Processing Systems (2011)
17. Lin, M., Chen, Q., Yan, S.: Network in network. arXiv:1312.4400 (2013)
18. Jetley, S., Lord, N.A., Lee, N., Torr, P.H.S.: Learn to pay attention. arXiv:1804.02391 (2018)
19. Vaswani, A., et al.: Attention is all you need. In: Advances in Neural Information Processing Systems (2017)
20. Ramachandran, P., Parmar, N., Vaswani, A., Bello, I., Levskaya, A., Shlens, J.: Stand-alone self-attention in vision models. arXiv:1906.05909 (2019)
21. Zhou, B., Khosla, A., Lapedriza, A., Oliva, A., Torralba, A.: Learning deep features for discriminative localization. In: Proceedings of the IEEE Conference on Computer Vision and Pattern Recognition (2016)
22. Kingma, D., Ba, J.: Adam: a method for stochastic optimization. In: International Conference on Learning Representations (2014)
23. Chen, D., Li, J., Xu, K.: AReLU: attention-based rectified linear unit. arXiv:2006.13858 (2020)
24. Dai, Y., Oehmcke, S., Gieseke, F., Wu, Y., Barnard, K.: Attention as activation. arXiv:2007.07729 (2020)
25. Eelbode, T., et al.: Optimization for medical image segmentation: theory and practice when evaluating with Dice score or Jaccard index. IEEE Trans. Med. Imaging **39**(11), 3679–3690 (2020)
26. Nida, N., Irtaza, A., Javed, A., Yousaf, M.H., Mahmood, M.T.: Melanoma lesion detection and segmentation using deep region based convolutional neural network and fuzzy C-means clustering. Int. J. Med. Inform. **124**, 37–48 (2019)
27. Bisla, D., Choromanska, A., Berman, R.S., Stein, J.A., Polsky, D.: Towards automated melanoma detection with deep learning: data purification and augmentation. In: Proceedings of the IEEE/CVF Conference on Computer Vision and Pattern Recognition Workshops (2019)
28. Adekanmi, A.A., Viriri, S.: Deep learning-based system for automatic melanoma detection. IEEE Access **8**, 7160–7172 (2019)
29. Adekanmi, A.A., Viriri, S.: Deep learning techniques for skin lesion analysis and melanoma cancer detection: a survey of state-of-the-art. Artif. Intell. Rev. **54**(2), 811–841 (2021)
30. Codella, N., et al.: Skin lesion analysis toward melanoma detection 2018: a challenge hosted by the international skin imaging collaboration (ISIC). arXiv:1902.03368 (2019)

An Efficient One-Stage Detector for Real-Time Surgical Tools Detection in Robot-Assisted Surgery

Yu Yang[1], Zijian Zhao[1(✉)], Pan Shi[1], and Sanyuan Hu[2]

[1] School of Control Science and Engineering, Shandong University, Jinan 250061, China
zhaozijian@sdu.edu.cn
[2] Department of General Surgery, First Affiliated Hospital of Shandong First Medical University, Jinan 250014, China

Abstract. Robot-assisted surgery (RAS) is a type of minimally invasive surgery which is completely different from the traditional surgery. RAS reduces surgeon's fatigue and the number of doctors participating in surgery. At the same time, it causes less pain and has a faster recovery rate. Real-time surgical tools detection is important for computer-assisted surgery because the prerequisite for controlling surgical tools is to know the location of surgical tools. In order to achieve comparable performance, most Convolutional Neural Network (CNN) employed for detecting surgical tools generate a huge number of feature maps from expensive operation, which results in redundant computation and long inference time. In this paper, we propose an efficient and novel CNN architecture which generate ghost feature maps cheaply based on intrinsic feature maps. The proposed detector is more efficient and simpler than the state-of-the-art detectors. We believe the proposed method is the first to generate ghost feature maps for detecting surgical tools. Experimental results show that the proposed method achieves 91.6% mAP on the Cholec80-locations dataset and 100% mAP on the Endovis Challenge dataset with the detection speed of 38.5 fps, and realizes real-time and accurate surgical tools detection in the Laparoscopic surgery video.

1 Introduction

Laparoscopic surgery has been widely used in clinical surgery because it causes less pain and allows a faster recovery rate [1]. With the development of computer technology, deep learning and modern medical technology, robot-assisted minimally invasive surgery (RMIS) systems have a large number of applications in clinical surgerys and achieve amazing results. Surgical tools tracking is very important for detecting the presence of surgical tools and then localizing them in the image [2]. Hence, in this paper we focus on real-time and accurate detection of 2D position of surgical tools in endoscopic surgery.

Many methods have been proposed for surgical tools detection. Most of the current methods are based on deep learning and take fully advantage of very deep networks to detect [3–5] and track [6–8] surgical tools. For example, Luis C Garca-Peraza-Herrera et al. [9] proposed to employ fully convolutional neural network (FCN) for surgical

© Springer Nature Switzerland AG 2021
B. W. Papież et al. (Eds.): MIUA 2021, LNCS 12722, pp. 18–29, 2021.
https://doi.org/10.1007/978-3-030-80432-9_2

tool tracking in 2016. Bareum Choi et al. [1] proposed to apply YOLO [10] Network to detect surgical tools in laparoscopic robot-assisted surgery in 2017. Mishra K, et al. proposed to apply improved LeNet to track surgical tools in minimally invasive eye surgery in 2017. Current state-of-the-art surgical tools detectors are composed of two components: backbone used for extracting low-level features and detection head used to predict classes and bounding boxes of surgical tools. Recently, many methods insert some layers (we named it detection neck) between backbone and detection head, these layers are usually used to collect and reuse several feature maps from different stages [11]. P. Shi et al. [12] proposed to use attention-guided CNN to detect surgical tools in 2020.

These detectors have shown excellent performance on surgical tools detection, but redundant feature map is an important characteristic of these successful methods' backbone.

Inspired by ZFNet [13] and GhostNet [14], we visualize the low-level feature maps (Fig. 1) from the first convolutional layer of Single Shot MultiBox Detector (SSD) [15] and put insight into the relationship between them. We found some feature maps which come from the same layer are similar. In other words, we used expensive operation (ordinary convolution) in order to achieve comparable performance but got similar hidden output.

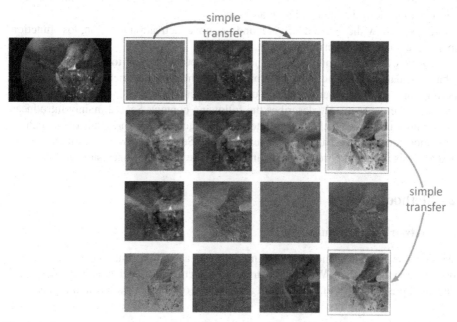

Fig. 1. We visualize 16 feature maps randomly selected from 64 feature maps generated from first convolutional layer in SSD's backbone. We marked two similar feature maps with the same color's boxes. One feature map can be simply transform from another similar feature map.

We diagnose the potential problem of this model and then propose the method to refine backbone network for performing feature extraction efficiently. Specifically, we generate some similar feature maps based on original feature maps through simple transformation instead of ordinary convolution. At the same time, we follow Cross Stage Partial Network (CSPNet) [16] to design the detection backbone and use Mish activation function [17]. In terms of detection neck, We employ architecture like U-Net [18] to fuse features from different stages and use spatial pyramid pooling (SPP) [19] after the last convolutional layer of backbone. In terms of detection head, we employ the Yolov3 [20] detection head on three different layers and decode the prediction result to get the bounding boxes and classes of surgical tools. In terms of loss function, we use Complete Intersection over Union (CIoU) Loss, [21] considering the overlapping area, the aspect ratio of bounding boxes, and the distance between center points simultaneously. This loss function can achieve better accuracy on the bounding box regression problem. We evaluated the proposed method's performance on the Cholec80-locations [3] dataset and EndoVis Challenge dataset [22], and the experimental results proved that the proposed method is better than the other three state-of-the-art methods in terms of detection accuracy and speed.

The main contributions of the proposed method can be summarized as follows:

(1) We visualize the feature maps from SSD and diagnose the potential problem – redundant feature maps and propose to employ Ghost module to generate feature maps cheaply for the fast and accurate surgical tools detection in Robot-assisted surgery. The Ghost module is designed to decrease the amount of parameters, increase the speed of surgical tools detection and maintain accuracy.

(2) We follow the spirit of CSPNet and use novel activation function, loss function to achieve comparable performance.

(3) We evaluate the proposed method for detecting surgical tools on the EndoVis Challenge dataset and Cholec80-location dataset. The proposed method shows superior performance over the other three state-of-the-art methods.

Next, the paper mainly describes the following content: Sect. 2 mainly introduces the proposed method, including architecture of detection network and tricks such as activation function, loss function for learning etc. Section 3 shows experiment results and analysis of the proposed method, and finally, we draw the conclusions in Sect. 4.

2   Methodology

2.1   Network Architecture

We describe the Ghost module, CSP module and activation function of the network backbone in this section. We then illustrate the detection head and explain the details of decoding prediction result. The overall architecture of network is shown in Fig. 2.

2.1.1   Ghost Module

Inspired by GhostNet [14], we designed an efficient backbone to solve the problem of redundant feature maps and maintain comparable accuracy.

As we can see in Fig. 3, Ghost module adopts ordinary convolution (kernel = 3×3, stride = 2) to generate some original feature maps firstly and then employ depthwise

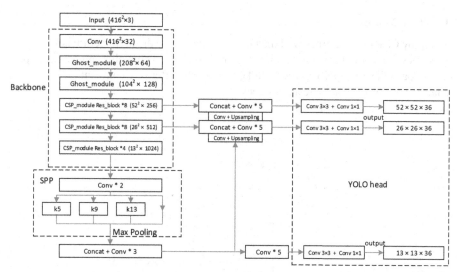

Fig. 2. The overall network structure of the proposed method including detection backbone, neck and head. The number behind the symbol * denotes the number of repeat times of corresponding module. In terms of SPP module, k5, k9, k13 denotes the size of pooling kernel, this module can increase the receptive field dramatically.

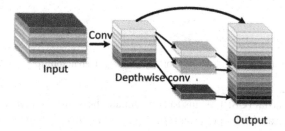

Fig. 3. An illustration of the Ghost module, and the output feature map is twice as many as input

convolution channel by channel. Lastly, we concatenate the output of depthwise convolution and intrinsic feature maps. Note that we implement the simple transfer mentioned in Fig. 1 by depthwise convolution because it is already well supported by current hardware and computational cost of GPU is much lower than the traditional convolution. We replace traditional convolution with the Ghost modules and the speed-up ratio is:

$$r_s = \frac{c \cdot k \cdot k \cdot n \cdot h \cdot w}{\frac{n}{2} \cdot c \cdot k \cdot k \cdot h \cdot w + \frac{n}{2} \cdot d \cdot d \cdot h \cdot w} = \frac{2 \cdot c \cdot k \cdot k}{c \cdot k \cdot k + d \cdot d} \approx 2 \quad (d = k = 3)$$

where n, h, w denote the dimension of output feature maps. c denotes the number of input feature map channels (always 128, 256…), k denotes the kernel size of ordinary convolution and d is the kernel size of depthwise convolution. The compression ratio of parameter is:

$$r_c = \frac{c \cdot k \cdot k \cdot n}{\frac{n}{2} \cdot c \cdot k \cdot k + \frac{n}{2} \cdot d \cdot d} \approx 2$$

2.1.2 CSP Module

We employ CSP modules followed the Ghost modules, and the architecture is shown in Fig. 4. As we can see, the input feature maps pass through a convolution layer and split into two parts. The former is directly linked to the end after one convolution layer, but the later goes through many residual blocks.

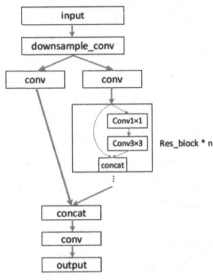

Fig. 4. An illustration of CSP module, a shortcut connection span multiple layers in this module.

Cross Stage Partial (CSP) means to concatenate the feature maps generated by two parts and then do transition operation [16]. By utilizing this strategy, we can drop computation cost significantly and reuse a large amount of gradient information for updating weights. The number of residual blocks used in CSP module is illustrated in Fig. 2. Note that the activation function in the backbone is:

$$Mish = x \times \tanh(\ln(1 + e^x))$$

2.1.3 Detection Head

The proposed one-stage surgical tools detection head is the same as YOLOv3 [20], predicting many bounding boxes and classes of these boxes at the same time. Specifically, we add convolutional layers on multi-scale feature maps and these layers decrease the number of feature map channels but maintain the size of feature map. We divide input image into $S \times S$ grid ($S = 13, 26, 52$). Each grid is responsible for predicting surgical tools whose center points falls in this grid [10]. Each grid predicts 3 default bounding boxes, so the output tensor of each scale is $S \times S \times [(4 + 1 + 7) \times 3]$ for 4 bounding box offsets relative to grid, 1 objectness confidence and 7 surgical tool classes (bipolar, grasper, hook, scissors, irrigator, clipper and specimen bag are shown in Fig. 5) in the

Cholec80-locations dataset. We predict bounding boxes at 3 different scales. 32 times downsampling feature maps have the largest receptive field and suitable for detecting large surgical tools, 8 times downsampling feature maps are suitable for detecting small surgical tools. Before obtaining the final result, objectness confidence score sorting and non-maximum suppression (NMS) [23] must be performed, because a large number of boxes are generated from the proposed method.

2.2   Loss Function for Learning

The network predicts a three dimension tensor to encode bounding box (BBox), objectness confidence, and class on three different scale feature maps. In terms of BBox, the network predict 4 coordinates for each BBox: p_x, p_y, p_w, p_h. Now we suppose the offset of the grid relative to the upper left corner is (o_x, o_y) and the default BBox width is d_w, height is d_h. The real prediction result relative to the original image is:

$$r_x = o_x + \sigma(p_x)$$

$$r_y = o_y + \sigma(p_y)$$

$$r_w = d_w \cdot e^{p_w}$$

$$r_h = d_h \cdot e^{p_h}$$

Note that the output tensor grid cell has 3 default prior boxes, we predict on 3 scale, so we have 9 prior boxes: (12×16), (19×36), (40×28), (36×75), (76×55), (72×146), (142×110), (192×243), (409×401).

BBox regression is the crucial step in surgical tools detection. In terms of loss function, we use CIoU loss [21] considering three geometric factors: overlap area, aspect ratio and central point distance. There are two reasons for not using Intersection over Union (IoU) loss function: (1) IoU is the concept of ratio, which is not sensitive to the scale of surgical tools (2) This loss function only works when the prediction BBox and ground truth BBox have overlap, and don't offer any gradient information for non-overlapping cases. An alternative loss function is Generalized IoU (GIoU) [24] loss which is proposed to solve the problem of gradient vanishing when two bounding boxes don't have overlapping area, but this loss function converges slowly and relies on IoU loss heavily. In comparison, we use the CIoU loss with faster convergence and comparable regression accuracy, the concept of CIoU is as follows:

$$CIoU = IoU - \frac{\rho^2\left(b, b^{gt}\right)}{c^2} - \alpha\gamma$$

where $\rho(b, b^{gt})$ is the distance between the prediction box's center and ground truth box's center. c denotes the diagonal distance of the smallest closed box covering the prediction box and ground truth box.

γ denotes the consistency of aspect ratio of two boxes:

$$\gamma = \frac{4}{\pi^2}\left(arctan\frac{w^{gt}}{h^{gt}} - arctan\frac{w}{h}\right)^2$$

α denotes trade-off parameter:

$$\alpha = \frac{\gamma}{(1 - IoU) + \gamma}$$

Then the CIoU loss function can be defined as:

$$L_{CIoU} = 1 - IoU + \frac{\rho^2\left(b, b^{gt}\right)}{c^2} + \alpha\gamma$$

3 Experiment and Results

3.1 Dataset

Fig. 5. Seven surgical tools in the Cholec80-locations dataset. (a) Grasper; (b) Hook; (c) Clipper; (d) Bipolar; (e) Scissors; (f) Irrigator; (g) Specimen bag.

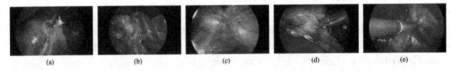

Fig. 6. Disturbing factors of Cholec80-locations dataset. (a–b) obscured by other surgical tools or organs; (c) smoke occlusion; (d) motion blurring; (e) high deformation.

We used the Cholec80-locations dataset and this dataset contains 80 videos in chole-cystectomy surgeries scene, and these surgeries were performed by 13 surgeons coming from University Hospital of Strasbourg [3]. There are 7 classes of surgical tools in this dataset and shown in Fig. 5. This dataset is downsampled to 1fps by taking the first frame from every 25 frames to reduce redundancy. The resolution of each frame is 716 × 402. We selected 11 video clips and labeled 4000 frames manually and we labeled surgical tool in a frame if half or more of the tip of surgical tool is visible. This dataset is very challenging because it comes from a real surgery scene, the surgical tools and camera moved freely in a small space. There are some disturbing factors of this dataset shown in Fig. 6: (1) obscured by other surgical tools or organs; (2) smoke occlusion; (3) motion blurring; (4) high deformation. To train the proposed model, we divided the entire dataset into three subparts: 2400 frames for training, 800 frames for validating and the leftover 800 frames for testing. The division of dataset is random. To further validate the proposed network, we evaluated the method on the EndosVis Challenge dataset containing 1083 frames [21]. Each frame has surgical tool annotations and the resolution is 720×576. This dataset was divided into two part: training set (800 frames) and test set (283 frames). The EndosVis Challenge dataset is not from the real surgery scene, the types of surgical tools and disturbing factors are few, so this dataset is much easier than the Cholec80-locations dataset.

3.2 Experiment Settings

We implemented the efficient one-stage networks based on Pytorch1.7 deep learning framework and operating system of Ubuntu 18.04 LTS. We used CUDA 10.0, CUDNN 7.3 and Python 3.8. We used a setting of multiple 1.60 GHz CPU processors, two TITAN $\times$ GPU were used to accelerate for training. The input resolution was fixed to 416×416. We used many data augment methods such as flipping, random scaling, cropping and cut mix before training. The learning rate was initialized at 0.001 for all convolutional layers, and declined by a factor of 10 at 120, 135 epoch respectively. We trained the proposed network for 150 epochs. In order to ensure the absolute fairness of comparative experiments, we downloaded source code and pre-trained weights to test the detection speed and accuracy for each model on the same environment and machine.

3.3 Results

We evaluated the proposed architecture on the Cholec80-locations dataset and EndosVis Challenge dataset. We compared the proposed method with three other state-of-the-art detection networks. We selected one-stage method (Yolov3 and Yolov4) and two-stage method (Faster R-CNN). We show the surgical tools bounding boxes of network output in Fig. 7 and show three other state-of-the-art network outputs on the same picture as a comparison. The proposed method is in red (the possible categories of tools are shown on the top-left corners of BBox), Yolov4 is in white, Yolov3 is in blue and the ground truth BBox is in green. Because of the limited drawing space, we did not show the output of Faster R-CNN, which is worse than Yolov3.

To evaluate the accuracy of the proposed network, the evaluation method we used is as follows: if the intersection over union (IoU) of the predicted BBox of network and the ground truth is bigger than 0.5, we consider that the surgical tools to be successfully detected in this frame. We use mean average precision (mAP) to measure the accuracy of detection. As shown in Table 1, mAP1 represents the accuracy on the Cholec80-locations dataset and mAP2 denotes the accuracy on EndoVis Challenge dataset. The proposed method achieved the mAP of 91.6% for the Cholec80-locations dataset and 100% for the EndosVis Challenge dataset. We can see that the accuracy of the proposed network is slightly lower than yolov4 but exceeds Yolov3 and Faster R-CNN by a significant margin on the Cholec80-locations dataset. Because the surgical tools move slowly and the posture doesn't change in EndosVis Challenge dataset, all methods can achieve good results on this dataset.

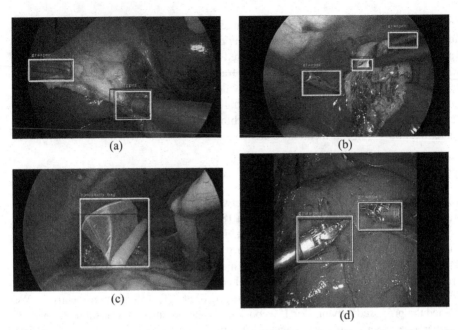

(a)   (b)

(c)

(d)

Fig. 7. Output BBoxes of proposed network. (a–c) based on the Cholec80-locations dataset, (d) based on Endovis Challenge dataset (Color figure online).

To understand the performance of the proposed network in more detail, we also assessed these methods based on a distance evaluation approach as follows: If the distance between the predicted BBox's center and the ground truth BBox's center is smaller than a threshold in the input image coordinates, we consider the surgical tool to be successfully detected in this frame.

The experimental results are shown in Fig. 8.

Table 1. Detection accuracy on the Cholec80-locations dataset and EndosVis Challenge datasets.

Method	Backbone	mAP1	mAP2
Faster R-CNN [24]	VGG16 [25]	86.37%	100%
Yolov3	DarkNet53	88.25%	100%
Yolov4	CSPDarkNet53	91.58%	100%
The proposed method	Ghost module + CSP module	91.60%	100%

We tested and compared the detection speed of the proposed method with the other three state-of-the-art detectors on two datasets respectively. The results are shown in Table 2. The proposed network has real-time performance and only need 0.026 s (over 25 FPS) per frame. The detection speed of the proposed method is faster than Yolov4 and

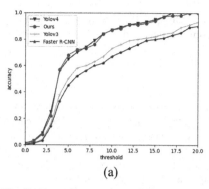
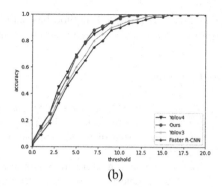

(a)   (b)

Fig. 8. Detection accuracy of distance evaluation approach. (a) based on the Cholec80-locations dataset, (b) based on Endovis Challenge dataset.

Yolov3, and is 4.6 times faster than Faster R-CNN. This demonstrates the proposed network is potential for online surgical tools detection. The considerable improvement of the proposed network is largely attributed to Ghost module to generate feature maps cheaply and uses three top feature maps to predict both confidences for multiple categories of surgical tools and bounding boxes [10].

Table 2. Speed comparison on two datasets.

Method	Backbone	Speed (seconds)	FPS
Faster R-CNN [24]	VGG16 [25]	0.12	8.3
Yolov3	DarkNet53	0.034	29.4
Yolov4	CSPDarkNet53	0.039	25.7
The proposed method	Ghost module+CSP module	0.026	38.5

The proposed method achieved comparable results in terms of accuracy and speed, but there are potential limitations about frame by frame detection. For example, if large part of surgical tool is obscured by other surgical tools or organs, the proposed method cannot locate them or locate the surgical tool but the category is wrong. Based on the above considerations, we will pay more attention to utilizing temporal information in the future, because there is richer information in the time domain.

4   Conclusion

We proposed an efficient and accurate one-stage detector for real-time surgical tools detection in robot-assisted surgery. The proposed architecture employs efficient Ghost module to generate feature maps cheaply and quickly and the proposed method utilizes novel activation function and loss function. We believe the proposed method is the first to employ Ghost module to speed up inference time and maintain accuracy for surgical tool detection. The proposed method has achieved excellent speed and accuracy to realize real-time surgical tools detection in the Laparoscopic surgery videos.

References

1. Choi, B., Jo, K., Choi, S., Choi, J.: Surgical-tools detection based on convolutional neural network in laparoscopic robot-assisted surgery. In: 2017 39th Annual International Conference of the IEEE Engineering in Medicine and Biology Society (EMBC), vol. 2017, pp. 1756–1759 (2017)
2. Liu, Y., Zhao, Z., Chang, F., Hu, S.: An anchor-free convolutional neural network for real-time surgical tool detection in robot-assisted surgery. IEEE Access. **PP**(99), 1 (2020)
3. Twinanda, A.P., Shehata, S., Mutter, D., Marescaux, J., De Mathelin, M., Padoy, N.: Endonet: a deep architecture for recognition tasks on laparoscopic videos. IEEE Trans. Med. Imaging **36**(1), 86–97 (2017)
4. Hajj, H.A., Lamard, M., Conze, P.H., Cochener, B., Quellec, G.: Monitoring tool usage in surgery videos using boosted convolutional and recurrent neural networks. Med. Image Anal. **47**, 203–218 (2018)
5. Sahu, M., Moerman, D., Mewes, P., Mountney, P., Rose, G.: Instrument state recognition and tracking for effective control of robotized laparoscopic systems. Int. J. Mech. Eng. Robot. Res. **5**(1), 33–38 (2016)
6. Du, X., et al.: Combined 2D and 3D tracking of surgical instruments for minimally invasive and robotic-assisted surgery. Int. J. Comput. Assist. Radiol. Surg. **11**(6), 1109–1119 (2016)
7. Zhao, Z., Voros, S., Chen, Z., Cheng, X.: Surgical tool tracking based on two CNNs: from coarse to fine. The J. Eng. **2019**(14), 467–472 (2019)
8. Nwoye, C.I., Mutter, D., Marescaux, J., Padoy, N.: Weakly supervised convolutional lstm approach for tool tracking in laparoscopic videos. Int. J. Comput. Assist. Radiol. Surg. **14**(6), 1059–1067 (2019)
9. Garc︮la-Peraza-Herrera, L.C., et al.: Real-time segmentation of non-rigid surgical tools based on deep learning and tracking. In: International Workshop on Computer-Assisted and Robotic Endoscopy (2016)
10. Redmon, J., Divvala, S., Girshick, R., Farhadi, A.: You only look once: unified, real-time object detection. In: Computer Vision and Pattern Recognition (2016)
11. Bochkovskiy, A., Wang, C.Y., Liao, H.Y.M.: Yolov4: optimal speed and accuracy of object detection (2020)
12. Shi, P., Zhao, Z., Hu, S., et al.: Real-time surgical tool detection in minimally invasive surgery based on attention-guided convolutional neural network. IEEE Access **PP**(99), 1–1 (2020)
13. Zeiler, M.D., Fergus, R.: Visualizing and understanding convolutional neural networks. In: European Conference on Computer Vision (2013)
14. Han, K., Wang, Y., Tian, Q., Guo, J., Xu, C.: Ghostnet: more features from cheap operations. In: 2020 IEEE/CVF Conference on Computer Vision and Pattern Recognition (CVPR) (2020)
15. Liu, W., et al.: SSD: single shot multibox detector (2016)
16. Wang, C.Y., Liao, H.Y.M., Wu, Y.H., Chen, P.Y., Yeh, I.H.: CSPNet: a new backbone that can enhance learning capability of CNN. In: 2020 IEEE/CVF Conference on Computer Vision and Pattern Recognition Workshops (CVPRW) (2020)
17. Misra, D.: Mish: a self regularized non-monotonic activation function (2019)
18. Ronneberger, O., Fischer, P., Brox, T.: U-net: convolutional networks for biomedical image segmentation (2015)
19. He, K., Zhang, X., Ren, S., Sun, J.: Spatial pyramid pooling in deep convolutional networks for visual recognition. IEEE Trans. Pattern Anal. Mach. Intell. **37**(9), 1904–1916 (2014)

20. Redmon, J., Farhadi, A.: Yolov3: an incremental improvement. arXiv e-prints (2018)
21. Zheng, Z., Wang, P., Liu, W., Li, J., Ren, D.: Distance-IoU loss: faster and better learning for bounding box regression. In: AAAI Conference on Artificial Intelligence (2020)
22. Du, X., et al.: Articulated multi-instrument 2-d pose estimation using fully convolutional networks. IEEE Trans. Med. Imaging **37**(5), 1276–1287 (2018)
23. Neubeck, A., Gool, L.J.V.: Efficient Non-Maximum Suppression (2006)
24. Rezatofighi, H., Tsoi, N., Gwak, J.Y., Sadeghian, A., Savarese, S.: Generalized intersection over union: a metric and a loss for bounding box regression. In: 2019 IEEE/CVF Conference on Computer Vision and Pattern Recognition (CVPR) (2019)
25. Ren, S., He, K., Girshick, R., Sun, J.: Faster r-CNN: towards real-time object detection with region proposal networks. IEEE Trans. Pattern Anal. Mach. Intell. **39**(6), 1137–1149 (2016)
26. Simonyan, K., Zisserman, A.: Very deep convolutional networks for large-scale image recognition. Comput. Sci. (2014)

A Comparison of Computer Vision Methods for the Combined Detection of Glaucoma, Diabetic Retinopathy and Cataracts

Jarred Orfao [ID] and Dustin van der Haar[(✉)] [ID]

University of Johannesburg, Kingsway Avenue and University Rd, Auckland Park, Johannesburg 2092, South Africa
216082337@student.uj.ac.za, dvanderhaar@uj.ac.za

Abstract. This paper focuses on the accurate, combined detection of glaucoma, diabetic retinopathy, and cataracts, all using a single computer vision pipeline. Attempts have been made in past literature; however, they mainly focus on only one of the aforementioned eye diseases. These diseases must be identified in the early stages to prevent damage progression. Three pipelines were constructed, of which 12 deep learning models and 8 Support Vector Machines (SVM) classifiers were trained. Pipeline 1 extracted Histogram of Oriented Gradients (HOG) features, and pipeline 2 extracted Grey-Level Co-occurrence Matrix (GLCM) textural features from the pre-processed images. These features were classified with either a linear or Radial Basis Function (RBF) kernel SVM. Pipeline 3 utilised various deep learning architectures for feature extraction and classification. Two models were trained for each deep learning architecture and SVM classifier, using standard RGB images (labelled as Normal). The other uses retina images with only the green channel present (labelled as Green). The Inception V3 Normal model achieved the best performance with accuracy and an F1-Score of 99.39%. The SqueezeNet Green model was the worst-performing deep learning model with accuracy and an F1-Score of 81.36% and 81.29%, respectively. Although it performed the worst, the model size is 5.03 MB compared to the 225 MB model size of the top-performing Inception V3 model. A GLCM feature selection study was performed for both the linear and RBF SVM kernels. The RBF SVM that extracted HOG features on the green-channel images performed the best out of the SVMs with accuracy and F1-Score of 76.67% and 76.48%, respectively. The green-channel extraction was more effective on the SVM classifiers than the deep learning models. The Inception V3 Normal model can be integrated with a computer-aided system to facilitate examiners in detecting diabetic retinopathy, cataracts and glaucoma.

Keywords: Glaucoma · Diabetic retinopathy · Cataract · Computer vision · Convolutional neural network · Deep learning · GLCM · HOG · CAD

1 Introduction

Cataracts, diabetic retinopathy (DR) and glaucoma are prevalent eye diseases that lead to vision loss if treatment is not sought after. Currently, a complete eye examination

© Springer Nature Switzerland AG 2021
B. W. Papież et al. (Eds.): MIUA 2021, LNCS 12722, pp. 30–42, 2021.
https://doi.org/10.1007/978-3-030-80432-9_3

conducted by a medical professional is used to detect all these diseases. However, various factors may prevent individuals from obtaining an eye examination. The longer the delay of diagnosis, the more damage a disease can cause, leading to partial vision loss, or in severe cases, complete vision loss. A potential solution to these factors can be achieved using computer vision. In this paper, we compare computer vision methods for the automated detection of DR, Cataracts and Glaucoma using retina fundus images.

The research problem and environment will be discussed in Sect. 2. A literature review regarding computer vision and eye-disease detection will be discussed in Sect. 3. Section 4 will discuss the experiment setup. The implemented models are illustrated in Sect. 5. Section 6 highlights the results achieved, followed by a result analysis in Sect. 7. Finally, the conclusion will be provided in Sect. 8.

2 Research Problem and Environment

Blindness and impaired vision are global challenges. According to [1], 285 million people in 2010 were estimated by the World Health Organisation to have some form of visual impairment. Cataracts, DR and glaucoma, are highly ranked eye diseases known to cause blindness or vision impairment.

Glaucoma is a group of conditions that damage the optic nerve. The damage is often caused by abnormally high eye pressure [2]. Fast fluid production or slow fluid drainage are the culprits for the increased eye pressure. As the optic nerve deteriorates, irreparable blind spots unwittingly develop; the risk of developing glaucoma increases with age.

DR is an impediment attributed to diabetes [2]. Exorbitant blood-sugar levels produce occlusions in the retina blood vessels. New blood vessels may develop to nourish the retina, but they are feeble and leak easily. There are no symptoms for this disease, and its progression may result in visual impairments such as blurred vision, dark spots and vision loss. The risk of developing DR increases with the duration an individual has diabetes.

A cataract is an eye disease that obscures the usually clear eye lens [2]. Ageing and eye-tissue damage are factors that spur cataract development. As the disease progresses, the cataract enlarges and distorts the light passing through it. As a result, vision may be blurred or dim. An eye operation can be performed to remove a cataract and restore an individual's vision.

Further damage progression can be prevented if these diseases are detected and treatment is sought after. However, access to medical care by some individuals is limited. According to [3, 4], 50% of blindness cases are related to cataracts, 2.5% to 5.3% of people older than 40 years are affected by glaucoma, and between 5% and 10% of people have DR in South Africa. Early detection of these diseases will have a significant impact on developing countries.

Thus, this research paper aims to compare computer vision methods that focus on detecting glaucoma, diabetic retinopathy, and cataracts in a timely and affordable manner using only one classifier. This research paper aims to answer the following research question: Can computer vision accurately distinguish between a cataract, glaucoma, diabetic retinopathy, and standard retina image with a single pipeline? The next section will discuss similar work regarding computer vision and disease detection.

3   Literature Review

Computer vision is frequently used for tasks such as image classification, interpretation, and object detection. The remainder of this section will discuss work that has utilised computer vision for disease classification.

Acharya et al. [5] present a method for the automated diagnosis of glaucoma using textural and higher-order spectral features. Sixty fundus images were obtained from the Kasturba Medical College in India. The subjects range from 20 to 70 years of age. The images were captured with a fundus camera and are 560×720 pixels in resolution. Pre-processing involved performing histogram equalisation to improve the image contrast. Higher-order spectra (HOS) features were extracted using a radon transform, which contains the amplitude and phase information for a given signal. Textural features were extracted using a grey-level co-occurrence matrix. These two features combined produced a low p-value which means there is more chance that these features have different values for normal and abnormal classes that can better discriminate the two classes. Four classifiers were investigated: Support Vector Machine (SVM), Sequential Minimal Optimization (SMO), random forest and Naïve Bayesian (NB). A Random-forest classifier, combined with z-score normalisation and feature-selection methods, performed the best with an accuracy of 91.7%. The advantages of this approach are high accuracy and being able to incorporate this approach into existing infrastructure.

Guo et al. [6] present a computer-aided system for cataract classification and grading. Feature extraction was performed using sketch-based methods and wavelet transforms. The feature extraction utilised the intuitive approach that a fundus image with a cataract will show less detail of the optic nerve. The wavelet transforms provide high contrast between the background and blood vessels. The Haar wavelet transform was used for computing efficiency, and the number of coefficients in the amplitude regions were used as features for the classification and grading. A larger number of different amplitude regions correspond to a higher frequency component, indicating less of a cataract. The features were classified into a non-cataract or cataract class and graded as mild, moderate or severe using the Multi-Fisher algorithm. The dataset contained 445 fundus images of varying intensity of cataract grading. The wavelet transforms approach achieved 90.9% and 77.1% accuracy for cataract classification and grading, respectively. The sketch-based method approach achieved 86.1% and 74% accuracy for cataract classification and grading, respectively. The advantage of this approach is the intuitive way to classify and grade cataracts. However, details regarding the pre-processing of the images were omitted along with the dataset characteristics.

Singh et al. [7] present a method for the automated detection and grading of diabetic retinopathy. The dataset was created by combining images from the Diabetic Retinopathy Detection dataset from Kaggle and the Messidor dataset. The input image was resized to 150×150 pixels during pre-processing for both models. The researchers developed a custom convolutional neural network architecture for the binary classification of diabetic retinopathy. This model was trained with 9600 images and tested with 4080 images for 100 epoch and yielded an accuracy of 71%. Another custom architecture was implemented for the grading of diabetic retinopathy. This model was trained on a set of 24000 images and tested on a set of 4800 images and yielded an accuracy of 56% after 100 epochs. The advantage of this approach is that the models can make predictions in real-time. Furthermore, a Convolutional Neural Network (CNN) can learn the features associated with abnormal and rare cases. The disadvantage of this approach is that both models achieved a low accuracy.

Panse et al. [8] produced a method for retinal disease diagnosis using image mining. The datasets used were DIARETDB0, DIARETDB1 and DRIONS DB resulting in 329 colour fundus images in total. The retina image is resized to a standard size during pre-processing and converted from Red Green Blue (RGB) to YCbCr format to increase the variation in pigment colours. The yellow channel was separated, and Low pass filtering was applied to remove the noise pigments in the image. Canny edge detection was used to detect the edges in the image. Binarisation segmentation and skeletonisation was applied to detect the small, round, dark-red and yellow dots in the image that correspond to exudates, haemorrhages and microaneurysms. A discrete cosine transform was applied to extract features classified as either normal, diabetic retinopathy or glaucoma, using a K-Nearest Neighbour (KNN) classifier. This approach achieved an accuracy of 97.36%. The advantage of this approach is the high accuracy that was achieved. Furthermore, both glaucoma and diabetic retinopathy could be detected. The disadvantage of this approach is that the accuracy decreases as the number of disease classes to be classified increases. Thus, this approach will only be accurate for a small number of diseases.

4 Experiment Setup

This section will discuss the data sampling, data population and technical aspects utilised in the model implementation.

For the experiment, the publicly available Ocular Disease Intelligent Recognition (ODIR) structure ophthalmic database [9] was utilised. This database consists of left and right human retina fundus RGB images of 5000 patients. Due to the dataset attempting to mimic actual patient data collected from hospitals, the images vary in resolution. Trained human readers annotated the images with quality control management. The images are classified into eight categories. Of the eight categories, the experiment utilises four, namely: Normal, Diabetic Retinopathy, Glaucoma and Cataract, consisting of 2819, 931, 200 and 262 images, respectively. From each category, 150 images were randomly selected to prevent a class imbalance from occurring. Dataset augmentation was performed by randomly rotating and zooming in on each of the images, which increased the dataset to 1649 images in each category and 6596 images, all 224×224 pixels. Furthermore, each category was randomly divided into 80% training images and 20% testing images. This produced 1319 training images and 330 testing images for each category. This split was performed to ensure that fair model metrics were generated to evaluate each model's performance accurately. The data population is composed of both male and females that are aged between 1 and 91 years.

In terms of the technical aspects of the implementation, the Scikit Learn library was used for pre-processing, model selection, the SVM implementation and model metrics. The Pandas library was used for reading and writing CSV files. The NumPy library was used for matrix manipulation for extracting GLCM features, and the skimage library was used for HOG feature extraction. Tensorflow was used for implementing the various CNN architectures. Early stopping and model checkpoints from the Keras Callbacks library in TensorFlow were used to monitor the training and save the top-performing model.

5   Model

Three pipelines have been implemented and are depicted in Fig. 1. The various components of the pipelines and the algorithms used to implement them will be discussed.

5.1   Image Acquisition

In this step, the user provides the image that will be classified. The implemented mobile application allows a user to select a picture from an Android or iOS mobile device's gallery or capture an image using the device's camera with a retina camera attachment. The acquired image is sent to a service that hosts the models.

5.2   The Composition of the Trained Models and Classifiers

In total, twelve deep learning models and eight SVM classifiers were trained. The composition of the various SVM classifiers is depicted in Table 1. Two versions of each model were trained to compare the effect of the green channel extraction with the standard RGB images. The models trained using the resized Red Green Blue (RGB) images

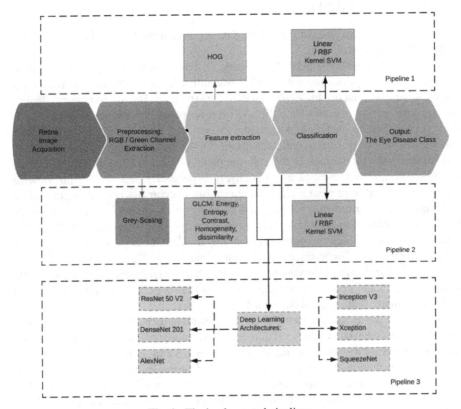

Fig. 1. The implemented pipelines.

are labelled as "Normal", and the other that was trained using the images consisting of only the green channel is labelled as "Green".

5.3 Pre-processing

In this step, noise is reduced, and features are enhanced in the acquired image. The acquired image is resized to 224 × 224 pixels to minimise memory overhead. The Green trained models pre-process the selected image by extracting only the green channel from the RGB image. Xu et al. [10] state that the green channel is less sensitive to noise, and Imran et al. [11] show the green channel achieves better feature highlighting. The effects of the green channel extraction can be viewed in Fig. 2.

Pipeline 2 grayscaled the RGB image by computing the RGB value of a pixel and dividing it by 3 to prepare the feature extraction algorithm that follows. The SVM pipelines applied feature-scaling to the extracted textural features. The deep learning pipelines normalised the pixels to be in a range of [−1, 1].

Cataract   Diabetic   Glaucoma   Normal
            Retinopathy

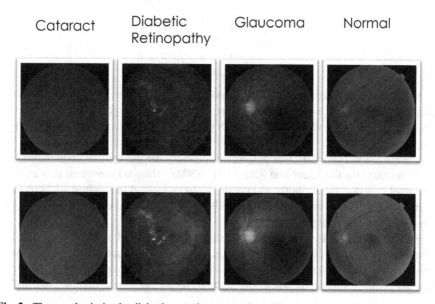

Fig. 2. The emphasis in detail is due to the green channel extraction on a cataract, diabetic retinopathy, glaucoma, and normal image. (Color figure online)

5.4 Feature Extraction and Classification

In this step, features that can classify an image are extracted. An SVM aims to find a hyperplane in an n-dimensional feature space such that the distance between the data points of all the classes is maximised [12]. How the vectors are mapped is controlled by the kernel function. The kernel can be linear, Gaussian, polynomic and many other types. The implementation utilises a linear kernel SVM and a radial basis function (RBF) kernel SVM separately.

Pipeline 1 extracts HOG Features [13] from the processed image. Pipeline 2 utilises a Gray-Level Co-occurrence Matrix (GLCM) to extract textural features from the processed image. Texture can be captured from images through the size of the area where the contrast change occurs; the contrast of the pixels; and the direction or lack of direction of this change. The GLCM was used to derive energy, contrast, entropy, dissimilarity, and homogeneity [14] features for pipeline 2. The following equations depict the calculation of the textural features using the GLCM:

$$\text{Contrast} = \sum_{x,y=0}^{N-1} Pxy(x - y)^2 \tag{1}$$

$$\text{Dissimilarity} = \sum_{x,y=0}^{N-1} Px, y|x - y| \tag{2}$$

$$\text{Homogeneity} = \sum_{x,y=0}^{N-1} \frac{Pxy}{1 + (x - y)^2} \tag{3}$$

$$\text{Energy} = \sum_{x,y=0}^{N-1} (Pxy)^2 \tag{4}$$

$$\text{Entropy} = \sum_{x,y=0}^{N-1} -\ln(Pxy)Pxy \tag{5}$$

In all the above equations, Pxy is located at row x and column y of the normalised, symmetrical GLCM. N is the number of grey levels in the image. A Linear and RBF kernel SVMs are used independently in both pipelines 1 and 2 to classify the extracted features.

A feature selection study was performed to determine the optimal combination of GLCM features for the linear and RBF kernel SVMs. The study revealed that a combination of contrast, dissimilarity and energy features yielded the highest F1-Score for the linear SVM. However, the RBF kernel SVM yielded the highest F1-Score when all of the GLCM features were combined.

Pipeline 3 utilises deep learning architectures to extract and classify features. One of the deep learning architectures used in pipeline 3 is the SqueezeNet architecture. SqueezeNet is renowned for being three times faster and 50 times smaller than AlexNet while maintaining similar accuracy. The SqueezeNet architecture is depicted in Fig. 3.

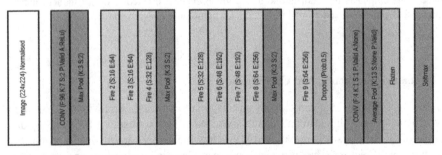

Fig. 3. The SqueezeNet architecture implemented in pipeline 3

5.5 Displaying the Results

When the classification is complete, the service hosting the models sends the category that the acquired image was classified as back to the mobile application. The application then displays the results to the user.

6 Results

The performance of the pipelines was evaluated using a confusion matrix [15] which was constructed by comparing the classifier's prediction with the data sample's ground-truth label. This result fell into one of four categories, namely true positive (TP), true negative (TN), false positive (FP) and false-negative (FN).

Table 1. The composition of SVM classifiers.

Classifier name	Pre-processing	Feature extraction	Kernel
RBF SVM green (HOG)	Green channel extraction	HOG features	RBF
RBF SVM (HOG)	Normal RGB	HOG features	RBF
Linear SVM (HOG)	Normal RGB	HOG features	Linear
Linear SVM green (HOG)	Green channel extraction	HOG features	Linear
RBF SVM (GLCM)	Normal RGB grey-scaled	GLCM	RBF
RBF SVM green (GLCM)	Green channel extraction	GLCM	RBF
Linear SVM green (GLCM)	Green channel extraction	GLCM	Linear
Linear SVM (GLCM)	Normal RGB grey-scaled	GLCM	Linear

7 Result Analysis

The performance of the deep learning models and SVM classifiers are depicted in Table 2. The Inception V3 Normal model achieved the best performance with a weighted F1-Score of 99.39%. Despite the SqueezeNet models performing the worst out of the deep learning models, its model size was 5.03 MB compared to the 225MB model size of the best performing Inception V3 model. The RBF SVM that extracted HOG features from the green channel images was the top-performing SVM classifier with an F1-Score of 76.48%. Although there is a minor gap of 0.52% between the RBF SVM that extracted HOG features from the RGB images, it is evident that the green channel extraction was more effective than a standard RGB image. The effectiveness of the green channel extraction is seen in the F1-Score difference of 3.09% between the two linear SVM classifiers that utilise GLCM features. The green channel extraction was not effective for the deep learning models as it performed poorly compared to the RGB

Fig. 4. Ground truth label is glaucoma. ResNet Normal predicted it to be a cataract.

Fig. 5. Ground truth label is a cataract. ResNet green predicted it to be glaucoma. (Color figure online)

version of the model. Thus, the green channel extraction is better suited for the SVM pipelines.

Based on the results achieved, it is evident that deep learning models are better suited to this problem than SVM classifiers. When analysing the misclassifications, there was evidence of more than one disease present in the retina images, as depicted in Fig. 4 and 5. In both figures, the retina is primarily opaque, which indicates the image could contain a cataract retina. Furthermore, the size of the optic disc present suggests that the retina image could contain glaucoma. These multi-disease retina images have had a massive impact on the metrics for each pipeline, especially in pipelines 1 and 2. The SVMs utilise textural features obtained from the GLCM matrix and HOG features for training and testing. If there is more than one disease present in the retina, the GLCM matrix will produce textural features that encompass all the present diseases. However, these features will affect the placement of the support vectors, which ultimately produces an unclear decision boundary. Thus, the dataset needs to be re-evaluated to omit the multi-class retina images.

Even with the dataset problem, the Inception V3 Normal produced comparable results to the methods highlighted in the literature review section as depicted in Table 4. The metrics for the classification of each eye-disease class by the Inception V3 Normal is displayed in Table 3. The Inception V3 Normal model achieved an accuracy of 99.62%, 99.85% and 99.77% for glaucoma, cataract and diabetic retinopathy classification, respectively.

Table 2. A comparison of the validation results achieved by the pipelines. "Normal" means standard RGB images, and "Green" means images containing only the green channel. Deep learning architectures used: Inception V3 [17], Xception [18], AlexNet [19], DenseNet [20], ResNet 50 V2 [21], SqueezeNet [22].

Models	Accuracy %	Weighted precision %	Weighted recall %	Weighted F1-score %
Inception V3 normal	**99.39**	**99.40**	**99.39**	**99.39**
Xception normal	99.17	99.17	99.17	99.17
Inception V3 green	99.02	99.02	99.02	99.02
AlexNet normal	98.86	98.87	98.86	98.86
DenseNet normal	98.86	98.87	98.86	98.86
AlexNet green	97.58	97.57	97.58	97.57
DenseNet green	97.50	97.50	97.50	97.50
ResNet 50 V2 normal	97.42	97.42	97.42	97.42
Xception green	97.00	98.00	97.00	97.00
ResNet 50 V2 green	95.00	95.05	95.00	94.99
SqueezeNet normal	84.47	84.57	84.47	84.45
SqueezeNet green	81.36	82.34	81.36	81.29
RBF SVM green (HOG)	76.67	76.45	76.67	76.48
RBF SVM (HOG)	76.14	76.13	76.14	75.96
Linear SVM (HOG)	60.15	59.66	60.15	59.79
Linear SVM green (HOG)	60.23	59.43	60.23	59.61
RBF SVM (GLCM)	53.71	54.51	53.71	53.63
RBF SVM green (GLCM)	53.64	53.91	53.64	52.97
Linear SVM green (GLCM)	51.29	50.44	51.29	50.42
Linear SVM (GLCM)	50.76	51.10	50.76	47.33

Memari et al. [16] highlight the need for computer-aided systems (CADs) for screening individuals that potentially have diabetes. Manual screening by an examiner is often subjective and dependent on their level of experience. Thus, CAD systems are needed to eliminate human bias and aid in isolated or over-populated screening areas. With the results achieved by the Inception V3 Normal model, we could easily integrate it with a CAD system to facilitate the screening process for diabetic retinopathy, cataracts and glaucoma detection.

Table 3. The metrics achieved by the inception V3 normal model for each disease class.

Metric	Glaucoma	Diabetic retinopathy	Cataract	Normal
Accuracy %	99.62	99.77	99.85	99.55
Precision %	99.09	99.40	99.40	99.69
Recall %	99.39	99.70	100.00	98.48
F1-Score %	99.24	99.55	99.70	99.09

Table 4. The accuracy achieved by researchers mentioned in the literature review section. Classification accuracy is the binary accuracy achieved by classifying the image as the respective disease or not the respective disease.

Research articles	Dataset used	Sample size	Classification accuracy %
Acharya et al. [5] automated diagnosis of glaucoma	A primary dataset collected from the Kasturba Medical College in India	60	91.7
Guo et al. [6] cataract classification and grading	A primary dataset that is not publicly available	445	90.90
Singh et al. [7] diabetic retinopathy detection and grading	A secondary dataset composed of selected images from the diabetic retinopathy detection dataset and the Messidor dataset	13680	71.00
Panse et al. [8] glaucoma and diabetic retinopathy detection	A secondary dataset composed of DIARETDB0, DIARETDB1 and DRIONS DB	329	97.36
Proposed inception V3 normal model for the detection of cataracts, glaucoma and diabetic retinopathy	A secondary dataset composed of selected images from the ODIR structure ophthalmic database	6596	**99.39**

8   Conclusion

This paper compared the combined detection of glaucoma, cataract and diabetic retinopathy using computer vision. Three pipelines were constructed of which, 12 deep learning models and 8 SVM classifiers were trained. Pipelines 1 and 2 extracted HOG and GLCM textural features from the pre-processed image, respectively. These features were then classified with either a linear or RBF kernel SVM. Pipeline 3 utilised various deep learning architectures for feature extraction and classification. The effect of using images with only the green channel being present was investigated and was shown to be only effective with the support vector machine pipelines compared to the standard RGB images. The Inception V3 Normal model achieved the best performance with accuracy and an F1-Score of 99.39%. The Xception Normal model came second with accuracy and an F1-Score of 99.17%. The SqueezeNet Green model was the worst-performing deep learning model with accuracy and an F1-Score of 81.36% and 81.29%, respectively. Although it performed the worst, the model size is 5.03 MB compared to the 225 MB model size of the top-performing Inception V3 model. A GLCM feature selection study was performed for both the linear and RBF SVM kernels. The linear kernel achieved its highest F1-Score of 47.33% and an accuracy of 50.76% when only contrast, dissimilarity, and energy features were used. The RBF kernel achieved its highest F1-Score of 53.63% when all the GLCM features were used. The SVM classifiers that extracted HOG features outperformed the SVMs that extracted GLCM features. The RBF SVM extracted HOG features on the green-channel images performed the best out of the SVMs with accuracy and F1-Score of 76.67% and 76.48%, respectively. The Linear SVM that utilised GLCM features extracted from the standard RGB images performed the worst out of all deep learning models and SVM classifiers with accuracy and an F1-Score of 50.76% and 47.33%, respectively.

The ODIR dataset was used to train and test the various models and classifiers. However, retina images labelled as only containing one eye disease showed evidence of containing multiple eye diseases. This dataset issue has impacted the results achieved, especially by the pipelines utilising an SVM. The dataset needs to be re-evaluated to omit the multi-disease images present in a single retina image. Despite the dataset issue, Inception V3 produced comparable results to similar research focused on glaucoma, cataract, and diabetic retinopathy detection. The performance achieved indicates that it is possible to detect glaucoma, cataract and diabetic retinopathy using a single pipeline.

References

1. Ettore Giardini, M.: The portable eye examination kit: mobile phones can screen for eye disease in low-resource settings. IEEE Pulse **6**, 15–17 (2015)
2. Kazi, A., Ajmera, M., Sukhija, P., Devadkar, K.: Processing retinal images to discover diseases. In: 2018 International Conference on Current Trends towards Converging Technologies (ICCTCT) (2018)
3. Labuschagne, M.: Glaucoma: what should the general practitioner know? S. Afr. Fam. Pract. **55**, 134–141 (2013)
4. Thomas, R., et al.: Incidence and progression of diabetic retinopathy within a private diabetes mellitus clinic in South Africa. J. Endocrinol. Metab. Diab. S. Afr. **20**, 127–133 (2015)
5. Acharya, U., Dua, S., Du, X., Sree, S.V., Chua, C.: Automated diagnosis of glaucoma using texture and higher order spectra features. IEEE Trans. Inf. Technol. Biomed. **15**, 449–455 (2011)
6. Guo, L., Yang, J., Peng, L., Li, J., Liang, Q.: A computer-aided healthcare system for cataract classification and grading based on fundus image analysis. Comput. Ind. **69**, 72–80 (2015)

7. Singh, T., Bharali, P., Bhuyan, C.: Automated detection of diabetic retinopathy. In: 2019 Second International Conference on Advanced Computational and Communication Paradigms (ICACCP) (2019)
8. Panse, N., Ghorpade, T., Jethani, V.: Retinal fundus diseases diagnosis using image mining. In: 2015 International Conference on Computer, Communication and Control (IC4) (2015)
9. Larxel: Ocular Disease Recognition. https://www.kaggle.com/andrewmvd/ocular-disease-recognition-odir5k
10. Xu, Y., MacGillivray, T., Trucco, E.: Computational Retinal Image Analysis. Elsevier Science & Technology (2019)
11. Imran, A., Li, J., Pei, Y., Akhtar, F., Yang, J., Wang, Q.: Cataract detection and grading with retinal images using SOM-RBF neural network. In: 2019 IEEE Symposium Series on Computational Intelligence (SSCI) (2019)
12. Jothi, A.K., Mohan, P.: A comparison between KNN and SVM for breast cancer diagnosis using GLCM shape and LBP features. In: 2020 Third International Conference on Smart Systems and Inventive Technology (ICSSIT) (2020)
13. Dalal, N., Triggs, B.: Histograms of oriented gradients for human detection. In: 2005 IEEE Computer Society Conference on Computer Vision and Pattern Recognition (CVPR 2005) (2005)
14. Haralick, R., Shanmugam, K., Dinstein, I.: Textural features for image classification. IEEE Trans. Syst. Man Cybern. **3**, 610–621 (1973)
15. Patro, V., Patra, M.: A novel approach to compute confusion matrix for classification of n-class attributes with feature selection. Trans. Mach. Learn. Artif. Intell. **3**(2), 52 (2015)
16. Memari, N., Abdollahi, S., Ganzagh, M., Moghbel, M.: Computer-assisted diagnosis (CAD) system for diabetic retinopathy screening using color fundus images using deep learning. In: 2020 IEEE Student Conference on Research and Development (SCOReD) (2020)
17. Szegedy, C., Vanhoucke, V., Ioffe, S., Shlens, J., Wojna, Z.: Rethinking the inception architecture for computer vision. In: 2016 IEEE Conference on Computer Vision and Pattern Recognition (CVPR) (2016)
18. Chollet, F.: Xception: deep learning with depthwise separable convolutions. In: 2017 IEEE Conference on Computer Vision and Pattern Recognition (CVPR) (2017)
19. Krizhevsky, A., Sutskever, I., Hinton, G.: ImageNet classification with deep convolutional neural networks. Commun. ACM **60**, 84–90 (2017)
20. Huang, G., Liu, Z., Van Der Maaten, L., Weinberger, K.: Densely connected convolutional networks. In: 2017 IEEE Conference on Computer Vision and Pattern Recognition (CVPR) (2017)
21. He, K., Zhang, X., Ren, S., Sun, J.: Identity mappings in deep residual networks. In: Leibe, B., Matas, J., Sebe, N., Welling, M. (eds.) ECCV 2016. LNCS, vol. 9908, pp. 630–645. Springer, Cham (2016). https://doi.org/10.1007/978-3-319-46493-0_38
22. Iandola, F., Han, S., Moskewicz, M., Ashraf, K., Dally, W., Keutzer, K.: SqueezeNet: AlexNet-level accuracy with 50× fewer parameters and <0.5MB model size, http://arxiv.org/abs/1602.07360

Prostate Cancer Detection Using Image-Based Features in Dynamic Contrast Enhanced MRI

Liping Wang[1(✉)], Yuanjie Zheng[1], Andrik Rampun[2], and Reyer Zwiggelaar[3]

[1] School of Information Science and Engineering, Shandong Normal University, Jinan, China
`{aberwlp,yjzheng}@sdnu.edu.cn`
[2] School of Medicine, Department of Infection, Immunity and Cardiovascular Disease, Sheffield University, Sheffield, UK
`y.rampun@sheffield.ac.uk`
[3] Department of Computer Science, Aberystwyth University, Aberystwyth, UK
`rrz@aber.ac.uk`

Abstract. Dynamic Contrast Enhanced Magnetic Resonance Imaging (DCE MRI) provides valuable information in prostate cancer detection. Existing computer-aided detection methods focus on estimating the DCE curves as pharmacokinetic models and directly calculating the perfusion-related measurements from the DCE signals. Substantial image content contained in DCE MRI series, which captures the spatio-temporal pattern receives less attention. This work aims to investigate the performance of the image-based features extracted from DCE MRI on prostate cancer detection. Various image-based features are extracted from DCE MRI series. Their performance on prostate cancer detection is compared with features extracted from the pharmacokinetic models and the perfusion-related measurements. Features are concatenated and feature selection is applied to reduce the feature dimensionality and improve cancer detection performance. Evaluation is based on a publicly available dataset. Using image-based features outperforms using either the features extracted from the pharmacokinetic models or the perfusion-related measurements. By applying feature selection to the aggregation of all features, the performance of prostate cancer detection achieves 0.821, for the area under the receiver operating characteristics curve. This study demonstrates that compared with the commonly used pharmacokinetic models and the perfusion-related features, image-based features provide an additional contribution to prostate cancer detection and can potentially be used as an alternative approach to model DCE MRI.

Keywords: Prostate cancer detection · Image-based features · DCE MRI

1 Introduction

Prostate cancer is the fourth most common cancer at a worldwide scale [1]. In the United States, aside from skin cancer, it remains the most frequently

B. W. Papież et al. (Eds.): MIUA 2021, LNCS 12722, pp. 43–55, 2021.
https://doi.org/10.1007/978-3-030-80432-9_4

diagnosed cancer affecting men and the second leading cause of death from cancer among men [2]. Prostate-specific antigen (PSA) testing and transrectal ultrasound (TRUS) guided biopsy have been widely used for prostate cancer screening. However, these techniques suffer from low accuracy, invasiveness and side effects [3].

Magnetic Resonance Imaging (MRI) has been used for non-invasive assessment of the prostate cancer since the 1980s [4]. In clinical practise, the Prostate Imaging – Reporting and Data System (PI-RADS) has been adopted for scoring the aggressiveness of the prostate cancer based on the anatomical, functional and physiologic characteristics provided by multi-parametric MRI (mpMRI), including T2-weighted (T2W), diffusion-weighted imaging (DWI) and its derivative apparent-diffusion coefficient (ADC) maps, dynamic contrast enhanced (DCE) MRI and magnetic resonance spectroscopic imaging (MRSI) [5]. In 2015, PI-RADS Version 2 was developed to promote global standardisation and diminish variation in prostate mpMRI examinations [6]. Studies have shown that integrating MRI into the standard TRUS approach benefits the tumour localisation and aggressiveness assessment from biopsy [7]. However, this process requires substantial human interaction, is time consuming and suffers from observer variability. Therefore, a computer aided prostate cancer detection system based on MRI can improve the repeatability and accuracy in carrying out biopsy, diagnosis and treatment planning.

It is notable that in PI-RADS Version 2, the role of DCE MRI becomes less important than in PI-RADS Version 1, especially for the assessment of the prostate tumour in the transition zone [6]. Nevertheless, DCE MRI possesses advantages over the other modalities – it allows image analysis in both the spatial and time domains. Studies have also shown that DCE MRI has better discrimination between cancerous and normal tissue compared to conventional T2W MRI [8]. Hence it is essential to further explore the potential of DCE MRI in prostate cancer detection in clinical practice.

In computer aided detection, DCE MRI images were quantitatively analysed by using parametric and nonparametric approaches [9,10]. Parametric approaches aim to estimate kinetic parameters by fitting pharmacokinetic models to the concentration curves. The pharmacokinetic models proposed in the literature include Brix [11], Tofts [12], Hoffmann [13] and the phenomenological universalities (PUN) [14]. Nonparametric approaches calculate perfusion-related measurements that characterise the shape and structure of the concentration curves. The commonly used measurements include the onset time, the maximum signal intensity, wash-in rate, wash-out rate, integral under the curve and others [9,10].

Both parametric and nonparametric approaches model DCE series by a few features/parameters for prostate cancer detection, which inevitably cause information loss. By contrast, image-based features could model substantial image content in DCE MRI images and capture both spatial and temporal enhancement patterns, which potentially benefit the prostate cancer detection. In contrast to the analysis of T2W, DWI and ADC maps, for which various image-based features were often extracted such as the edges, statistical features, filter responses, texture descriptors and others [15,16], the large amount of image content in DCE MRI series has not been fully taken into account for prostate cancer detection. In the literature, there have been few studies which extract image-based features from DCE MRI images for prostate cancer diagnosis. Niaf et al. extracted features such as the intensity values, texture and gradient features for prostate cancer diagnosis [17]. However, the features were extracted from only one series (i.e. one volume) of DCE MRI images and no explanation was given about why that particular series was selected. Therefore, in this study, we would like to investigate the performance of the image-based features extracted from DCE MRI on prostate cancer detection and compare the results with the commonly used parametric and non-parametric approaches.

2   Materials and Methods

2.1   Dataset

This work is based on the Initiative for Collaborative Computer Vision Benchmarking (I2CVB) dataset, which is publicly available at http://i2cvb.github. io/. It provides multi-parametric MRI data including T2W, DCE, DWI and MRSI data, acquired using a 3T whole body MRI scanner (Siemens Magnetom Trio TIM, Erlangen, Germany). The dataset consists of the MRI images taken for 17 patients who have biopsy proven prostate cancer. In our work, we only used the DCE MRI images for the prostate cancer detection experiments. T2W MRI was used for preprocessing DCE MRI images, which will be described in the following subsection. The size of the T2W MRI images ranges from $308 \times 384 \times 64$ to $368 \times 448 \times 64$ with the voxel resolution ranging between $0.68\,mm \times 0.68\,mm \times 1.25\,mm$ and $0.79\,mm \times 0.79\,mm \times 1.25\,mm$. The DCE MRI data of each patient include 40 series taken over approximately 5 min; each series has 16 slices with 3.5 mm slice thickness; the image size of each slice is 192×256 or 200×256 with the in-plane pixel resolution ranging between $1.09\,mm \times 1.09\,mm$ and $1.37\,mm \times 1.37\,mm$. The dataset also includes the delineations provided by an experienced radiologist. For T2W MRI, the prostate gland, the tumour region, the peripheral zone and the transition zone were annotated; for DCE MRI, only the prostate gland was annotated. Figure 1 illustrates examples of T2W and DCE MRI data from I2CVB.

2.2   Preprocessing

In order to propagate the tumours annotated on T2W to DCE images, registration was applied to align the DCE and T2W images. Normalisation was performed to reduce the inter-patient signal intensity variations due to the image acquisition process of DCE MRI. For both the registration and normalisation steps, we used the methods proposed by Lemaître [18] and the publicly available code on GitHub (https://github.com/I2Cvb/mp-mri-prostate).

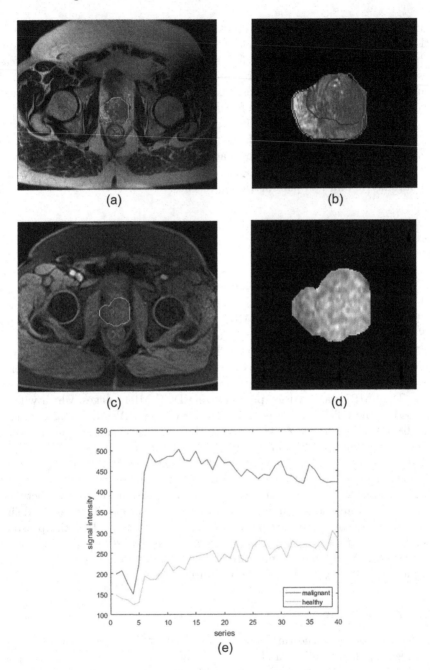

Fig. 1. (a): Example slice of T2W MRI with the prostate gland, the peripheral zone and the tumour annotated in white, blue and red contours, respectively. (b): Zoomed-in region of the prostate gland in (a). (c): Matching slice of DCE MRI with the prostate gland annotated in a white contour. (d): Zoomed-in region of the prostate gland in (c). (e): Variation of signal intensities of cancerous and healthy tissue voxels over time series in DCE MRI. The figure is better viewed in colour. (Color figure online)

2.3   Feature Extraction

Pharmacokinetic Models. Pharmacokinetic models estimate a set of parameters that reflect the physiological exchanges between vessels and extravascular extracellular space (EES) [19]. These have obtained considerable attention owing to their simplicity and small number of parameters to be estimated [9]. Brix has been one of the most commonly used models, which calculates three parameters from the DCE signal: the contrast media exchange rate k_{ep}; the elimination rate from the plasma compartment k_{el}; and an arbitrary constant simulating the tissue properties A [11]. The Hoffmann model was derived from the Brix model [12]. It redefines the constant A and the three parameters k_{ep}, k_{el} and A can also be computed. Tofts is another commonly used model, from which three parameters were calculated: the forward transfer constant of the contrast media diffusing from the blood plasma K_{trans}; the reverse constant of the contrast media returning to the blood plasma K_{ep}; and the plasma volume fraction v_p [13]. For the Tofts model, the patient-based arterial input function (AIF) signal was estimated by selecting the most enhanced voxels from the femoral and iliac arteries [20]. In addition, for the PUN model, three parameters β, a_0 and r can be calculated, where β and a_0 control the growth rate of the curve in its first part and r determines the behaviour and the speed of change of the curve in the second part [14]. The performance of the four models for prostate cancer detection were evaluated in this work.

Perfusion-Related Measurements. Nonparametric approaches calculate the empirical perfusion-related measurements that correlate with the physiology of the organ. These features have been commonly used because they are straightforward in definition and simple to compute [9]. We categorised these into four sets: time features, signal intensity features, derivative and integral features and ratio features as listed in Table 1.

Image-Based Features. DCE MRI consists of a set of T1-weighted MRI images acquired over time and hence contains substantial image information. Various image-based features can be extracted from all series of DCE MRI images such as intensity, statistical features, gradient-based features, edges, filter responses and texture descriptors.

Intensity. The signal intensities of the whole time series of DCE MRI data are the most basic image information. For each voxel, the intensity values across all DCE series were used as the feature.

Statistical Features. The statistical features capture the distribution of intensities within a local patch centered at each voxel. These features include mean, median, variance, standard deviation, mean of absolute deviation, median of absolute deviation, skewness, kurtosis, local contrast, local probability, 25th percentile, 75th percentile and others [15].

Table 1. Perfusion-related features calculated using nonparametric approaches.

Category	Symbol	Description
Time features	t_0	Onset time of the enhancement curve
	t_{max}	Time corresponding to the maximum signal intensity
	τ	Exponential time constant
Signal intensity features	S_0	Intensity at the onset of the enhancement
	S_{max}	Maximum signal intensity
	$S_{95\%}$	95% of the maximum signal intensity
	S_{end}	Signal intensity at the final time point
	$S_{max} - S_0$	Difference between the peak and the baseline intensities
Derivative and integral features	Wash-in rate	Signal slope from t_0 to t_{max}
	Wash-out rate	Signal slope from t_{max} to the final time point
	AUC	Area under the curve between t_0 and the final time point
	IAUC	Initial area under the curve between t_0 and t_{max}
	average plateau	Average signal change during the wash-out phase (from t_{max} to the final time point)
Ratio features	PER	Peak enhancement ratio calculated as $(S_{max} - S_0)/S_0$
	MITR	Maximum intensity time ratio calculated as $(S_{max} - S_0)/t_{max}$
	nMITR	Normalised MITR calculated as $(S_{max} - S_0)/(S_0 t_{max})$

Gradient-Based Features. Gradient-based features are able to detect the signal intensity changes and characterise micro-textures. The gradients of the image intensity computed along three dimensions, the magnitude, the gradient azimuth and elevation all belong to this category.

Edges. Edges were detected by convolving edge operators like Sobel, Scharr, Prewitt and Kirsch with the original image. Phase congruency measures the significance of image features in the frequency domain [21]. It detects contrast invariant features such as the maximum moment of phase congruency covariance, the orientation image and the local weighted mean phase angle at each point in the image.

Filter Responses. Gabor filters capture specific patterns of the image by tuning the scale, orientation and frequency of the kernels. The filter responses generated by convolving the Gabor filter bank with the original image were used as features. The eight maximum response (MR8) filter bank extract rotation invariant features [22]. It consists of two anisotropic filters (an edge and a bar filter at six orientations and three scales) and two isotropic filters (one Gaussian and one Laplacian of Gaussian filter). The dimensionality of the features was reduced from 38 to 8 by taking the maximum responses of the two anisotropic filters across different orientations at each scale.

Texture Descriptors. Local binary patterns (LBP) describe image texture by local spatial patterns and gray level contrast [23]. It is invariant to monotonic

Table 2. Image-based features.

Category	Features	Dimensionality
Intensity	Signal intensities	1×40
Statistical features	Mean, median, variance, standard Deviation, mean of absolute deviation, Median of absolute deviation, skewness, Kurtosis, local contrast, local probability, 25th percentile, 75th percentile	12×40
Gradient-based features	Gradients in three dimensions; Magnitude, gradient azimuth, gradient elevation	6×40
Edges	Sobel	3×40
	Scharr	3×40
	Prewitt	3×40
	Kirsch	2×40
	Phase congruency	3×40
Filter responses	Gabor filters	6×40
	MR8 filters	8×40
Texture descriptors	LBP	6×40
	Texton	1×40
	Tamura contrast	1×40
	Haar-like features	2×40

gray level changes and efficient in both computation and texture classification. In the texton-based approach, a texton dictionary was generated by applying clustering algorithms to the patches extracted from all images in the dataset. Each voxel in the image was assigned a texton identification (ID) corresponding to the closest texton in the dictionary [22]. In our work, we used texton IDs across all DCE series as the feature for each voxel [24]. Tamura contrast measures the variation of the intensity in a local region [25]. Haar-like features were calculated as the mean intensity of any cubical region or the mean intensity difference of any two random asymmetric regions within a local patch [25].

All the image-based features described above and their corresponding dimensionalities are listed in Table 2. We extracted most of the features using the parameters described in Lemaître's work [18].

Anatomical Features. Five anatomical features were computed: the relative distance to the centre of the prostate, the relative distance to the contour of the prostate, the relative position in the Euclidean and cylindrical coordinate system and the zone location (i.e. the peripheral zone or the transitional zone) the voxel is in.

2.4   Classification

Leave-one-patient-out cross validation was performed in our work. A random forest classifier was adopted to classify all voxels into cancer or non-cancer. There

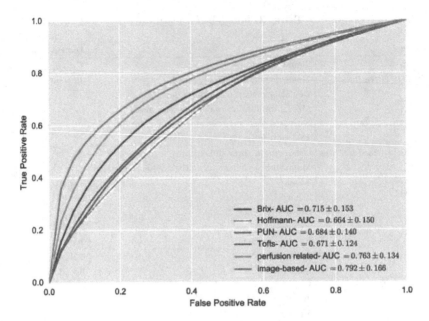

Fig. 2. Comparisons of the prostate cancer detection performance using image-based features, pharmacokinetic features and perfusion-related measurements extracted from DCE MRI. The classification performance was measured by the mean AUC value and the standard deviation over the dataset. The figure is better viewed in colour. (Color figure online)

were 100 classifiers in the forest; the minimum number of features required to split an internal node was 2; the bootstrap samples were used when building the trees. The classification performance was evaluated by conducting a receiver operating characteristics (ROC) analysis and calculating the area under the curve (AUC) for each patient. By applying the random forest as the classifier, the importance of the features can be computed using the Gini impurity across all classifiers in the forest and the most discriminant features can be selected by setting a threshold.

3   Results

3.1   Comparison of Image-Based, Pharmacokinetic and Perfusion-Related Features

We compared the prostate cancer detection performance using different categories of features extracted from DCE MRI: the image-based features, the pharmacokinetic features estimated from various models and the perfusion-related measurements. The anatomical features were always included in the evaluation of each category of features. The results shown in Fig. 2 demonstrate that the Brix model outperforms the other three pharmacokinetic models in the prostate

Table 3. Classification results of using selected features from the aggregation of all features extracted from DCE MRI.

Percentile (%)	1	5	10	15	20	30	100
AUC	0.790	0.816	0.819	0.821	0.816	0.813	0.802
	± 0.157	± 0.152	± 0.159	± 0.153	± 0.157	± 0.165	± 0.161

cancer detection; using perfusion-related features produces better results than using the features estimated from the pharmacokinetic models; and using the image-based features results in the best classification performance.

3.2 Aggregation of All Features

All extracted features were concatenated to perform prostate cancer detection. Subsequently, percentile thresholds were set to select the most important features based on the Gini impurity values calculated in the random forest classification.

Table 3 lists the classification results generated by setting different percentiles. By contrast to the results shown in Fig. 2, it can be noted that using the aggregation of all features (when the percentile equals to 100) outperforms using each individual category of features. By applying feature selection, the classification performance was further improved and the best result was obtained when 15% of the features were selected (AUC = 0.821 ± 0.153).

Figure 3 illustrates the probability maps of a few cases produced by selecting the most often selected features in the leave-one-patient-out cross validation. By setting the percentile as 15%, a set of features were selected for each fold in the cross validation; all the selected features were ranked according to the times being selected during the entire cross validation process; then the features with higher rankings, which also accounts for 15% of the aggregation of all features, were used to generate the results. It can be observed that the proposed method accurately identifies the prostate cancer for cases for which the tumour exists in the peripheral zone only (a, b), in the transitional zone only (c) and in both regions (d, e). However, for the case shown in (f), the performance of prostate cancer detection was poor with lots of false positives detected.

The features used to generate the results are listed in Table 4. Most of them were selected in all folds (i.e. 17 times) in the cross validation; a few were selected in 16 or 15 folds. Features of all four categories were selected. Among all the image-based features, the statistical features and the filter responses contribute the most to the prostate cancer detection; the intensity, the gradient-based features, the edges and the texture descriptors were hardly selected.

4   Discussion

Although image-based features have been widely adopted in prostate cancer detection using MRI images like T2W, ADC and DWI, they are rarely used

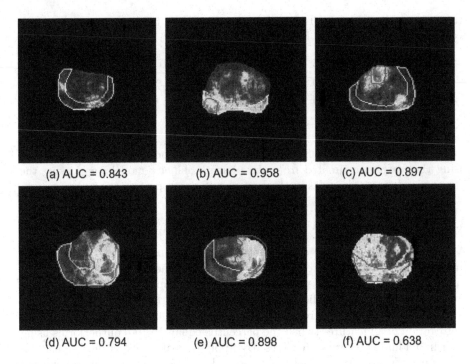

Fig. 3. Probability maps of the prostate cancer detection produced by selecting 15% of the aggregation of all features extracted from DCE MRI. Only the region of the prostate gland has been shown in these example slices of 6 different cases. The annotations of the peripheral zone and the tumour are depicted in white and red contours, respectively. The jet overlap represents the probability map of the cancer detection, where high and low probabilities of being cancer are indicated by red and blue colours, respectively. The AUC value corresponding to each case is also displayed. The figure is better viewed in colour. (Color figure online)

to model the spatial-temporal characteristics of DCE MRI series. Experimental results have shown that compared to the commonly used approaches modelling DCE MRI series, using image-based features achieves superior performance, which indicates the additional contribution image-based features provide in prostate cancer detection.

Because all DCE MRI series were used, it was time consuming to extract various image-based features, which resulted in a high dimensionality. To mitigate this problem, a number of the most important features can be selected by applying feature selection techniques. We extracted the features with high discriminability and used them for the cancer detection.

The I2CVB dataset used in this work contains limited number of cases. However, it is the only publicly available DCE MRI prostate dataset that provides tumour segmentations. The PROSTATEx and PROSTATEx-2 challenge datasets also include DCE MRI. But these two datasets do not provide the

Table 4. Selected features from the aggregation of all features.

Category	Selected features	Dimensionality
Pharmacokinetic features	Brix model: A	1
	PUN: β	1
	Tofts: K_{trans}, K_{ep}	2
Perfusion-related features	t_{max}, S_{max}, $S_{95\%}$, S_{end}	7
	Wash-in rate, IAUC, MITR	
Image-based features	Intensity	2
	Statistical features	179
	Gradient-based features	2
	Edges	1
	Filter responses	142
	Texture descriptors	10
Anatomical features	Distance to the contour	1
	Relative Euclidean position	2
	Relative cylindrical position	3

original image series but only the K_{trans} images calculated from them. And for the tumour ground truth, only the scanner coordinate position of the tumour region is provided, which makes the datasets unsuitable to investigate the performance of the spatio-temporal features contained in DCE MRI series for prostate tumour segmentation in our work.

As future work, we would like to analyse the impacts of the parameters used in feature extraction on the performance of prostate cancer detection and validate our approach on a larger dataset. We can also validate the effectiveness of the image-based features extracted from DCE MRI series on other tasks such as prostate cancer staging and Gleason grading. Based on the voxel-level classification results generated in this work, a region-level lesion segmentation and classification can be applied to remove false positives and further improve cancer detection accuracy. Moreover, instead of feeding the manually engineered features into a conventional classifier, the classification can be achieved by extracting the hierarchical features by applying deep learning algorithms. Besides of the DCE MRI, multi-parametric MRI data, such as T2W, DWI and MRSI data, can be incorporated into the approach to benefit prostate cancer detection.

5   Conclusions

This work investigated the performance of image-based features extracted from DCE MRI series on prostate cancer detection. The experimental results have demonstrated that using the image-based features outperforms other widely used approaches, such as estimating DCE signals using pharmacokinetic models and extracting perfusion-related measurements, because for image-based features substantial image information was taken into account and they can model both the spatial and temporal characteristics of the DCE series. Selecting the most discriminant features from all categories of features reduced the feature redundancy, removed noisy information and produced excellent performance for prostate cancer detection. It has been demonstrated that apart from the commonly used features in the literature, image-based features, especially the statistical features and the filter responses, provide additional contribution for prostate cancer detection. The post-processing steps, the hierarchical features, more advanced machine learning algorithms and multi-parametric MRI can be investigated in the future to further improve prostate cancer detection.

References

1. Siegel, R., Ma, J., Zou, Z., Jemal, A.: Cancer statistics. CA: Can. J. Clin. **64**(1), 9–29 (2014)
2. Siegel, R., Naishadham, D., Jemal, A.: Cancer statistics. CA: Can. J. Clin. **63**(1), 11–30 (2013)
3. Chou, R., Croswell, J.M., Dana, T., et al.: Screening for prostate cancer: a review of the evidence for the US preventive services task force. Ann. Internal Med. **155**(11), 762–771 (2011)
4. Hricak, H., Dooms, G.C., McNeal, J.E., et al.: MR imaging of the prostate gland: normal anatomy. Am. J. Roentgenol. **148**(1), 51–58 (1987)
5. Barentsz, J.O., Richenberg, J., Clements, R., et al.: ESUR prostate MR guidelines 2012. Eur. Radiol. **22**(4), 746–757 (2012)
6. Weinreb, J.C., Barentsz, J.O., Choyke, P.L., et al.: 2016. PI-RADS prostate imaging-reporting and data system: 2015, version 2. Eur. Urol. **69**(1), 16–40 (2016)
7. Moore, C.M., Ridout, A., Emberton, M.: The role of MRI in active surveillance of prostate cancer. Curr. Opin. Urol. **23**(3), 261–267 (2013)
8. Kim, C.K., Park, B.K., Kim, B.: Localization of prostate cancer using 3T MRI: comparison of T2-weighted and dynamic contrast-enhanced imaging. J. Comput. Assist. Tomogr. **30**(1), 7–11 (2006)
9. Khalifa, F., Soliman, A., El-Baz, A., et al.: Models and methods for analyzing DCE-MRI: A review. Med. Phys. **41**, 124301 (2014)
10. Sung, Y.S., Kwon, H.J., Park, B.W., et al.: Prostate cancer detection on dynamic contrast-enhanced MRI: computer-aided diagnosis versus single perfusion parameter maps. Am. J. Roentgenol. **197**(5), 1122–1129 (2011)
11. Brix, G., Semmler, W., Port, R., Schad, L.R., Layer, G., Lorenz, W.J.: Pharmacokinetic parameters in CNS Gd-DTPA enhanced MR imaging. J. Comput. Assis. Tomogr. **15**(4), 621–628 (1991)
12. Tofts, P.S., Kermode, A.G.: Measurement of the blood-brain barrier permeability and leakage space using dynamic MR imaging. 1. Fundamental concepts. Magn. Reson. Med. **17**(2), 357–367 (1991)
13. Hoffmann, U., Brix, G., Knopp, M.V., et al.: Pharmacokinetic mapping of the breast: a new method for dynamic MR mammography. Magn. Reson. Med. **33**(4), 506–514 (1995)
14. Gliozzi, A.S., Mazzetti, S., Delsanto, P.P., Regge, D., Stasi, M.: Phenomenological universalities: a novel tool for the analysis of dynamic contrast enhancement in magnetic resonance imaging. Phys. Med. Biol. **56**(3), 573 (2011)

15. Rampun, A., Zheng, L., Malcolm, P., Tiddeman, B., Zwiggelaar, R.: Computer-aided detection of prostate cancer in T2-weighted MRI within the peripheral zone. Phys. Med. Biol. **61**(13), 4796–4825 (2016)
16. Trigui, R., Mitéran, J., Walker, P.M., Sellami, L., Hamida, A.B.: Automatic classification and localization of prostate cancer using multi-parametric MRI/MRS. Biomed. Sig. Process. Control **31**, 189–198 (2017)
17. Niaf, E., Rouvière, O., Mège-Lechevallier, F., Bratan, F., Lartizien, C.: Computer-aided diagnosis of prostate cancer in the peripheral zone using multiparametric MRI. Phys. Med. Biol. **57**(12), 3833 (2012)
18. Lemaître, G.: Computer-Aided Diagnosis for Prostate Cancer using Multi-Parametric Magnetic Resonance Imaging. Doctoral dissertation 2016, Universite de Bourgogne, Universitat de Girona, pp. 109–116 (2016)
19. Cover, T.M.: Elements of Information Theory, pp. 12–23. Wiley, New Delhi (2006)
20. Kybic, J., Thévenaz, P., Unser, M.: Multiresolution spline warping for EPI registration. In: Proceedings of the SPIE: Mathematical Imaging-Wavelet Applications in Signal and Image Processing 1999, Denver, Colorado, vol. 3813, pp. 571–579 (1999)
21. Scharr, H.: Optimal operators in digital image processing. Doctoral dissertation 2000. Repertus Carola University, Heidelberg, Germany (2000)
22. Kirsch, R.A.: Computer determination of the constituent structure of biological images. Comput. Biomed. Res. **4**(3), 315–328 (1971)
23. Kovesi, P.: Image features from phase congruency. Videre: J. Comput. Vis. Res. **1**(3), 1–26 (1999)
24. Gabor, D.: Theory of communication. Part 1: The analysis of information. J. Inst. Electr. Eng. Part III: Radio Commun. Eng. **93**(26), 429–441 (1946)
25. Ojala, T., Pietikäinen, M., Harwood, D.: A comparative study of texture measures with classification based on featured distributions. Pattern Recogn. **29**(1), 51–59 (1996)

Controlling False Positive/Negative Rates for Deep-Learning-Based Prostate Cancer Detection on Multiparametric MR Images

Zhe Min[1]([✉]), Fernando J. Bianco[2], Qianye Yang[1], Rachael Rodell[1,3], Wen Yan[1,4], Dean Barratt[1], and Yipeng Hu[1]

[1] Centre for Medical Image Computing and Wellcome/EPSRC Centre for Interventional and Surgical Sciences, University College London, London, UK
z.min@ucl.ac.uk
[2] Urological Research Network, Miami Lakes, FL, USA
[3] Focalyx Technologies, Miami, FL, USA
[4] City University of Hong Kong, Hong Kong, China

Abstract. Prostate cancer (PCa) is one of the leading causes of death for men worldwide. Multi-parametric magnetic resonance (mpMR) imaging has emerged as a non-invasive diagnostic tool for detecting and localising prostate tumours by specialised radiologists. These radiological examinations, for example, for differentiating malignant lesions from benign prostatic hyperplasia in transition zones and for defining the boundaries of clinically significant cancer, remain challenging and highly skill-and-experience-dependent. We first investigate experimental results in developing object detection neural networks that are trained to predict the radiological assessment, using these high-variance labels. We further argue that such a computer-assisted diagnosis (CAD) system needs to have the ability to control the false-positive rate (FPR) or false-negative rate (FNR), in order to be usefully deployed in a clinical workflow, informing clinical decisions without further human intervention. However, training detection networks typically requires a multi-tasking loss, which is not trivial to be adapted for a direct control of FPR/FNR. This work in turn proposes a novel PCa detection network that incorporates a lesion-level cost-sensitive loss and an additional slice-level loss based on a lesion-to-slice mapping function, to manage the lesion- and slice-level costs, respectively. Our experiments based on 290 clinical patients concludes that 1) The lesion-level FNR was effectively reduced from 0.19 to 0.10 and the lesion-level FPR was reduced from 1.03 to 0.66 by changing the lesion-level cost; 2) The slice-level FNR was reduced from 0.19 to 0.00 by taking into account the slice-level cost; (3) Both lesion-level and slice-level FNRs were reduced with lower FP/FPR by changing the lesion-level or slice-level costs, compared with post-training threshold adjustment using networks without the proposed cost-aware training. For the PCa application of interest, the proposed CAD system is capable of substantially reducing FNR with a relatively preserved FPR, therefore is considered suitable for PCa screening applications.

© Springer Nature Switzerland AG 2021
B. W. Papież et al. (Eds.): MIUA 2021, LNCS 12722, pp. 56–70, 2021.
https://doi.org/10.1007/978-3-030-80432-9_5

Keywords: Prostate cancer · Multi-parametric resonance images · Object detection · False negative reduction

1 Introduction

Prostate Cancer (PCa) is one major public health problem for males globally [12]. It is estimated that 191,930 cases have been newly diagnosed with PCa and 33,330 associate deaths in the United States in 2020 [12]. Multi-parametric Magnetic Resonance images (mpMR) has potential to play a part in every stage of prostate cancer patient management, including enabling targeted biopsy for early-to-medium stage cancer diagnosis and screening programmes for avoiding unnecessary biopsy [6,14]. However, reading mp-MR requires highly specialised radiologists and, for those experienced, it remains a challenging and arguably tedious task.

Automated computer-aided PCa detection not only can help significantly reduce the radiologist's time in examining the volumetric, multi-modality mpMR images, but also provides higher consistency over human interpreters with rivaling human performance at the same time [9]. Computer-aided diagnosis (CAD) of PCa using mpMR has therefore attracted growing attention and, in particular, modern machine learning methods have been proposed recently for the end-to-end, fully-automated CAD tasks, such as classification, detection and localisation. However, automating PCa detection has to overcome several challenges innate to several imaging and pathology characteristics specific in this application. For example, inherently high inter-patient variance in shape and size among cancerous regions; spatial misalignment between different MR sequences [16]; and similar imaging patterns exhibited between the benign prostatic hyperplasia (BPH) and high grade PCa, which subsequently leads to false positives (FPs) [9,15], for both CAD models and human observers, thus their labelling.

Scores based on Prostate Imaging and Reporting Data System (PI-RADS) [13] and Gleason groups based on biopsy or prostatectomy specimens are examples of radiological and histopathological labels. These two types of labels and their combinations are useful to train a CAD system. Sanford *et al.* utilized a ResNet-based network to assign specific PI-RADS scores to already delineated lesions [10]. Schelb *et al.* compared the clinical performance between PI-RADS and U-Net-based methods for classification and segmentation of suspicious lesions on T2w and diffusion MRI sequences, where the ground-truth is acquired by combined targeted and extended systematic MRI–transrectal US fusion biopsy [11]. While the radiological labels are limited by the challenges discussed above, histopathological labels are also subject to errors and bias in sampling, due to, for example, shift in patient cohort, localisation error in needle biopsy and variance in pathology report. Searching best gold-standard between the two is still an open question and may be beyond the scope of this study. In our work, we use the experienced radiologist PI-RADS scores as our prediction of interest - the training labels. See more details of the data in Sect. 3.

A CAD system for detecting PCa has been considered as a semantic segmentation problem. Many recent PCa segmentation algorithms adopted convolution neural networks (CNNs) [1]. Cao *et al.* has proposed a multiclass CNN called FocalNet to jointly segment the cancerous region and predict the Gleason scores on mpMR images [1]. Cao *et al.* adapted the focal loss (FL) to the PCa detection task that predicts the pixel-level lesion probability map on both the Apparent Diffusion Coefficient (ADC) and T2-Weighted (T2w) images, where the training concentrates more on the cancerous or suspicious pixels [2]. In addition, to account for the fact that the lesions may show different size or shapes across imaging modalities, an imaging component selector called selective dense conditional random field is designed to select the best imaging modality where the lesion is observable more clearly [2]. Finally, the predicted probability maps is refined into the lesion segmentation on that selected imaging component [2]. It should be noted that only slices with annotated lesions are included in both the training and validation in [1,2]. Yu *et al.* utilised a standalone false positive reduction network with inputs being the detected true positives (TPs) and false positives (FPs) from another U-net-based detection network [15].

Object detection algorithms have also been proposed for detecting and segmenting PCa from mpMR images, explicitly discriminating between different lesions through instance classification and segmentation. Multiple-staged object detection networks have been shown to have fewer false positives in challenging lesion detecting tasks, compared with segmentation methods such as U-Net [16]. Li *et al.* adapted the MaskRCNN to detect the presence of epithelial cells on histological images for predicting Gleason grading [5], with an addition branch classifying epithelial cell presence in and the MaskRCNN branch classifying, detecting (bounding boxes), and segmenting (into binary masks) the epithelial areas. Dai *et al.* investigated the performances of the MaskRCNN to segment prostate gland and the intra-prostatic lesions and reported consistent superior performances over the U-net [3]. Yu *et al.* also used the MaskRCNN in the PCa detection task, where an additional segmentation branch has been added to the original detection network [16].

Two- or multi-stage object detectors have been shown superior performance, compared with the one-stage counterparts [7]. However, existing two-stage object detection network in fields of computer vision, such as Mask-RCNN, optimise for overall accuracy, weighting false positive and false negative regions based on their respective prevalence, rather than the associated clinical and societal costs. In this work, we focus on real-world clinical scenarios, in which the CAD system is developed for, for example, assisting population screening or treatment referrals, by alleviating the need for further radiologist examining individual lesions or slices. These clinical applications mandate the developed CAD system to guarantee a low false negative rate and a low false positive rate at lesion or slice levels, in the two respective examples. Instead of thresholding the detection network post-training to achieve the desired sensitivity/specificity at either lesion or slice level, in this study, we aim to answer the research question: With a two-stage object detector, can more desirable FPR or FNR be controlled by changing their costs during training?

We explore the plausible answer to this question through formulating and incorporating two cost-sensitive classification losses at the lesion and slice levels respectively, which will give the flexibility of biasing towards reducing FPR or FNR during training. This is not trivial for a detection network training scheme that minimises a multi-tasking loss, as the following technical questions need to be addressed in this work: a) whether a cost-sensitive loss replacing the original instance-level classification loss is effective; b) how slice-level cost can be quantified and subsequently controlled; c) whether changing slice-level cost by the additional slice-level loss is effective; and d) how these two level costs can be combined during training to archive desirable levels of control of FPR/FNR at lesion or slice level, on test data set.

Our key contributions of this study are summarised as follows. (1) We modify the classification loss in the original detection network with the aim of controlling the lesion-level FP/FN for PCa detection. (2) We propose a novel slice-level classification loss with the aim of controlling the slice-level FP/FN for PCa detection. We investigate its utility in improving baseline sensitivity with lower FPR by incorporating the classifier into the overall detection network. (3) We study the effect of different weighting schemes in the two classifier branches on lesion-level and slice-level FP/FN reduction.

2   Methods

2.1   Problem Definition

In this work, PCa detection is formulated as an instance segmentation problem. The slices within mpMR images without annotated cancerous regions are regarded as background images. The multiple tasks in PCa detection include: classify whether one proposal region is a lesion or not; regress the coordinates of the bounding box (BB) surrounding the proposal region; segment the mask of the lesion.

The overall architecture of our proposed CAD system is depicted in Fig. 1. The network utilizes a Feature Pyramid Network (FPN) backbone on top of the ResNet architecture [4], to generate multi-scale features. The extracted features are shared by the following two modules: (a) a region proposal network (RPN) module that generates candidate object bounding boxes [8]; (b) a detection module that performs the bounding box regression, classification and the region of interest (RoI) mask prediction on the candidates boxes.

2.2   Overall Training Loss Function

As shown in Fig. 1, our multi-task loss consists of the following six terms

$$L_{total} = L_{rpn\_reg} + L_{rpn\_cls} + L_{box} + L_{mask} + L_{cost\_cls} + L_{slice\_cls} \qquad (1)$$

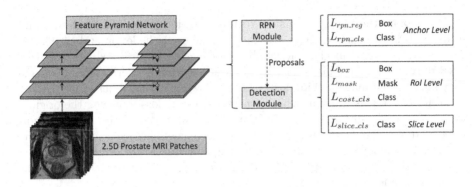

Fig. 1. Illustration of the overall architecture based on the MaskRCNN. ROI: region of interest, RPN: region proposal network. L_{rpn_reg}: RPN regression loss, L_{rpn_cls}: RPN classification loss, L_{box}: the regression loss at the proposal/RoI level, L_{mask}: the mask loss at the RoI level, L_{cost_cls}: the lesion-level (i.e., RoI level) cost-sensitive classification loss, L_{slice_cls}: the slice-level cost-sensitive classification loss.

where L_{rpn_reg} and L_{rpn_cls} are the smoothed bounding box regression loss based on L^1-norm and the cross entropy classification loss, at the anchor level, respectively; L_{box}, L_{mask} and L_{cost_cls} are the L^1-norm smoothed bounding box regression loss, the binary cross entropy loss and the weighted cross entropy classification loss, at the RoI level, respectively; and L_{slice_cls} is the weighted cross entropy classification loss at the slice level. Among all the loss terms in Eq. (1), L_{rpn_reg}, L_{rpn_cls}, L_{box}, and L_{mask} are the same as those in the original Mask-RCNN framework. The rationale of L_{slice_cls} is to evaluate whether the model can classify the category of a slice being cancerous or not. The inputs to L_{slice_cls} and L_{slice_cls} are the class probabilities of the proposals being cancerous.

2.3 Lesion-Level Cost-Sensitive Classification Loss

To control the cost of mis-classification of individual lesions, the lesion-level (RoI level) cost sensitive classification loss $L_{cost\_cls}(p_i, p_i^{\star})$ is defined as follows

$$L_{cost\_cls}(p_i, p_i^{\star}) = \underbrace{-\alpha_{lesion}p_i^{\star} \log p_i}_{L_{cost\_cls}^{positive}} \underbrace{-\beta_{lesion}(1 - p_i^{\star}) \log (1 - p_i)}_{L_{cost\_cls}^{negative}} \tag{2}$$

where $p_i^{\star} = 1$ if the ith region proposal is positive, $p_i^{\star} = 0$ if negative, $p_i \in [0, 1]$ is the predicted class probability (by the classification branch in MaskRCNN) of the region proposal i being an cancerous region, α_{lesion} and β_{lesion} are the weights associated with the positive and negative regions. In this study, three different combinations of α_{lesion} and β_{lesion} are tested as follows. (i) $\alpha_{lesion} > 1$ and $\beta_{lesion} = 1$, during training, the network emphasizes more on regions with positive labels; (ii) $\alpha_{lesion} = 1$ and $\beta_{lesion} > 1$, the network emphasizes more on the regions with negative labels; (iii) $\alpha_{lesion} = 1$ and $\beta_{lesion} = 1$, the network weights them equally. In other words, in the above first two cases, the network will penalise more on the (1) false negatives (FNs); (2) false positives (FPs) respectively at the lesion level. In the third case, $L_{cost\_cls}(p_i, p_i^\star)$ degenerates to the binary cross entropy loss when $\alpha_{lesion} = 1$ and $\beta_{lesion} = 1$.

Positive Slices. In the slices where there are GT lesions, the training loss associated with that slice is defined as

$$L_{total} = L_{rpn\_reg} + L_{rpn\_cls} + L_{box} + L_{mask} + L_{cost\_cls}. \tag{3}$$

Negative Slices. In the slices where there is no GT lesion, the training loss associated with that slice is defined as

$$\begin{aligned} L_{total} &= L_{rpn\_cls} + L_{cost\_cls}^{negative} \\ &= L_{rpn\_cls} - \beta_{lesion}(1 - p_i^\star) \log(1 - p_i). \end{aligned} \tag{4}$$

2.4   Slice-Level Cost-Sensitive Classification Loss

Let us suppose there are N proposal regions or region of interest (ROI) in one slice. The slice-level cost-sensitive classification loss is defined as the weighted cross entropy as follows

$$L_{slice\_cls} = \underbrace{-\alpha_{slice} p_{slice}^\star \log p_{slice}}_{L_{slice\_cls}^{positive}} - \underbrace{\beta_{slice}(1 - p_{slice}^\star) \log(1 - p_{slice})}_{L_{slice\_cls}^{negative}}, \tag{5}$$

where $p_{slice}^\star \in \{0, 1\}$ and $p_{slice} \in [0, 1]$ is given by

$$p_{slice}^\star = max(p_1^\star, ..., p_N^\star), \tag{6}$$

$$p_{slice} = max(p_1, ..., p_N), \tag{7}$$

where $p_i^\star$ and p_i are the GT and predicted probability that i^{th} region being cancerous. More specifically, $p_{slice}^\star = 1$ indicates that there is at least one cancerous region in the interested slice, p_{slice} is the largest predicted probability of one detected region being cancerous. The rational behind the lesion-to-slice mapping function, for computing $p_{slice}^\star$ and p_{slice}, is that (1) for GT labels, one slice is considered to be a 'cancerous' slice if there exists at least one 'positive' region (i.e., $p_i^\star = 1$ for at least one i); (2) for predictions, the probability of one slice being 'cancerous' is the largest predicted probability of one detected region being 'positive' in the interested slice. Like the function of α_{lesion} and β_{lesion} in $L_{cost\_cls}(p_i, p_i^\star)$, α_{slice} and β_{slice} weight the loss L_{slice_cls} in an adversarial manner: Whilst $\alpha_{slice} > 1$ and $\beta_{slice} = 1$, the network penalises FNs more heavily at the slice level, $\alpha_{slice} = 1$ and $\beta_{slice} > 1$, the network penalises FPs more.

Positive Slices. In the slices where there are GT lesions, the overall training loss remains L_{total}, defined in Eq. (1), and can be expanded as follows

$$L_{total} = L_{rpn\_reg} + L_{rpn\_cls} + L_{box} + L_{mask} + L_{cost\_cls} - \alpha_{slice} p_{slice}^\star \log p_{slice}. \tag{8}$$

Negative Slices. In the slices where there is no GT lesion, the overall training loss is therefore given by

$$L_{total} = L_{rpn\_cls} + L_{cost\_cls}^{negative} + L_{slice\_cls}^{negative}$$
$$= L_{rpn\_cls} - \beta_{lesion}(1 - p_i^\star)\log(1 - p_i) - \beta_{slice}(1 - p_{slice}^\star)\log(1 - p_{slice}). \tag{9}$$

where only the classification losses at the anchor, lesion/region, and slice levels are included.

3   Experiments and Evaluation

3.1   Data Set and Implementation Details

Our data sets consist of 290 clinical prostate cancer patients with approved Institutional Review Board (IRB) protocol. The ground-truth labels (including cancerous masks) have been acquired based on the Prostate Imaging Reporting and Data System (PI-RADS) scores reported by radiologists with more than 15 years of experience. PIRADS ≥ 3 annotated lesions are regarded as clinically significant and are considered positive in this work. The ratios of number of patients in the training, validation and test sets are 8:1:1. The inputs to our proposed detection include the T2-Weighted (T2w), the Apparent Diffusion Coefficient (ADC), and the Diffusion-Weighted Images (DWI) b-2000 images. ADC and DWI b-2000 images were spatially aligned with corresponding T2w images using the rigid transformation based on the coordinate information stored in the imaging files. All slices were cropped from the center to be 160×160 and the intensity values were normalized to $[0, 1]$. Our networks were constructed with 2D convolutional layers, with a so-called 2.5D input bundle which concatenated two neighboring slices for each of the T2, ADC and DWI b-2000 image slices at the slice of interest, i.e. resulting in a nine-channel input as denoted in Fig. 1.

The proposed method was implemented with the TensorFlow framework. Each network was trained for 100 epochs with the stochastic gradient descent (SGD) optimizer and the initial learning rate was set to be 0.001. Random affine transformations were applied for data augmentation during training. If not otherwise specified, the parameter threshold[1] was set to 0.7 and the maximum number of lesions in one slice being 6 was configured at both the training and test stages.

3.2   Evaluation Metrics

We evaluate the methods with descriptive statistics at both the lesion and slice levels. The slice-level false positive rate (FPR) and false negative rate (FNR) are defined as follows FPR $= \frac{FP}{FP+TN} = 1 -$ specificity, FNR $= \frac{FN}{FN+TP} = 1 -$ sensitivity, ACC $= \frac{TP+TN}{TP+TN+FP+FN}$, where FP, TN FN and TP are numbers of

---

[1] We use threshold to denote parameter DETECTION_MIN_CONFIDENCE in the original MaskRCNN codes, for brevity.

Table 1. The false positive rate (FPR) and false negative rate (FNR) on the test data sets, where L_{cost_cls} was used in the training process. With $\alpha_{lesion} = 3$, $\beta_{lesion} = 1$ in Eq. (2), both the lesion-level and slice-level FNRs were considerably reduced, compared to the case where $\alpha_{lesion} = 1$, $\beta_{lesion} = 1$. With $\alpha_{lesion} = 1$, $\beta_{lesion} = 3$ in Eq. (2), the lesion-level FPs and slice-level FPRs were lower, compared to the case where $\alpha_{lesion} = 1$, $\beta_{lesion} = 1$.

	$\alpha_{lesion}/\beta_{lesion} = 1$	$\alpha_{lesion} = 3$, $\beta_{lesion} = 1$	$\alpha_{lesion} = 1$, $\beta_{lesion} = 3$
Lesion-level FP	1.0327	2.0218	**0.6567**
Lesion-level FNR	0.1941	**0.1013**	0.4118
Slice-level FPR	0.5878	0.8434	**0.5049**
Slice-level FNR	0.0097	**0.0028**	0.0736
ACC	0.5744	0.3924	**0.6161**

Table 2. The false positive rate (FPR) and false negative rate (FNR) on the test data sets where $L_{cost\_cls}(\alpha_{lesion}/\beta_{lesion} = 1)$ and L_{slice_cls} were incorporated into the training. With $\alpha_{slice} = 3$, $\alpha_{slice} = 1$ in Eq. (5): (a) the lesion-level FNR was reduced; (b) the slice-level FNR remained close to zero with reduced slice-level FPR, compared to the case where $\alpha_{slice} = 1$, $\beta_{slice} = 1$. With $\alpha_{slice} = 1$, $\beta_{slice} = 3$ in Eq. (5), the slice-level FPR was reduced, compared to the case where $\alpha_{slice} = 1$, $\beta_{slice} = 1$.

	$\alpha_{slice} = 1$, $\beta_{slice} = 1$	$\alpha_{slice} = 3$, $\beta_{slice} = 1$	$\alpha_{slice} = 1$, $\beta_{slice} = 3$
Lesion-level FP	**1.7202**	1.9493	1.7965
Lesion-level FNR	0.1190	**0.0970**	0.1232
Slice-level FPR	0.8505	0.8234	0.8277
Slice-level FNR	**0.0000**	**0.0000**	0.0014
ACC	0.3882	**0.4076**	0.4041

false positive, true negative, false negative and true positive cases, respectively. It is noteworthy that the above definitions are defined and used at the slice level. At the lesion level, only the definition of FNR remains valid. Instead, we compute the mean FP per slice.

At the lesion level, a TP prediction requires the GT lesion has an Intersection of Union (IoU) greater than or equal to 0.2, between the GT bounding box (BB) and any predicted BB. A FP prediction means IoUs are smaller than 0.2 (including no overlap) between the predicted BB and all GT BBs. A GT lesion that has no TP prediction is counted as a FN. TN is not defined at the lesion level.

At the slice level, one slice with at least one GT annotated lesion mask is considered as a TP if there is any detected region at that slice. If there is no detection on the slices with GT lesion masks, the slice is counted as a FN. A TN slice means no lesion found in both prediction and GT. Any positive lesion predicted on a slice that has no GT lesion leads to a FP slice.

Table 3. The false positive rate (FPR) and false negative rate (FNR) on the test data sets where L_{cost_cls} and L_{slice_cls} were incorporated. With $\alpha = 3, \beta = 1$ in Eq. (2) and Eq. (5), (a) the lesion-level FNR was reduced; (b) the slice-level FNR remained to be 0, compared to the case where $\alpha = 1, \beta = 1$. With $\alpha = 1, \beta = 3$, the lesion-level FP and slice-level FPR were reduced, compared to those where $\alpha = 1, \beta = 1$.

	$\alpha = 1, \beta = 1$	$\alpha = 3\ \beta = 1$	$\alpha = 1, \beta = 3$
Lesion-level FP	1.7202	2.3827	**1.0982**
Lesion-level FNR	0.1190	**0.0734**	0.2262
Slice-level FPR	0.8505	0.9220	**0.6576**
Slice-level FNR	**0.0000**	**0.0000**	0.0014
ACC	0.3882	0.3367	**0.5265**

4 Results

4.1 Adjusting Mis-classification Cost at Lesion-Level

In this experiment, we study the impact on the lesion-level and slice-level performances, due to different L_{cost_cls} in the training, while the slice-level loss L_{slice_cls} in Eq. (5) is not included. More specifically, we compare the original MaskRCNN (i.e., $\alpha_{lesion} = 1, \beta_{lesion} = 1$ in Eq. (2)), with our proposed two variants where $\alpha_{lesion} = 3, \beta_{lesion} = 1$ and $\alpha_{lesion} = 1, \beta_{lesion} = 3$.

Table 1 summarises the comparative results with the case where $\alpha_{lesion} = 1, \beta_{lesion} = 1$. With $\alpha_{lesion} = 3, \beta_{lesion} = 1$, the lesion-level and slice-levels FNRs were reduced from 0.1941 to 0.1013, from 0.0097 to 0.0028, respectively. With $\alpha_{lesion} = 1, \beta_{lesion} = 3$, the lesion-level FP was reduced from 1.0327 to 0.6567 while the slice-level FPR was reduced from 0.5878 to 0.5049.

Figure 2 shows the examples where the FNs were reduced with $\alpha_{lesion} = 3, \beta_{lesion} = 1$, by comparing Fig. 2(c) with Fig. 2(a, b), and comparing Fig. 2(g) with Fig. 2(e, f). By comparing Fig. 2(g) with Fig. 2(c), the FP was reduced with a higher threshold. In contrast, more FNs can be found with larger threshold and $\alpha_{lesion} = 1, \beta_{lesion} = 3$ by comparing Fig. 2(f) with Fig. 2(b).

Figure 4 shows the example where the FPs were avoided/reduced with $\alpha_{lesion} = 1, \beta_{lesion} = 3$, by comparing Fig. 4(g) with Fig. 4(a, b, c, e, f). In the first row in Fig. 4(c), with relatively lower value of the parameter threshold, the FP still exists with $\alpha_{lesion} = 1, \beta_{lesion} = 3$. In contrast, by comparing Fig. 4(g) with Fig. 4(e, f), with larger value of the parameter threshold, the FP was avoided as shown in Fig. 4(g).

4.2 Adjusting Mis-classification Cost at Slice-Level

In this experiment, we study the effect of incorporating and changing L_{slice_cls} in the training loss whereas the weighting in L_{cost_cls} was fixed as $\alpha_{lesion} = 1, \beta_{lesion} = 1$. Table 2 includes the quantitative results with different settings of $\alpha_{slice}, \beta_{slice}$: (1) $\alpha_{slice} = 1, \beta_{slice} = 1$; (2) $\alpha_{slice} = 3, \beta_{slice} = 1$; (3) $\alpha_{slice} =$

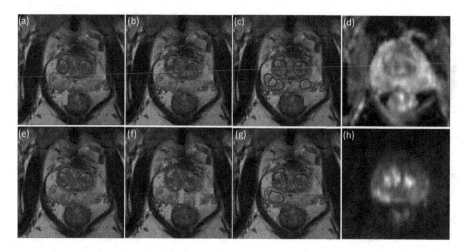

Fig. 2. In all figures in this paper, (1) the red circles denote the ground-truth (GT) lesion region, while the blue circles denote the predicted regions of interest; (2) a false positive (FP) predicted detection is denoted with the yellow arrow, while another false negative (FN) lesion is denoted with the green arrow. In this study, only the lesion-level classification loss L_{cost_cls} in the training process. All example sub-figures shown here correspond to the performances on one same slice in the test data set. In the first row, threshold = 0.7 while threshold = 0.95 in the second row. The first three columns from the left show the detected results with only L_{cost_cls} incorporated into the training loss. (a,e) $\alpha_{lesion} = 1, \beta_{lesion} = 1$; (b,f) $\alpha_{lesion} = 1, \beta_{lesion} = 3$; (c,g) $\alpha_{lesion} = 3, \beta_{lesion} = 1$. (d) Apparent Diffusion Coefficient (ADC) image; (h) Diffusion-Weighted Images (DWI) b-2000 image (Color figure online).

$1, \beta_{slice} = 3$. With $\alpha_{slice} = 3, \beta_{slice} = 1$, (a) the lesion-level FNR was reduced from 0.1190 to 0.0970; (b) the slice-level FNR remained to be 0.0000 while the slice-level FPR was also reduced from 0.8505 to 0.8234, compared to the case where $\alpha = 1, \beta = 1$. With $\alpha_{slice} = 1, \beta_{slice} = 3$, (a) the FPR was reduced from 0.8505 to 0.8277; (b) the lesion-level FP was increased from 1.7202 to 1.7965, compared to the case where $\alpha = 1, \beta = 1$.

By comparing the second column in Table 1 and the second column in Table 2, we can find that the lesion-level and slice-level FNRs were reduced from 0.1941 to 0.1190 and from 0.0097 to 0, respectively. Comparing the third column in Table 1 and the third column in Table 2, finds that lesion-level and slice-level FNRs were reduced from 0.1013 to 0.0970 and from 0.0028 to 0.0000 respectively while (1) the lesion-level FP was reduced from 2.0218 to 1.9493; (2) the slice-level FPR was reduced from 0.8434 to 0.8234. The improvements in both FPRs and FNRs, by incorporating and further changing the slice-level cost, indicate the benefits and the significance of using the slice-level cost-sensitive classification loss.

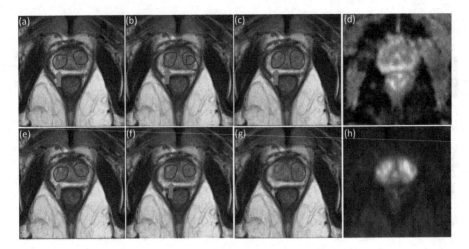

Fig. 3. This figure demonstrates that both the lesion-level and slice-level FNs were reduced by incorporating L_{slice_cls} into the training process. In all the ablation examples presented in this figure, threshold $= 0.95$. Example sub-figures shown here in the same row correspond to the same slice in the test data set. (a,e,i) depicts the detection results with only the lesion-level classification loss L_{cost_cls} incorporated, where $\alpha_{lesion}, \beta_{lesion}$ vary. (b,f,j) depicts the detection results with both L_{cost_cls} and L_{slice_cls} utilized in the training. The weighting schemes in this ablation study are as follows. (a) $\alpha_{lesion} = 1, \beta_{lesion} = 1$; (b) $\alpha = 1, \beta = 1$; (e) $\alpha_{lesion} = 1, \beta_{lesion} = 3$; (f) $\alpha = 1, \beta = 3$; (i) $\alpha_{lesion} = 3, \beta_{lesion} = 1$; (j) $\alpha = 3, \beta = 1$. (c,g,k) ADC images; (d,h,l) DWI b-2000 images.

4.3   Adjusting Mis-classification Cost at Both Levels

In this experiment, we study the effect of changing both L_{cost_cls} and L_{slice_cls} on the performance by varying α and β. Table 3 shows the corresponding results with three different settings of α and β: (a) $\alpha_{lesion/slice} = 1, \beta_{lesion/slice} = 1$; (b) $\alpha_{lesion/slice} = 3, \beta_{lesion/slice} = 1$; (c) $\alpha_{lesion/slice} = 1, \beta_{lesion/slice} = 3$. With $\alpha = 3, \beta = 1$, compared to the case where $\alpha = 1, \beta = 1$, (a) the lesion-level FNR was reduced from 0.1190 to 0.0734; (b) the slice-level FNR remained to be 0. With $\alpha = 1, \beta = 3$, compared to the case where $\alpha = 1, \beta = 1$, (a) the lesion-level FP was reduced from 1.7202 to 1.0982; (b) the slice-level FPR was reduced from 0.8505 to 0.6576.

By comparing the corresponding results in the same columns in Table 3 with those in Table 1 respectively, both the lesion-level and slice-level FNRs were substantially reduced by incorporating the slice-level classification loss L_{slice_cls} into training. By comparing corresponding results in the third column in Table 3 with those in Table 2, (1) the lesion-level FNR was reduced from 0.0970 to 0.0734; (2) the slice-level FNR remained to be 0. By comparing corresponding results in the last column in Table 3 with those in Table 2, it becomes clear that (1) the lesion-level FP was reduced from 1.7965 to 1.0982; (2) the slice-level FPR was reduced from 0.8277 to 0.6576.

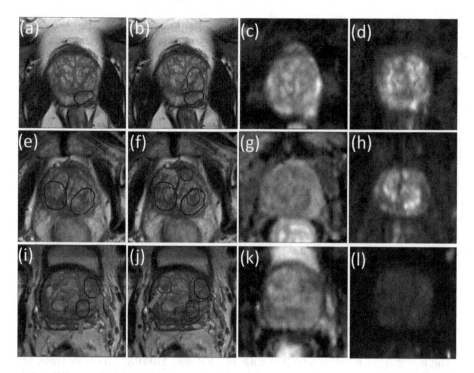

Fig. 4. This figure demonstrates the reduction of the lesion-level FPs by changing the lesion-level classification cost L_{cost_cls}. The same setting in training was adopted as that in Fig. 2, and all example sub-figures shown here correspond to the performances on one same slice in the test data set (but a different slice with that in Fig. 2. In the first row, threshold = 0.7 while threshold = 0.95 in the second row. The weighting schemes are summarised as follows: (a,e) $\alpha_{lesion} = 1, \beta_{lesion} = 1$; (b,f) $\alpha_{lesion} = 3, \beta_{lesion} = 1$; (c,g) $\alpha_{lesion} = 1, \beta_{lesion} = 3$. (d) ADC image; (h) DWI b-2000 image.

Figure 3 includes the three ablation examples where the slice-level FNs were reduced by incorporating the slice-level classification loss L_{slice_cls} into training. Three different slices are utilized to demonstrate the improvements in the three different rows in Fig. 3. Comparing Fig. 3(b) with Fig. 3(a), shows that the slice-level FN was reduced with the sacrifice of one more lesion-level FP. By comparing Fig. 3(f) with Fig. 3(e), we find that both lesion-level and slice-level FNs were reduced with one more lesion level FP. By comparing Fig. 3(j) with Fig. 3(i), we find that both lesion-level and slice-level FNs were reduced with the sacrifice of one more lesion-level FP.

4.4   Results Analysis

It should be noted that all the terms in the loss are weighted equally in this work. The effects of different weighting factors associated with different sub-tasks will be explored in the future. In addition, a wider range of α and β will be tested to find their optimal values. In this section, we quantitatively analyse the impact of changing the training-time cost-sensitive losses, compared with those where the threshold parameter was adjusted post-training. For brevity, in what follows, we use (1) $\alpha_{lesion}, \beta_{lesion}$ to refer to the case where only the cost-sensitive loss L_{cost_cls} was used in training; (2) $\alpha_{slice}, \beta_{slice}$ to refer to the case where $\alpha_{lesion} = 1, \beta_{lesion} = 1$ while the cost-sensitive slice-level loss L_{slice_cls} was also utilized in training, and the weights may vary; (3) α, β to refer to the case where both L_{cost_cls} and L_{slice_cls} were used in training, and the weights in the both losses can change.

We further group the interesting conclusions into positive and negative results, indicating the resulting impact difference to our specific PCa application. These, however, may not generalise to other clinical applications that adopt the same proposed cost-adjusting strategies.

Positive Results

1. With $\alpha_{lesion} = 1$, $\beta_{lesion} = 1$, by adjusting the post-training threshold, the lesion-level FNR was reduced to 0.1131 with the lesion-level FP being **5.9758**. In contrast, (1) with $\alpha_{lesion} = 3, \beta_{lesion} = 1$, the lesion-level FNR was 0.1013 while the FP was **2.0218**; (2) with $\alpha_{slice} = 1, \beta_{slice} = 1$, the lesion-level FNR was 0.1190 while the FP was **1.7202**; (3) with $\alpha_{slice} = 3, \beta_{slice} = 1$, the lesion-level FNR was 0.0970 with the FP was **1.9493**. To summarize, by choosing the appropriate loss during training, a considerable lower FP value can be achieved with comparable or reduced lesion-level FNs, compared to those from changing the threshold.
2. With $\alpha_{lesion} = 1, \beta_{lesion} = 1$, by adjusting the threshold, the slice-level FNR was reduced to be 0.0042 with the FPR being **0.6972**. In contrast, with $\alpha = 1, \beta = 3$, the slice-level FNR was 0.0014 while the FPR was **0.6576**.
3. With $\alpha_{slice} = 1, \beta_{slice} = 1$, by adjusting the threshold, the lesion-level FNR was reduced to 0.0987 while the FP was **2.0530**. In contrast, with $\alpha_{slice} = 3, \beta_{slice} = 1$, the lesion-level FNR and FP were 0.0970 and **1.9493**, respectively.
4. With $\alpha_{slice} = 3, \beta_{slice} = 1$, compared to the case where $\alpha_{slice} = 1, \beta_{slice} = 1$, the slice-level FPR was reduced to 0.8234 while the FNR remained to be 0.
5. With $\alpha = 1, \beta = 1$, by adjusting the threshold, the lesion-level FNR was reduced to be 0.0734 while the lesion-level FP was **5.4910**. In contrast, with $\alpha = 3, \beta = 1$, the lesion-level FP was **2.3827** while the FNR was 0.0734.
6. With $\alpha = 1, \beta = 1$, the slice-level FNR was reduced to be 0.014 with the slice-level FPR being **0.7161**. In contrast, with $\alpha = 1, \beta = 3$, the slice FNR and the slice-level FPR were 0.0014 and **0.6576**, respectively.

Comparing results in 1 and 5, at the lesion level, shows that the added FPs can be reduced in order to achieve a lower FNR by simply adding the classification loss at the slice level. The above results demonstrate the significant advantage of incorporating the cost-sensitive classification loss in reducing the lesion-level and slice-level FNRs.

Negative Results

1. With $\alpha = 1, \beta = 1$, by adjusting the threshold, the slice-level FNR was reduced to be 0 with the FPR being 0.9166, which is smaller than 0.9220 where $\alpha = 3, \beta = 1$.
2. With $\alpha_{lesion} = 1, \beta_{lesion} = 1$, by adjusting the threshold, the lesion-level FP was reduced to be **0.4774** with the FNR being **0.3662**. These two values are smaller than those where $\alpha_{lesion} = 1, \beta_{lesion} = 3$, respectively.
3. At the slice level where $\alpha = 1, \beta = 1$, the FNR was reduced to be 0.014 with FPR being **0.7161**. In contrast, in the case where $\alpha = 1, \beta = 3$, FNR was 0.0014 with FPR being **0.8277**.

In the training data set, the class imbalance problem was present where much more background objects/slices exist. Interestingly, this is the reason we believe that the so-called negative results originated in this application, in which, a greater weighting towards the majority class(es) would further reduce the biased (usually lower) prediction performance on the minority class(es), although the associated costs may have been correctly minimised. Further analysis for this phenomenon between prediction performance with- and without considering costs might warrant further investigation.

5   Conclusions

In this study, we explore the feasibility of controlling the false positives/negatives at the lesion or slice level during training, together with an in-depth analysis of the associated advantages and disadvantages. We conclude the quantitative results obtained from the clinical patient data set as follows: 1) Incorporating the proposed cost-sensitive classification losses at either lesion or slice level (or both) demonstrates the expected flexibility of controlling the false positive rate (FPR) and false negative rate (FNR); and 2) Incorporating the proposed cost-aware losses was able to reduce the FNRs while maintaining or further reducing the FPRs, which can be particularly useful for real-world clinical applications such as population screening for prostate cancer.

Acknowledgements. This work is supported by the Wellcome/EPSRC Centre for Interventional and Surgical Sciences (203145Z/16/Z). This work was supported by the International Alliance for Cancer Early Detection, a partnership between Cancer Research UK [C28070/A30912; C73666/A31378], Canary Center at Stanford University, the University of Cambridge, OHSU Knight Cancer Institute, University College London and the University of Manchester.

References

1. Cao, R., et al.: Joint prostate cancer detection and Gleason score prediction in mp-MRI via FocalNet. IEEE Trans. Med. Imaging **38**(11), 2496–2506 (2019)

2. Cao, R., et al.: Prostate cancer detection and segmentation in multi-parametric MRI via CNN and conditional random field. In: 2019 IEEE 16th International Symposium on Biomedical Imaging (ISBI 2019), pp. 1900–1904. IEEE (2019)

3. Dai, Z., et al.: Segmentation of the prostatic gland and the intraprostatic lesions on multiparametic magnetic resonance imaging using mask region-based convolutional neural networks. Adv. Radiat. Oncol. 5(3), 473–481 (2020)

4. He, K., Zhang, X., Ren, S., Sun, J.: Deep residual learning for image recognition. In: Proceedings of the IEEE Conference on Computer Vision and Pattern Recognition, pp. 770–778 (2016)

5. Li, W., et al.: Path R-CNN for prostate cancer diagnosis and Gleason grading of histological images. IEEE Trans. Med. Imaging 38(4), 945–954 (2018)

6. Litjens, G., Debats, O., Barentsz, J., Karssemeijer, N., Huisman, H.: Computer-aided detection of prostate cancer in MRI. IEEE Trans. Med. Imaging 33(5), 1083–1092 (2014)

7. Liu, L., et al.: Deep learning for generic object detection: a survey. Int. J. Comput. Vis. 128(2), 261–318 (2020)

8. Ren, S., He, K., Girshick, R., Sun, J.: Faster R-CNN: towards real-time object detection with region proposal networks. arXiv preprint arXiv:1506.01497 (2015)

9. Saha, A., Hosseinzadeh, M., Huisman, H.: End-to-end prostate cancer detection in bpMRI via 3D CNNs: effect of attention mechanisms, clinical priori and decoupled false positive reduction. arXiv preprint arXiv:2101.03244 (2021)

10. Sanford, T., et al.: Deep-learning-based artificial intelligence for PI-RADS classification to assist multiparametric prostate MRI interpretation: a development study. J. Magn. Reson. Imaging 52(5), 1499–1507 (2020)

11. Schelb, P., et al.: Classification of cancer at prostate MRI: deep learning versus clinical PI-RADS assessment. Radiology 293(3), 607–617 (2019)

12. Siegel, R.L., Miller, K.D., Jemal, A.: Cancer statistics, 2020. CA Cancer J. Clin. 70(1), 7–30 (2020)

13. Turkbey, B., Choyke, P.L.: PIRADS 2.0: what is new? Diagn. Interven. Radiol. 21(5), 382 (2015)

14. Wildeboer, R.R., van Sloun, R.J., Wijkstra, H., Mischi, M.: Artificial intelligence in multiparametric prostate cancer imaging with focus on deep-learning methods. Comput. Meth. Prog. Biomed. 189, 105316 (2020)

15. Yu, X., et al.: False positive reduction using multiscale contextual features for prostate cancer detection in multi-parametric MRI scans. In: 2020 IEEE 17th International Symposium on Biomedical Imaging (ISBI), pp. 1355–1359. IEEE (2020)

16. Yu, X., et al.: Deep attentive panoptic model for prostate cancer detection using biparametric MRI scans. In: Martel, A.L., et al. (eds.) MICCAI 2020. LNCS, vol. 12264, pp. 594–604. Springer, Cham (2020). https://doi.org/10.1007/978-3-030-59719-1_58

Optimising Knee Injury Detection with Spatial Attention and Validating Localisation Ability

Niamh Belton[1,2](✉)[iD], Ivan Welaratne[3][iD], Adil Dahlan[2][iD],
Ronan T. Hearne[4][iD], Misgina Tsighe Hagos[1,5][iD], Aonghus Lawlor[5,6][iD],
and Kathleen M. Curran[1,2][iD]

[1] Science Foundation Ireland Centre for Research Training in Machine Learning,
Dublin, Ireland
[2] School of Medicine, University College Dublin, Dublin, Ireland
niamh.belton@ucdconnect.ie, {adil.dahlan,kathleen.curran}@ucd.ie
[3] Department of Radiology, Mater Misericordiae University Hospital, Dublin, Ireland
[4] School of Electronic and Electrical Engineering, University College Dublin, Dublin,
Ireland
ronan.hearne@ucdconnect.ie
[5] School of Computer Science, University College Dublin, Dublin, Ireland
misgina.hagos@ucdconnect.ie, aonghus.lawlor@ucd.ie
[6] Insight Centre for Data Analytics, University College Dublin, Dublin, Ireland

Abstract. This work employs a pre-trained, multi-view Convolutional Neural Network (CNN) with a spatial attention block to optimise knee injury detection. An open-source Magnetic Resonance Imaging (MRI) data set with image-level labels was leveraged for this analysis. As MRI data is acquired from three planes, we compare our technique using data from a single-plane and multiple planes (multi-plane). For multi-plane, we investigate various methods of fusing the planes in the network. This analysis resulted in the novel 'MPFuseNet' network and state-of-the-art Area Under the Curve (AUC) scores for detecting Anterior Cruciate Ligament (ACL) tears and Abnormal MRIs, achieving AUC scores of 0.977 and 0.957 respectively. We then developed an objective metric, Penalised Localisation Accuracy (PLA), to validate the model's localisation ability. This metric compares binary masks generated from Grad-Cam output and the radiologist's annotations on a sample of MRIs. We also extracted explainability features in a model-agnostic approach that were then verified as clinically relevant by the radiologist.

Keywords: Deep learning · Musculoskeletal · Magnetic Resonance Imaging · Medical imaging · Spatial attention · Explainability

1 Introduction

Knee injuries are one of the most prevalent injuries in sporting activities [13]. Musculoskeletal (MSK) knee injuries can be detrimental to athletes' careers.

© Springer Nature Switzerland AG 2021
B. W. Papież et al. (Eds.): MIUA 2021, LNCS 12722, pp. 71–86, 2021.
https://doi.org/10.1007/978-3-030-80432-9_6

Such injuries require early intervention and appropriate rehabilitation. Magnetic Resonance Imaging (MRI) is the gold standard imaging modality for non-invasive knee injury diagnosis [21]. MRI is a volumetric imaging technique that is acquired from three planes, namely axial, coronal and sagittal. Machine Learning (ML) can be used to develop Computer Aided Detection (CAD) systems that automate tasks such as detecting injuries from MRIs. These systems have several benefits including reduced diagnosis time for patients and alleviating the ever-growing workload of radiologists by assisting diagnosis.

With the growing interest in CAD systems, there is a corresponding growing requirement for the CAD systems to be explainable. Explainability increases clinicians' confidence in the models and allows for smoother integration of these models into clinical workflows [6]. Validating the localisation ability and extracting features of automated injury detection systems aids their explainability. There has been significant research in the area of assessing the localisation ability of various saliency techniques [1, 7]. There is, however, a lesser focus on assessing a model's localisation ability. While object detection algorithms can be specifically trained to locate the site of the target task, they have increased model complexity and they require annotations of the complete data set. It is more often that saliency maps are employed to assess the localisation ability of a classification model. In this work, we validate the model's localisation ability by comparing the saliency map technique Grad-Cam [20] to annotations defined by the radiologist. This analysis found that existing segmentation metrics such as Intersection over Union (IoU) and Dice Coefficient are not suitable for quantifying the localisation ability of a model. This highlights the need for an interpretable metric that communicates an accurate representation of a model's localisation ability.

In this work, we propose a CNN with a ResNet18 architecture and integrated spatial attention block for MSK knee injury detection. The main aspects of this work are as follows.

1. We investigate if an injury can be accurately detected using data from a single-plane or if additional planes are required for optimal detection accuracy.
2. For cases where additional planes increase detection accuracy, we investigate methods of fusing planes in a network and develop a new multi-plane network, 'MPFuseNet', that achieves state-of-the-art performance for ACL tear detection.
3. We develop an objective metric, Penalised Localisation Accuracy (*PLA*), to validate the localisation ability of the model. This proposed metric communicates a more accurate representation of the model's localisation ability than typical segmentation metrics such as IoU and Dice.
4. We extract explainability features using a post-hoc approach. These features were then validated as clinically relevant.

2   Related Work

The research in the field of ML with MRI for MSK injury detection was accelerated by the release of the open-source MRNet data set [3]. This is a rich data set of knee MRIs with image-level labels acquired from the Stanford Medical Centre. Bien [3] implemented a pre-trained AlexNet architecture for each plane and trained separate models for detecting different types of injuries. Azcona [2] followed a similar approach and employed a ResNet18 with advanced data augmentation, while Irmakci [10] investigated other classic Deep Learning (DL) architectures. Tsai [25] employed the ELNet, a CNN trained end-to-end with additional techniques such as multi-slice normalisation and blurpool. This method was shown to have superior performance than previous methods. Other research studies have also demonstrated the effectiveness of using Deep Learning for detecting ACL tears on additional data sets [12].

Spatial attention is another area that has sparked interest in the ML and medical imaging community. Spatial attention is a technique that focuses on the most influential structures in the medical image. This technique has been shown to improve the performance of various medical imaging tasks such as image segmentation [19] and object detection [24]. Tao [24] implemented 3D spatial and contextual attention for deep lesion detection. We base our spatial attention block on this work.

The rapid growth of ML techniques for medical tasks has highlighted the requirement for DL models to be explainable. Several saliency map techniques have been proposed in recent years [4, 20, 22]. These techniques highlight regions of an image that are influential in the output of the model. There has also been extensive work in the area of using ML to extract features from models. Recently, Janik [11] used Testing with Concept Activation Vectors (TCAV), originally proposed by Kim [14], to extract underlying features for cardiac MRI segmentation. These features not only aid the explainability of the model but they can also discover previously unknown aspects of the pathology.

3   Materials

The MRNet data set first published by Bien [3] consists of 1,250 MRIs of the knee. Each MRI is labelled as having an Anterior Cruciate Ligament (ACL) tear, meniscus tear and/or being abnormal. The MRIs were acquired using varying MRI protocols as outlined in Table S1 in the original paper [3]. Each plane was made available in a different MRI sequence. The sequences for the sagittal, coronal and axial planes are T2-weighted, T1-weighted and Proton Density-weighted respectively [3]. The MRI data was previously pre-processed to standardise the pixel intensity scale. This was conducted using a histogram intensity technique [3, 18] to correct inhomogeneous pixel intensities across MRIs.

4   Method

4.1   Model Backbone

The backbone model is a 2D multi-view ResNet18. Azcona [2] demonstrated that a ResNet18 [8], pre-trained on the ImageNet data set [5] outperforms other classic architectures such as AlexNet for detecting knee injuries from MRI data. For this reason, we employ a ResNet18 architecture. The weights were initialised with the pre-trained model weights and all weights were subsequently fine-tuned.

MRI data is volumetric and therefore, each plane of an MRI is made up of several images known as 'slices'. A multi-view technique was employed to combine slices of the same MRI in the network. 'Views' in the multi-view network are equivalent to slices. The batch dimension of the multi-view CNN is equal to the number of slices in one MRI. The slices are input into the multi-view CNN and the output has dimensions (b, f) where b is the batch dimension, equal to the number of slices of an MRI and f is a one-dimensional (1D) vector that is representative of each slice. Su [23] demonstrated that an element-wise maximum operation was optimal for combining views. For this reason, we implemented an element-wise maximum operation across the batch dimension to obtain a final 1D vector of combined views (slices) of size $(1, f)$ that is representative of a stack of slices from one plane of an MRI. This vector is then passed through a fully connected layer to produce the final output.

4.2   Spatial Attention

A soft spatial attention mechanism is integrated into our base model. This follows the spatial attention method proposed by Tao [24] who integrated spatial attention into a VGG16 for deep lesion detection. We opt for a ResNet18 as its performance for this task has been demonstrated and it has approximately 127 million less parameters than the VGG16. The attention block is designed as follows.

We define our output volume D^l from a convolutional layer l of size (b, c, h, w) where b is number of slices, c is the number of channels (feature maps), and h and w are the height and width of the feature maps. In the attention block, volume D^l is input into a 1×1 convolution. This produces an output volume A^l of shape (b, c, h, w), where the dimensions are of identical size to D^l. A softmax is applied to each feature map of A^l so that each of the c feature maps of dimensions (h, w) sum to one. Small values are sensitive to the learning rate [24] and therefore, each feature map is normalised by dividing the values of the feature map by the maximum value in the feature map. The resulting attention mask $A^{l'}$ is then multiplied by the original convolutional volume D^l.

The volume $A^{l'}$ is equivalent to a 3D spatial weight matrix where each 2D feature map in volume $A^{l'}$ is an attention mask for each feature map in volume D^l. Attention masks are known to switch pixels in the original volume 'on' if they are important and 'off' if they are not important to the target task [17]. Standard attention mechanisms develop a single 2D attention map, while our method implements a 3D attention mechanism that creates a tailored attention mask for each feature map. Figure 1 uses Grad-Cam to illustrate three example cases where the addition of the spatial attention block correctly redirects the model to more precisely locate a knee injury and associated abnormalities. The examples shown are MRI slices from the sagittal plane that were annotated by a radiologist on our team (I.W.).

4.3 Single-Plane and Multi-plane Analysis

For the single-plane analysis, a separate model was trained for each plane and each task. Figure 2 shows the architecture of the single-plane model. All slices from one plane of one MRI are input into a 2D multi-view ResNet18 with a spatial attention block of output size $(s_{plane}, 512, 8, 8)$, where s_{plane} is the number of slices of an MRI from one plane. The model then has a Global Average Pooling layer of output size $(s_{plane}, 512)$, a fully connected layer of output size $(s_{plane}, 1000)$, an element-wise maximum operation of output size $(1, 1000)$, a final fully connected layer and a sigmoid function to get the final output.

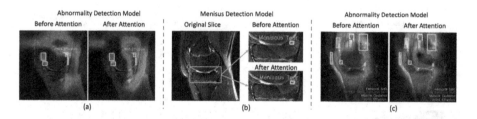

Fig. 1. The examples shown were generated from the same model before and after the spatial attention block was applied. (a) Spatial attention redirects the attention of the model to an abnormality (posterior joint effusion). Note that Grad-Cam overlaps with the annotations on the anterior knee in other slices. (b) A threshold is applied to Grad-Cam on both images to only colour-code the most influential regions. This then highlights how the spatial attention fine-tunes the localisation ability of the model. (c) Spatial attention removes focus from the healthy femur to focus on abnormalities.

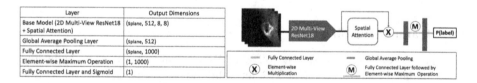

Layer	Output Dimensions
Base Model (2D Multi-View ResNet18 + Spatial Attention)	(splane, 512, 8, 8)
Global Average Pooling Layer	(splane, 512)
Fully Connected Layer	(splane, 1000)
Element-wise Maximum Operation	(1, 1000)
Fully Connected Layer and Sigmoid	(1)

Fig. 2. Output dimensions of each layer and architecture of the Single-Plane model.

A multi-plane analysis was then conducted in order to investigate if using data from multiple planes increases injury detection ability. Figure 3 visualises the multi-plane networks. These networks follow the same general architecture as the single-plane models. Three methods of fusing planes along this architecture were investigated. All of the multi-plane networks start with the Base Model (BM). The BM is the 2D multi-view ResNet18 with spatial attention block. The methods of fusing planes are described as follows.

1. The first multi-plane join is named MPFuseNet. This network fuses the output from the BM of each plane. The output volumes from each plane are of

dimensions $(s_{axial}, 512, 8, 8)$, $(s_{coronal}, 512, 8, 8)$ and $(s_{sagittal}, 512, 8, 8)$. These volumes are fused along the batch dimension, resulting in a fused volume of dimensions $((s_{axial} + s_{coronal} + s_{sagittal}), 512, 8, 8)$.

2. 'Multi-plane Join 2' (MP2) fuses planes after the first fully connected layer and element-wise maximum operation. This converts the volume of each plane from dimensions $(1, 1000)$ to a combined volume of dimensions $(1, 3000)$.

3. 'Multi-Plane Logistic Regression' (MPLR) trains each CNN separately and combines the predictions with a logistic regression model. This is the most common method in the literature [2,3].

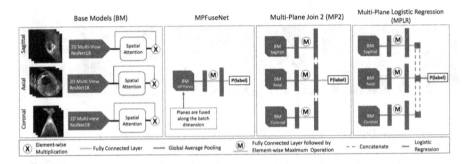

Fig. 3. The BM for each plane is a 2D multi-view CNN with ResNet18 architecture and integrated spatial attention. MPFuseNet fuses the planes after the BM. MP2 fuses planes after the first fully connected layer and element-wise maximum operation. MPLR fuses the final prediction from each plane.

4.4 Training Pipeline

The models were trained on the official MRNet training set using eight-fold cross-validation and the official MRNet validation set was used as an unseen test set. Affine data augmentation techniques such as flipping, translating and rotating were applied to slices to improve model robustness. We employed the Adam optimiser [15], with an initial learning rate of 1e−5 and a weighted cross entropy loss function. As the element-wise maximum operation for the multi-view technique is implemented over the batch dimension, the batch size is set to one to avoid combining slices of different MRIs.

5   Evaluation

5.1   Quantitative

Table 1 reports the Area Under the Curve (AUC) of all multi-plane models and the best performing single-plane models for each task on the validation data. The results demonstrate that although single-plane methods can accurately detect knee injuries, including data from additional planes further improves performance. However, the training time is three times longer for multi-plane models with the number of trainable parameters tripling in comparison to single-plane models. The trade-off between performance and model complexity should be considered.

In the case of the multi-plane methods, MPLR resulted in optimal performance for detecting abnormal MRIs. However, MPLR was detrimental to performance for detecting ACL and meniscus tears. This is because single-plane models that had a lower validation AUC had a higher training AUC than other models. This resulted in the logistic regression weighting the sub-optimal performing single-plane models higher than the best performing single-plane models, resulting in a performance degradation.

MPFuseNet performed optimally for detecting ACL and meniscus tears. This network reduces the risk of overfitting by performing the element-wise maximum operation over all planes. It does not perform optimally for abnormality detection. However, the nature of this task is substantially different to the other tasks. The models for ACL and meniscus tear detection determine whether a tear is present or not, while the abnormality detection model determines if there is a major abnormality or several minor abnormalities. Particularly in cases where the model has detected several minor abnormalities, it is possible that performing the maximum operation over all planes in MPFuseNet causes significant information loss and therefore, degrades the model's performance.

This analysis has demonstrated that MPFuseNet outperforms MPLR for ACL and meniscus tear detection and that MPLR can be detrimental to the model's performance. This is a significant finding given that MPLR is the most common method in the literature.

From the validation results in Table 1, we select the novel MPFuseNet for ACL and meniscus tear detection and MPLR for abnormality detection as our final models.

Table 1. Comparison of single-plane and multi-plane on validation data (AUC)

Model	ACL	Abnormal	Meniscus
Single-Plane (SP)	0.953_{Axial}	$0.923_{Coronal}$	$0.862_{Coronal}$
Multi-Plane Logistic Regression (MPLR)	0.923	**0.952**	0.862
MPFuseNet	**0.972**	0.916	**0.867**
Multi-Plane Join 2 (MP2)	0.948	0.903	0.853

Tsai [25] trained the ELNet using four fold cross-validation. Table 2 compares the performance of our selected ResNet18 + Spatial Attention models and the performance of ELNet and MRNet on unseen test data, as reported by Tsai [25]. The ELNet improved on the MRNet performance by 0.004 and 0.005 AUC for

ACL and abnormality detection respectively, while our proposed models for ACL and abnormality detection result in a substantial performance improvement. Although our proposed model for meniscus detection does not outperform the ELNet, it shows an improvement on the MRNet performance.

Table 2. Comparison with known methods on unseen test data (AUC)

Model	ACL	Abnormal	Meniscus
Proposed Models	$0.977_{MPFuseNet}$	0.957_{MPLR}	$0.831_{MPFuseNet}$
ELNet [25]	0.960	0.941	**0.904**
MRNet [3, 25]	0.956	0.936	0.826

5.2   Ablation Study

An ablation study was conducted to demonstrate the effects of adding the spatial attention block. A multi-view ResNet18 with no spatial attention was trained for this study. Figure 4 shows the Receiver Operator Curves (ROC) and the AUC for single-plane models with and without spatial attention on the test set. The addition of the spatial attention block results in increased performance, most notably for meniscus tear and abnormality detection.

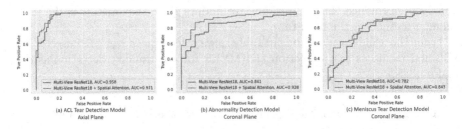

Fig. 4. ROC curves for ResNet18 with and without spatial attention.

6   Explainability

This section aims to validate the model by assessing the localisation ability and extracting features. This will aid the explainability of the model.

For this study, the radiologist annotated MRIs from 50 randomly sampled healthy and unhealthy MRIs. The radiologist used bounding boxes to annotate meniscus tears, ACL tears and significant abnormalities that can be associative or non-associative of the aforementioned tears. Examples of the annotated abnormalities include joint effusion and bone bruising. Out of the sample of annotated MRIs, MRIs where the radiologist annotated an ACL tear and the single-plane model correctly detected the ACL tear were used for this analysis, resulting in a sample of twelve MRIs. There were additional cases where an ACL tear was found by the radiologist on another plane but they were not visible to the radiologist on the plane in question due to the thickness of the slices. In some of these cases, the models confidently detected a tear on the plane where it was not visible. This indicates that there are other features that the model relies on or the model has found evidence of a tear that is not visible to the human eye.

6.1   Localisation Ability

As previously outlined, object detection models can be trained specifically to localise the tear. However, these models require localised annotations. The MRNet data set provides only image-level labels. Therefore, we use the output of Grad-Cam and the sample of MRIs that were annotated by the radiologist to quantify the model's localisation ability. This task requires a metric that meets the following two criteria.

Does Not Over-Penalise False Positive Regions: Grad-Cam can highlight excess area around the region of interest. This excess area outside the bounds of an annotation is a false positive region. This results in low IoU and Dice scores that over-penalise the model's localisation abilities. Furthermore, heavy penalisation of false positive regions is not ideal for assessing localisation ability for medical imaging tasks as it is likely that the area outside the bounds of the annotation may also be abnormal. For example, there could be visible joint effusion or tissue damage in the region around an ACL tear annotation.

Stand-Alone: Although IoU and Dice over-penalise the model's localisation abilities, they are still informative when comparing localisation abilities across models. However, a stand-alone metric can communicate an accurate representation of the localisation capabilities in the absence of any comparative information.

A metric that meets the two criteria will accurately reflect a model's localisation ability. Our proposed metric, PLA, meets the aforementioned criteria.

Localisation Accuracy and Penalised Localisation Accuracy. In order to calculate Localisation Accuracy (LA) and PLA, we compare the output of Grad-Cam to the annotations defined by the radiologist. Grad-Cam generates a matrix that assigns a feature importance score between zero and one to each pixel. This matrix is normally shown as a heat-map overlaid on the MRI slice in question (Fig. 5(a)). A Grad-Cam mask was generated based the pixel importance values. The masks had pixel values of one when the pixel importance value was above the threshold of 0.6 and had pixel values of zero when the pixel importance value was below the threshold of 0.6 (Fig. 5(b)). The threshold of 0.6 was chosen as pixel importance values above 0.5 have above average importance. However, 0.5 was found to generate large Grad-Cam masks and higher threshold values created Grad-Cam masks that were sometimes smaller than the annotation. Therefore, the optimal threshold value is 0.6 for this study. Similarly, an annotation mask was created for each annotated MRI slice where pixels within the bounds of the annotation have a value of one and pixels outside the bounds of the annotation have a value of zero (Fig. 5(d)). The Grad-Cam and annotation mask can then be used to calculate metrics such as LA to quantify the localisation ability. Figure 5 outlines this process for assessing ACL tear localisation. The metric LA is calculated as the percentage of the annotation that is covered by the Grad-Cam mask. The formula for LA is shown in Eq. 1. One of the limitations of this metric is that if the Grad-Cam mask covers a significantly large area, the annotation is likely to overlap with the Grad-Cam mask due to random chance. We therefore introduce an adjusted version of LA, PLA. This metric is equal to LA with a penalty for false positive regions. However, unlike IoU and Dice, it does not over-penalise false positive regions. It is calculated using the formulae shown in Eq. 2 and 3.

$$LA_x = \frac{overlap_x}{ann} \tag{1}$$

$$FPP_x = \frac{gc_x - overlap_x}{total} \tag{2}$$

$$PLA_x = MAX(LA_x - FPP_x, 0) \tag{3}$$

$overlap_x$ is the number of pixels that are both, within the bounds of the annotation and have a pixel value of one in the Grad-Cam mask. The subscript x means that the metric was calculated based on a Grad-Cam mask generated at pixel importance threshold x. As outlined earlier, we set x to equal the pixel importance threshold 0.6. ann is the number of pixels within the bounds of the radiologist's annotation. FPP_x is the False Positive Penalty (FPP). gc_x is the number of pixels that have a value of one in the Grad-Cam mask and $total$ is the total number of pixels in the image. Equation 3 also shows a maximum operation. This is to avoid negative PLA values. Without a maximum operation, PLA would have a negative value when there is no overlap between the Grad-Cam mask and the annotation mask.

PLA returns a value between zero and one where a score of one is perfect localisation. A score of one is achieved when the area of the Grad-Cam mask is equal to the area of the annotation mask and they have a perfect overlap. A score of zero is achieved if there is no overlap between the Grad-Cam mask and the annotation mask. PLA also effectively penalises false positive regions. For example, if the Grad-Cam mask covers 100% of the image and the annotation takes up 1% of the area, the value of LA would be one, while the value of PLA would be 0.01. Although PLA penalises false positive regions, it does not over-penalise them as is the case with IoU and Dice. This is demonstrated in a later

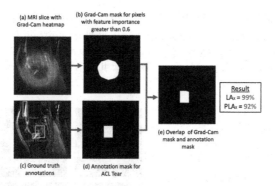

Fig. 5. Pipeline for calculating LA and PLA based on an ACL tear detection model.

paragraph. PLA is also a stand-alone metric, meaning it is interpretable and informative of the localisation ability without any comparative information.

Area Under the Curve for Localisation. The Area Under the Curve is another metric that can be used to assess a model's localisation ability that meets the previously outlined criteria. To calculate the AUC, the Grad-Cam output and annotation masks are flattened into 1D vectors. The Grad-Cam 1D vector is interpreted as the probability of the pixel showing the tear and the annotation 1D vector is the ground truth. The AUC is then calculated. This metric takes into account false positive regions and is advantageous over IoU and PLA as it does not require a Grad-Cam mask to be generated at a specific threshold. Instead, the direct Grad-Cam output can be used for the AUC calculation. However, it is not as easily interpreted as our proposed metric, PLA.

Metric Comparison. Figure 6(a) shows the value of each metric on a sample of MRIs. These scores are based on the ACL detection model that was trained on axial data. For each MRI, the metrics shown in the figure are based on the slice with the highest metric values as we interpret this to be the MRI's key slice. The metrics are shown on the same graph as all metrics have a range of possible values from zero to one. All metrics, with the exception of AUC, were calculated using a Grad-Cam mask generated at the 0.6 pixel importance threshold. AUC uses the direct Grad-Cam output. Figure 6(b) shows the Grad-Cam heatmap overlaid on the MRI slice and the corresponding radiologist annotations. As we are assessing the model's ability to localise an ACL tear, it is only the ACL tear annotations shown in blue that are of interest. Figure 6(b) also shows the values of the metrics for each example. The radiologist was satisfied with the model's localisation of the ACL tear in the examples shown. However, the IoU and Dice metrics have near zero values, indicating poor localisation ability. It can be seen from Fig. 6(a) that all samples have low IoU and Dice scores. The disagreement between these metrics and the radiologist's opinion demonstrates the ineffectiveness of IoU and Dice as stand-alone, interpretable metrics that can be used to assess localisation ability. This is due to their over-penalisation of false positive regions. A further cause for concern is that neither IoU or Dice show a dis-improvement in their scores for the problematic MRI Case 4. In this MRI case, the Grad-Cam mask overlaps with only 40% of the ACL tear annotation.

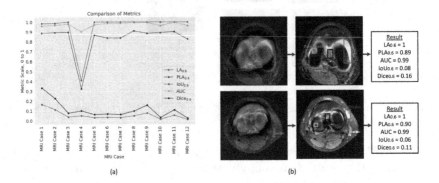

Fig. 6. (a) The $LA_{0.6}$, $PLA_{0.6}$, AUC, $IoU_{0.6}$ and $Dice_{0.6}$ scores for sample MRIs. (b) Grad-Cam overlaid on axial plane MRI slices and the corresponding radiologist annotations for MRI Case nine (top) and eleven (bottom). These metrics are based on the model's ability to locate the ACL tear (ACL tear annotations are shown in blue). (Color figure online)

Figure 6(b) shows that LA achieves a perfect score for almost all samples. However, LA will not accurately communicate the localisation ability of a model in cases where there are significantly large false positive regions. It can be concluded from the figure that our proposed PLA and AUC give the most accurate representation of the model's localisation ability. They both give high scores when the model has effectively localised the tear but still penalise false positive regions. An advantage of AUC is that it does not require a Grad-Cam mask to be generated based on a pixel importance threshold. It uses the direct Grad-Cam output. However, it appears to under-penalise the model in cases. MRI Case four, as previously outlined, only overlaps with 40% of the annotation. However, this case achieves a high AUC score of 0.9. Therefore, PLA can be considered to be advantageous over AUC. Moreover, our proposed metric, PLA is akin to a straight-forward accuracy calculation with an adjustment for false positive regions and therefore, it is more interpretable than AUC.

Aggregated Localisation Results. To quantify the localisation ability over the entire sample, we can calculate the percentage of cases where the model accurately located the tear based on the PLA scores. Value a_x^s is the PLA value of slice s of an MRI that overlaps with the Grad-Cam mask segmented at feature importance threshold x where x is equal to 0.6 for this study. The slice with the best a_x^s score is selected from each MRI as we interpret this slice to be the MRI's key slice. For each MRI, a_x^s is subsequently transformed to a binary outcome of

[0, 1] using a threshold value k. The value k can be varied in the range (0.5, 1). For each value of k, a_x^s is assigned a value of one if k is less than a_x^s and zero otherwise. Once a binary outcome has been obtained for each MRI, the accuracy can be calculated over the sample for each threshold value k. We report that 91.7% of the sample accurately localised the tear at k values ranging from 0.5 to 0.85. It is evident that our proposed model is correctly localising ACL tears and the output of the model is based on the site of the ACL tear. Moreover, all MRI cases have an AUC greater than 0.9. This further demonstrates the model's localisation ability.

6.2   Features

The radiologist's annotated abnormalities were used to extract features. The purpose of annotating abnormalities in addition to the ligament tears is to determine if there are additional structures present in the MRI that influence whether a model detects a ligament tear or not. Such 'structures' can be thought of as features. Figure 7 visualises the most common abnormalities that the ACL tear model detected. The table shows the number of MRIs where the feature was present and the detection rate. The detection rate is calculated based on number of times the Grad-Cam mask generated at the 0.6 pixel importance threshold covered over 60% of the annotation. The results show that joint effusion and bone bruising are the most common features of an ACL tear. The radiologist verified that these features are frequently seen in the setting of acute ACL tears.

It was also found from an analysis of the ACL detection model trained on sagittal data that there was a correlation between abnormal cases and cases where growth plates were visible on the MRI. Growth plates are found in children and adolescents. Grad-Cam highlighted these normal regions to indicate that they are influential in the model's output. Therefore, the model learned an incorrect relationship. This is an example of the model falling victim to the correlation versus causation fallacy. This further justifies the requirement of extracting features and clinically verifying them.

6.3   Limitations

Although the experiment quantified the localisation ability of the model and extracted features, there were a number of limitations. The annotations were limited to bounding boxes. This could result in excess area being included in the annotation if the abnormality is not rectangular. The radiologist minimised the effects of this by using multiple annotations for irregularly shaped abnormalities. An example of this is the joint effusion annotation in Fig. 7. The radiologist also noted that the slice thickness of some MR protocols made it difficult to detect and annotate tears. Grad-Cam has some known limitations such as not capturing the entire region of interest [4]. A variation of visualisation techniques could be employed in future to overcome this limitation.

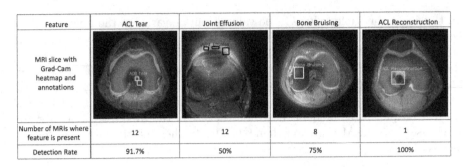

Fig. 7. Table demonstrating the number of MRIs where the feature is present and the feature detection rate.

7   Conclusion

We have demonstrated how spatial attention corrects the model's focus to salient regions. In our analysis of single and multi-plane, it was found that including data from additional planes increases the detection ability. However, the training time for multi-plane methods and the number of parameters is tripled and thus, the trade-off between model complexity and performance should be considered. We developed the multi-plane network, MPFuseNet, that outperforms the common MPLR method for ACL and meniscus tear detection. Furthermore, our proposed methods achieve state-of-the-art results for ACL tear and abnormality detection.

We developed an objective metric, PLA for quantifying and validating the localisation ability of our model. Using this metric, we verified that the ACL detection model accurately located the ACL tear in 91.7% of a sample of MRIs. We then extracted features from the model that were then verified by the radiologist. Validating the localisation ability and extracting features for explainability will improve clinical trust in ML systems and allow for smoother integration into clinical workflows. This work will be extended to consider the opinion of several radiologists and a larger data set of MRIs. Future work will also elaborate on the single-plane and multi-plane analysis by considering additional planes. Oblique MRI planes have previously been shown to improve ACL tear detection [16]. The quality of the extracted explainability features will also be assessed using the System Causability Scale [9] and by means of a user evaluation study with clinicians.

Code for this work will be made publicly available at https://git.io/J3pA5.

Acknowledgements. This work was funded by Science Foundation Ireland through the SFI Centre for Research Training in Machine Learning (18/CRT/6183). This work is supported by the Insight Centre for Data Analytics under Grant Number SFI/12/RC/2289_P2. This work was funded by Enterprise Ireland Commercialisation Fund under Grant Number CF 201912481. We would also like to acknowledge and thank Professor Kevin McGuinness from Dublin City University (DCU) for his guidance on the research study.

References

1. Arun, N., et al.: Assessing the (un)trustworthiness of saliency maps for localizing abnormalities in medical imaging (2020)
2. Azcona, D., McGuinness, K., Smeaton, A.F.: A comparative study of existing and new deep learning methods for detecting knee injuries using the MRNet dataset. arXiv preprint arXiv:2010.01947 (2020)
3. Bien, N., et al.: Deep-learning-assisted diagnosis for knee magnetic resonance imaging: development and retrospective validation of MRNet. PLOS Med. **15**(11), 1–19 (2018). https://doi.org/10.1371/journal.pmed.1002699
4. Chattopadhay, A., Sarkar, A., Howlader, P., Balasubramanian, V.N.: Grad-CAM++: generalized gradient-based visual explanations for deep convolutional networks. In: 2018 IEEE Winter Conference on Applications of Computer Vision (WACV), pp. 839–847 (2018). https://doi.org/10.1109/WACV.2018.00097
5. Deng, J., Dong, W., Socher, R., Li, L., Li, K., Fei-Fei, L.: ImageNet: a large-scale hierarchical image database. In: 2009 IEEE Conference on Computer Vision and Pattern Recognition, pp. 248–255 (2009). https://doi.org/10.1109/CVPR.2009.5206848
6. Floruss, J., Vahlpahl, N.: Artificial intelligence in healthcare: acceptance of AI-based support systems by healthcare professionals (2020)
7. Graziani, M., Lompech, T., Müller, H., Andrearczyk, V.: Evaluation and comparison of CNN visual explanations for histopathology (2020)
8. He, K., Zhang, X., Ren, S., Sun, J.: Deep residual learning for image recognition. In: Proceedings of the IEEE Conference on Computer Vision and Pattern Recognition, pp. 770–778 (2016). https://doi.org/10.1109/CVPR.2016.90
9. Holzinger, A., Carrington, A., Müller, H.: Measuring the quality of explanations: the system causability scale (SCS). KI - Künstliche Intelligenz **34**(2), 193–198 (2020). https://doi.org/10.1007/s13218-020-00636-z
10. Irmakci, I., Anwar, S.M., Torigian, D.A., Bagci, U.: Deep learning for musculoskeletal image analysis. In: 2019 53rd Asilomar Conference on Signals, Systems, and Computers, pp. 1481–1485. IEEE (2019). https://doi.org/10.1109/IEEECONF44664.2019.9048671
11. Janik, A., Dodd, J., Ifrim, G., Sankaran, K., Curran, K.: Interpretability of a deep learning model in the application of cardiac MRI segmentation with an ACDC challenge dataset. In: Išgum, I., Landman, B.A. (eds.) Medical Imaging 2021: Image Processing, vol. 11596, pp. 852–863. International Society for Optics and Photonics, SPIE (2021). https://doi.org/10.1117/12.2582227
12. Awan, M.J., Rahim, M.S.M., Salim, N., Mohammed, M.A., Garcia-Zapirain, B., Abdulkareem, K.H.: Efficient detection of knee anterior cruciate ligament from magnetic resonance imaging using deep learning approach. Diagnostics **11**(1) (2021). https://doi.org/10.3390/diagnostics11010105
13. Kennedy, M., Dunne, C., Mulcahy, B., Molloy, M.: The sports' clinic: a one year review of new referrals. Ir. Med. J. **86**(1), 29–30 (1993)
14. Kim, B., et al.: Interpretability beyond feature attribution: quantitative testing with concept activation vectors (TCAV) (2018)
15. Kingma, D.P., Ba, J.: Adam: a method for stochastic optimization (2017)
16. Kosaka, M., et al.: Oblique coronal and oblique sagittal MRI for diagnosis of anterior cruciate ligament tears and evaluation of anterior cruciate ligament remnant tissue. Knee **21**(1), 54–57 (2014). https://doi.org/10.1016/j.knee.2013.04.016

17. Mader, K.: Attention on pretrained-vgg16 for bone age (2018). https://www.kaggle.com/kmader/attention-on-pretrained-vgg16-for-bone-age
18. Nyúl, L.G., Udupa, J.K.: On standardizing the MR image intensity scale. Magn. Reson. Med. **42**(6), 1072–1081 (1999)
19. Oktay, O., et al.: Attention U-Net: learning where to look for the pancreas (2018)
20. Selvaraju, R.R., Cogswell, M., Das, A., Vedantam, R., Parikh, D., Batra, D.: Grad-CAM: visual explanations from deep networks via gradient-based localization. Int. J. Comput. Vision **128**(2), 336–359 (2019). https://doi.org/10.1007/s11263-019-01228-7
21. Skinner, S.: MRI of the knee. Aust. Fam. Physician **41**(11), 867–869 (2012)
22. Smilkov, D., Thorat, N., Kim, B., Viégas, F., Wattenberg, M.: SmoothGrad: removing noise by adding noise (2017)
23. Su, H., Maji, S., Kalogerakis, E., Learned-Miller, E.: Multi-view convolutional neural networks for 3D shape recognition. In: Proceedings of the IEEE International Conference on Computer Vision, pp. 945–953 (2015). https://doi.org/10.1109/ICCV.2015.114
24. Tao, Q., Ge, Z., Cai, J., Yin, J., See, S.: Improving deep lesion detection using 3d contextual and spatial attention (2019). https://doi.org/10.1007/978-3-030-32226-7_21
25. Tsai, C.H., Kiryati, N., Konen, E., Eshed, I., Mayer, A.: Knee injury detection using MRI with efficiently-layered network (ELNet). In: Proceedings of the Third Conference on Medical Imaging with Deep Learning, Montreal, QC, Canada. Proceedings of Machine Learning Research, vol. 121, pp. 784–794. PMLR, 06–08 July 2020. http://proceedings.mlr.press/v121/tsai20a.html

Improved Artifact Detection in Endoscopy Imaging Through Profile Pruning

Ziang Xu[1], Sharib Ali[1,2], Soumya Gupta[1], Numan Celik[1], and Jens Rittscher[1]($\boxtimes$)

[1] Department of Engineering Science, Big Data Institute, University of Oxford, Oxford, UK
{ziang.xu,sharib.ali,jens.rittscher}@eng.ox.ac.uk
[2] Oxford NIHR Biomedical Research Centre, Oxford, UK

Abstract. Endoscopy is a highly operator dependent procedure. During any endoscopic surveillance of hollow organs, the presence of several imaging artifacts such as blur, specularity, floating debris and pixel saturation is inevitable. Artifacts affect the quality of diagnosis and treatment as they can obscure features relevant for assessing inflammation and morphological changes that are characteristic to precursors of cancer. In addition, they affect any automated analysis. It is therefore desired to detect and localise areas that are corrupted by artifacts such that these frames can either be discarded or the presence of these features can be taken into account during diagnosis. Such an approach can largely minimise the amount of false detection rates. In this work, we present a novel bounding box pruning approach that can effectively improve artifact detection and provide high localisation scores of diverse artifact classes. To this end, we train an EfficientDet architecture by minimising a focal loss, and compute the Bhattacharya distance between probability density of the pre-computed instance specific mean profiles of 7 artifact categories with that of the predicted bounding box profiles. Our results show that this novel approach is able to improve commonly used metrics such as mean average precision and intersection-over-union, by a large margin.

Keywords: Endoscopy imaging · Artifact detection · Deep learning · Artifact profile

1 Introduction

Endoscopy is a widely used clinical procedure for assessing inflammation in hollow organs and performing subsequent treatments. Every endoscope typically consists of a light source and a camera connected to an external monitor to help clinicians navigate through the organ of interest in real-time. However, many factors such as lighting conditions, unstable motion, and organ structure result in

© Springer Nature Switzerland AG 2021
B. W. Papież et al. (Eds.): MIUA 2021, LNCS 12722, pp. 87–97, 2021.
https://doi.org/10.1007/978-3-030-80432-9_7

presence of inevitable imaging artifacts that hinder the endoscopic surveillance and affects imaging data analysis. In clinical endoscopic surveillance of gastrointestinal organs, more than 60% of endoscopic video frames can be affected by artifacts [1]. A large number of artifacts can significantly affect the lesion detection algorithms, surface-based 3D reconstruction and other automated analysis methods [2,3]. It is therefore extremely vital to find and locate these artifacts for a detailed endoscopic surveillance. While artifacts can be of various shapes, sizes and nature, accurate detection and precise localisation is difficult to achieve. We utilise deep learning-based object detectors to generate a list of potential artifacts, and then exploit the nature of each artifact instance to eliminate the false positives thereby enabling improved accuracy and localisation.

Widely used object detection methods can be categorised into a single-stage and two-stage detectors. Single-stage detectors, such as You Only Look Once (YOLO-v2) [4] and Single Shot multiBox Detector (SSD) [5], refer to detectors that directly regress the category probability and position coordinates, and only perform a single pass on the data. RetinaNet introduced by Lin et al. [6] is another single-stage detector where the authors introduced focal loss that puts a focus on hard examples to boost performance. Subramanian et al. [7] used RetinaNet model using ResNet101 for the detection of artifacts. Oksuz et al. [8] utilised the RetinaNet model and focal loss to detect artifacts in artifact detection challenge EAD2019 [1].

On the other hand, multi-stage detectors first generate the region proposal, and then perform object detection for the selected object known as *region proposals*. Among them, two-stage architecture such as R-CNN [9] and Faster R-CNN [10] have become the basis of many successful models in the field of object detection [1,11,12]. Polat et al. [11] combined Faster R-CNN, Cascade R-CNN and RetinaNet. The prediction of their model is based on class agnostic non-maximum suppression and false-positive elimination. Yang et al. [13] improved Cascade R-CNN model by combining feature pyramid networks (FPN). Zhang et al. [12] implemented a mask-aided R-CNN model with a flexible and multi-stage training protocol to deal with artifact detection tasks. Two-stage detectors tend to have better results than single-stage detectors, but single-stage detectors have faster detection speed.

To improve on standard detection metrics, such as mean average precision (mAP) and intersection-over-union (IoU), while maintaining the test speed, we propose a pruning-based mechanism. We compare the intensity profiles of each predicted instance from the detector network with the pre-computed mean profile of that instance. To our knowledge this is the first work, that exploits the nature of each detected bounding box to prune detection by first computing probability density distribution of their center line gray-values and then comparing them with the computed mean profiles of 7 artifact instances. Here, we use EfficientDet [14], one of the state-of-the-art single-stage detector, with EfficientNet [15] backbone as our detector network. Our results demonstrate the effectiveness of our approach and shows improvement over both mAP and IoU metrics.

2   Proposed Method

We propose a comprehensive artifact detection method that comprises of: a) Bounding box prediction using EfficientNet with a bidirectional feature pyramid layer, b) NMS suppression to optimise the predictions, and c) proposed artifact profile-based pruning of the artifact instance boxes. Relevant details are presented in the following sections.

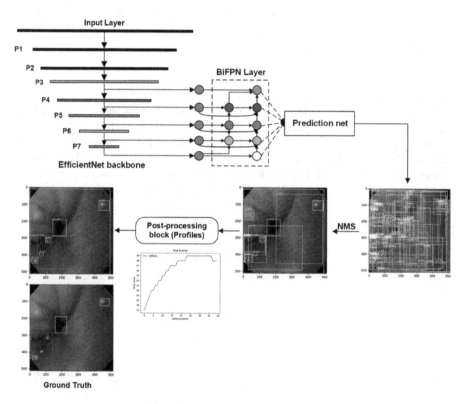

Fig. 1. The architecture of proposed method. The overall process is divided into two parts, the artifact detection network and post-processing block based on artifact instance profiles.

2.1   Artifact Detection

We use EfficientNet [15] as the backbone for the EfficientDet [14] and stacked bidirectional feature pyramid network (BiFPN) layers to obtain high-level fusion features. The design of BiFPN layers comes from path aggregation network (PANet) [16]. It removes the node with only one input and connects the input and output nodes of the same level. The entire BiFPN layers are used as a base layer and are repeatedly connected three times as a feature network. It takes P3 to P7 features from the backbone network, and repeated two-way feature fusion. Both the bounding box and class prediction networks can share features and produce bounding box predictions and object classes, respectively. We modified the class prediction network for classifying 8 artifact classes. Figure 1 illustrates the architecture of EfficientDet.

Our goal is to minimise the total weighted loss $Loss_{total}$ during training. Focal loss [6] with $\alpha = 0.25$ and $\gamma = 1.5$ and smooth $L1$ [10] loss L_{reg} are applied to the classification task and regression tasks, respectively. The regression loss is scaled by multiplication factor of 50 to balance the two loss functions, λ is set to 0.5. The resulting total loss L_T can be written as

$$L_T = (1 - \lambda)L_{cls} + 50 \times \lambda L_{reg} , \tag{1}$$

In addition, non-maximum suppression (NMS) [17] is used to optimize our results. As shown in Fig. 1, the direct prediction result of the model generates a large number of bounding boxes. The NMS selects the bounding box with the highest confidence in each category and calculates the IoU between the highest confidence bounding boxes and other bounding boxes one by one. If the score is greater than the set threshold, it is removed. It can be observed from the Fig. 1 that after NMS, the detection and localisation of artifacts has been greatly improved. However, compared with the ground truth, there are still erroneous predictions and multiple overlapping predictions.

2.2   Novel Pruning Method Using Instance Profiles

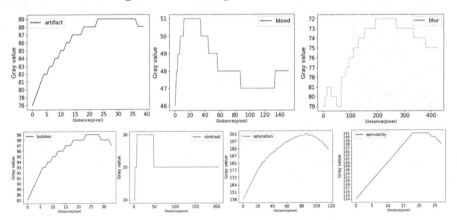

Fig. 2. Atlas or mean probability density function for center-line of 7 different artifact classes created from ground truth samples. From left to right on the top row are artifact, blood, blur and on the bottom row are instances bubbles, contrast, saturation and specularity. Evidently, the gray value distribution of each category has a significant difference.

Through detailed inspection of the dataset, we found that there are obvious differences in the profiles between different categories. It is specifically expressed as the probability density distribution (PDF) of gray values corresponding to the horizontal center-line pixels of the bounding boxes. For example, the characteristic of saturation profile is that the gray value quickly rises to the vertex (about 200) and then drops quickly. On the other hand, the line profile of contrast is extremely stable, and the gray value has been maintained at around 25. We also found that most of the artifacts are located in the center of the bounding boxes and the same artifacts have similar bounding boxes size (see Fig. 3). Additionally, since instrument can be part of the imaging unlike other artefacts, so we have discarded instrument class in our work for profile pruning.

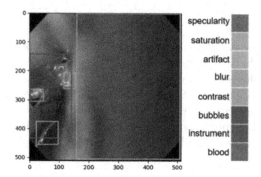

Fig. 3. Examples of artifact locations in the ground truth image. The size of bounding boxes belonging to artifact class (green) are analogous and they are all on or close to the center line of bounding boxes. (Color figure online)

Profile Atlases. We first computed the mean probability density distribution along the horizontal line profiles of each category in the ground truth annotations present in all training samples. For this, the gray value distribution of the horizontal center line is calculated in each bounding boxes. Among all the profiles in each category, we calculate the median length of them as the average length of this category. Then we select a threshold of 20 pixels to cut and filter out all profiles. If the length of the center line is greater than the average length, the profiles with the size equal to the average length are intercepted from the middle of the center line. If the length of the center line is less than the average length and more than 20 pixels, then we discard this profile. If the length of the center line is less than the average length and no more than 20 pixels, we insert the zeros to the gray value distribution on both sides of the profiles. After processing, all profiles in the same category have the same length. The mean gray value of each pixel is then calculated, and finally, we obtain the mean probability density distribution of each category as shown in Fig. 2.

Detection Box Pruning. The prediction results obtained by the EfficientDet model include prediction categories, confidence scores and bounding boxes information. Profiles of the same style are extracted in each prediction result and are either shortened or elongated by adding zeros such that they could be compared with the profile atlases. Then the processed profiles and mean probability density distribution are compared to optimize the prediction results. We calculate the Bhattacharyya distance [18] between the two PDFs say p and q to reflect the difference between them. In statistics, the Bhattacharyya distance [19] is used to measure the similarity of two probability distributions and can be written as:

$$D_B(p,q) = \frac{1}{4}\ln\left(\frac{1}{4}\left(\frac{\sigma_p^2}{\sigma_q^2} + \frac{\sigma_q^2}{\sigma_p^2} + 2\right)\right) + \frac{1}{4}\left(\frac{(\mu_p - \mu_q)^2}{\sigma_p^2 + \sigma_q^2}\right), \tag{2}$$

here σ_q^2 and μ_q represent the variance of the q-class distribution and the mean of the q-class distribution respectively.

In addition to the confidence score, we obtain another evaluation score, the Bhattacharyya distance score. Smaller Bhattacharyya distance means that the probability density distributions in two profiles are more similar. We keep the artifact bounding box predictions whose Bhattacharyya distance score is less than 4, and those greater than four are dropped out.

3 Results

3.1 Dataset

The dataset used is from the Endoscopy Artifact Detection Challenge (EAD2020) [1,2]. There were eight different artifact categories, i.e. saturation, specularity, bubbles, instrument, contrast, blood, blur and imaging artifacts. EAD2020 contains of 2200 single endoscopic frames and 232 sequential frames[1]. Each frame included an image frame and its corresponding bounding box annotations and category labels. We have used single endoscopic frames for training (80% of them are used for training set and the remaining 20% are used as validation set), and sequential frames are used as the out-of-sample test set. Table 1 provides the total number of artifacts present in the training data which split by classes.

3.2 Evaluation Metrics

In this work, we have computed widely used detection metrics described below:

- Intersection-over-union (IoU): This metric quantifies the area overlap between two spatial regions using the intersection-over-union between reference or ground-truth (denoted R) and predicted bounding boxes and segmentations (denoted S).

$$IoU(R, S) = \frac{R \cap S}{R \cup S}. \tag{3}$$

---

[1] http://dx.doi.org/10.17632/c7fjbxcgj9.3.

Table 1. Statistics of the training data.

Class	Number of instances
Specularity	8596
Artifacts	7688
Bubbles	4627
Contrast	1590
Saturation	1184
Blur	671
Instrument	446
Blood	446
Total	25248

– Mean average precision (mAP): This metric measures the ability to accurately retrieve all ground truth bounding boxes. mAP calculates the average precision (AP) of each category based on different IoU thresholds, and then the average of all categories of AP are estimated. AP refers to the calculation of the area under curve (AUC) of the precision-recall curve after obtaining the maximum Precision value.

3.3   Experimental Setup

We implemented EfficientDet using PyTorch[2]. In the training process, we first resized the training data to 512×512. We used Adam optimiser with a weight decay of 1e−4 and learning rate of 1e−3 for training the model. In the loss function, we used $Loss_{total}$ to include Focal loss with $\alpha = 0.25$ and $\gamma = 1.5$ and a smooth $L1$-loss regularisation. The model was trained for 1500 epochs on NVIDIA Quadro RTX 6000. The results of network prediction were optimized by NMS with set threshold of 0.5. We additionally set a confidence threshold of 0.2 to further eliminate false positive predictions. The results were then fed to the post-processing block for further pruning and optimization. As described in the method section, we got our final result by calculating the distance and discarding all scores greater than 4, which is empirically set based on our experimental validation of improvement.

---

[2] https://pytorch.org.

Table 2. Experimental results on testing set. Each model has 16 million parameters.

Test model	mAP_{25}	mAP_{50}	$mAP_{overall}$	IoU
EfficientDet-d0	30.67%	20.93%	20.29%	5.17%
EfficientDet-d0-NMS	29.56%	19.77%	19.93%	11.60%
EfficientDet-d0-NMS-PDF	**31.69%**	**19.98%**	**21.33%**	**19.01%**

3.4 Quantitative Results

Table 2 summarises the result of artifacts detection on test set. It can be seen that EfficientDet-d0 model got overall mAP of 20.29% and an IoU of 5.17%. This means the model over predicted the class instances with poor localisation. However, after the NMS, overall mAP dropped to 19.93% with increased IoU by 6.43% (i.e., IoU of 11.60%). However, by using our proposed elimination technique, (EfficientDet-d0-NMS-PDF), where we compare distance between the predicted PDF and the mean PDF, the results are improved. It can be observed that with technique we achieve an overall mAP of 21.33% and an IoU of 19.01% (an increase by nearly 15%). Similarly, we observed improvements in mAPs at other IoU thresholds (e.g., mAP_{25} of 31.69%, mAP_{50} of 19.98%).

3.5 Qualitative Results

As shown in Fig. 4, the results after NMS processing have been significantly improved, but there is large number of wrong bounding box predictions. Our PDF-based post-processing method further optimises the results and reduces the incorrect predicted bounding boxes. In the original prediction of the model, there is a large number of erroneous results between different artifacts, especially in specularity and artifact, as well as prediction bounding boxes of different sizes for the same artifact area, such as contrast. Our proposed method prunes these predictions by calculating the Bhattacharyya distance-based score. From the Qualitative Results point of view, our results have fewer and more accurate annotation bounding boxes which can be observed to be closer to the ground truth annotations.

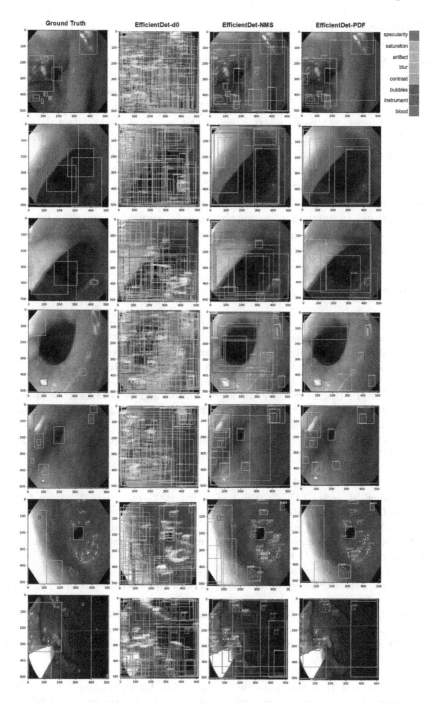

Fig. 4. Qualitative Results of EfficientDet-d0, EfficientDet-NMS and EfficientDet-NMS-PDF. Different colored bounding boxes represent different classes. (Color figure online)

4   Discussion and Conclusion

In this paper, we proposed a novel method of eliminating wrong predictions based on distances between the horizontal line profile of the predicted bounding boxes with that of the pre-computed atlas line profiles. The proposed method achieves the purpose of bounding box pruning by calculating the Bhattacharya distance between instances. State-of-the-art EfficientDet is trained for artifact detection tasks. The experimental results show that our method effectively improves the mean average precision and intersection-over-union. Compared with the widely used non-maximum suppression technique, our method has a higher metric scores. It can be also seen from our qualitative experiment that the results of the proposed PDF-based method are closer to the ground truth bounding box annotations. In particular, our method can be applied to any deep learning-based object detection model. In the future, we plan to integrate proposed profile pruning technique within end-to-end learning framework.

References

1. Ali, S., et al.: An objective comparison of detection and segmentation algorithms for artefacts in clinical endoscopy. Sci. Rep. **10**(1), 1–15 (2020)
2. Ali, S., et al.: Deep learning for detection and segmentation of artefact and disease instances in gastrointestinal endoscopy. Med. Image Anal. **70**, 102002 (2021)
3. Sánchez, F.J., Bernal, J., Sánchez-Montes, C., de Miguel, C.R., Fernández-Esparrach, G.: Bright spot regions segmentation and classification for specular highlights detection in colonoscopy videos. Mach. Vis. Appl. **28**(8), 917–936 (2017). https://doi.org/10.1007/s00138-017-0864-0
4. Redmon, J., Farhadi, A.: YOLO9000: better, faster, stronger. In: Proceedings of the IEEE Conference on Computer Vision and Pattern Recognition, pp. 7263–7271 (2017)
5. Liu, W., et al.: SSD: single shot MultiBox detector. In: Leibe, B., Matas, J., Sebe, N., Welling, M. (eds.) ECCV 2016. LNCS, vol. 9905, pp. 21–37. Springer, Cham (2016). https://doi.org/10.1007/978-3-319-46448-0_2
6. Lin, T.-Y., Goyal, P., Girshick, R., He, K., Dollár, P.: Focal loss for dense object detection. In: Proceedings of the IEEE International Conference on Computer Vision, pp. 2980–2988 (2017)
7. Subramanian, A., Srivatsan, K.: Exploring deep learning based approaches for endoscopic artefact detection and segmentation. In: EndoCV@ ISBI, pp. 51–56 (2020)
8. Oksuz, I., Clough, J.R., King, A.P., Schnabel, J.A.: Artefact detection in video endoscopy using retinanet and focal loss function. In: Proceedings of the 2019 Challenge on Endoscopy Artefacts Detection (EAD2019), Venice, Italy, 8th April 2019, vol. 2366 (2019)
9. Girshick, R., Donahue, J., Darrell, T., Malik, J.: Rich feature hierarchies for accurate object detection and semantic segmentation. In: Proceedings of the IEEE Conference on Computer Vision and Pattern Recognition, pp. 580–587 (2014)
10. Ren, S., He, K., Girshick, R., Sun, J.: Faster R-CNN: towards real-time object detection with region proposal networks. arXiv preprint arXiv:1506.01497 (2015)

11. Polat, G., Sen, D., Inci, A., Temizel, A.: Endoscopic artefact detection with ensemble of deep neural networks and false positive elimination. In: EndoCV@ ISBI, pp. 8–12 (2020)
12. Zhang, P., Li, X., Zhong, Y.: Ensemble mask-aided R-CNN. In: ISBI, pp. 6154–6162 (2018)
13. Yang, S., Cheng, G.: Endoscopic artefact detection and segmentation with deep convolutional neural network. In: Proceedings of the 2019 Challenge on Endoscopy Artefacts Detection (EAD2019), Venice, Italy, 8th April 2019, vol. 2366 (2019)
14. Tan, M., Pang, R., Le, Q.V.: EfficientDet: scalable and efficient object detection. In: Proceedings of the IEEE/CVF Conference on Computer Vision and Pattern Recognition, pp. 10781–10790 (2020)
15. Tan, M., Le, Q.: EfficientNet: rethinking model scaling for convolutional neural networks. In: International Conference on Machine Learning, pp. 6105–6114. PMLR (2019)
16. Liu, S., Qi, L., Qin, H., Shi, J., Jia, J.: Path aggregation network for instance segmentation. In: Proceedings of the IEEE Conference on Computer Vision and Pattern Recognition, pp. 8759–8768 (2018)
17. Neubeck, A., Van Gool, L.: Efficient non-maximum suppression. In: 18th International Conference on Pattern Recognition (ICPR 2006), vol. 3, pp. 850–855. IEEE (2006)
18. Kailath, T.: The divergence and Bhattacharyya distance measures in signal selection. IEEE Trans. Commun. Technol. **15**(1), 52–60 (1967)
19. Coleman, G.B., Andrews, H.C.: Image segmentation by clustering. Proc. IEEE **67**(5), 773–785 (1979)

Automatic Detection of Extra-Cardiac Findings in Cardiovascular Magnetic Resonance

Dewmini Hasara Wickremasinghe[1][(✉)] [iD], Natallia Khenkina[3,4] [iD],
Pier Giorgio Masci[1,2] [iD], Andrew P. King[1] [iD], and Esther Puyol-Antón[1] [iD]

[1] School of Biomedical Engineering and Imaging Sciences, King's College London,
London, UK
dewmini.wickremasinghe@kcl.ac.uk
[2] Guy's and St Thomas' Hospital, London, UK
[3] Luigi Sacco University Hospital, Milan, Italy
[4] Postgraduate School of Diagnostic and Interventional Radiology, Università degli
Studi di Milano, Milan, Italy

Abstract. Cardiovascular magnetic resonance (CMR) is an established, non-invasive technique to comprehensively assess cardiovascular structure and function in a variety of acquired and inherited cardiac conditions. In addition to the heart, a typical CMR examination will also image adjacent thoracic and abdominal structures. Consequently, findings incidental to the cardiac examination may be encountered, some of which may be clinically relevant. We compare two deep learning architectures to automatically detect extra cardiac findings (ECFs) in the HASTE sequence of a CMR acquisition. The first one consists of a binary classification network that detects the presence of ECFs and the second one is a multi-label classification network that detects and classifies the type of ECF. We validated the two models on a cohort of 236 subjects, corresponding to 5610 slices, where 746 ECFs were found. Results show that the proposed methods have promising balanced accuracy and sensitivity and high specificity.

Keywords: Deep learning classification · Extra cardiac findings · Cardiovascular magnetic resonance

1 Introduction

Cardiovascular Magnetic Resonance (CMR) is a valuable, non-invasive, diagnostic option for the evaluation of cardiovascular diseases as it allows the assessment of both cardiac structure and function [21]. CMR is an advantageous image modality due to its wide field of view, accuracy, reproducibility and ability to scan in different planes. Also, due to its lack of exposure to ionizing radiation,

---

A. P. King and E. Puyol-Antón—Joint last authors.

© Springer Nature Switzerland AG 2021
B. W. Papież et al. (Eds.): MIUA 2021, LNCS 12722, pp. 98–107, 2021.
https://doi.org/10.1007/978-3-030-80432-9_8

CMR can be widely employed, except for patients with implanted electronic devices such as pacemakers [9,15,16]. The importance of the CMR imaging modality is evident from the fact that in the time period ranging from 2008 to 2018 CMR use increased by 573%, growing by the 14.7% in the year 2017–2018 alone [10].

A typical CMR acquisition begins with a gradient echo 'scout' image of several slices in the coronal, sagittal, and axial planes, followed by axial imaging of the entire chest, conventionally using a Half-Fourier Acquisition Single-shot Turbo spin Echo (HASTE) sequence [8]. The acquired images contain the heart, as well as significant portions of the upper abdomen and thorax. These regions may present extra-cardiac irregularities, which are defined as incidental Extra Cardiac Findings (ECFs) [8].

The importance of investigating the presence of incidental ECFs in CMR scans has been shown in previous studies [2,8,17,22], and acknowledged by the European Association of Cardiovascular Imaging, which includes it as part of the European CMR certification exam [14]. In particular, Dunet et al. [2] reported a pooled prevalence of 35% of ECFs in patients undergoing CMR, 12% of which could be classified as major findings, i.e. requiring further investigation. ECFs are important for the early diagnosis of unknown diseases but can also be useful to determine the primary cardiac pathology which is being examined, due to the fact that some cardiac conditions have a multi-systemic environment [17]. In addition, when an ECF is identified, the important clinical question is whether the abnormality represents a benign or malignant lesion [17]. Two key examples are breast and lung cancer, since mammary and pulmonary tissue can be visualized on axial cross sectional imaging at the time of CMR. Previous works have shown that incidental breast lesions are identified in 0.1–2.5% of CMR studies and over 50% of these lesions are clinically significant [14–16, 52,]; similarly, the incidence of significant pulmonary abnormalities found in CMR examinations are up to 21.8% [2]. Another important factor to consider is that, depending on the institution, CMR examinations may be reported by cardiologists, radiologists or a combination thereof. A recent study showed that the highest accuracy to assess prevalence and significance of ECF in clinical routine CMR studies was reported when cardiologist and radiologist were working together [3], but this is not possible at all institutions. We believe that a computer-aided ECF detection tool could be beneficial in a clinical setting, especially when reporting is performed by only one specialist or by inexperienced operators.

The application of artificial intelligence (AI) in healthcare has great potential, for example by automating labour-intensive activities or by supporting clinicians in the decision-making process [19]. Liu et al. [12] compared the performance of clinicians to deep learning models in disease detecting tasks. They showed that AI algorithms performed equivalently to health-care professionals in classifying diseases from medical imaging. Although there is room for improvement, these results confirm the positive impact that AI could have in healthcare. Deep learning models have been employed in the automated detection of incidental findings in computed tomography (CT) and obtained promising results [1,18,23].

An automated pipeline able to detect the presence of incidental ECFs in CMR would not only be beneficial to the investigation of primary conditions and possible unknown diseases of the patient but could also reduce burden on overworked clinicians. In this paper, we investigate the feasibility of using deep learning techniques for the automatic detection of ECFs from the HASTE sequence.

2   Materials

This is a retrospective multi-vendor study approved by the institutional ethics committee and all patients gave written informed consent. A cohort of 236 patients (53.7 ± 15.7 years, 44% female) who underwent clinical CMR was manually reviewed to specifically assess the prevalence and importance of incidental ECFs. CMR image acquisitions were acquired with scanners of different magnetic field strengths and from different vendors with the following distributions: 70 subjects with 1.5T Siemens, 86 subjects with 1.5T Phillips and 80 subjects with 3.0T Phillips. From the CMR acquisitions, the HASTE sequence was used to detect any abnormal finding located outside the pericardial borders and the great vessels (aortic and pulmonary). ECFs were classified by anatomical location (i.e. neck, lung, mediastinum, liver, kidney, abdomen, soft tissue and bone) and by severity (i.e. major for findings that warrant a further investigation, new treatment, or a follow up e.g. lymphadenopathy or lung abnormalities; minor for findings that are considered benign conditions and don't require further investigations, follow up or treatment) [2,13]. Of the 236 studies analysed, that correspond to 5610 slices, 746 ECFs were found. The distribution of ECFs by location and severity is shown in Fig. 1.

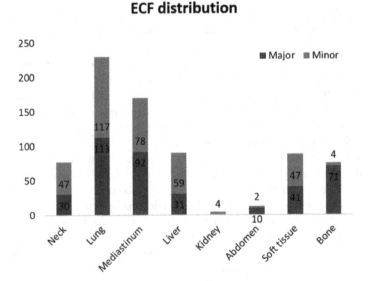

Fig. 1. Distribution of the ECFs split by location and severity.

3   Methods

The proposed framework for automatic detection of ECFs from the CMR HASTE images is summarized in Fig. 2, and each step is described below.

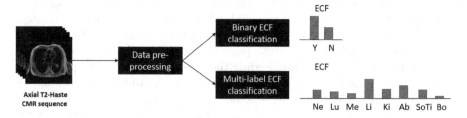

Fig. 2. Overview of the proposed framework for automatic ECF detection/ classification.

3.1   Data Pre-processing

To correct for variation in acquisition protocols between vendors, all images were first resampled to an in-plane voxel size of 1.25×1.25 mm and cropped to a standard size of 256×256 pixels. The cropping was based on the centre of the images and the standard size was selected as the median size of all the images in the database All DICOM slices were converted to numpy arrays and the pixel values were normalised between 0 and 1.

3.2   Binary ECF Classification

The first strategy aims to detect if any of the slices of the HASTE sequence has an ECF and we frame the problem as a binary classification task. We trained and evaluated seven state-of-the-art convolutional neural network (CNN) architectures: AlexNet [11], DenseNet [6], MobileNet [5], ResNet [4], ShuffleNet [24], SqueezeNet [7] and VGG [20].

3.3   Multi-label ECF Classification

The second strategy aims to not only detect the presence of ECFs but also identify to which class the ECF belongs. The chosen classes represent eight different areas of the body, namely neck, lung, mediastinum, liver, kidney, abdomen, soft tissue and bone. In this paper, we focus on the identification of ECFs and their subsequent classification based on the classes mentioned above, rather than their major/minor classification. As each slice can contain more than one ECF, we have framed the problem as a multi-label classification, which means that the output of the deep learning classifier supports multiple mutually non-exclusive classes. We extended the previous seven state-of-the-art CNN architectures to multi-label classification by using the number of classes as the number of nodes in the output layer and adding a sigmoid activation for each node in the output layer.

3.4 Training

The manually classified data were divided as follows: 80% were used for training and validation of the classification networks and 20% were used for testing. Data were split at patient level and bounded to a specific set, either training, validation or testing, in order to maintain data independence. Each network was trained for 200 epochs with binary cross entropy with a logit loss function. During training, data augmentation was performed on-the-fly using random translations (± 30 pixels), rotations ($\pm 90°$), flips (50% probability) and scalings (up to 20%) to each mini-batch of images before feeding them to the network. The probability of augmentation for each of the parameters was 50%. Additionally, we implemented an adaptive learning rate scheduler, which decreases the learning rate by a constant factor of 0.1 after every 5 epochs, stopping at a plateau on the validation set (commonly known as ReduceLRonPlateau). This step was added as it improves training when presented with unbalanced datasets.

3.5 Statistics

The performance of the models was evaluated using a receiver operating characteristic (ROC) curve analysis, and based on this the balanced accuracy (BACC), sensitivity, specificity, positive predictive value (PPV) and negative predictive value (NPV) were computed for the optimal classifier selected using the weighted Youden index. Sensitivity, also known as the true positive rate, is defined as the proportion of ground truth positively labelled examples that are identified as positive by the model; specificity, also known as true negative rate, is defined as the proportion of ground truth negatives that are identified as negative; PPV is defined as the proportion of identified positives that have a ground truth positive label; NPV is defined as the proportion of identified negatives with a ground truth negative label. For the multi-label classification algorithm, we extended this analysis to two conventional methods, namely micro-averaging and macro-averaging. Micro-averaging calculates metrics globally by counting the total true positives (TPs), true negatives (TNs), false positives (FPs) and false negatives (FNs), while macro-averaging calculates metrics for each label and finds their unweighted mean.

4 Results

4.1 Binary ECF Classification

Table 1 summarises the statistics computed from the results of the binary classification for each of the employed state-of-the-art networks.

Table 1. Mean sensitivity, specificity, positive predictive value (PPV), negative predictive value (NPV) and balanced accuracy (BACC) for the different binary ECF classifiers. Bold font highlights the best results.

	ALEX NET	DENSE NET	MOBILE NET	RES NET	SHUFFLE NET	SQUEEZE NET	VGG
Sensitivity	0.34	0.36	0.17	0.44	0.45	0.00	**0.55**
Specificity	0.80	0.77	0.93	0.79	0.72	1.00	**0.76**
PPV	0.19	0.18	0.25	0.23	0.18	NaN	**0.24**
NPV	0.90	0.90	0.89	0.91	0.91	0.88	**0.93**
BACC	0.57	0.57	0.55	0.62	0.59	0.50	**0.66**

The computed sensitivity values range between 0.34 and 0.55, except for MobileNet and SqueezeNet that obtained 0.17 and 0.00. Specificity and NPV, which are respectively the proportion of the correctly identified negative labels and the chance of the assigned label to be correct if identified as negative, obtained results close to 1. PPV values range between 0.18 and 0.25, although SqueezeNet, the only outlier, had NaN. BACC values fluctuate around 0.55. SqueezeNet obtained the lowest value (0.50) and VGG obtained 0.66, which is the highest computed BACC. It is noticeable that SqueezeNet performed poorly compared to the other networks and achieved the lowest computed values. On the other hand, the best performing network was VGG, which obtained the best sensitivity and BACC values.

4.2   Multi-label ECF Classification

Table 2 summarises the statistics computed from the results of the multi-label classification for each of the employed state-of-the-art networks.

As stated before, micro-averaging computes the metrics by calculating the total numbers of TPs, TNs, FPs and FNs. The best sensitivity value of 0.62 was obtained with AlexNet. Specificity and NPV, similarly to the binary classification task, have high values, often close to 1. The only outliers are AlexNet and ShuffleNet, which had specificity respectively 0.85 and 0.83. PPV had low results for all the networks and obtained the lowest values for AlexNet and ShuffleNet, which got respectively 0.21 and 0.20. It is noteworthy that although these networks obtained high values for the other metrics, they obtained the lowest values for PPV. The computed BACC values, in parallel to the ones computed in the binary case, oscillate between 0.50 and 0.74. The best BACC values were computed from AlexNet and ShuffleNet.

Macro-averaging calculates metrics per label and then finds their unweighted mean. In this case, sensitivity values are lower than the ones mentioned above. For macro averaging they range between 0.28 and 0.51, except for MobileNet and SqueezeNet that obtained the lowest sensitivities, respectively 0.09 and 0.00. As before, specificity and NPV values are close to the maximum. PPV obtained low results, between 0.13 and 0.34, except for SqueezeNet that obtained NaN. The NaN values are caused when the network predicts all the cases as negatives

Table 2. Mean sensitivity, specificity, positive predictive value (PPV), negative predictive value (NPV) and balanced accuracy (BACC) for the different multi-label ECF classifiers. Micro-averaging calculates metrics globally by counting the total true positives, true negatives, false positives and false negatives, while macro-averaging calculates metrics for each label and finds their unweighted mean. Bold font highlights the best results.

	ALEX NET	DENSE NET	MOBILE NET	RES NET	SHUFFLE NET	SQUEEZE NET	VGG
Micro sensitivity	**0.62**	0.44	0.12	0.40	**0.57**	0.00	0.32
Micro specificity	**0.85**	0.94	0.99	0.91	**0.83**	1.00	0.94
Micro PPV	**0.21**	0.32	0.41	0.23	**0.20**	NaN	0.28
Micro NPV	**0.97**	0.96	0.94	0.96	**0.96**	0.94	0.95
Micro BACC	**0.74**	0.69	0.56	0.65	**0.70**	0.50	0.63
Macro sensitivity	**0.41**	0.28	0.09	0.31	**0.51**	0.00	0.32
Macro specificity	**0.84**	0.93	0.99	0.91	**0.83**	1.00	0.94
Macro PPV	**0.13**	0.21	0.34	0.18	**0.19**	NaN	0.14
Macro NPV	**0.97**	0.96	0.94	0.95	**0.96**	0.96	0.95
Macro BACC	**0.62**	0.60	0.54	0.61	**0.67**	0.50	0.63

and therefore there are no true positives or false positives. This reflects the poor performance of the SqueezeNet network. The highest computed BACC was computed for ShuffleNet, while the other networks obtained values around 0.60. Again, MobileNet and SqueezeNet obtained the lowest values (0.54 and 0.50).

Visual results from the multi-label ECF classifier are shown in Fig. 3. The top row shows five images containing different ECFs (i.e. location and severity) which were correctly classified and the bottom row shows five images that have been misclassified by the network. Overall, it is apparent that the size and shape of the ECF can significantly vary. The performance of the network seems to be strongly influenced by the size of the ECF as well as the number of cases of that class of ECF in the training database.

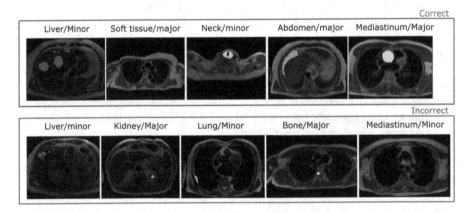

Fig. 3. Example results for proposed multi-label ECF classifier: top row shows correct cases and bottom row shows cases that have been misclassified.

5 Discussion and Conclusion

In a CMR examination, a careful assessment of non-cardiac structures may also detect relevant non-cardiac diseases. During a CMR acquisition, the inferior neck, entire thorax and upper abdomen are routinely imaged, particularly in the initial multi-slice axial and coronal images. Correctly identifying and reporting ECFs is beneficial to the patient and can prevent unnecessary over-investigation whilst ensuring that indeterminate or potentially important lesions are investigated appropriately. Cardiovascular diseases also often have systemic effects and the identification of ECFs can help with the interpretation of the primary cardiac pathology. In this paper, we have proposed for the first time a deep learning-based framework for the detection of ECFs from the HASTE sequence.

We approached the problem following two strategies: the first one consisted of a binary classification task that aimed to identify the presence of ECFs in each slice of the HASTE sequence; the second one consisted of a multi-label classification task which, in addition to the identification of ECFs, aimed to classify the ECFs based on their location.

For the first approach results showed that the best performing network was VGG, with BACC, sensitivity and specificity respectively equal to 66%, 55% and 76%. For the second approach the best performing networks were AlexNet and ShuffleNet, with a micro BACC higher than 70%, micro specificity above 83% and micro sensitivity values respectively of 62% and 57%. Macro metrics show that when computing the unweighted mean obtained from each label, performance decreases.

Our vision is that deep learning models could be used in clinical workflows to automate the identification of ECFs from CMR exams, thus reducing clinical workloads. This would require a high sensitivity to ensure that potential ECFs are not missed (i.e. minimise false negatives). False positives are less important as they can be eliminated by a subsequent cardiologist review. Therefore, performance is not currently sufficient for clinical needs. We believe that the main reasons for the low sensitivity are the limited amount of training data, the variation in image appearance due to the multi-vendor nature of the study and the large variation in the size, appearance and position of ECFs, as shown in Fig. 3. Obtaining more data, with less class imbalance would likely improve performance in future work. However, this work represents the first AI-based framework for automated detection and localization of ECFs in CMR images and therefore serves as a proof-of-principle. In future work, we will aim to gather more training data and develop novel techniques to improve the sensitivity of our models.

A limitation of the current framework is that we do not differentiate between major and minor ECFs and this is important in clinical practice to decide which ECF should be treated and which could be considered benign. We plan to address this in future work. We will also aim to combine a classification network with a segmentation network to allow localisation and differentiation of the different ECFs.

In conclusion, we have demonstrated the feasibility of using deep learning for the automatic screening of HASTE images for identifying potential ECFs.

Further work is required to improve the sensitivity of the technique and fully evaluate its role and utility in clinical workflows.

Acknowledgements. This work was supported by the EPSRC (EP/R005516/1 and EP/P001009/1) and the Wellcome EPSRC Centre for Medical Engineering at the School of Biomedical Engineering and Imaging Sciences, King's College London (WT 203148/Z/16/Z).

References

1. Blau, N., Klang, E., Kiryati, N., Amitai, M., Portnoy, O., Mayer, A.: Fully automatic detection of renal cysts in abdominal CT scans. Int. J. Comput. Assist. Radiol. Surg. **13**(7), 957–966 (2018)
2. Dunet, V., Schwitter, J., Meuli, R., Beigelman-Aubry, C.: Incidental extracardiac findings on cardiac MR: systematic review and meta-analysis. J. Magn. Resonan. Imaging **43**(4), 929–939 (2016)
3. Greulich, S., et al.: Extra cardiac findings in cardiovascular MR: why cardiologists and radiologists should read together. Int. J. Cardiovasc. Imaging **30**(3), 609–617 (2014)
4. He, K., Zhang, X., Ren, S., Sun, J.: Deep residual learning for image recognition. In: Proceedings of the IEEE Conference on Computer Vision and Pattern Recognition, pp. 770–778 (2016)
5. Howard, A.G., et al.: MobileNets: efficient convolutional neural networks for mobile vision applications. arXiv preprint arXiv:1704.04861 (2017)
6. Huang, G., Liu, Z., Van Der Maaten, L., Weinberger, K.Q.: Densely connected convolutional networks. In: Proceedings of the IEEE Conference on Computer Vision and Pattern Recognition, pp. 4700–4708 (2017)
7. Iandola, F.N., Han, S., Moskewicz, M.W., Ashraf, K., Dally, W.J., Keutzer, K.: SqueezeNet: AlexNet-level accuracy with 50x fewer parameters and <0.5 MB model size. arXiv preprint arXiv:1602.07360 (2016)
8. Irwin, R.B., et al.: Incidental extra-cardiac findings on clinical CMR. Eur. Heart J.-Cardiovasc. Imaging **14**(2), 158–166 (2013)
9. Karamitsos, T.D., Francis, J.M., Myerson, S., Selvanayagam, J.B., Neubauer, S.: The role of cardiovascular magnetic resonance imaging in heart failure. J. Am. Coll. Cardiol. **54**(15), 1407–1424 (2009)
10. Keenan, N., et al.: UK national and regional trends in cardiovascular magnetic resonance usage-the British society of CMR survey results. Eur. Heart J. **41**(Supplement_2), ehaa946-0200 (2020)
11. Krizhevsky, A., Sutskever, I., Hinton, G.E.: ImageNet classification with deep convolutional neural networks. Adv. Neural Inf. Process. Syst. **25**, 1097–1105 (2012)
12. Liu, X., et al.: A comparison of deep learning performance against health-care professionals in detecting diseases from medical imaging: a systematic review and meta-analysis. Lancet Digit. Health **1**(6), e271–e297 (2019)
13. Mantini, C., et al.: Prevalence and clinical relevance of extracardiac findings in cardiovascular magnetic resonance imaging. J. Thoracic Imaging **34**(1), 48–55 (2019)
14. Petersen, S.E., et al.: Update of the European association of cardiovascular imaging (EACVI) core syllabus for the European cardiovascular magnetic resonance certification exam. Eur. Heart J.-Cardiovasc. Imaging **15**(7), 728–729 (2014)

15. Peterzan, M.A., Rider, O.J., Anderson, L.J.: The role of cardiovascular magnetic resonance imaging in heart failure. Cardiac Failure Rev. **2**(2), 115 (2016)
16. Rajiah, P., Desai, M.Y.: Cardiac magnetic resonance imaging-role and applications. Eur. Cardiol. **8**(1), 17–22 (2012)
17. Rodrigues, J.C., et al.: Extra-cardiac findings in cardiovascular magnetic resonance: what the imaging cardiologist needs to know. J. Cardiovasc. Magn. Resonan. **18**(1), 1–21 (2016)
18. Shenkman, Y., et al.: Automatic detection and diagnosis of sacroiliitis in CT scans as incidental findings. Med. Image Anal. **57**, 165–175 (2019)
19. Siegersma, K., Leiner, T., Chew, D., Appelman, Y., Hofstra, L., Verjans, J.: Artificial intelligence in cardiovascular imaging: state of the art and implications for the imaging cardiologist. Netherlands Heart J. **27**(9), 403–413 (2019)
20. Simonyan, K., Zisserman, A.: Very deep convolutional networks for large-scale image recognition. arXiv preprint arXiv:1409.1556 (2014)
21. Stacey, R.B., Hundley, W.G.: The role of cardiovascular magnetic resonance (CMR) and computed tomography (CCT) in facilitating heart failure management. Curr. Treat. Options Cardiovasc. Med. **15**(4), 373–386 (2013)
22. Wyttenbach, R., Médioni, N., Santini, P., Vock, P., Szucs-Farkas, Z.: Extracardiac findings detected by cardiac magnetic resonance imaging. Eur. Radiol. **22**(6), 1295–1302 (2012)
23. Yan, K., Wang, X., Lu, L., Summers, R.M.: DeepLesion: automated mining of large-scale lesion annotations and universal lesion detection with deep learning. J. Med. Imaging **5**(3), 036501 (2018)
24. Zhang, X., Zhou, X., Lin, M., Sun, J.: ShuffleNet: an extremely efficient convolutional neural network for mobile devices. In: Proceedings of the IEEE Conference on Computer Vision and Pattern Recognition, pp. 6848–6856 (2018)

Brain-Connectivity Analysis to Differentiate Phasmophobic and Non-phasmophobic: An EEG Study

Suhita Karmakar[1], Dipayan Dewan[1(✉)], Lidia Ghosh[1], Abir Chowdhury[1], Amit Konar[1], and Atulya K. Nagar[2]

[1] Artificial Intelligence Laboratory, Department of Electronics and Telecommunication Engineering, Jadavpur University, Kolkata, India
amit.konar@jadavpuruniversity.in
[2] Department of Mathematics and Computer Science, Liverpool Hope University, Liverpool, UK
nagara@hope.ac.uk

Abstract. Brain-connectivity refers to a pattern of functional or effective connectivity of distinct modules of human brain due to interactions between them. In this paper, the authors have attempted to conduct a brain connectivity based analysis to study the brain circuitry in subjects from their electroencephalographic (EEG) data, while they are engaged in playing a horror video game. The main motive of our work is to understand the differences in the effective connectivity among phasmophobic and non-phasmophobic subjects. In the present analysis, we propose a modified version of the causality test, named as Convergent Cross Mapping (CCM) to perform the analysis. The proposed CCM improves the performance of the standard CCM with an added feature of finding the possible direction of causation in terms of conditional entropy or maximum information transfer among the brain signal-sources. Experimental results and statistical analysis show that the proposed method shows superior efficacy in estimating the directed brain-connectivity as compared to the very well-known classical Granger Causality, classical CCM and other off-the-shelf brain-connectivity algorithms.

Keywords: Brain connectivity · Phasmophobia · Convergent cross mapping · EEG · Entropy

1 Introduction

Phasmophobia is a psychological condition where people experience irrational fear of supernatural things. Many people undergo fear of supernatural things from a young age. The onset age of any specific phobia depends on the type of phobia, any traumatic situation in the past or genetic tendency of the person [8]. For some people, mild development of phobia at their young age might disappear as they move into adolescence, but for others the fears and the anxieties persist and the severity of phobia increases gradually. It may even worsen

ⓒ Springer Nature Switzerland AG 2021
B. W. Papież et al. (Eds.): MIUA 2021, LNCS 12722, pp. 108–122, 2021.
https://doi.org/10.1007/978-3-030-80432-9_9

into a chronic and potentially lead to several other forms of severe psychological disorders like trauma, intense anxiety, panic attacks, debilitating enough for a diagnosis. In [26], the authors demonstrate overall clinical and psychological case study of six patients having phobia of supernaturals. According to the study, the patients were diagnosed with persistent fear of supernatural things, intense anxiety, panic attacks, developing other types of phobia like nyctophobia (fear of darkness) or autophobia (fear of being alone), insomnia, dementia, and mild epilepsy. Although there exist plenty of literature [12,14,27,36], which involve research outcomes of psychologists, who investigated different types of phobia and attempted to provide solutions for the diagnosis and prognosis of such phobias, knowledge based on the analysis of brain connectivity measures in specific phobia is still limited. The present paper fills the void by aiming at investigating the brain dynamics of phasmophobic and non-phasmophobic subjects with respect to the functional connectivity of the brain activation region while playing a horror video game.

In the past few years, a considerable amount of works [9,19,22] have been done focusing on functional neuro-imaging in specific phobia, which has revealed new perspectives into the neuro-biological mechanism underlying the fear response when exposed to a phobic stimulus. According to the existing literature [2,3], neural analysis underpinning specific phobia are mostly based on animal subtype, particularly snake or spider phobia subjects. These studies show evidence of hyper-activation in the brain region involved in fear evaluation structures, including the amygdala, dorsal anterior cingulate cortex (ACC), thalamus and insula. The comparative study [15,20] based on neural response analysis reveals that there exist two subtypes of specific phobia, named as animal subtype and blood-injection-injury (BII) sub-type. Elevated BOLD responses are observed in insula, thalamus, cingulate gyrus, supplementary motor area, superior frontal gyrus, parietal cortex, superior temporal gyrus and the cerebellum in case of snake phobia (animal subtype) whereas in dental phobia (BII subtype) only prefrontal and orbitofrontal cortex activates. The literature, [6,10] enhances the understanding of neural connectivity mechanisms of social phobia and panic disorder by analysing brain activation regions through emotional face perception. Functional connectivity analysis of patients suffering from agoraphobia (fear of situation/places) [18] shows decrease in inferior frontal gyrus activity at advanced stage with depression and panic disorder: altered prefrontal cortex connectivity and intensifying connectivity in fear producing regions, low connectivity in two brain networks in patients with sub-clinical agoraphobia [16].

Additionally, there exists literature [2,3,6,41], which includes fMRI and PET studies investigating functional connectivity of the whole brain for phobic patients. The fMRI studies reveal that there is a significant increase in cortical thickness in the insula region, and increase or decrease in cortical thickness in ACC region of phobic patients. It is also found that the volume depletion in the amygdala may occur for the patients with animal phobia, which can in turn cause severe vomiting, diarrhea, kidney failure, etc. The PET study in [1] presents the neural correlates of phobic patients which shows that para-hippocampal memory processes are influenced by amygdala-related arousal.

However, there exist only a handful of studies assessing functional connectivity of specific phobia using functional Near Infrared Spectroscopy (fNIRS) and electroencephalography (EEG). The study [11] using fNIRS shows that the activation of left inferior frontal gyrus (IFG) decreases over time in response to the semantic stimuli during phobia-relevant emotional Stroop paradigm. The analysis [33] of in-situ cortical blood oxygenation using fNIRS of arachnophobic (extreme fear of spiders) patient shows promising results for the application of the technique in the field of psychotherapy. In [33], the authors identified the severity of acrophobia using EEG. The literature [32] utilizes both the fNIRS and EEG devices to study the neural correlates for arachnophobic patients. Although there exist numerous research works based on various types of phobia but understanding the brain functional connectivity changes in phasmophobic patient is a new research era.

The novelty of the present study lies in understanding the difference in brain functional connectivity of phasmophobic and non-phasmophobic subjects from their EEG signals by applying a modified version of Convergent cross mapping (CCM) technique [35]. CCM technique has the ability to overcome the limitations provided by Granger Causality [13,17]. CCM is a statistical test of detecting the causal relationships among the time-series data (here, brain signals) of a unidirectionally connected chaotic system (here, brain) and thus provides a new era of research in brain-connectivity analysis.

Our approach is to experimentally determine the information transfer among brain regions, participating in a specific cognitive task (here, playing horror game) and to detect the causal dependence of the time-varying signals collected from multiple brain lobes. The source signals can be acquired by means of a brain-rhythm capturing device. EEG consists of several number of metal electrodes which can be mounted over the scalp of the subject to capture the source signals. Determining functional brain connectivity from the EEG data is a challenging problem as they have wide variations within and across experimental sessions. Although there exist a few interesting approaches like transfer entropy [39], Probabilistic Relative Correlation Adjacency Matrix (PRCAM) [17], several forms of Granger Causality analysis [13,17,25,31], none of these shows promising results in determining the causal relationships between the pairs of active EEG electrodes [4].

CCM can fortunately provide the solutions to the drawbacks experienced by the existing techniques and thus successfully has been utilized to understand the causal dependence of the time-series data [5,21,24]. Thereby, we undertake an extended version of classical CCM to test inter-connectivity between pairs of brain modules. In this paper, the possible causal connectivity from brain region X to Y is checked by the following 2 steps. First, the acquired EEG time-series signals, acquired from the brain regions X and Y are checked for the causality by employing the classical CCM technique and then the direction of causation between the observation variables (here, EEG signals) has been inferred by calculating the amount of information transfer among them using conditional entropy of probability distributions of the residuals of CCM output.

The paper is divided into 4 sections. In Sect. 2, the principle of classical CCM and its extension in brain-connectivity analysis is introduced. Section 3 provides a framework for the experimental protocol design along with results and interpretations. Performance analysis and statistical validation are also undertaken in this section. Conclusions are listed in Sect. 4.

2 Principles and Methodologies

This section provides a brief outline about the existing classical CCM algorithm and the proposed extension to determine the brain-connectivity from the acquired EEG signals for a given cognitive task.

2.1 Classical CCM

CCM [35] is a causality-detection algorithm, which can be effectively utilized to obtain the causal relationship between two time-series variables, say X and Y. The principle, that CCM uses, includes reconstruction of system states from X and Y using specific values of a few pre-determined parameters and then to quantify the relationship values between them using a nearest neighbor algorithm.

The mathematical basis of CCM algorithm are constituted using the concept of Takens' Theorem [37], which is one of the very well-known time-delay embedding theorems. In [37], Taken states that "an attractor manifold can be reconstructed from a set of observation variables of a dynamical system". Let, $X(t)$ and $Y(t)$ be the two discrete time-varying signals of a dynamical system (here, brain as shown in Fig. 1), that are sharing a common attractor manifold, denoted by M. Following the Taken's principle, one can easily reconstruct the D-dimensional shadow manifolds, denoted by M_X and M_Y for the temporal signals $X(t)$ and $Y(t)$ respectively, with the time delay τ between the successive steps. Numerically, one can represent the time-points on the shadow manifold M_X as: $\underline{x}(t) =< X(t), X(t - \tau), X(t - 2\tau), ..., X(t - (D-1)\tau) >$, where D is the embedding dimension. The optimal values of the parameters D and τ can be obtained using Simplex Projection technique [36]. It is followed from the Taken's theorem that, the time-points on the shadow manifold M_X have one-to-one correspondence to that of the original manifold M [35]. In an alternate way, it can be said that M_X is diffeomorphic reconstruction of M, as shown in Fig. 2. The other shadow manifold M_Y, which is also diffeomorphic to M can also be easily constructed in similar manner.

CCM actually aims at discovering the degree of correspondence between the local neighbors of M_X and M_Y [35], i.e., how well the local neighbors of M_X correspond to M_Y, as shown in Fig. 2. Therefore, the next step is to find $D + 1$ nearest neighbors by a suitable nearest-neighbor algorithm. Now let, $t_1, ..., t_{D+1}$ be the time-indices of the $D+1$ nearest neighbors of $\underline{x}(t)$ on M_X. These time-indices can be utilized to construct the cross mapping of $\hat{Y}(t)$ using Eq. (1).

$$\hat{Y}(t)|M_X = \sum w_i Y(t_i) \,, \tag{1}$$

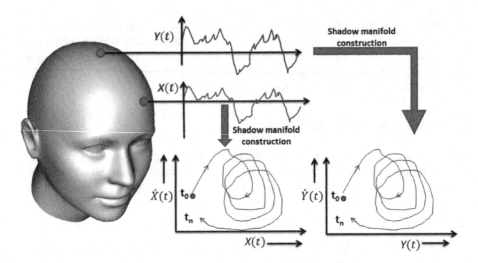

Fig. 1. Shadow manifold of two brain signals

where, i belongs to 1 to $D+1$ and w_is are the weights corresponding to the distance between $\underline{x}(t)$ and each of the $D+1$ nearest neighbors.

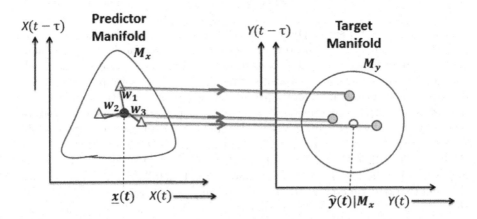

Fig. 2. Nearest Neighbors of M_X and M_Y

The weights can be estimated by

$$w_i = \frac{u_i}{\sum u_j}, \tag{2}$$

where, $j = 1, ..., D+1$ and u_i is given by

$$u_i = e^{-\frac{R(\underline{x}(t), \underline{x}(t_i))}{R(\underline{x}(t), \underline{x}(t_1))}}. \tag{3}$$

In (3), $R(\underline{x}(t), \underline{x}(t_i))$ denotes the Euclidean distance between $\underline{x}(t)$ and $\underline{x}(t_i)$. Similarly, $\hat{X}(t)|M_Y$ can be obtained. Next, Pearson's Correlation Coefficient [29], ρ (rho), between the estimated $(\hat{Y}(t)|M_X)$ and observed values of $Y(t)$ can be computed using the following transformation.

$$\rho = \frac{N\sum XY - (\sum X \sum Y)}{\sqrt{[N\sum x^2 - (\sum x)^2][N\sum y^2 - (\sum y)^2]}} \,. \tag{4}$$

2.2 Estimating the Direction of Causation Using Conditional Entropy

In the classical CCM algorithm, the causation is unidirectional, i.e., either $X \rightarrow Y$ or $Y \rightarrow X$. The implication relation $X \rightarrow Y$ denotes that textitX can be estimated from Y, but the converse is not hold. To find the direction of causation one can compare the correlation coefficient values $\rho_{x,y}$, obtained between X and $\hat{X}|M_Y$, and $\rho_{y,x}$, obtained between Y and $\hat{Y}|M_X$. If $\rho_{x,y}$ is greater than $\rho_{y,x}$, then one may conclude that $X \rightarrow Y$. However, this is a contradiction of Granger's Intuitive Scheme [35].

One approach to find the direction of causation is to compare the probability distribution of the euclidean distance between manifold M_X and manifold $M_{\hat{X}}|M_Y$ and the same between manifold M_Y and manifold $M_{\hat{Y}}|M_X$. Let, R_x be the euclidean distance between M_X and $M_{\hat{X}}|M_Y$, and R_y be the euclidean distance between M_Y and $M_{\hat{Y}}|M_X$. It is observed from Fig. 3 that the probability distributions of R_x and R_y has negligible difference. A suitable way to enhance the inference of the direction of causality is to determine the amount of information transfer among the probability distributions of R_x and R_y [30] by computing Conditional Entropy of and combine them with the corresponding correlation coefficients $\rho_{x,y}$ and $\rho_{y,x}$.

Let $P(R_x)$ and $P(R_y)$ be the probability distribution function of R_x and R_y. The Conditional entropy of R_y for given R_x is determined as

$$H(R_y|R_x) = -\sum P(R_x) \sum P(R_y|R_x) log(P(R_y|R_x)). \tag{5}$$

The entropy of R_y conditioned on R_x actually measures the uncertainty in R_y given R_x. Similarly, $H(R_x|R_y)$ is computed. Then the mutual dependence measures between the two variables, denoted by $C_{X|Y}$ and $C_{Y|X}$ are derived by

$$C_{X|Y} = \frac{\rho_{x,y}}{H(R_x|R_y)}, \ and \ C_{Y|X} = \frac{\rho_{y,x}}{H(R_y|R_x)} \tag{6}$$

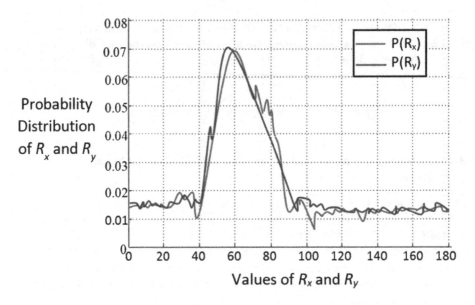

Fig. 3. Probability distributions of R_x and R_y

These are the measures of *Information Quality Ratio* (IQR), which quantifies the amount of information transferred from one variable to the other against the total uncertainty. The direction of the causality is then determined based on the higher IQR values. Thus, for all possible combinations of m number of EEG electrode-pairs, we obtain m^2 IQR values, which are represented in a matrix form. We name the matrix as $\mathbf{W}_{ECM}$ which has the dimension $m \times m$. The entire experimental workflow is illustrated in Fig. 4.

2.3   Classification Using Kernelized Support Vector Machine

To understand the performance of the proposed brain-connectivity algorithm, the outputs obtained from the proposed CCM are treated as features and fed to a classifier. To classify the two classes: phasmophobic and non-phasmophobics from the EEG responses of the subjects, W_{ECM} are first reshaped to form $(1 \times m^2)$ vector and then used as features to train and test a kernelized Support Vector Machine (KSVM) [7] classifier with polynomial kernel function.

3   Experiments and Results

This section describes the experimental results obtained using the principles introduced in Sect. 2 for analysing functional connectivity in phasmophobic and non-phasmophobic subjects.

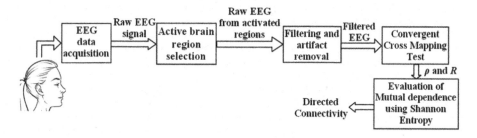

Fig. 4. Block diagram of the proposed framework

3.1   Experimental Setup

The experiment has been performed in the Artificial Intelligence Laboratory of Jadavpur University, Kolkata, India. A 14-channel EMOTIV EPOC+ EEG device is used to capture the electrical response of the brain. 24 subjects: 13 phasmophobics and 11 non-phasmophobics, in the age group 15–27 years participated in the said experiment. Subjects are asked to play a horror video game, namely Amnesia: The Dark Descent and during that time, EEG signals were extracted from them. To obtain data with minimum possible noise, the subjects had to relax and concentrate on the screen before the experiment could begin and the experiment was conducted in a dark and quiet room. The game was played on a computer. The 14 electrodes used to acquire EEG data are AF3, F7, F3, FC5, T7, P7, O1, O2, P8, T8, FC6, F4, F8, AF4. Additionally, P3 and P4 electrodes are used as reference electrodes to collect the signals properly. The electrode placement diagram is shown in Fig. 5.

The EMOTIV EPOC+ EEG device has built-in characteristics, which assists us in properly fitting the headset by directing that which EEG channel reflects bad quality signal (indicated by red light) and which one reflects good quality signal (indicated by green light). Here, the term 'good quality' emphasizes the signal with less noise and the term 'bad' indicates the channel with more noise. To obtain the good quality signal to the maximum extent, subjects are asked to close their eyes for 20 s in a rest position and then to open it, and then again to close their eyes for 10 s. In this manner, we were able to locate and fix the faulty electrodes by tracking the EEG channels. We didn't start the experiment until every EEG channel was properly fit to the scalp of the subjects and was in good working order. To capture a wide range of variations in the brain responses, the experiment was conducted over 3 days for each subject. Each day, the subjects were asked to play the horror computer game for 10 min. After each trial session, the subjects were asked to rate their level of scariness in a scale of 1–10. Based on this rating we further divided the dataset into 2 clusters to create 2 different categories:

1. rating 0–5 – Non-Phasmophobic
2. rating 6–10 – Phasmophobic

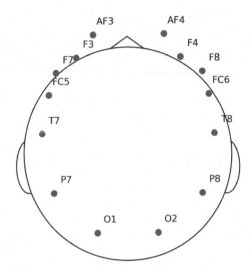

Fig. 5. EEG electrode placement on the scalp

3.2 Data Preprocessing

The EEG signals, thus obtained, are then passed through various signal pre-processing steps to remove the noise or artifacts. The sampling frequency of the recording was set 128 Hz. Sample length of each EEG data is 76,800. The built-in filter is present in the EMOTIV EEG headset device, which is the 5th order sinc filter. The frequency bandwidth of 0.1–40 Hz is set to record the EEG signals in a particular window and this is the first step of preprocessing of the raw signal. EEG signals are contaminated with various forms of artifacts. Three most common forms of artifacts that need special mention include 1) step artifacts, 2) spike artifacts, and 3) physiological artifacts [23]. The step artifacts come into play, when there is a change in the surrounding environment. The step artifact can be removed by minimizing the variation of the external light and instrumental noise. The spike/motion artifacts are related to decoupling between the electrodes and their assigned positions due to head or muscle movement. They result in abrupt changes in the amplitude of the received signals. For example, a sudden change in the ambient light intensity results in a spike-like noise. Another important artifact is physiological artifact, which may include the artifacts due to eye-blinking [34], respiration, heart-beat, blood pressure fluctuations and Mayer wave [40]. To remove most of the EEG artifacts, the raw EEG signal is passed through an Elliptical type bandpass filter of order 10 and frequency band of 0.1–40 Hz.

3.3 Active Brain Region Selection Usings LORETA

In our study, the selection of activated brain regions of the subjects playing horror video games is done using sLORETA [28] software. The experiment is

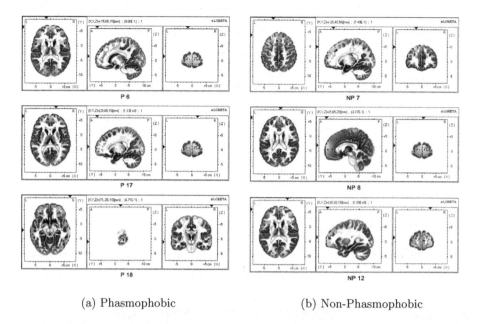

(a) Phasmophobic   (b) Non-Phasmophobic

Fig. 6. sLORETA activation topographical map for 3 phasmophobic (P) and 3 non-phasmophobic (NP) subjects.

carried out by analyzing and estimating the electrical activity of the intra-cortical distribution from the EEG signal acquired during gameplay using sLORETA software. From the sLORETA solutions, it is found that the prefrontal, frontal, temporal and occipital regions have the greater activation for the phasmophobics as compared to the non-phasmophobics, as indicated in Fig. 6.

3.4   Effective Connectivity Estimation by CCM Algorithm

To get the effective functional connectivity between the active brain lobes, first of all we have checked for the causality between the electrodes by CCM method. To do so, the shadow manifold of each electrode is reconstructed by computing the optimum embedding dimension (D) and time delay (τ) by simplex projection. For each electrode the computed value of D and τ are 2 and 1 respectively. Hence from the reconstructed manifold we check for the cross map skill for every possible combination of electrode pair. Next to find the direction of causation, we have employed the conditional entropy based CCM approach described in Sect. 2. The directed brain-connectivity obtained by the proposed algorithm is depicted in Fig. 7 (a)-(f) for phasmophobic and non-phasmophobic subjects. Among 11 non-phasmophobics (denoted as NP1, NP2,. . . NP11) and 13 phasmophobics (denoted as P1, P2, . . . , P13), the brain-connectivity of only 3 subjects from each group is given in Fig. 7. It is observed from the figure that the phasmophobics have strong connectivity between left frontal and right temporal region, whereas the connections become weaker for non-phasmophobics.

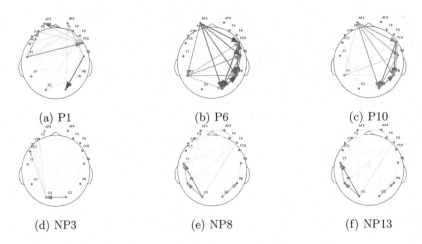

Fig. 7. The directed brain connectivity obtained by the proposed CCM for 3 non-phasmophobic (NP) and 3 phasmophobic (P) subjects.

3.5 Statistical Analysis Using One-Way ANOVA Test

We performed ANOVA tests to i) evaluate and determine the effect of stimuli in different brain regions and ii) to test the efficacy of the proposed connectivity algorithm. For the former case, we performed one way ANOVA test for each electrode individually which shows difference in impact of fear stimulus in the brain activation region between phasmophobic and non-phasmophobic subjects. As mentioned above, 13 subjects are phasmophobic and 11 are non- phasmophobic. We compared mean value of all the subjects from two groups (phasmophobic and non-phasmophobic) for each electrode of EEG data using F-distribution, initially considering null hypothesis i.e., two means are equal. The one-way ANOVA test shows statistically significant results, i.e., the probability (p-value) is less than the specific significant level ($p < 0.05$), which rejects the null hypothesis. The box-plot of the one-way ANOVA test is depicted in Fig. 8a for each of the occipital, frontal and temporal regions. Here, means of the high activation region are unequal between two groups: phasmophobic and non-phasmophobic. Brain Regions: (Occipital Lobe: O2 ($p = 0.0009$), Frontal region: F4 ($p = 0.02$), F7 ($p = 0.007$), F8 ($p = 0.006$), FC5 ($p = 0.007$), FC6 ($p = 0.004$), AF3($p = 0.002$)) and Temporal region: T7 ($p = 0.005$), T8 ($p = 0.04$).

Next, to test the differences in effective connectivity among phasmophobic and non-phasmophobic, obtained by the proposed CCM, again we utilize oneway ANOVA test. The significance level is set at $p < 0.05$. The result is depicted in Fig. 8b, which emphasizes that there is increased connectivity for phasmophobic as compared to non-phasmophobic.

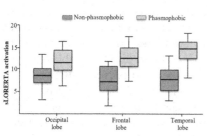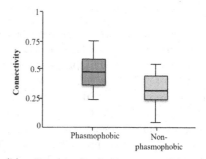

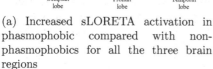

(a) Increased sLORETA activation in phasmophobic compared with non-phasmophobics for all the three brain regions

(b) Increased brain-connectivity in phasmophobics as compared to non-phasmophobics

Fig. 8. Box plot for ANOVA statistical analysis

3.6   Relative Performance Analysis of the Proposed CCM

The proposed CCM algorithm for brain-connectivity analysis is compared here with 5 well-known techniques of functional connectivity analysis. They are Cross-correlation technique [38], probabilistic relative correlation adjacency matrix (PRCAM) [17], Granger Causality [13], standard CCM [4] and Transfer entropy based analysis [39]. The relative performance of the proposed algorithm with the existing algorithms, has been evaluated on the basis of three performance metrics of classifier: classifier accuracy, sensitivity and specificity. The results of the study are given in Table 1. It is apparent from the table that the proposed technique outperforms the existing techniques in all the cases.

Table 1. Relative performance analysis of the proposed CCM

Brain-connectivity algorithms	Classifier accuracy (%)	Sensitivity	Specificity
Cross Correlation technique	80.18	0.89	0.84
PRCAM	80.35	0.90	0.87
Granger Causality	87.80	0.82	0.90
Standard CCM	86.03	0.85	0.86
Transfer entropy	85.35	0.89	0.87
Proposed CCM	**88.91**	**0.96**	**0.92**

4   Conclusion

The paper introduces a novel approach for brain-connectivity analysis for phasmophobic and non-phasmophobic subjects using conditional entropy based CCM technique in EEG-based Brain Computer Interfacing paradigm. The method is advantageous for its inherent potential to detect directional causality in brain-connectivity, which can not be carried out by traditional CCM or Granger causality techniques. The proposed approach has successfully been employed to determine brain-connectivity of phasmophobics and non-phasmophobics. Experiments undertaken reveal that phasmophobics possess brain-connectivity with weaker strength between left frontal and right temporal regions as compared to the non-phasmophobics. Thus the proposed method can easily segregate phasmophobics from their non-phasmophobic counterparts. This technique is useful for diagnosis of the subjects with phobia of supernatural at its early stage, which can in turn, will suggest an early treatment to reduce associated disorders like sleeplessness, anxiety, etc.

References

1. Åhs, F., et al.: Arousal modulation of memory and amygdala-parahippocampal connectivity: a pet-psychophysiology study in specific phobia. Psychophysiology **48**(11), 1463–1469 (2011)
2. Åhs, F., et al.: Disentangling the web of fear: amygdala reactivity and functional connectivity in spider and snake phobia. Psychiatry Res. Neuroimaging **172**(2), 103–108 (2009)
3. Britton, J.C., Gold, A.L., Deckersbach, T., Rauch, S.L.: Functional MRI study of specific animal phobia using an event-related emotional counting stroop paradigm. Depression Anxiety **26**(9), 796–805 (2009)
4. Chowdhury, A., Dewan, D., Ghosh, L., Konar, A., Nagar, A.K.: Brain connectivity analysis in color perception problem using convergent cross mapping technique. In: Nagar, A.K., Deep, K., Bansal, J.C., Das, K.N. (eds.) Soft Computing for Problem Solving 2019. AISC, vol. 1138, pp. 287–299. Springer, Singapore (2020). https://doi.org/10.1007/978-981-15-3290-0_22
5. Clark, A.T., et al.: Spatial convergent cross mapping to detect causal relationships from short time series. Ecology **96**(5), 1174–1181 (2015)
6. Danti, S., Ricciardi, E., Gentili, C., Gobbini, M.I., Pietrini, P., Guazzelli, M.: Is social phobia a "mis-communication" disorder? Brain functional connectivity during face perception differs between patients with social phobia and healthy control subjects. Front. Syst. Neurosci. **4**, 152 (2010)
7. Das, S., Halder, A., Bhowmik, P., Chakraborty, A., Konar, A., Nagar, A.: Voice and facial expression based classification of emotion using linear support vector machine. In: 2009 Second International Conference on Developments in eSystems Engineering, pp. 377–384. IEEE (2009)
8. De Vries, Y.A., et al.: Childhood generalized specific phobia as an early marker of internalizing psychopathology across the lifespan: results from the world mental health surveys. BMC Med. **17**(1), 1–11 (2019)
9. Del Casale, A.: Functional neuroimaging in specific phobia. Psychiatry Res. Neuroimaging **202**(3), 181–197 (2012)
10. Demenescu, L., et al.: Amygdala activation and its functional connectivity during perception of emotional faces in social phobia and panic disorder. J. Psychiatric Res. **47**(8), 1024–1031 (2013)

11. Deppermann, S., et al.: Functional co-activation within the prefrontal cortex supports the maintenance of behavioural performance in fear-relevant situations before an itbs modulated virtual reality challenge in participants with spider phobia. Behav. Brain Res. **307**, 208–217 (2016)
12. Eaton, W.W., Bienvenu, O.J., Miloyan, B.: Specific phobias. Lancet Psychiatry **5**(8), 678–686 (2018)
13. Granger, C.W.: Investigating causal relations by econometric models and cross-spectral methods. Econometrica J. Econometric Soc. **37**, 424–438 (1969)
14. Grös, D.F., Antony, M.M.: The assessment and treatment of specific phobias: a review. Curr. Psychiatry Rep. **8**(4), 298–303 (2006)
15. Hilbert, K., Evens, R., Maslowski, N.I., Wittchen, H.U., Lueken, U.: Neurostructural correlates of two subtypes of specific phobia: a voxel-based morphometry study. Psychiatry Res. Neuroimaging **231**(2), 168–175 (2015)
16. Indovina, I., Conti, A., Lacquaniti, F., Staab, J.P., Passamonti, L., Toschi, N.: Lower functional connectivity in vestibular-limbic networks in individuals with subclinical agoraphobia. Front. Neurol. **10**, 874 (2019)
17. Kar, R., Konar, A., Chakraborty, A., Nagar, A.K.: Detection of signaling pathways in human brain during arousal of specific emotion. In: 2014 International Joint Conference on Neural Networks (IJCNN), pp. 3950–3957. IEEE (2014)
18. Kunas, S.L., et al.: The impact of depressive comorbidity on neural plasticity following cognitive-behavioral therapy in panic disorder with agoraphobia. J. Affect. Disord. **245**, 451–460 (2019)
19. Lange, I., et al.: Functional neuroimaging of associative learning and generalization in specific phobia. Prog. Neuro-Psychopharmacol. Biol. Psychiatry **89**, 275–285 (2019)
20. Lueken, U., Kruschwitz, J.D., Muehlhan, M., Siegert, J., Hoyer, J., Wittchen, H.U.: How specific is specific phobia? Different neural response patterns in two subtypes of specific phobia. NeuroImage **56**(1), 363–372 (2011)
21. Luo, C., Zheng, X., Zeng, D.: Causal inference in social media using convergent cross mapping. In: 2014 IEEE Joint Intelligence and Security Informatics Conference, pp. 260–263. IEEE (2014)
22. Linares, I.M., Chags, M.H.N., Machado-de Sousa, J.P., Crippa, J.A.S., Hallak, J.E.C.: Neuroimaging correlates of pharmacological and psychological treatments for specific phobia. CNS Neurol. Disord. Drug Targets (Formerly Curr. Drug Targets-CNS Neurol. Disord.) **13**(6), 1021–1025 (2014)
23. Maulsby, R.L.: Some guidelines for assessment of spikes and sharp waves in EEG tracings. Am. J. EEG Technol. **11**(1), 3–16 (1971)
24. McCracken, J.M., Weigel, R.S.: Convergent cross-mapping and pairwise asymmetric inference. Phys. Rev. E **90**(6), 062903 (2014)
25. Nolte, G., Bai, O., Wheaton, L., Mari, Z., Vorbach, S., Hallett, M.: Identifying true brain interaction from EEG data using the imaginary part of coherency. Clin. Neurophysiol. **115**(10), 2292–2307 (2004)
26. de Oliveira-Souza, R.: Phobia of the supernatural: a distinct but poorly recognized specific phobia with an adverse impact on daily living. Front. Psychiatry **9**, 590 (2018)
27. Pachana, N.A., Woodward, R.M., Byrne, G.J.: Treatment of specific phobia in older adults. Clin. Interv. Aging **2**(3), 469 (2007)
28. Pascual-Marqui, R.D., et al.: Standardized low-resolution brain electromagnetic tomography (sLORETA): technical details. Methods Find. Exp. Clin. Pharmacol. **24**(Suppl D), 5–12 (2002)

29. Pearson, K.: Vii. Note on regression and inheritance in the case of two parents. Proc. R. Soc. London **58**(347–352), 240–242 (1895)
30. Pukenas, K.: An algorithm based on the convergent cross mapping method for the detection of causality in uni-directionally connected chaotic systems. Math. Models Eng. **4**(3), 145–150 (2018)
31. Rathee, D., Cecotti, H., Prasad, G.: Estimation of effective fronto-parietal connectivity during motor imagery using partial granger causality analysis. In: 2016 International Joint Conference on Neural Networks (IJCNN), pp. 2055–2062. IEEE (2016)
32. Rosenbaum, D., et al.: Neuronal correlates of spider phobia in a combined fNIRS-EEG study. Sci. Rep. **10**(1), 1–14 (2020)
33. Rosenbaum, D., et al.: Cortical oxygenation during exposure therapy-in situ fNIRS measurements in arachnophobia. NeuroImage Clin. **26**, 102219 (2020)
34. Shoker, L., Sanei, S., Latif, M.A.: Removal of eye blinking artifacts from EEG incorporating a new constrained BSS algorithm. In: Processing Workshop Proceedings, 2004 Sensor Array and Multichannel Signal, pp. 177–181. IEEE (2004)
35. Sugihara, G., et al.: Detecting causality in complex ecosystems. Science **338**(6106), 496–500 (2012)
36. Sugihara, G., May, R.M.: Nonlinear forecasting as a way of distinguishing chaos from measurement error in time series. Nature **344**(6268), 734–741 (1990)
37. Takens, F.: Detecting strange attractors in turbulence. In: Rand, D., Young, L.-S. (eds.) Dynamical Systems and Turbulence, Warwick 1980. LNM, vol. 898, pp. 366–381. Springer, Heidelberg (1981). https://doi.org/10.1007/BFb0091924
38. Toyama, K., Kimura, M., Tanaka, K.: Cross-correlation analysis of interneuronal connectivity in cat visual cortex. J. Neurophysiol. **46**(2), 191–201 (1981)
39. Vicente, R., Wibral, M., Lindner, M., Pipa, G.: Transfer entropy–a model-free measure of effective connectivity for the neurosciences. J. Comput. Neurosci. **30**(1), 45–67 (2011)
40. Winkler, I., Haufe, S., Tangermann, M.: Automatic classification of artifactual ICA-components for artifact removal in EEG signals. Behav. Brain Functions **7**(1), 1–15 (2011)
41. Zilverstand, A., Sorger, B., Sarkheil, P., Goebel, R.: fMRI neurofeedback facilitates anxiety regulation in females with spider phobia. Front. Behav. Neurosci. **9**, 148 (2015)

Image Registration, and Reconstruction

Virtual Imaging for Patient Information on Radiotherapy Planning and Delivery for Prostate Cancer

Miguel Martínez-Albaladejo[1]([✉]), Josep Sulé-Suso[1,2], David Lines[3], James Bisson[1], Simon Jassal[1], and Craig Edwards[1]

[1] Cancer Centre, University Hospitals of North Midlands, Stoke-on-Trent, UK
{Miguel.Martinez,Josep.SuleSuso}@uhnm.nhs.uk
[2] Guy Hilton Research Centre, Keele University, Stoke-on-Trent, UK
[3] Proton Beam Therapy Centre, The Christie NHS Trust, Manchester, UK
David.Lines@christie.nhs.uk

Abstract. The provision of information on radiotherapy (RT) planning and delivery for patient with cancer is a vital issue in improving not only patients' satisfaction but also patients' compliance to treatment. In the present study, patients with prostate cancer receiving RT were randomised to have the provision of information on RT planning and delivery using Virtual Reality (VR) prior to RT or on the last day of RT. Bladder and rectal volumes and separations were measured prior to starting RT, weekly during RT and on the last day of RT. A quality of life questionnaire was completed prior to RT, halfway through RT, at the end of RT and at 3 and 6 months after RT in order to assess side effects. Bladder volumes decreased towards the end of RT but no differences were seen between patients in either cohort. The scores of the quality of life questionnaire showed only a statistical difference at 3 months after RT with lower scores (less intense side effects) for patients in Cohort 1. Although this study failed to show that provision of information to patients receiving RT using VR improves compliance, VT tools were very welcomed by patients as a means of providing information on RT planning and delivery. In the future, this study may provide cancer centres with an useful tool to audit local protocols regarding patient preparations and training for prostate RT.

Keywords: Virtual environment for radiotherapy training · Compliance to prostate radiotherapy · Bladder volume · Computed tomography

1 Introduction

The planning of radiotherapy (RT) for prostate cancer has to find a fine balance between the dose to the prostate and to neighbouring organs. The aim of achieving highest tumour control with minimal side effects is not always an easy task.

© Springer Nature Switzerland AG 2021
B. W. Papież et al. (Eds.): MIUA 2021, LNCS 12722, pp. 125–139, 2021.
https://doi.org/10.1007/978-3-030-80432-9_10

This is made even more difficult when patients' input (i.e., maintaining a full bladder) is required such as in RT for prostate cancer. The two main organs at risk close to the prostate are the bladder and the rectum. It is accepted that in order to reduce side effects from RT, prostate cancer needs to be treated with a full bladder and empty rectum. A full bladder will reduce the dose to the whole bladder and push bowel away from the treatment area. While this is the desirable outcome, it is also recognised that patients find it difficult to keep a full bladder the more RT sessions they receive [1,2]. Moreover, it is widely accepted that the bladder volume consistently decreased towards the end of the RT when compared to that measured on the RT planning Computed Tomography (CT) scan. This could obviously have an implication on side effects caused by RT such as bladder and bowel toxicity [3], being the latter those with the greatest impact on quality of life [4]. Thus, patient education is an important element in cancer care when patients undergo a wide range of examinations and treatments [5]. In RT planning and delivery, providing clear information to patients is paramount not only on shared decision making but also to comply with the treatment plan, improve patient positioning [5], and making them more aware of potential side-effects [6]. Furthermore, it has been described that well informed patients report better quality of life [7]. In the case of RT for prostate cancer, patient information and preparation prior to RT is paramount as organ motion could affect patient set-up. In fact, it is anticipated that an informed patient is more likely to adhere to instructions [8].

However, it is not clear yet whether providing as much information as possible to patients with prostate cancer about the reasons behind having a full bladder improves patients' compliance. Moreover, it is not clear either whether improved patients' compliance would translate also into a reduction of side effects. This is made more difficult by the fact that RT, with its complex procedures, treatment delivery and highly specialised equipment, is an abstract thing to explain to patients which could be challenging to comprehend for some of them [9].

Table 1. Patients characteristics.

	Cohort 1 (pre-VERT)	Cohort 2 (post-VERT)
Age		
Mean	69	69
Range	54 - 80	53 - 82
Stage		
IIA	2	2
IIB	4	3
IIC	6	2
IIIB	16	26
IIIC	17	6
IVA	2	4
IVB	0	2
PSA		
0-10	15	17
>10-20	19	17
>20	13	11

Over the last few years, there has been the possibility of explaining RT planning and delivery using virtual reality (VR). An example is the Virtual Environment Radiotherapy (VERT) system. Although it was initially used for teaching purposes [10–13], it has been exploited more recently as a tool to provide information to patients and relatives on RT planning and delivery [5,8,9,14–16]. A dedicated patient information module known as Patient Education and Radiotherapy Learning (PEARL) has also been successfully used for patient education [17]. We were the first to show that the VERT system improved not only the provision of information on a one to one basis (clinician with patient, but sometimes including their relatives) to cancer patients receiving RT, but also patients and relatives' satisfaction [14]. Interestingly, during the preliminary study, some radiographers treating these patients felt that patients with prostate cancer receiving RT who were informed about RT planning and delivery using the VERT system were more compliant during their treatment, i.e., holding their water better than those patients who did not receive this information using the VERT system.

On this basis, we carried out the present randomised study to assess whether providing information on RT planning and delivery using the VERT system, patients with prostate cancer receiving RT would be more compliant to treatment. This compliance was assessed by measuring bladder and rectal volumes at different time points. Furthermore, a quality of life (QOL) questionnaire was also used to measure early side effects caused by RT in order to assess whether better compliance translated in improved quality of life.

2   Materials and Methods

2.1   Study Design

This research was approved by the human research ethics committee of West Midlands, Staffordshire. Patients were identified by an oncologist who explained to them the aims of the study and what it entailed. Patients were given an information sheet and allowed 24 h before taking a decision to participate and signing a consent form.

96 patients with the histological diagnosis of adenocarcinoma of the prostate treated with RT between 2016 and 2017 were included in this study. Four patients declined to participate in the study as they thought that enough information had 50 already been provided to them, thus, the data from 92 patients are presented here. The study was open to prostate cancer patients undergoing RT with curative intent.

The patients' characteristics are summarised in Table 1. Patients were randomised to receiving either information on RT planning and delivery using the VERT system prior to starting RT (Cohort 1, Pre-VERT, 47 patients) or after the last dose of RT (Cohort 2, Post-VERT, 45 patients). The information was given on a one-to-one basis (radiographer and patient and his relatives and/or friends). Patients in Cohort 2 received the standard information on planning and delivery of their RT prior to starting their treatment. This consisted of written

 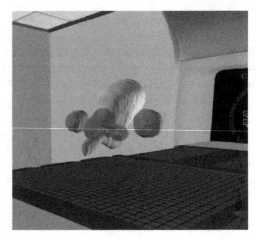

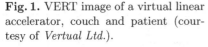

Fig. 1. VERT image of a virtual linear accelerator, couch and patient (courtesy of *Vertual Ltd.*).

Fig. 2. VERT image showing the bladder (yellow), rectum (pink), both femoral heads (orange) and the PTV (red) for a patient receiving RT for a cancer of the prostate. (Color figure online)

information explaining how to do their preparation to ensure consistency in emptying bowels and bladder filling. Information about planning and delivery was given to them by radiographers. Patients were invited to bring relatives or friends during the VERT session. The limitation encountered was to have enough staff to show patients how RT is planned and delivered by means of the VERT system (if patients had been randomised to Cohort 1) between the time of completing the CT planning for RT and the start of RT. This limitation was also present in our previous study [14].

For each study patient, their CT Scan images and RT plan were uploaded onto the VERT system (Fig. 1 and Fig. 2) using the Digital Imaging and Communication in Medicine (DICOM) standard which has been developed to enable standardised storage and transfer of medical images with associated RT planning detail overlay [18]. This means that a database for each patient can be created. The way information on RT planning and delivery was given to patients using the VERT system has been previously described in detail and does not need to be iterated [14]. The uploading of the CT scan images onto the VERT 3D system took around 10 min for each individual patient. The time needed to explain using VR how RT is planned and given and answer patients and relatives' questions took around 30 min.

Patients were asked to fill in the QOL questionnaire EORTC QLQ-PR25 prior to starting RT, half way through the RT, after the last session of RT and at 3 months and 6 months after RT. The scores to the different questions were labelled as 1 (not at all), 2 (a little), 3 (quite a bit) or 4 (very much). This questionnaire collated information on bladder and bowel function. The questionnaires at 3 and 6 months were sent to patients with a self-addressed envelope in order for them to send the completed questionnaires back to the Cancer Centre.

While every effort was carried out to obtain the completed questionnaires at 3 and 6 months' time frame after completing RT, only 32 questionnaires were completed for all time points for patients randomised in Cohort 1 (47 patients, 68.0%) and 24 for patients randomised in Cohort 2 (45 patients, 53.3%). The remaining patients in both groups did not returned all the 22 completed questionnaires at 3 and/or 6 months.

2.2 Eligibility and Exclusion Criteria

Patients with the histological diagnosis of prostate cancer attending the Cancer Centre for RT were included in this pilot study. The only exclusion criteria were patients unable to fully understand the nature of the study and therefore, unable to provide a signed consent form.

2.3 Radiotherapy

Radiotherapy was performed using 10 MV photons delivered from Varian 2100iX linear accelerators. Treatment planning was carried out in Eclipse Treatment Planning System using CT scans. All patients received RT using Volumetric Modulate Arc Therapy (VMAT), with a prescribed dose of 74 Gy in 37 fractions to the mean of the Planning Target Volume (PTV) [PTV-3 = prostate + 0.5 cm/0 cm posterior margin and avoiding rectum], with additional PTVs treated to 70.9 Gy [PTV-2 = PTV-3 + 0.5 cm] and 59.2 Gy [PTV-1 = prostate + seminal vesicles + 1.0 cm]. All patients completed the prescribed treatment, delivering the dose 5 d a week for 8 weeks. However, during this study, the results of the CHHiP trial were published [19]. This meant that a change in practice for RT for prostate cancer took place at the Cancer Centre where patients with prostate cancer started to be treated using 20 fractions rather than 37 fractions. Therefore, it was decided to stop the study when it became difficult to recruit patients receiving RT over 37 fractions.

2.4 Bladder and Rectal Measurements

Cone beam CT (CBCT) scans used for position verification were performed with slice thickness of 2.5 mm, however these scans did not always include the whole bladder. It was not possible in these cases to repeat the CBCT scan in order to include the full bladder as this would pose further delays in an already busy department and unnecessary additional radiation dose. For all patients, the whole bladder could be seen in the planning CT Scan (pCT).

   Data on bladder and bowel volumes were measured in Eclipse Treatment Planning System (TPS) using the patients' pCT scan, their CBCT scans on days 1, 2 and 3, and from one CBCT scan each week until the end of their treatment. If the patient had more scans available (i.e., imaged daily) then one scan per week was selected, typically the scan occurring on the same day of the week and before the correction of the patient's position in order to have more reliable information regarding compliance to RT. Therefore, no patient required extra CBCT Scans. For each of the scans used in the study, the following information was acquired (Fig. 3, 4 and 5):

* CBCT (or pCT) scan number and date.
* Contoured bladder volume (if the whole bladder could be seen in the scan).
* The bladder anterior/posterior (A/P) distance at the expected central superior/inferior (SUP/INF) slice of the bladder (a).
* Half of the bladder SUP/INF distance (b).
* The bladder left/right (L/R) distance (at the central SUP/INF slice of the bladder (c).
* Location of central SUP/INF slice of the prostate.
* Rectum A/P and L/R distances at the central slice of the prostate.
* Rectum A/P and L/R distances at 3 cm SUP to the central slice of the prostate.
* Rectum A/P and L/R distances at 3 cm INF to the central slice of the prostate.

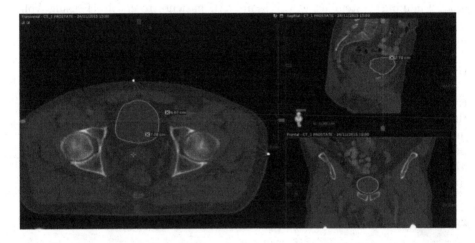

Fig. 3. Representative example of the measurement of the bladder volume in Eclipse TPS (left hand screen) and location to measure A/P, L/R and SUP/INF bladder distances (a, b, c, respectively).

Bladder, prostate and rectum were outlined in the CBCT scans by one of the authors (MMA) to minimise inter-user variability. The solid bladder volumes were quantified directly in Eclipse if whole volume was visible. The other advantage is that the z-slice of Eclipse is useful for determining the 3 cm SUP/INF locations for the rectum.

2.5 Bladder Volume Model

In those cases where it was not possible to delineate the whole bladder in the CBCT (partial scans), the bladder volumes were estimated by an in-house model applied to the volumes obtained from the whole bladder ($V_{bladder}$) assuming an ellipsoidal shape ($V_{ellipsoid}$), by means of the following fitting process:

$$V_{bladder} = k_1 \cdot (V_{ellipsoid}) + k_2 \quad cm^3 \tag{1}$$

$$V_{ellipsoid} = \pi/3 \cdot a \cdot b \cdot c \quad cm^3 \tag{2}$$

where k_1, k_2 are constants and a, b, c the variables above mentioned [20,21].

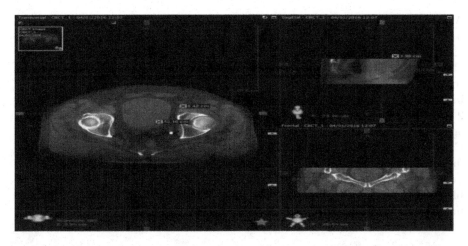

Fig. 4. Representative example of the measurement of the bladder volume in Eclipse TPS (left hand screen) and location to measure A/P, L/R and SUP/INF bladder distances (a, b, c, respectively), when the bladder is not completely seen on a CBCT.

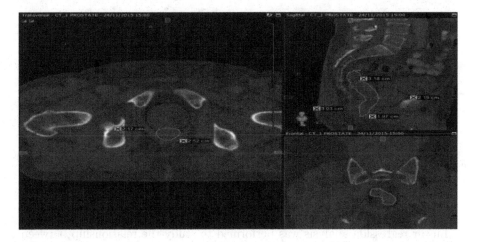

Fig. 5. Locations to measure rectum separations. Measurements were carried out on a transverse CT slice (left screen). A measurement example is shown in the sagittal slice.

Differences between the estimated and real bladder volumes were also calculated in the cohort of CBCT where the whole bladder was completely seen in
order to validate the model.

2.6 Statistical Analysis

Descriptive analysis (frequencies and percentages) were generated from QOL
questionnaires and bladder volumes. The p-values are estimated by two sample
t-tests.

3 Results

As mentioned before, it was important that this work would fit within the daily
practice at the Cancer Centre without causing extra work or disruption in an
already busy department apart from extra patient radiation. This meant that
CBCT which did not include the whole bladder volume were not obtained again
to acquire images with the whole bladder volume as this would have put delays
in our daily practice. This was solved by applying the mathematical Eqs. 1 and
2 described in the previous section. However, it was important to confirm the
agreement between the bladder volumes when the whole bladder was seen in the
CBCT images and the calculated total bladder volume when the CBCT images
did not provide the whole bladder volume. To this purpose, we took the CBCT
in which the whole bladder was seen. At this point, we calculated the whole
bladder volume by delineating the bladder contour but also, we used the above
mentioned formulae to calculate the whole bladder volume.

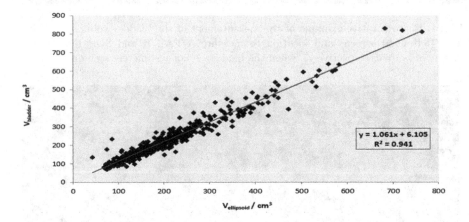

Fig. 6. Linear regression showing $V_{bladder}$ (the bladder volume given by Eclipse) vs.
$V_{ellipsoid}$ (bladder volume calculated assuming ellipsoidal shape) using those CBCT
where the whole bladder could be seen.

Linear regression analysis was performed to study the relationship between
the estimated bladder volume using the formula above and the real bladder

volume by means of the statistical environment MATLAB v11 (*Mathworks*) and Microsoft Office. The model which estimates the bladder volume results from studying the correlation between the bladder volumes determined by Eclipse and the ones assuming ellipsoids in the subpopulation of scans for which the bladder is completely seen. The number of images used for this purpose has been 498, almost 50% (49.2%) of the total of analysed images, taking into account both pCT and CBCT. The accuracy of this volume was validated by calculating relative differences in the whole control group. As can be seen in Fig. 6, there was a good correlation between the 2 volumes thus obtained, with a regression coefficient of 0.941 and an average difference model-Eclipse of -0.4%. In this scenario, the possible impact of measuring the formulae parameters (a, b, c) seems minor. The mean bladder volume prior to starting RT for patients in Cohort 1 was (331 ± 48) ml and for patients in Cohort 2 was (328 ± 46) ml. For both groups, Fig. 7 shows that there was a decrease in bladder volume as the patient treatment progressed.

The mean bladder volumes at week 8 were (254 ± 28) ml for patients in Cohort 1 and (267 ± 42) ml for patients in Cohort 2. The bladder volumes when compared to the pre-RT bladder volume were $(81.8 \pm 7.3)\%$ and $(80.2 \pm 11.4)\%$ respectively at week 4, and $(84.7 \pm 10.3)\%$ and $(76.5 \pm 10.3)\%$ respectively on the last week of RT. There were no statistical differences between patients in Cohort 1 and patients in Cohort 2 on the bladder volumes at the first CBCT (p = 0.063), 4 weeks (p = 0.196) and 8 weeks (p = 0.078) after starting RT. Furthermore, there were no differences in the number of total CBCT scans between Cohort 1 (39.0 per patient) and Cohort 2 (38.5 per patient). With regards to the rectal separations, there were no significant differences in the product of distances at each of the three anatomical points measured prior to starting RT and after the last fraction of RT (Fig. 8A, B and C).

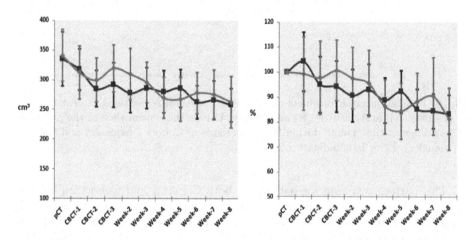

Fig. 7. Population mean bladder volume estimated and when badder was seen (left side) and mean of the percentage of the bladder volume in relation to each pCT (right side) at different time points for patients in Cohort 1 (squares) and Cohort 2 (triangles). Error bars indicate ± 2 SD_{mean}.

The analysis of the quality of life data showed in Table 2 and Table 3, for bladder side effects, an increase of the scores at weeks 4 and 8 indicating the presence of side effects caused by RT. The scores decreased at 3 months and 6 months after RT indicating a reduction in the intensity of the side effects.

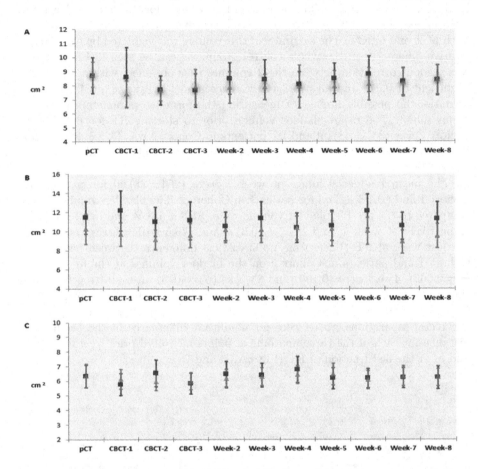

Fig. 8. Population mean products (A/P x R/L) of rectal separations at central slice of prostate (A), at 3 cm SUP (B) and at 3 cm INF to the central slice of the prostate (C) at different time points during RT for patients in Cohort 1 (squares) and Cohort 2 (triangles). Error bars indicate ±2 SD_{mean}.

This pattern was seen for patients in both Cohort 1 and Cohort 2 (Table 2 and Table 3). For bowel symptoms, the scores gradually increased at weeks 4 and 8 but did not revert to initial pre-RT values at 6 months as in the case of the scores for the bladder side effects (Table 2 and Table 3). This pattern was also seen for patients in both Cohort 1 and Cohort 2. Statistical analysis of the scores linked to bladder side effects showed a marginal statistical difference ($p = 0.041$) between patients in Cohort 1 and patients in Cohort 2 prior to starting RT. This could be due to the small number of cases included in the analysis. However, it can introduce a bias on the statistical analysis of scores at different time points.

Table 2. Quality of life questionnaire for patients in Cohort 1 (pre-VERT). Values are the means of the scores for each question.

	Pre-RT	Week 4	Week 8	3 months post RT	6 months post RT
Have you had to urinate frequently during the day	1.84	2.19	2.34	1.83	1.75
Have you had to urinate frequently at night	2.25	2.75	2.81	2.07	2.09
When you felt the urge to pass urine, did you have to hurry to get to the toilet?	1.75	2.00	2.06	1.67	1.63
Was it difficult for you to get enough sleep, because you needed to get up frequently at night to urinate?	1.56	1.94	2.00	1.67	1.69
Have you had difficulty going out of the house because you needed to be close to a toilet?	1.03	1.22	1.31	1.13	1.16
Have you had any unintentional release (leakage) of urine?	1.34	1.19	1.34	1.17	1.22
Did you have pain when you urinated?	1.03	1.38	1.47	1.13	1.16
Did you pass blood when you urinated?	1.03	1.03	1.06	1.00	1.00
Have your daily activities been limited by your urinary problems?	1.07	1.17	1.26	1.17	1.22
Mean urinary symptoms	**1.43**	**1.65**	**1.74**	**1.43**	**1.43**
SD (±)	**0.44**	**0.59**	**0.59**	**0.38**	**0.37**
Have your daily activities been limited by your bowel problems?	1.06	1.06	1.34	1.23	1.22
Did you have pain when opening your bowels?	1.06	1.03	1.25	1.17	1.16
Have you had any unintentional release (leakage) of stools?	1.09	1.03	1.19	1.17	1.22
Have you had blood in your stools?	1.00	1.13	1.09	1.00	1.03
Did you have a bloated feeling in your abdomen?	1.34	1.25	1.28	1.43	1.44
Mean bowel symptoms	**1.11**	**1.10**	**1.23**	**1.20**	**1.21**
SD (±)	**0.13**	**0.09**	**0.10**	**0.16**	**0.15**

Table 3. Quality of life questionnaire for patients in Cohort 2 (post-VERT). Values are the means of the scores for each question.

	Pre-RT	Week 4	Week 8	3 months post RT	6 months post RT
Have you had to urinate frequently during the day	2.46	2.58	2.63	2.09	2.08
Have you had to urinate frequently at night	2.63	2.92	2.83	2.59	2.50
When you felt the urge to pass urine, did you have to hurry to get to the toilet?	1.83	1.79	2.17	1.77	1.83
Was it difficult for you to get enough sleep, because you needed to get up frequently at night to urinate?	2.04	2.58	2.54	2.23	1.92
Have you had difficulty going out of the house because you needed to be close to a toilet?	1.25	1.42	1.58	1.27	1.33
Have you had any unintentional release (leakage) of urine?	1.29	1.38	1.38	1.41	1.46
Did you have pain when you urinated?	1.04	1.50	1.54	1.14	1.13
Did you pass blood when you urinated?	1.00	1.00	1.00	1.00	1.00
Have your daily activities been limited by your urinary problems?	1.13	1.21	1.42	1.18	1.26
Mean urinary symptoms	**1.63**	**1.82**	**1.89**	**1.63**	**1.61**
SD (±)	**0.62**	**0.69**	**0.65**	**0.56**	**0.49**
Have your daily activities been limited by your bowel problems?	1.13	1.13	1.21	1.23	1.21
Did you have pain when opening your bowels?	1.08	1.09	1.08	1.18	1.21
Have you had any unintentional release (leakage) of stools?	1.00	1.09	1.13	1.36	1.33
Have you had blood in your stools?	1.04	1.17	1.13	1.14	1.21
Did you have a bloated feeling in your abdomen?	1.42	1.35	1.50	1.68	1.50
Mean bowel symptoms	**1.13**	**1.17**	**1.21**	**1.32**	**1.29**
SD (±)	**0.17**	**0.11**	**0.17**	**0.22**	**0.13**

The analysis for bladder side effects showed only a statistical difference (p = 0.02) between patients in Cohort 1 and patients in Cohort 2 at 3 months after starting RT.

There were no statistical differences between patients in Cohort 1 and patients in Cohort 2 on the scores at 4 weeks (p = 0.093), 8 weeks (p = 0.166) and 6 months (p = 0.068) after starting RT. Regarding scores for bowel side effects, no statistical differences were seen at baseline (prior to starting RT) or at the time points (4 weeks, 8 weeks, 3 months and 6 months) after starting RT.

4 Discussion

The provision of information to patients helps them not only to better understand the treatment and tolerate side effects but also to be more compliant during the delivery of RT. This is paramount in the treatment of prostate cancer. Furthermore, the involvement of relatives in this exercise is also an important factor for these patients. While our previous work showed that VERT system improves patients and relatives' satisfaction [14], further work was needed to assess whether this translated, in the case of prostate cancer, into improved patients' compliance and reduced side effects from RT.

In the treatment of prostate cancer, to have a full bladder during RT is important as a decrease in bladder volume affects both bladder dose volume and the position of adjacent organs (the prostate, seminal vesicles, small intestine, sigmoid colon, and rectum) [2]. It has been stated that a fixed drinking protocol does not eliminate all variations in the bladder volume [22], in part due to significant individual variations in velocity of bladder filling [2]. On the other hand, it has been reported that patients are able to accurately judge their bladder filling state and it has even been suggested that subjective patient assessments should be taken into account to better control bladder volume [22]. The mean bladder volume prior to starting RT was (331 ± 48) ml for patients in Cohort 1 and (328 ± 46) ml for patients in Cohort 2. The volumes on the last day of RT were (254 ± 28) ml for patients in Cohort 1 and (267 ± 42) ml for patients in Cohort 2 (Fig. 5). These values represent a decrease in bladder volume of 23.5% for patients in Cohort 1 and 23.0 % for patients in Cohort 2. This decrease in bladder volume is less when compared to other studies reporting a decrease in bladder volume between 30% and 50% [1,2]. In the present study, there were no differences in the reduction of bladder volume regardless of whether patients had been shown using the VERT system how RT is planned and given prior to RT or after the last fraction of RT. It is possible that no difference was found because not enough patients were included in the study due to it ending sooner than planned. Additionally, it is also possible that even providing information using VR tools might not be sufficient to help patients to better understand the importance of having a full bladder towards the end of RT. Further work is thus needed to clarify this.

The other organ at risk during RT for prostate cancer is the rectum. Due to the close anatomic relation between the rectum, anal canal and the prostate, it is impossible to completely spare these organs [23]. In the present study, we did not see any differences between the rectal sections at each of the three anatomical slices measured (described above) prior to and on the last fraction of RT. Furthermore, there were no differences between patients allocated to Cohort 1 or Cohort 2 (Fig. 8A, B and C). It has been shown that rectal volume decreases at 6 months after RT but it was not measured on the last day of RT [24].

In our cohort of patients, side effects from RT, as measured with a QOL questionnaire, were present at week 4 and became higher at week 8 of treatment. This is in accordance to previous published data stating that symptoms gradually increase over the 2–3 weeks after the start of RT, rising towards a peak at the end of the RT [25,26]. After the completion of RT, acute symptoms (bladder frequency, nocturia and urgency, and diarrhoea, bleeding, cramps or bowel urgency) will start to disappear within 4–8 weeks [26]. Furthermore, scores measuring bladder side effects start to decrease at 1 month after the completion of VMAT, although the scores remain significantly higher than those at baseline [25]. Again, our data follows this pattern as the scores decreased at 3 months after RT but still remained higher when compared to scores prior to RT. This was the case for both patients in Cohort 1 and in Cohort 2. Statistical analysis did not show any statistical difference in the scores between patients in Cohort 1 and Cohort 2 after starting RT except at 3 months after RT with patients in Cohort 1 having lower scores. No statistical difference was seen when comparing bowel scores between patients in Cohort 1 and Cohort 2 at any of the study time-points.

Finally, as our previous study indicated, all patients welcomed the provision of information on RT using the VERT system. In fact, patients in Cohort 2 stated that they would have wished to have seen how RT is planned and given using the VERT system prior to starting RT.

5   Conclusion

Previous studies have shown that most patients receiving RT welcome VR tools to better understand how RT is planned and given. Specifically, providing information on RT using the 3D VERT imaging system rather than others 2D helped patients and relatives to better understand the complexity of RT planning and delivery, as well as to reduce their anxieties. However, it is not clear yet whether this leads to improved patients' compliance to RT which could also translate to reduced side effects from RT. Despite the improvement of patients' experience and satisfaction, the present randomised study did not show any differences in patients' compliance regardless of whether patients were showed how RT is planned and given using the VERT system prior to RT or on the last day of RT. Nevertheless, this paper reiterates the wish of patients to receive information on RT planning and delivery using VR tools. Future work is still needed with perhaps bigger cohort of patients. In the future, this study may provide cancer centres with an useful tool to audit local protocols regarding patient preparations and training for prostate RT.

Acknowledgements. This research did not receive any specific grant from funding agencies in the public, commercial or not-for-profit sectors.

References

1. Stam, M.R., van Lin, E.N.J.T., van der Vight, L.P., Kaander, J.H.A.M., Visser, A.G.: Bladder filling variation during radiation treatment of Prostate cancer: can the use of a bladder ultrasound scanner and biofeedback optimize bladder filling. Int. J. Radiat. Oncol. Biol. Phys. **65**, 371–377 (2006). https://doi.org/10.1016/j.ijrobp.2005.12.039
2. Nakamura, N., et al.: Variability in bladder volumes of full bladders in definitive radiotherapy for cases of localized prostate cancer. Strahlenther Onkol **36**(186), 637–42 (2010). https://doi.org/10.1007/s00066-010-2105-6
3. Bonet, M., Cayetano, L., Núñez, M., Jovell-Fernández, E., Aguilar, A., Ribas, Y.: Assessment of acute bowel function after radiotherapy for prostate cancer: is it accurate enough? Clin. Transl. Oncol. **20**(5), 576–583 (2017). https://doi.org/10.1007/s12094-017-1749-4
4. Bacon, C.G., Giovannucci, E., Testa, M., Glass, T.A., Kawachi, I.: The association of treatment-related symptoms with quality-of-life outcomes for localized prostate carcinoma patients. Cancer **94**, 862–71 (2002). https://doi.org/10.1002/cncr.10248
5. Hansen, H., Nielsen, B.K., Boejen, A., Vestergaard, A.: Teaching cancer patients the value of correct positioning during radiotherapy using visual aids and practical exercises. J. Cancer Educ. **33**(3), 680–685 (2016). https://doi.org/10.1007/s13187-016-1122-2
6. Berger, O., Grønberg, B.H., Loge, J.H., Kaasa, S., Sand, K.: Cancer patients' knowledge about their disease and treatment before, during and after treatment: a prospective, longitudinal study. BMC Cancer **18**, 381 (2018). https://doi.org/10.1186/s12885-018-4164-5
7. Mallinger, J.B., Griggs, J.J., Shields, C.G.: Patient-centered care and breast cancer survivors' satisfaction with information. Patient Educ. Couns. **57**, 342–9 (2005). https://doi.org/10.1016/j.pec.2004.09.009
8. Stewart-Lord, A., Brown, M., Noor, S., Cook, J., Jallow, O.: The utilisation of virtual images in patient information giving sessions for prostate cancer patients prior to radiotherapy. Radiography **22**, 269–273 (2016). https://doi.org/10.1016/j.radi.2016.05.002
9. Jimenez, Y.A., Wang, W., Stuart, K., Cumming, S., Thwaites, D., Lewis, S.: Breast cancer patients' perceptions of a virtual learning environment for pretreatment education. J. Cancer Educ. **33**(5), 983–990 (2017). https://doi.org/10.1007/s13187-017-1183-x
10. Phillips, R, Ward, J.W., Beavis, A.W.: Immersive visualization training of radiotherapy treatment. Stud. Health Technol. Inf. **111**, 390–396 (2011). https://europepmc.org/article/med/15718766
11. Boejen, A., Grau, C.: Virtual reality in radiation therapy training. Surg. Oncol. **20**, 185–8 (2011). https://doi.org/10.1016/j.suronc.2010.07.004
12. Jimenez, Y.A., Lewis, S.J.: Radiation therapy patient education using VERT: combination of technology with human care. J. Med. Radiat. Sci. **65**, 158–62 (2018). https://doi.org/10.1002/jmrs.282
13. Kane, P.: Simulation-based education: a narrative review of the use of VERT in radiation therapy education. J. Med. Radiat. Sci. **65**, 131–136 (2018). https://doi.org/10.1002/jmrs.276

14. Sulé-Suso, J., et al.: Pilot study on virtual imaging for patient information on radiotherapy planning and delivery. Radiography **21**, 273–7 (2015). https://doi.org/10.1016/j.radi.2015.02.002

15. Marquess, M., et al.: A pilot study to determine if the use of a virtual reality education module reduces anxiety and increases comprehension in patients receiving radiation therapy. J. Radiat. Oncol. **6**(3), 317–322 (2017). https://doi.org/10.1007/s13566-017-0298-3

16. Jimenez, Y.A., Cumming, S., Wang, W., Stuart, K., Thwaites, D.I., Lewis, S.J.: Patient education using virtual reality increases knowledge and positive experience for breast cancer patients undergoing radiation therapy. Support. Care Cancer **26**(8), 2879–2888 (2018). https://doi.org/10.1007/s00520-018-4114-4

17. Chapman, K., James, S.: A review of results from patient experience surveys during the introduction of group pre-radiotherapy patient information sessions. Radiography **22**, 237–43 (2016). https://doi.org/10.1016/j.radi.2016.01.003

18. Patel, D., Muren, L.P., Mehus, A., Kvinnsland, Y., Ulvang, D.M., Villanger, K.P.: A virtual reality solution for evaluation of radiotherapy plans. Radiother. Oncol. **82**, 218–21 (2007). https://doi.org/10.1016/j.radonc.2006.11.024

19. Dearnaley, D., Syndikus, I., Mossop, H., et al.: Conventional versus hypofractionated high-dose intensity-modulated radiotherapy for prostate cancer: 5-year outcomes of the randomised, non-inferiority, phase 3 CHHiP trial. Lancet **17**, 1047–60 (2016). https://doi.org/10.1016/s1470-2045(16)30102-4

20. Wilson, A.J.: Volume of n-dimensional ellipsoid. Sci. Acta Xaveriana **1**, 101–106 (2010). http://oaji.net/articles/2014/1420-1415594291.pdf

21. Chan, E., et al.: Tri-axial ellipsoid volume calculation: a method for bladder volume estimation. J. Med. Imag. Rad. Sci. **44**, 46–7 (2013). https://doi.org/10.1016/j.jmir.2012.12.011

22. O'Doherty, Ú.M., et al.: Variability of bladder filling in patients receiving radical radiotherapy to the prostate. Radiother. Oncol. **79**, 335–40 (2006). https://doi.org/10.1016/j.radonc.2006.05.007

23. Krol, R., McColl, G.M., Hopman, W.P.M., Smeenk, R.J.: Anal and rectal function after intensity-modulated prostate radiotherapy with endorectal balloon. Radiother. Oncol. **128**, 364–8 (2018). https://doi.org/10.1016/j.radonc.2018.03.032

24. Stasi, M., et al.: Emptying the rectum before treatment delivery limits the variations of rectal dose-volume parameters during 3DCRT of prostate cancer. Radiother. Oncol. **80**, 363–70 (2006). https://doi.org/10.1016/j.radonc.2006.08.007

25. Nakai, Y., et al.: Quality of life worsened the most severely in patients immediately after intensity-modulated radiotherapy for prostate cancer. Res. Rep. Urol. **10**, 169–80 (2018). https://doi.org/10.2147/RRU.S168651

26. Nakamura, K., Konishi, K., Komatsu, T., Ishiba, R.: Quality of life after external beam radiotherapy for localized prostate cancer: comparison with other modalities. Int. J. Urol. **26**, 950–4 (2019). https://doi.org/10.1111/iju.14026

Data-Driven Speed-of-Sound Reconstruction for Medical Ultrasound: Impacts of Training Data Format and Imperfections on Convergence

Farnaz Khun Jush[1,2(✉)] ⓘ, Peter Michael Dueppenbecker[1],
and Andreas Maier[2] ⓘ

[1] Technology Excellence, Siemens Healthcare GmbH, Erlangen, Germany
{farnaz.khunjush,peter.dueppenbecker}@siemens-healthineers.com
[2] Pattern Recognition Lab, Friedrich-Alexander-University, Erlangen, Germany
andreas.maier@fau.de

Abstract. B-mode imaging is a qualitative method and its interpretation depends on users' experience. Quantitative tissue information can increase precision and decrease user ambiguity. For example, Speed-of-Sound (SoS) in tissue is known to carry diagnostic information. Studies showed the possibility of SoS reconstruction from ultrasound raw data (a.k.a., RF data) using deep neural networks (DNNs). However, many ultrasound systems are designed to process demodulated data (i.e., IQ data) and often decimate data in early stages of acquisition. In this study we investigated the impacts of input data format and decimation on convergence of the DNNs for SoS reconstruction. Our results show that fully data-driven SoS reconstruction is possible using demodulated ultrasound data presented in Cartesian or Polar format using an encoder-decoder network. We performed a study using only amplitude and only phase information of ultrasound data for SoS reconstruction. Our results showed that distortion of the phase information results in inconsistent SoS predictions, indicating sensitivity of the investigated approach to phase information. We demonstrated that without losing significant accuracy, decimated IQ data can be used for SoS reconstruction.

Keywords: Speed-of-Sound · Deep Neural Network · Ultrasound

1 Introduction

B-mode imaging is a non-ionising, real-time and portable imaging modality which is vastly used for medical diagnostics [22]. However, b-mode imaging is a qualitative approach and outcomes highly depend on operator's expertise. In b-mode imaging, different tissue types can result in very similar image impressions. This complicates image interpretation which can lead to missed lesions or unnecessary biopsies [24].

ⓒ Springer Nature Switzerland AG 2021
B. W. Papież et al. (Eds.): MIUA 2021, LNCS 12722, pp. 140–150, 2021.
https://doi.org/10.1007/978-3-030-80432-9_11

Additional quantitative information (e.g., density, elasticity, Speed-of-Sound (SoS), attenuation) can increase ultrasound specificity and ease image interpretation. Specialized ultrasound devices exist that provide quantitative information side by side with b-mode images, e.g., shear-wave elastography techniques are available on commercial devices which estimate Young's modulus in tissues. Young's modulus indicates stiffness and can be used for differentiating pathological tissues [8]. However, shear-wave elastography is only limited to high-end devices [4].

SoS carries similar diagnostic information as shear wave elastography which can be used as a biomarker for tissue characterization [6,9,15,18,19]. Multiple studies have shown the possibility of estimating SoS using conventional ultrasound setups [1,17]. The most common practice is based on the idea that if the SoS chosen for b-mode beamforming matches the actual SoS in the tissue being examined, the quality of the b-mode image is maximized [1]. However, there exist studies in which the SoS is reconstructed by either formulating the forward problem based on deformation of the wavefront in the reflection pattern [21] or solving an ill-posed inverse problem [20].

Fundamentally, acoustic properties of tissues including SoS information is available in raw data but extracting this information is a non-trivial task. Conventional methods for SoS reconstruction are often either iterative or optimization-based methods and therefore can be computationally expensive with varying run times. Deep learning solutions can be used to offload the computational burden to the training phase and often have predictable run times, thus, they can be used for real-time processing. In recent years, deep learning-based techniques were successfully applied to solve inverse problems which often resulted in superior solutions compared to analytical approaches [14,16]. There are studies which show the possibility of SoS reconstruction using DNNs [2,4,5,10,24].

The idea of a deep learning based algorithm for SoS reconstruction without going through b-mode beamforming is first introduced in [4]. However, in many available setups access to RF data is very limited because ultrasound systems produce huge amount of raw data and often the bandwidth required to transfer the data from the probe to the processing system exceeds the capacities of data links. Therefore, demodulation and decimation of data is often applied in early stages of the acquisition and many ultrasound data processing pipelines are designed to work with decimated IQ demodulated data.

This work focuses on investigating the feasibility and impact of data representation format for data-driven SoS reconstruction. As the reference method, we set up a DNN which takes RF data as input and estimates the SoS in output based on [5]. We proposed a network to reconstruct SoS from demodulated data and compared its performance with the network which uses RF data. Having demodulated complex data, phase and amplitude information can be easily extracted. We presented the amplitude and phase information to a DNN and investigated the contribution of phase and amplitude on SoS estimations. We further examined the impacts of decimation on the estimated SoS map.

2   Methods

2.1   Data Simulation

Performance and capability of generalisation of deep learning-based techniques strongly depend on huge amount of diverse training data [3]. Measuring and labeling sufficient data for training DNNs is often time consuming and costly. Additionally, in some cases measuring the ground truth (GT) is difficult or impossible, for example, acquiring GT data for SoS imaging is a challenging task, because neither a proper reference method capable of creating GT from ultrasound reflection data is available nor sufficient number of reference phantoms with known SoS exist [4]. Therefore, in this study we use simulated data for training.

For the simulation we use the k-Wave toolbox (version 1.3) [23]. K-Wave toolbox benefits from CUDA accelerated code which offers decent run time on Nvidia GPUs and is suitable for simulation of diverse heterogeneous mediums.

We simulated a linear probe geometry with 128 elements. During acquisition, only the 64 central elements of the probe are activated. The transmit pulse is a tone burst pulse with rectangular envelope and a center frequency of 5 MHz. Inspired by [4], we use single plane-wave and a simplified geometric model for inclusions. The medium size is 3.8×3.8 cm and the probe head is placed in the center line of the medium. A uniform SoS of 1535 m/s is chosen in the background. One to five inclusions are placed randomly in the medium with SoS contrast from background. Inclusions are mimicked as elliptical shapes with randomly chosen SoS values in range [1300–1700] m/s. Studies suggest that SoS values in human soft tissue is in range [1400–1600] m/s [13,22], but we used a greater range for better generalization. The acoustic attenuation is set to the constant value of 0.75 dB/MHz.cm [22]. The mass density is set to 1020 kg/m^3 [22]. Random speckle particles are added with $\pm3\%$ variation and mean distribution of 2 speckles per λ^2. The central section of the medium which is placed directly under the transducer is recovered, a region of size 1.9×3.8 cm. In order

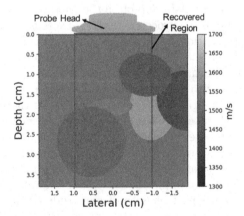

Fig. 1. Sample simulated medium of size 3.8×3.8 cm, probe head is placed in center line and the region inside the red box is recovered for training. (Color figure online)

to include out of the plane reflections, the size of the simulated medium is twice the size of the recovered region. Figure 1 shows a sample simulated medium.

2.2   Network Setup

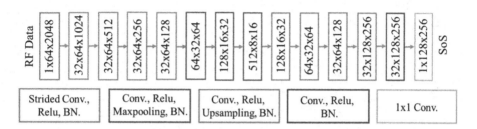

Fig. 2. RF-Net architecture with encoder-decoder structure for mapping RF data (input) to SoS map (output).

Prior Work, RF Data. SoS reconstruction from RF data using DNNs is already addressed in [4,5,10]. For the sake of comparison, we set up a network similar to the network proposed in [5]. The network architecture is shown in Fig. 2. The network is a fully convolutional neural network with encoder and decoder paths which maps the ultrasound RF data directly to SoS map created by simulation.

The encoder or contracting path consists of seven layers. In first four layers Strided Convolution (1×2), Relu [25] and BatchNormalization (BN) [7] is applied. Last three layers consist of Convolution, Relu, Maxpooling and BN.

The decoder or expanding path consists of six layers. In first four layers Convolution, Relu, Upsampling and BN is applied. The fifth layer consists of Convolution, Relu and BN and the last layer is a 1×1 convolution which results in SoS map.

Hereafter, this network will be referred as RF-Net. The input of the network is RF data (64 channels and 2048 samples, sampled at 40 MHz, 64-bit floating point). The output of the network is a 256×128 matrix of SoS map which translates to the recovered section of the medium of size 1.9×3.8 cm.

Complex Data. The data format used in image processing pipelines is manufacture specific. Since ultrasound data is a band-pass signal and the bandwidth of the transducer is usually between [50–70]% [12], often IQ demodulation and decimation is done in early stages of data acquisition. This data is usually presented as complex data. In order to handle complex data, we propose a network as shown in Fig. 3(a), hereafter known as IQ-Net. IQ-Net has two encoders and one decoder. IQ data is fed to the two parallel encoders. First six layers of each encoder is similar to the RF-Net. In seventh layer features from two encoders are

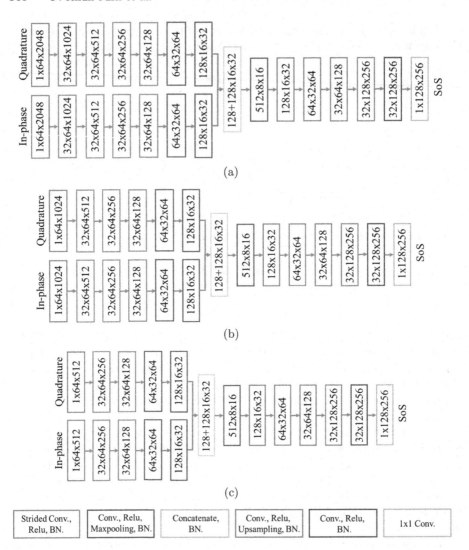

Fig. 3. a) IQ-Net consists of two paths, each path takes 64 channels sampled at 40 MHz, (b) Modified IQ-Net I for decimated data by factor of 2, (c) Modified IQ-Net II for decimated data by factor of 4.

concatenated. Eventually a Convolution, Relu, Maxpooling and BN is applied. The decoder of IQ-Net has the same architecture as in RF-Net.

Cartesian Representation: First of all, we investigate the possibility of SoS reconstruction using IQ demodulated data sampled at the same sampling frequency (F_s) as RF data (40 MHz). Since the data consists of real and imaginary parts, the encoder consists of two parallel paths. The real (a.k.a., in-phase) and the imaginary (a.k.a., quadrature) part of the data is fed to two parallel channels as input. Features are extracted simultaneously in parallel paths and are eventually concatenated at latent space as demonstrated in Fig. 3(a). The output of the decoder is the SoS map similar to the RF-Net.

Polar Representation: Having IQ demodulated data, phase and amplitude of the signal can easily be extracted. In order to have a comprehensive investigation on how the format of the data affects SoS estimations, we present the data with polar representation to IQ-Net as well. Meaning, instead of in-phase and quadrature parts, amplitude and phase of the data is fed to each encoding paths as inputs.

Additionally, phase and amplitude can be separately presented to RF-Net. This can be used to examine how phase or amplitude of the data contribute to SoS estimations. Therefore, in addition to RF data, we trained RF-Net once with phase only and once with amplitude only.

Decimation. In order to inspect the impacts of decimation, we modified IQ-Net by removing one top layer of each encoding path by each decimation step, as shown in Fig. 3(b) and (c). The sampling rate is reduced by factor of 2 and 4 (F_s: 20 and 10 MHz).

3   Results and Discussion

Based on [4,10] we simulated 6700 cases and split the dataset to 6000 cases for train set (90% for training, 10% for validation) and 700 cases for test set. The mean square error (MSE) is used as loss function and Adam [11] is used as optimizer. Prior to feeding, the input data is normalized, time gain compensation with 0.75 dB/MHz.cm at SoS of 1535 m/s is applied and a uniformly distributed quantization noise is added to each channel. For regularization, during training, a random Gaussian noise with standard deviation of 0.5 is added to the input data. This setup is consistent for all following sections.

3.1   Data Format

In this section, the data is only demodulated but the decimation step is not applied. First of all, we examine the possibility of SoS reconstruction using different data formats. Essentially, the aim is to determine whether the networks are sensitive to the format of the input data. To this end, we compared the IQ-Net trained with demodulated data sampled at 40 MHz with the RF-Net trained with RF data sampled at the same frequency. Demodulated data is presented to IQ-Net once in Cartesian and once in Polar representation.

Additionally, we extracted amplitude and phase components and trained RF-Net once with amplitude only and once with phase only. After 150 epochs the networks converged to the Root Mean Square Error (RMSE) and Mean Absolute Error (MAE) shown in Table 1.

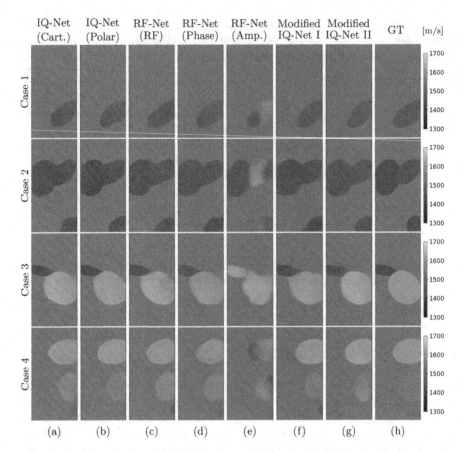

Fig. 4. SoS estimations for four cases each case is placed in one row, columns demonstrate: (a): IQ-Net Cartesian representation, (b): IQ-Net Polar representation, (c): RF-Net using RF data, (d): RF-Net using phase only, (e): RF-Net using amplitude only, (f): Modified IQ-Net I, (g): Modified IQ-Net II and (h): GT. Image sizes in lateral and axial directions are same as red box demonstrated in Fig. 1. (Color figure online)

The lowest RMSE and MAE is achieved using IQ-Net trained with demodulated data presented in Cartesian form. However, the RMSE and MAE corresponding to RF-Net using RF data and IQ-Net using Polar representation show only low variations which based on the standard deviations between multiple runs shown in Table 1 are statistically insignificant. Thus, we can conclude that SoS estimation is feasible using both RF and demodulated data.

Qualitative results are presented in Fig. 4 column (a), (b) and (c). Sample cases, shown in each row, are including mediums with single or multiple inclusions with lower SoS compared to the background (Case 1 and 2), mediums with multiple inclusions with both SoS lower and higher than the background (Case 3 and 4). Figure 4 column (h) shows the GT.

The predicted SoS maps in Fig. 4 column (a), (b) and (c) are very similar to GT. Only slight variations are present which mostly appear as under/overestimations of SoS on the edges of the inclusions. This can be seen as deformation in shapes of ellipses compared to the GT image.

Table 1. RMSE and MAE for multiple runs, $\pm$ shows the standard deviation (σ) for multiple runs.

	RMSE (m/s)	MAE (m/s)
IQ-Net (Cartesian)	12.34 ± 0.38	4.66 ± 0.26
IQ-Net (Polar)	13.14 ± 0.24	5.25 ± 0.30
RF-Net (RF data)	12.92 ± 0.22	5.10 ± 0.18
RF-Net (Phase)	14.94 ± 0.58	5.96 ± 0.33
RF-Net (Amplitude)	22.93 ± 1.27	9.29 ± 1.04

Based on Table 1, networks trained with only phase or only amplitude are prone to higher errors. After 150 epochs, RF-Net trained with phase data converges to RMSE of 14.94 ± 0.58 m/s and RF-Net trained with amplitude data converges to RMSE of 22.93 ± 1.27 m/s. Since using amplitude only results in high error rates, it can be concluded that the phase information is more significant for the DNN to estimate SoS. This is more obvious in qualitative results shown Fig. 4 column (d) and (e). Comparing the predicted SoS map with GT, the network trained with phase input could correctly estimate shape, location and SoS values while under/overestimations are only present at the edges of the inclusions. On the other hand, the network trained with amplitude could only estimate the shape and the location of the inclusions. However, SoS values inside inclusions are not uniform (Fig. 4 column (e) Case 1, 2 and 4) or completely inverted (Fig. 4 column (e) Case 3).

In order to demonstrate the sensitivity of the network to phase information, we extracted phase component of two samples shown in Fig. 5, column (f). After IQ demodulation based on [12], the phase component is shifted in range $[0-180]°$ with steps of $45°$ and then the IQ data with shifted phase components are reconstructed back to RF form [12]. The RF data with shifted phase is then fed to the RF-Net trained with RF data with original phase information. The SoS predictions are shown in Fig. 5 column (a–e). It can be seen that the estimated SoS values for each inclusion is directly related to the phase information of the input RF data while the shape and location of inclusions are relatively intact and therefore related to amplitude of the RF data. The data with zero degree shift is included, Fig. 5 column (a), to show that the process of conversion from RF to IQ and back to RF does not introduce distortion in relevant information for the network and the only altering component which affects the respective SoS estimations is the phase shift.

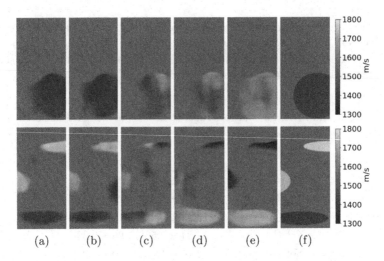

Fig. 5. SoS predictions for phase shifted RF data, where phase shift is: (a) 0°, (b) 45°, (c) 90°, (d) 135°, (e) 180°, (f) GT. Image sizes in lateral and axial directions are same as red box demonstrated in Fig. 1. (Color figure online)

3.2   Decimation

After demodulation, decimation is done and the Modified IQ-Nets shown in Fig. 3(b) and (c) are trained for 150 epochs.

Table 2 shows the corresponding RMSE and MAE for IQ-Nets. Modified IQ-Net I converges to similar RMSE and MAE as IQ-Net. Modified IQ-Net II has higher RMSE and MAE in comparison. However, since researches suggest that measurement accuracy up to 30 m/s is useful for clinical relevance [4], here the loss of accuracy with decimation can be considered insignificant (RMSE <3 m/s and MAE <1.5 m/s).

Figure 4 column (f) and (g) show the qualitative results for modified IQ-Nets. Compared to GT, it can be seen that in Fig. 4 column (f) and (g), Case 1, 2 and 3 SoS map matches the GT (except as previous networks at the edges of the inclusions). However, in Fig. 4 column (g), Case 4, there is a small inclusion in the bottom right with lower SoS compared to background which is missing. Based on our observations on this case and other cases not included here, the networks trained with decimated data can miss small inclusions (smaller than 1 mm radius) more often than other networks.

Nevertheless, the amount of data to transfer and save reduced by factor of 4 for each acquisition for Modified IQ-Net II compared to IQ-Net. Network training time reduced from 2.75 h to 1 h on Nvidia Quadro RTX 8000. On the same GPU, IQ-Net and modified IQ-Net II perform at 233 and 700 fps, respectively. Comparing modified IQ-Net II and RF-Net trained with fully sampled data, the training time is reduced from 1.6 h to 1 h on Nvidia Quadro RTX 8000. On the same GPU RF-Net performs at 350 fps. Therefore, in the trade-off between speed and storage and slight loss of accuracy, Modified IQ-Net II with decimated data can be used for SoS reconstruction.

Table 2. RMSE and MAE for IQ-Net and Modified IQ-Nets, $\pm$ shows the standard deviation (σ) for multiple runs.

	RMSE (m/s)	MAE (m/s)
IQ-Net	12.34 ± 0.30	4.66 ± 0.36
Modified IQ-Net I	12.99 ± 0.37	5.75 ± 0.41
Modified IQ-Net II	14.98 ± 0.40	6.20 ± 0.58

4 Conclusion

In this study, we investigated the possibility of fully data-driven SoS reconstruction from various data representation using DNNs. We demonstrated that between real and complex data format used for training DNNs, the format in which the data is presented to the network is not a crucial aspect and networks with different input data formats eventually converge to similar RMSE and MAE. However, training networks using only amplitude and only phase information showed that phase information is a significant aspect for the network and providing only phase information it would be possible to reconstruct SoS in tissues, while introducing phase distortion or phase shifts has a high impact on the estimated SoS values. Furthermore, in a trade-off between speed and storage and losing slight accuracy, demodulated decimated data can also be used for SoS reconstruction.

Disclaimer

The information in this paper is based on research results that are not commercially available.

References

1. Benjamin, A., et al.: Surgery for obesity and related diseases: I. A novel approach to the quantification of the longitudinal speed of sound and its potential for tissue characterization. Ultrasound Med. Biol. **44**(12), 2739–2748 (2018)
2. Bernhardt, M., Vishnevskiy, V., Rau, R., Goksel, O.: Training variational networks with multidomain simulations: speed-of-sound image reconstruction. IEEE Trans. Ultrason. Ferroelectr. Freq. Control **67**(12), 2584–2594 (2020)
3. Cho, J., Lee, K., Shin, E., Choy, G., Do, S.: How much data is needed to train a medical image deep learning system to achieve necessary high accuracy? arXiv preprint arXiv:1511.06348 (2015)
4. Feigin, M., Freedman, D., Anthony, B.W.: A deep learning framework for single-sided sound speed inversion in medical ultrasound. IEEE Trans. Biomed. Eng. **67**(4), 1142–1151 (2019)
5. Feigin, M., Zwecker, M., Freedman, D., Anthony, B.W.: Detecting muscle activation using ultrasound speed of sound inversion with deep learning. In: 2020 42nd Annual International Conference of the IEEE Engineering in Medicine & Biology Society (EMBC), pp. 2092–2095. IEEE (2020)

6. Hachiya, H., Ohtsuki, S., Tanaka, M.: Relationship between speed of sound in and density of normal and diseased rat livers. Jpn. J. Appl. Phys. **33**(5S), 3130 (1994)

7. Ioffe, S., Szegedy, C.: Batch normalization: accelerating deep network training by reducing internal covariate shift. arXiv preprint arXiv:1502.03167 (2015)

8. Jeong, W.K., Lim, H.K., Lee, H.K., Jo, J.M., Kim, Y.: Principles and clinical application of ultrasound elastography for diffuse liver disease. Ultrasonography **33**(3), 149 (2014)

9. Khodr, Z.G., et al.: Determinants of the reliability of ultrasound tomography sound speed estimates as a surrogate for volumetric breast density. Med. Phys. **42**(10), 5671–5678 (2015)

10. Khun Jush, F., Biele, M., Dueppenbecker, P.M., Schmidt, O., Maier, A.: DNN-based speed-of-sound reconstruction for automated breast ultrasound. In: 2020 IEEE International Ultrasonics Symposium (IUS), pp. 1–7. IEEE (2020)

11. Kingma, D.P., Ba, J.: Adam: A method for stochastic optimization. arXiv preprint arXiv:1412.6980 (2014)

12. Kirkhorn, J.: Introduction to IQ-demodulation of RF-data (1999)

13. Li, C., Duric, N., Littrup, P., Huang, L.: In vivo breast sound-speed imaging with ultrasound tomography. Ultrasound Med. Biol. **35**(10), 1615–1628 (2009)

14. Lucas, A., Iliadis, M., Molina, R., Katsaggelos, A.K.: Using deep neural networks for inverse problems in imaging: beyond analytical methods. IEEE Sig. Process. Mag. **35**(1), 20–36 (2018)

15. Matsuhashi, T., Yamada, N., Shinzawa, H., Takahashi, T.: An evaluation of hepatic ultrasound speed in injury models in rats: correlation with tissue constituents. J. Ultrasound Med. **15**(8), 563–570 (1996)

16. Ongie, G., Jalal, A., Baraniuk, C.A.M.R.G., Dimakis, A.G., Willett, R.: Deep learning techniques for inverse problems in imaging. IEEE J. Sel. Areas Inf. Theor. **1**, 39–56 (2020)

17. Qu, X., Azuma, T., Liang, J.T., Nakajima, Y.: Average sound speed estimation using speckle analysis of medical ultrasound data. Int. J. Comput. Assist. Radiol. Surg. **7**(6), 891–899 (2012)

18. Sak, M., et al.: Using speed of sound imaging to characterize breast density. Ultrasound Med. Biol. **43**(1), 91–103 (2017)

19. Sanabria, S., et al.: Breast-density assessment with hand-held ultrasound: a novel biomarker to assess breast cancer risk and to tailor screening? Eur. Radiol. **28**(8), 3165–3175 (2018). https://doi.org/10.1007/s00330-017-5287-9

20. Sanabria, S.J., Rominger, M.B., Goksel, O.: Speed-of-sound imaging based on reflector delineation. IEEE Trans. Biomed. Eng. **66**(7), 1949–1962 (2018)

21. Stähli, P., Kuriakose, M., Frenz, M., Jaeger, M.: Forward model for quantitative pulse-echo speed-of-sound imaging. arXiv preprint arXiv:1902.10639 (2019)

22. Szabo, T.L.: Diagnostic Ultrasound Imaging: Inside Out. Academic Press (2004)

23. Treeby, B.E., Cox, B.T.: k-wave: Matlab toolbox for the simulation and reconstruction of photoacoustic wave fields. J. Biomed. Opt. **15**(2), 021314 (2010)

24. Vishnevskiy, V., Sanabria, S.J., Goksel, O.: Image reconstruction via variational network for real-time hand-held sound-speed imaging. In: Knoll, F., Maier, A., Rueckert, D. (eds.) MLMIR 2018. LNCS, vol. 11074, pp. 120–128. Springer, Cham (2018). https://doi.org/10.1007/978-3-030-00129-2_14

25. Xu, B., Wang, N., Chen, T., Li, M.: Empirical evaluation of rectified activations in convolutional network. arXiv preprint arXiv:1505.00853 (2015)

Selective Motion Artefact Reduction via Radiomics and k-space Reconstruction for Improving Perivascular Space Quantification in Brain Magnetic Resonance Imaging

Jose Bernal[1]([⊠]) [iD], William Xu[1] [iD], Maria d. C. Valdés-Hernández[1] [iD],
Javier Escudero[2] [iD], Angela C. C. Jochems[1] [iD], Una Clancy[1] [iD], Fergus N. Doubal[1] [iD],
Michael S. Stringer[1] [iD], Michael J. Thrippleton[1] [iD], Rhian M. Touyz[3] [iD],
and Joanna M. Wardlaw[1] [iD]

[1] Centre for Clinical Brain Sciences, The University of Edinburgh, Edinburgh, UK
jose.bernal@ed.ac.uk
[2] Institute for Digital Communications, The University of Edinburgh, Edinburgh, UK
[3] Institute of Cardiovascular and Medical Sciences, University of Glasgow, Glasgow, UK

Abstract. Current evidence points towards perivascular spaces playing a key role in cerebral haemodynamics and waste clearance. Hence, their precise quantification may become a powerful tool for assessing brain health and further establishing their relationship with neurological diseases. Large strides have been made towards developing automatic tools to computationally assess the burden of perivascular spaces in MRI in recent years. However, their applicability depends to a large extent on the quality of the images. In this paper, we propose a pipeline to improve perivascular space quantification by means of radiomics-based image quality control and selective motion artefacts reduction. We demonstrate our method on a sample of patients with mild stroke (n = 60) with different extents of small vessel disease features and image quality. We show our proposal can differentiate high- and low-quality scans (AUROC = 0.98) and reduce imaging artefacts, which leads to greater correlations between visual and computational measurements, especially in the centrum semiovale (polyserial correlation: 0.86 [95% CI 0.85, 0.88] and 0.17 [95% CI 0.14, 0.21] with and without our proposal, respectively). Our preliminary results demonstrate the potential of our proposal for retaining clinically relevant information while reducing imaging artefacts.

Keywords: Perivascular spaces · Cerebral small vessel disease · Image enhancement · Imaging artefact reduction · Brain magnetic resonance imaging

1 Introduction

Perivascular spaces (PVS) in the brain are the cavities surrounding perforating cerebral microvessels, serving as drainage conduits through which interstitial fluid clearance is facilitated [1]. If enlarged or dilated, they appear hyperintense on T2-weighted MRI

© Springer Nature Switzerland AG 2021
B. W. Papież et al. (Eds.): MIUA 2021, LNCS 12722, pp. 151–164, 2021.
https://doi.org/10.1007/978-3-030-80432-9_12

sequences [2]. Although first described in the mid-1800s, their potential significance as indicator of brain health has only emerged recently following advancements in imaging technology [3]. Nonetheless, their association and involvement alone with neurological risk factors and disease is still currently debated [3], hence the importance of assessing PVS quantitatively to validate their use as a neuroimaging feature.

The majority of studies on PVS rely solely on clinical visual ratings [1, 4]. Although qualitative PVS assessments are quick, easily interpreted, and replicable, they are limited by their discrete nature, as with many qualitative scoring methods. Attempts to automate the quantification of PVS, while partly successful, are limited by imaging artefacts [5]. The incidence of imaging artefacts impeding automatic PVS segmentation has peaked at 20% in large-scale studies [5]. Thus, reduction of imaging artefacts during pre-processing prior to PVS quantification is necessary to reduce research waste and increase reliability of study results. Nonetheless, the application of image enhancement techniques requires distinguishing between high- and low-quality input scans as filtering high-quality scans is detrimental to PVS sensitivity [6].

The assessment of image quality is a long standing and challenging problem. Even though image analysts could potentially visually inspect all acquisitions, the task is tedious, time-consuming and error-prone due to inter- and intra-observer variability, especially in large-scale studies [7]. In the field of PVS segmentation, automatic quality control has been successfully conducted via texture analysis [6]. However, the presence of brain pathologies may limit the effectiveness of such approaches as textures vary with the burden of neuroimaging features and imaging artefacts [8, 9].

In this work, we propose a framework that identifies and enhances scans distorted by motion artefacts via radiomics and k-space analyses prior to PVS quantification, thus reducing imaging artefacts compromising computational assessments while retaining clinically relevant patterns on high-quality acquisitions. We tested our proposal on a well-phenotyped cohort of patients with a history of mild stroke presenting various neuroimaging features of small vessel disease.

2   Materials and Methods

As illustrated in Fig. 1, we determine the quality of the input image automatically and, if it is distorted by motion, we enhance its quality before PVS segmentation.

2.1   Subjects, Magnetic Resonance Imaging and Clinical Visual Scores

T2-w scans were obtained from a sample of an ongoing prospective study (The Mild Stroke Study 3: ISCTRN 12113543) of patients with a recent mild stroke (n = 60; 24 women; median age 69 years [IQR 58–75]; age range 40 to 85 years) with a varied burden of neuroimaging features of small vessel disease. We used data from the first and second visit (60 and 35 scans, respectively, were available at the time). Approval for this study was obtained from South East Scotland Research Ethics Committee (Ref 18/SS/0044) and NHS Lothian Research & Development (Ref 2018/0084).

T2-w MRI was performed on a 3T MRI scanner (MAGNETOM Prisma, Siemens Healthcare, Erlangen, Germany) and a 3D axial SPACE T2-w imaging sequence (TR/TE = 3200/408 ms, 0.94 × 0.94 × 0.90 mm acquired resolution, 24.0 × 24.0 cm field of view)[1]. An experienced neuroradiologist provided visual clinical ratings for PVS in the basal ganglia and centrum semiovale for the entire sample following the Potter scale [1]. The distribution of scores in the sample can be found in Table 1.

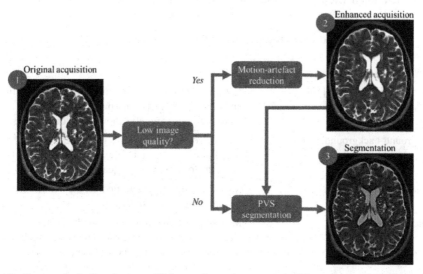

Fig. 1. Proposed pipeline for quantifying perivascular spaces in T2-w scans. We evaluate the quality of the scans, selectively correct frequencies that appear corrupted by motion, and segment perivascular spaces (detected perivascular spaces in green). (Color figure online)

Table 1. Distribution of clinical visual ratings for PVS in the basal ganglia and centrum semiovale in the MSS3 subsample. We report both frequencies and relative frequencies.

Rating	Basal ganglia	Centrum semiovale
0	0 (0%)	0 (0%)
1	16 (27%)	9 (15%)
2	26 (43%)	16 (27%)
3	18 (30%)	27 (45%)
4	0 (0%)	8 (13%)

---

[1] Full details of the study protocol and image acquisition are provided in [24].

2.2 Image Quality Assessment

Image quality metrics specifically designed for quantifying distortion due to noise and motion based on the analysis of background signal are suitable for image quality control as ghosting artefacts may be prevalent in this region and their values are unlikely to be indicative of the status of the brain. We tested four metrics accounting for noise and motion artefacts: the entropy-focus criterion, foreground-background energy ratio, mean background intensity, and signal-to-noise ratio (Table 2).

Table 2. Considered metrics for quantifying motion and noise in T2-w scans.

Image quality metric	Description
Entropy-focus criterion [10]	Entropy of voxel intensities as a measure of ghosting and blurring artefacts caused by motion. Lower entropy values may reflect lower motion artefacts
Foreground-background energy ratio [7]	Variance of voxel intensities in the intracranial region divided by that in background. The lower the ratio, the lower the image quality
Mean background intensity	Mean intensity in background. Higher values may be indicative of motion artefacts
Signal-to-noise ratio	Mean intensity in brain tissue divided by the standard deviation of intensities in background. The higher the ratio, the higher the image quality

The classification steps are as follows. Firstly, we extracted the aforementioned descriptors from each input image using the automatic MRI Quality Control tool[2] [7]. Secondly, we entered the resulting values in a logistic regression model to predict image quality (high-quality vs motion-corrupted). This model was trained using data from the second visit as described in Sect. 2.1 (9 motion-corrupted and 29 high-quality scans).

We validated our image quality control step against the classification of a trained analyst blind to our automatic assessment.

2.3 Motion Artefact Reduction

Once we separate high-quality and motion-corrupted images, we aim to reduce imaging artefacts on low-quality images, especially motion distortion.

Motion correction has been widely investigated [11, 12] and remains an unresolved challenge, although many partial solutions exist [11]. The applicability of motion reduction methods is dependant to a large extent on the availability of motion-tracking devices, prior information on the patient's movement, and the stage at which signal post-processing is applied (i.e., prospective or retrospective motion correction). In this particular case, we operate on the 'k-space' data obtained via Fourier transformation given the lack of the original k-space data or patient movement information.

---

[2] The MRIQC documentation can be found in mriqc.readthedocs.io.

Considering this, we decided to identify and compensate for inconsistencies in the 'k-space' data obtained via Fourier transformation of the T2-w images- sawtooth-like patterns depicted in Fig. 2. Assuming we can identify them appropriately (e.g., Fig. 2C, but not Fig. 2D), we can formulate motion reduction as a regression problem in which we estimate missing Fourier coefficients that increase image quality. In this work, we manually inspected k-spaces to select segments affected by sawtooth patterns.

Let $y : \mathbb{R}^3 \to \mathbb{R}$ be a motion-corrupted image. If we just nullify segments of the k-space displaying sawtooth patterns using an ideal low-pass filter M, the resulting image will be of greater visual quality in-plane, but of a poorer resolution in the superior-inferior direction compared to the original one. Instead, regress a new k-space $x^* : \mathbb{R}^3 \to \mathbb{R}$ that minimises the following expression

$$x^* = \underset{x}{\mathrm{argmin}} \, \gamma \|x - y\|_2^2 + \|M \star (x - y)\|_2^2 + \omega \|x\|_{TV}, \qquad (1)$$

where $\star$ denotes the convolution operator, ω, $\gamma \in \mathbb{R}$ weighting parameters and $\|\cdot\|_{TV}$ the total variation semi-norm. Each term in the expression ensures that (i) the resulting image x^* appears similar to the original, (ii) frequencies that are not visually affected by motion artefacts are retained, and (iii) potential "ringing" artefacts are reduced, in that order. Following qualitative analysis, we found $\omega = 0.5$ and $\gamma = 0.01$ balanced improving image quality and minimising blurring. We used the package developed by Lustig et al. [13] for finding an x^* by means of conjugate gradient and line-search.

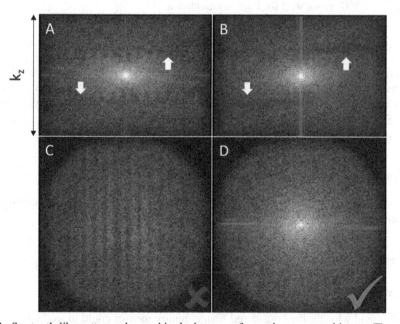

Fig. 2. Sawtooth-like patterns observed in the k-space of a motion-corrupted image. These patterns are visible in the slice direction kz in A and B (indicated by the white arrows) and in plane in C. Image D shows a slice visually free of artefacts.

2.4 PVS Segmentation

Most techniques segmenting PVS are based on eigenanalysis of the Hessian matrix as it encapsulates geometric properties used for distinguishing tubular structures (e.g., local curvature and eccentricity) [14]. The Hessian matrix H of a continuous and twice differentiable function $f : \mathbb{R}^3 \rightarrow \mathbb{R}$ is expressed as $(Hf)_{ij} = \partial^2 f/\partial x_i \partial x_j$. Digital images are typically convolved with multiple Gaussian kernels to ensure continuity and differentiability, and to enable multi-scale detection.

The eigenvalues of the Hessian matrix λ_1, λ_2, λ_3, with $|\lambda_1| \leq |\lambda_2| \leq |\lambda_3|$, characterises PVS in T2-w sequences since regions fulfilling $|\lambda_1| \approx 0$ and $\lambda_2 \approx \lambda_3 \ll 0$ are hyperintense tubular structures [14]. Moreover, complementary analysis permits targeting tubular objects with specific properties. For instance, the Frangi filter examines three additional properties $R_b = |\lambda_1|/\sqrt{|\lambda_2 \lambda_3|}$, $R_a = |\lambda_2|/|\lambda_3|$, and $S = \sqrt{\lambda_1^2 + \lambda_2^2 + \lambda_3^2}$ to filter blobs, lines, and low contrast structures, respectively [14, 15]. These three aspects are jointly evaluated in the vessel likelihood response as follows:

$$
V = \begin{cases} 0 \text{ if } \lambda_2 > 0 \text{ or } \lambda_3 > 0, \\ \left(1 - e^{-R_a^2/2\alpha^2}\right)\left(e^{-R_b^2/2\beta^2}\right)\left(1 - e^{-S^2/2c^2}\right) \text{ otherwise,} \end{cases}
$$

where α, β, and c control the sensitivity of the filter to the aforementioned properties. Parameter optimisation experiments have shown that default parameters ($\alpha = 0.5$, $\beta = 0.5$, and $c = 500$) work well for PVS segmentation [16].

2.5 Comparison Against a Relevant Framework

We compared each step of our proposal against that of a similar framework that attempts to evaluate image quality and correct imaging artefacts prior to PVS segmentation [6]. In that work, image quality was assessed via Haralick-based texture analysis; imaging artefacts corrected by means of the total variation denoising framework; and PVS segmentation using the Frangi filter.

2.6 Validation Against Clinical Parameters

We computed polyserial correlations to determine the strength of the relationship between visual and computational measurements of PVS.

3 Results

3.1 Image Quality Classification Results

To understand whether the considered image quality metrics could indeed help distinguish between high-quality and motion-corrupted image scans, we plotted quality scores for each group (Fig. 3). High-quality scans displayed significantly higher foreground-to-background energy ratios, higher signal-to-noise-ratios, and lower mean background intensities compared to motion-corrupted scans. Even though the entropy-focus criterion specifically targets motion artefacts, we found no significant differences between the two groups of images in this regard (high-quality: 0.48 [IQR 0.46, 0.50] vs motion-corrupted: 0.47 [IQR 0.46, 0.49]; p = 0.52).

We then used a logistic regression model trained on similar images to predict image quality, calculating an AUROC of approximately 0.98. Although entropy-focus criterion values were not significantly different for high-quality and motion-corrupted images, omitting it did not lead to a better model (AUCROC = 0.96). Only two scans out of the total 60 (one image of each quality group) were misclassified. Further visual inspection of these two cases revealed increased signal in the background region of the high-quality appearing scan compared to that in the apparently motion-corrupted scan (foreground-to-background energy ratio: 4403.45 vs 4779.42). This finding agrees with what can be perceived via visual inspection (Fig. 4), suggesting that the image labelled as high-quality was corrupted by motion artefacts.

We also compared our approach with one considering texture analysis of brain tissues, described in [6]. The use of motion and noise descriptors led to improved quality control performance when compared to that obtained using Haralick-based textures (AUROC = 0.98 with our proposal vs 0.94 with texture analysis). Hence, we used our logistic regression model using motion and noise descriptors to predict image quality.

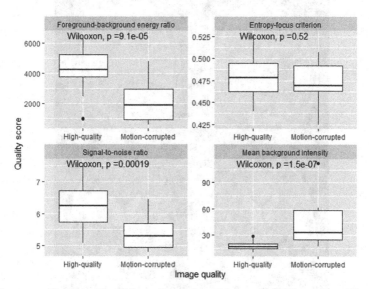

Fig. 3. Image quality scores for high-quality and motion-corrupted T2-w scans. We tested for differences in quality scores using the unpaired two-sample Wilcoxon test.

3.2 Motion Artefact Reduction

We applied the proposed motion artefact reduction method on motion-corrupted scans. The algorithm improves the visual quality of the T2-w images while retaining detail (Fig. 5); in some cases, these features become evident as noise is also reduced.

Low-quality appearing image   **High-quality appearing image**
Original scan   Thresholded image   Original scan   Thresholded image

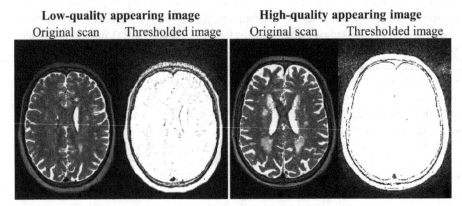

Fig. 4. Images that were misclassified using our logistic regression model. We thresholded intensities above 50 to show the background signal more clearly (thresholded images).

Original images   Enhanced images

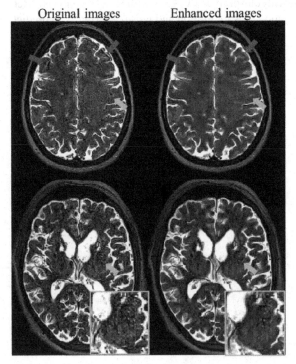

Fig. 5. Original and enhanced scans. Yellow arrows point to regions displaying enlarged perivascular spaces and red ones to evident motion artefacts corrected after image enhancement. (Color figure online)

3.3 Relationship Between Computational Measures and Clinical Visual Scores

We calculated polyserial correlations to assess the relationship between visual and computational PVS measures. We compared correlations obtained when assessing PVS on original scans, those enhanced using the total variation denoising framework [6], and those using our proposed motion artefact reduction (Table 3). Overall, correlation scores increased consistently after selective filtering. The effect was particularly evident in the centrum semiovale where the correlation between visual scores and volume increased from $\rho = 0.17$ (95% CI 0.14, 0.21) to $\rho = 0.29$ (95% CI 0.26, 0.32) and $\rho = 0.86$ (95% CI 0.85, 0.88) using the total variation framework and proposed method, respectively. Furthermore, the relationship between quantitative and qualitative scores was positive once we filtered motion-corrupted images.

Table 3. Polyserial correlations between quantitative and visual measures of perivascular spaces before and after filtering. The higher the correlation, the stronger the relationship between computational and visual scores. We consider the count and volume as quantitative measurements of the presence and enlargement of perivascular spaces. CI: confidence interval. BG: basal ganglia. CSO: centrum semiovale.

	Variable	Original			Total variation in image space [6]			Our proposal		
		ρ	95% CI		ρ	95% CI		ρ	95% CI	
BG	Count	0.38	0.35	0.41	0.43	0.40	0.46	0.50	0.48	0.53
	Volume	0.63	0.61	0.65	0.69	0.67	0.71	0.72	0.71	0.74
CSO	Count	− 0.10	− 0.13	− 0.06	− 0.08	− 0.05	− 0.01	0.34	0.31	0.37
	Volume	0.17	0.14	0.21	0.29	0.26	0.32	0.86	0.85	0.88

Further examination of the computational PVS measurements stratified by clinical visual score and image quality (Fig. 6) suggested the application of the k-space analysis led to more similar estimates between high-quality and image-enhanced scans compared to when no filtering was considered or when filtering in the image space.

4 Discussion

Imaging artefacts limit the applicability of computational solutions, especially those quantifying PVS. This is primarily due to their small size (less than 3 mm) [1] and limitations in imaging technology. We present a computational pipeline for assessing image quality via radiomics analysis and reducing motion artefacts selectively, thus improving PVS segmentation and quantification.

Motion artefacts are a common problem in clinical studies using MRI [11]. Whilst motion may happen occasionally, medical conditions – e.g., strokes [17] – may also cause movement disorders. Alarmingly, the prevalence of motion artefacts limiting computational PVS segmentation has been reported to be as high as 20% [5], a similar proportion to that in our sample. Recovering this data that would otherwise be unusable is therefore

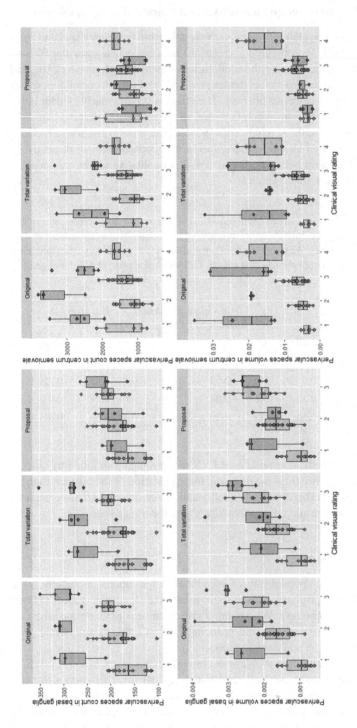

Fig. 6. Perivascular space count and volume quantified in high-quality (pink) and motion-corrupted T2-w scans (green) stratified by clinical visual ratings. The count and volume are computed on original scans as well as those enhanced using the total variation framework and the proposed autofocusing method. We anticipate intra-group similarity irrespective of image quality, i.e., patients with the same visual rating should exhibit similar PVS count and volume and this should not be subject to image quality. (Color figure online)

crucial in the assessment of PVS and further validate their use as biomarker of brain health.

A key step in the proposed pipeline involves separating low- and high-quality scans to avoid blurring the latter unnecessarily as it would otherwise lead to reduced PVS sensitivity [6]. Previous work has shown that this step can be carried out via texture analysis. However, textures extracted from brain tissues reflect both the quality of the scans but also the presence of brain pathologies [9]. Instead, we opted for image quality metrics using background information only or ratios between background and foreground signal to avoid this problem. We tested whether the entropy-focus criterion, foreground-background energy ratio, mean background intensity, and signal-to-noise ratio were appropriate for characterising imaging artefacts in the considered sample.

Our experimental results suggest that the use of image quality metrics permit distinguishing high-quality scans from low-quality ones (AUROC = 0.98) better than texture-based proposals (AUROC = 0.94). Moreover, we observed an image that was originally classified as high-quality was actually corrupted by motion artefacts, supporting the claim that some artefacts evade human detection due to their subtlety or visual fatigue [7]. Automatic image quality control can therefore allow the timely detection of imaging issues that may compromise subsequent processes.

Once we automatically segregated motion-corrupted scans, we proceeded with artefact reduction. Previous work showed filtering in the image space can help to reduce imaging artefacts [6], however, these approximations do not necessarily tackle the primary problem: missing or corrupted k-space measurements. We hypothesise that analysis of the k-space measurements is essential to improve the quality of the images without dispensing with clinically relevant information. In this particular case, T2-w scans were acquired using an elliptical k-space filling trajectory, a technique which oversamples the centre of the k-space [11]. Thus, motion reduction can be achieved by nullifying segments of the k-space displaying, for example, pie-slice [12] or sawtooth patterns. However, these segments need to be re-estimated to prevent a loss in resolution. As a proof of concept, we manually nullified segments of the k-space evidencing these patterns and regressed them using the total variation optimisation framework. We confirmed these segments were indeed linked with motion artefacts since their correction led to increased image quality. Future work should consider automatic detection of these motion-corrupted k-space regions.

Our work has some limitations. First, while nullifying segments of the k-space may work for scans acquired using elliptical trajectories, the same strategy may not be suitable for other types of acquisitions. Approximations using artificial intelligence – e.g., deep learning – may be explored in the future to compensate for these problems in a more generalised way. Moreover, a digital reference object containing PVS and a computational model mimicking image acquisition – similar to the work in [18] – may help to train a model with heterogeneous yet realistic cases, thus preventing potential generalisability issues [19]. Second, we also evaluated our proposal on a relatively small sample of an ongoing prospective study of mild stroke patients. While our results seem encouraging, further validation on a larger dataset containing T1w and T2w imaging sequences acquired with multiple imaging protocols is necessary to determine the suitability of the proposed pipeline for improving PVS quantification. We expect relevant image quality metrics to differ from those considered in this work, especially if scans are acquired with different k-space trajectories. The outcomes of this experiment could later be compared to descriptors that were found relevant for general automatic quality control [7, 20–23]. Third, statistical analysis of the relationship between computational estimates of PVS burden with demographics and risk factors could complement current analysis, helping to further validate PVS as a biomarker of brain health. Fourth, our proposal offers a partial solution to PVS quantification in motion-corrupted images: reduced false positive rates due to imaging artefact reduction but also reduced true positive rates as k-space extrapolation inevitably blurs some small PVS (e.g., those located in the medial part of the basal ganglia in Fig. 5). Although sharpening filters may improve visual appearance, they do not address the underlying problem: data loss. Motion correction leveraging Fourier transform properties may help to tackle this problem [11], but adjustments may be needed to ensure optimal results in scans acquired using non-Cartesian trajectories.

In conclusion, we have developed a pipeline for evaluating image quality automatically and correct motion artefacts, if needed, to improve PVS quantification. Experimental results suggests our proposal reduces imaging artefacts successfully and leads to a higher correlation between computational and visual measurements of PVS burden. Our development is practical since it helps to recover data otherwise useless and timely given the growing interest in PVS as a potential biomarker of brain dysfunction.

Acknowledgements. This work is supported by: MRC Doctoral Training Programme in Precision Medicine (JB - Award Reference No. 2096671); the UK Dementia Research Institute which receives its funding from DRI Ltd, funded by the UK MRC, Alzheimer's Society and Alzheimer's Research UK; the Foundation Leducq Network for the Study of Perivascular Spaces in Small Vessel Disease (16 CVD 05); Stroke Association 'Small Vessel Disease-Spotlight on Symptoms (SVD-SOS)' (SAPG 19\100068); The Row Fogo Charitable Trust Centre for Research into Aging and the Brain (MVH) (BRO-D.FID3668413); Stroke Association Garfield Weston Foundation Senior Clinical Lectureship (FND) (TSALECT 2015/04); NHS Research Scotland (FND); British Heart Foundation Edinburgh Centre for Research Excellence (RE/18/5/34216); a British Heart Foundation Chair award (RMT) (CH/12/4/29762); NHS Lothian Research and Development Office (MJT); European Union Horizon 2020, PHC-03-15, project No666881, 'SVDs@Target' (MS); Chief Scientist Office of Scotland Clinical Academic Fellowship (UC) (CAF/18/08); Stroke Association Princess Margaret Research Development Fellowship (UC) (2018); Alzheimer Nederland (ACCJ). The Research MR scanners are supported by the Scottish Funding Council through the Scottish Imaging Network, A Platform for Scientific Excellence (SINAPSE) Collaboration; the 3T scanner is funded by the Wellcome Trust (104916/Z/14/Z), Dunhill Trust (R380R/1114), Edinburgh and Lothians Health Foundation (2012/17), Muir Maxwell Research Fund, and the University of Edinburgh. We thank the participants, their families, radiographers at Edinburgh Imaging Facility Royal Infirmary of Edinburgh, and the Stroke Research Network at the University of Edinburgh.

References

1. Potter, G.M., Chappell, F.M., Morris, Z., Wardlaw, J.M.: Cerebral perivascular spaces visible on magnetic resonance imaging: development of a qualitative rating scale and its observer reliability. Cerebrovasc. Dis. **39**(3–4), 224–231 (2015)

2. Wardlaw, J.M., Smith, C., Dichgans, M.: Small vessel disease: mechanisms and clinical implications. Lancet Neurol. **18**(7), 684–696 (2019)
3. Wardlaw, J.M., et al.: Perivascular spaces in the brain: anatomy, physiology and pathology. Nat. Rev. Neurol. **16**(3), 137–153 (2020)
4. del Maria, C., Hernández, V., Piper, R.J., Wang, X., Deary, I.J., Wardlaw, J.M.: Towards the automatic computational assessment of enlarged perivascular spaces on brain magnetic resonance images: a systematic review: computational assessment of perivascular spaces. J. Magn. Reson. Imaging **38**(4), 774–785 (2013). https://doi.org/10.1002/jmri.24047
5. Ballerini, L., et al.: Computational quantification of brain perivascular space morphologies: associations with vascular risk factors and white matter hyperintensities. A study in the Lothian Birth Cohort 1936. NeuroImage Clin. **25**(2019), 102120 (2020)
6. Bernal, J., et al.: A framework for jointly assessing and reducing imaging artefacts automatically using texture analysis and total variation optimisation for improving perivascular spaces quantification in brain magnetic resonance imaging. In: Papież, B.W., Namburete, A.I.L., Yaqub, M., Noble, J.A. (eds.) MIUA 2020. CCIS, vol. 1248, pp. 171–183. Springer, Cham (2020). https://doi.org/10.1007/978-3-030-52791-4_14
7. Esteban, O., Birman, D., Schaer, M., Koyejo, O.O., Poldrack, R.A., Gorgolewski, K.J.: MRIQC: advancing the automatic prediction of image quality in MRI from unseen sites. PLoS ONE **12**(9), 1–21 (2017)
8. Valdés Hernández, M.d.C., et al.: Application of texture analysis to study small vessel disease and blood–brain barrier integrity. Front. Neurol. **8**, 327 (2017)
9. Bernal, J., et al.: Analysis of dynamic texture and spatial spectral descriptors of dynamic contrast-enhanced brain magnetic resonance images for studying small vessel disease. Magn. Reson. Imaging **66**, 240–247 (2020)
10. Atkinson, D., Hill, D.L.G., Stoyle, P.N.R., Summers, P.E., Keevil, S.F.: Automatic correction of motion artifacts in magnetic resonance images using an entropy focus criterion. IEEE Trans. Med. Imaging **16**(6), 903–910 (1997)
11. Zaitsev, M., Maclaren, J., Herbst, M.: Motion artifacts in MRI: a complex problem with many partial solutions. J. Magn. Reson. Imaging **42**(4), 887–901 (2015)
12. Godenschweger, F., et al.: Motion correction in MRI of the brain. Phys. Med. Biol. **61**(5), R32–R56 (2017)
13. Lustig, M., Donoho, D., Pauly, J.M.: Sparse MRI: the application of compressed sensing for rapid MR imaging. Magn. Reson. Med. **58**(6), 1182–1195 (2007)
14. Lamy, J., et al.: Vesselness filters: a survey with benchmarks applied to liver imaging. In: International Conference on Pattern Recognition (2020)
15. Frangi, A.F., Niessen, W.J., Vincken, K.L., Viergever, M.A.: Multiscale vessel enhancement filtering. In: Wells, W.M., Colchester, A., Delp, S. (eds.) MICCAI 1998. LNCS, vol. 1496, pp. 130–137. Springer, Heidelberg (1998). https://doi.org/10.1007/BFb0056195
16. Ballerini, L., et al.: Perivascular spaces segmentation in brain MRI using optimal 3D filtering. Sci. Rep. **8**(1), 1–11 (2018)
17. Handley, A., Medcalf, P., Hellier, K., Dutta, D.: Movement disorders after stroke. Age Ageing **38**(3), 260–266 (2009)
18. Bernal, J., et al.: A four-dimensional computational model of dynamic contrast-enhanced magnetic resonance imaging measurement of subtle blood-brain barrier leakage. Neuroimage **230**, 117786 (2021). https://doi.org/10.1016/j.neuroimage.2021.117786
19. Billot, B., Robinson, E., Dalca, A.V., Iglesias, J.E.: Partial volume segmentation of brain MRI scans of any resolution and contrast. In: Martel, A.L., et al. (eds.) MICCAI 2020. LNCS, vol. 12267, pp. 177–187. Springer, Cham (2020). https://doi.org/10.1007/978-3-030-59728-3_18
20. Magnotta, V.A., Friedman, L.: Measurement of signal-to-noise and contrast-to-noise in the fBIRN multicenter imaging study. J. Digit. Imaging **19**(2), 140–147 (2006)

21. Kellman, P., McVeigh, E.R.: Image reconstruction in SNR units: a general method for SNR measurement. Magn. Reson. Med. **54**(6), 1439–1447 (2005)
22. Dietrich, O., Raya, J.G., Reeder, S.B., Reiser, M.F., Schoenberg, S.O.: Measurement of signal-to-noise ratios in MR images: influence of multichannel coils, parallel imaging, and reconstruction filters. J. Magn. Reson. Imaging **26**(2), 375–385 (2007)
23. Mortamet, B., et al.: Automatic quality assessment in structural brain magnetic resonance imaging. Magn. Reson. Med. **62**(2), 365–372 (2009)
24. Clancy, U., et al.: Rationale and design of a longitudinal study of cerebral small vessel diseases, clinical and imaging outcomes in patients presenting with mild ischaemic stroke: mild stroke study 3. Eur. Stroke J. **6**(1), 81–88 (2020)

Mass Univariate Regression Analysis for Three-Dimensional Liver Image-Derived Phenotypes

Marjola Thanaj[1]([✉])[ID], Nicolas Basty[1][ID], Yi Liu[2][ID], Madeleine Cule[2][ID],
Elena P. Sorokin[2][ID], E. Louise Thomas[1][ID], Jimmy D. Bell[1][ID],
and Brandon Whitcher[1][ID]

[1] Research Centre for Optimal Health, School of Life Sciences,
University of Westminster, London, UK
m.thanaj@westminster.ac.uk
[2] Calico Life Sciences LLC, South San Francisco, CA, USA

Abstract. Image-derived phenotypes of abdominal organs from magnetic resonance imaging reveal variations in volume and shape and may be used to model changes in a normal versus pathological organ and improve diagnosis. Computational atlases of anatomical organs provide many advantages in quantifying and modeling differences in shape and size of organs for population imaging studies. Here we made use of liver segmentations derived from Dixon MRI for 2,730 UK Biobank participants to create 3D liver meshes. We computed the signed distances between a reference and subject-specific meshes to define the surface-to-surface (S2S) phenotype. We employed mass univariate regression analysis to compare the S2S values from the liver meshes to image-derived phenotypes specific to the liver, such as proton density fat fraction and iron concentration while adjusting for age, sex, ethnicity, body mass index and waist-to-hip ratio. Vertex-based associations in the 3D liver mesh were extracted and threshold-free cluster enhancement was applied to improve the sensitivity and stability of the statistical parametric maps. Our findings show that the 3D liver meshes are a robust method for modeling the association between anatomical, anthropometric, and phenotypic variations across the liver. This approach may be readily applied to different clinical conditions as well as extended to other abdominal organs in a larger population.

Keywords: Registration · Surface-to-surface · Morphology · Magnetic resonance imaging

1 Introduction

Magnetic resonance imaging (MRI) has become the benchmark for clinical research in the study of body composition, particularly for measurements of visceral adipose tissue, liver and pancreatic fat content. The incidence of chronic

© Springer Nature Switzerland AG 2021
B. W. Papież et al. (Eds.): MIUA 2021, LNCS 12722, pp. 165–176, 2021.
https://doi.org/10.1007/978-3-030-80432-9_13

conditions such as type-2 diabetes, cardiovascular disease and non-alcoholic fatty liver disease are rising rapidly, which reflects the increasing prevalence of obesity in society [27]. Organ and tissue MRI measurements, referred to as image-derived phenotypes (IDPs) have the potential to enhance our understanding of the precise phenotypic changes underlying these conditions [25].

The UK Biobank is a population-based prospective study that has recruited over 500,000 volunteers, aged 40–69 years old, with the goal of advancing our understanding of health and disease [10]. A subset of 100,000 participants has been invited for a medical imaging assessment that includes a standardised abdominal acquisition protocol. The UK Biobank abdominal imaging protocol produces several MRI datasets that focus on basic structure and composition measurements in the thorax, abdomen and pelvis [16]. Specifically, the abdominal protocol includes a two-point Dixon sequence [12] with neck-to-knee coverage, as well as a multiecho single-slice acquisition of the liver. This latter acquisition enables non-invasive estimation of tissue composition including proton density fat fraction (PDFF) and iron concentration. Together these sequences enable accurate, quantitative analysis of multiple liver IDPs.

Performing semantic segmentation on abdominal organs using deep learning methodology is now widely established. The basic IDP obtained from organ segmentation is total volume, a single number that is informative but does not capture the complex morphology of the underlying physical structure of an organ. Computational image analysis, by which machine learning is used to annotate and segment the images, is gaining traction as a means of representing detailed three-dimensional (3D) mesh-derived phenotypes related to shape variations at thousands of vertices in a standardised coordinate space. One approach to inference is to transform the spatially correlated data into a smaller number of uncorrelated principal components [4], while the modes from PCA are useful in exploratory data analysis they do not provide an explicit model relating 3D shape to other phenotypic measures. A more powerful approach may be to estimate parameters at each vertex of the 3D surface mesh, hence creating a so-called statistical parametric map (SPM), a concept widely used in functional neuroimaging [20] and cardiac imaging [9]. A recent study in the cardiovascular imaging domain, implementing a mass univariate framework, showed the ability of this technique in identifying interactions between genetic variation related to hypertrophic cardiomyopathy and a 3D cardiac phenotype [18]. On that note, scientific questions of interest, such as, how organ shape is associated with disease state (e.g., type-2 diabetes, cardiovascular disease, non-alcoholic fatty liver disease) or gene expression (e.g., a single-nucleotide polymorphism or SNP), may be quantified by selecting a cohort of cases and controls from an available population and including common covariates in a linear regression framework.

In this paper we extend techniques developed in neuroimaging and cardiovascular imaging fields to liver imaging by implementing a mass univariate framework that maps associations within the phenotypic variation. Such an approach would provide overly conservative inferences without considering spatial dependencies in the underlying data, so we evaluated threshold-free cluster-enhancement with respect to liver phenotypes for the sensitive detection of coherent signals in order to understand their efficacy in identifying phenotypic interactions in liver shape and structure. A cohort of UK Biobank participants are analyzed using our methodology, with the aim to investigate the associations between both image-derived and non-image-derived phenotypes and local liver morphology and establish anatomical structures in the liver. We further explore the feasibility of SPMs in comparing groups of subjects using mesh-based shape analysis. These 3D derived morphometric phenotypes will significantly contribute to our understanding of the relationship between form and function in health and disease.

2   Materials and Methods

2.1   Data

In this work, we used liver segmentations predicted from the Dixon MRI acquisitions in the UK Biobank Imaging Study. The Dixon acquisition contains six separate series for each subject. We performed basic pre-processing to assemble the six series into a single volume [17]. Briefly, the six series were resampled to the same resolution ($2.23 \times 2.23 \times 3.0$ mm^3), bias-field correction was performed and the six series were blended together resulting in a final volume of size ($224 \times 174 \times 370$ voxels). Segmentations were performed using a convolutional neural network based on the U-net architecture [22], trained using 120 manual annotations performed by a team of radiographers. The segmentation model achieved a dice score coefficient of 0.932 in out-of-sample data [17].

For the mass univariate regression analysis, we randomly selected 2,730 participants from the UK Biobank Imaging Study to cover a broad range of age, gender and body compositions. Detailed descriptions of the full cohort can be found in the section below. A randomly selected sub-cohort of 20 participants (70% females; age range: 50–78 years; average age: 56.8 years) was used to construct the template liver mesh.

2.2   Image Analysis and Mesh Construction

Image registration was performed in two stages to construct the liver template: affine registration to account for translation, rotation, scaling and shearing, and non-rigid registration to account for local deformation using the symmetric image normalization (SyN) method with cross-correlation as the similarity metric [1,3]. The algorithm maps a group of images to a common space by finding the template and set of transformations with the lowest dataset parameterization. The size of the parameterization, here, is given by the SyN implementation which measures the image similarity metric and diffeomorphism maps [3]. Here, we performed the image registration with no initial template as input instead, the template is derived completely from twenty images. In particular, we transformed the liver segmentations of twenty subject-specific volumes to the template space using the non-rigid transformation and computed a template in four iterations. This computes the Euclidean distance as a metric for average shape which corresponds to the average liver across all subjects. A surface mesh was then constructed from the average template using the marching cubes algorithm and Laplacian smoothing in the VTK library (Kitware Inc.), and a template was created in a coordinate space [5]. Figure 1, illustrates a scheme summarising the pipeline for the template mesh construction.

Once the template was available we performed a registration step where all subject-specific segmentations were rigidly aligned to a template segmentation. The subject meshes were then constructed enabling the further computation of 3D mesh-derived phenotypes. The liver meshes encode the 3D mesh-derived phenotype variation for the study cohort, in particular, for each subjects' mesh, surface-to-surface (S2S) values were measured by computing the nearest-neighbour signed distance between the template surface and a subjects' surface for each vertex. After a manual quality control process investigating for outliers on the surface-to-surface values, all values were no larger than the range ± 65 mm.

The template construction was performed using Advanced Neuroimaging Tools (ANTs) [1–3] using cross correlation (CC) as the similarity metric, the B-spline non-rigid transformation model (BSplineSyN) and default parameters otherwise. The subject's registration was performed using the rigid and affine transform type (a) and the rest of the default parameters from ANTs. The 3D mesh-derived phenotype was computed using the packages **FNN** [7] and **Rvcg** [23] in R 3.6.1 [21].

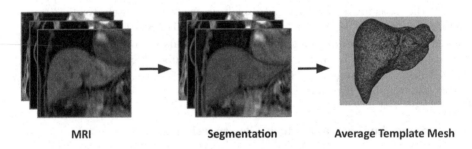

MRI   **Segmentation**   **Average Template Mesh**

Fig. 1. Average template mesh construction. Dixon MRI volumes from UK Biobank abdominal protocol (left) are used to produce subject-specific 3D liver segmentations (middle), then images are registered to a common space and combined to produce an average template mesh.

2.3   Mass Univariate Regression Analysis

The association between the 3D mesh-derived phenotype and anthropometric variables is modeled using a linear regression framework. Given n_s subjects from a sample of the population under study, the linear regression model was expressed as

$$Y = X\beta + \epsilon, \tag{1}$$

where Y is a $n_s \times n_v$ matrix containing (n_v is the number of voxels in the mesh), for example, the S2S values of all the n_s subjects at each vertex of the 3D liver mesh, X is the $n_s \times p$ design matrix of p known covariates (including the intercept) and the clinical variables for each subject, such as age and sex, used to model the hypothesis under investigation. X is related with Y by the vector of the regression coefficients β. In this way, Y may be associated with each of the columns of X adjusted for the other covariates. Finally ϵ is a $n_s \times n_v$ matrix which is independent and identical distributed across the subjects and is assumed to be a zero-mean Gaussian process [14]. The estimated regression coefficients $\hat{\beta}$ and their related p-values at each vertex in the mesh may be displayed on the whole 3D liver anatomy, providing the spatially-distributed associations. We applied threshold-free cluster-enhancement (TFCE) to enhance areas of the signal that exhibit spatial contiguity and better discriminate the estimated parameters between noise and spatially-correlated signal [9,24]. The mass univariate regression model for deriving associations between clinical parameters and a 3D phenotype is outlined in Fig. 2.

The TFCE statistic at each vertex v of the 3D mesh under study is given by

$$\text{TFCE}(v) = \int_{h=h_0}^{h_v} e(h)^E h^H dh, \tag{2}$$

where h is the value of the corresponding t-statistic and is raised from zero (h_0) up to the height of the vertex v (h_v), $e(h)$ is the extent of the cluster with threshold h that contains v vertices, and E and H are two hyperparameters empirically defined to be 0.5 and 2 [24].

The derived p-values were corrected to control the false discovery rate (FDR) using the Benjamini-Hochberg procedure [6] as it has been shown to provide the optimal p-values and areas of significance [9]. Together TFCE and permutation testing were applied to compute a new set of p-values at each mesh vertex v, by sampling the data and computing the TFCE N times over the obtained statistical parametric maps. The permutation testing was performed to estimate the null distribution for the univariate statistics. In particular, we permuted the data N times obtaining TFCE scores where the test statistics are summed forming a cluster mass statistic. The permutation testing identifies the largest cluster among those permutations [9]. Here, we used the Freedman-Lane technique as a permutation framework, as it provides powerful permutation and optimal control of false positives [13,26]. The mass univariate regression analysis was performed from a refined version of the package **mutools3D** [8] in R 3.6.1 [21].

3   Results

We analysed liver MRI data from 2,730 participants in UK Biobank from which 97.7% were Caucasian and aged between 46 to 80 years old. The main cohort

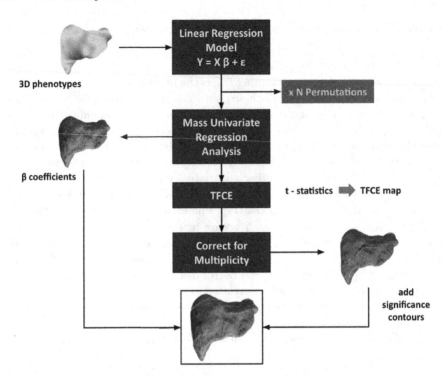

Fig. 2. Flow diagram for the mass univariate regression analysis of three-dimensional phenotypes. Phenotypes are used to construct the linear regression model, where mass univariate regression analysis produces parameter estimates ($\hat{\beta}$) and the null distribution via permutation. TFCE is applied to the t-statistics from the regression analysis to produce the significance threshold. The associated p-values are corrected for multiple comparisons and mapped on to the mesh for visualisation. This diagram was modified from [9].

characteristics are shown in Table 1. To assess the associations between S2S and anthropometric covariates as well as liver IDPs, we performed mass univariate regression analysis adjusting for age, sex, ethnicity, body mass index (BMI) and waist-to-hip ratio (WHR). The TFCE algorithm was applied to the t-statistics and on the permuted t-statistics ($N = 1,000$ times) for each analysis. Correction for multiplicity via the FDR was applied for the number of vertices and the number of anthropometric covariates/IDPs tested.

A summary of the regression models for the whole cohort, representing the significance area on the liver is shown in Table 2. S2S values is shown to increase with age with a positive association on 39 out of 57% of the significance area, while WHR and BMI showed a decline with increased S2S values. To determine the nature of the liver modeling we examined the effects of the liver IDPs such as iron concentration and PDFF. Liver iron was positively associated with S2S values on 36% of vertices, and negatively associated on 2% of the vertices whereas liver PDFF showed a positive association on 48 out of 55% of the vertices.

Table 1. Baseline characteristics and liver IDPs. Values are presented as mean ± SD for continuous variables and counts for discrete variables.

	Full cohort	Men	Women
N	2,730	1,368	1,362
N Caucasian	2,669	1,337	1,332
Age (years)	62.8 ± 7.2	63.4 ± 7.3	62.2 ± 7.1
BMI (kg/m^2)	26.3 ± 4.1	26.7 ± 3.9	25.8 ± 4.3
WHR	0.88 ± 0.09	0.93 ± 0.06	0.82 ± 0.07
Liver Iron (mg/g)	1.21 ± 0.26	1.22 ± 0.27	1.20 ± 0.24
Liver PDFF (%)	4.8 ± 4.7	5.6 ± 5.1	4.0 ± 4.1

Table 2. Significance areas for covariates in the mass univariate regression model. The significance area is the percentage of vertices on the liver mesh where the regression coefficients are statistically significant ($p < 0.05$) after adjustment for multiple comparisons. The total area has been split into areas of positive and negative associations.

	Significance area	$\hat{\beta} < 0$	$\hat{\beta} > 0$
Age	57%	18%	39%
BMI	87%	54%	33%
WHR	55%	29%	26%
Liver Iron	38%	2%	36%
Liver PDFF	55%	7%	48%

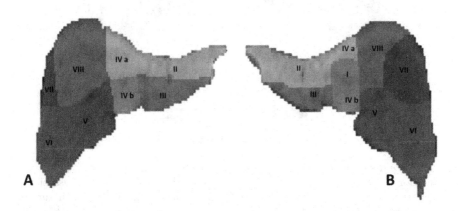

Fig. 3. Segments of the liver as described in the Couinaud classification, overlaid on the liver template. Projections are anterior (A) and posterior (B).

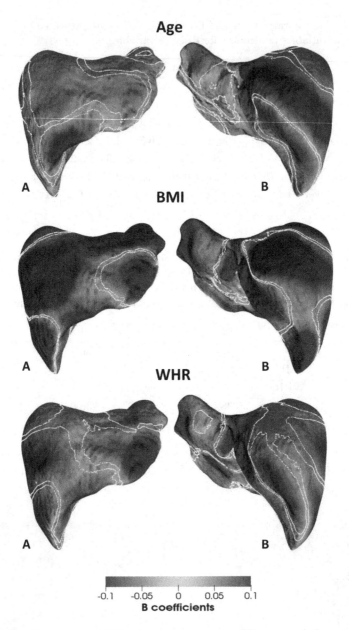

Fig. 4. Three-dimensional statistical parametric maps of liver morphology, projections are anterior (A) and posterior (B). The SPMs show the local strength of association between age, body mass index (BMI) and waist-to-hip ratio (WHR) with surface-to-surface values. Yellow contour lines indicate significant regions ($p < 0.05$) after correction for multiple testing, with positive associations in bright red and negative associations in bright blue. (Color figure online)

Liver Iron

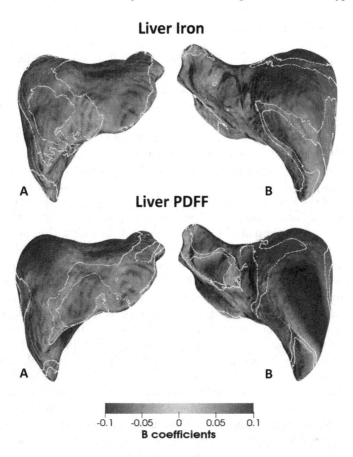

Fig. 5. Three-dimensional statistical parametric maps of liver morphology, projections are anterior (A) and posterior (B). The SPMs show the local strength of association between liver iron concentration and proton density fat fraction (PDFF) with surface-to-surface values. Yellow contour lines indicate significant regions ($p < 0.05$) after correction for multiple testing, with positive associations in bright red and negative associations in bright blue. (Color figure online)

The Couinaud classification [11] of liver segments has been applied to the liver template for reference (Fig. 3). All the significant associations between S2S values and the three anthropometric covariates are shown in Fig. 4 and the association between S2S values and the liver IDPs are in Fig. 5. Interestingly, the statistical parametric maps with associations between S2S values and liver iron concentration and PDFF appear to show regional differences congruent with different segments of the liver proposed by Couinaud. Changes in liver morphology associated with liver PDFF appeared to be most pronounced in segments II & III and VI & VII, while changes in liver morphology associated with liver iron concentration were strongest in segments II & III. BMI and WHR were associated with the most pronounced positive S2S values in segments III & VI and part of IV & VII with negative S2S values in segments II & VIII.

4   Discussion and Conclusions

In this paper, we constructed surface meshes from liver segmentations of 2,730 subjects in the UK Biobank. Based on the vertices of the mesh, we were able to compute a 3D phenotype related to local shape variations and perform mass univariate regression analysis to model the associations with anthropometric and phenotypic traits.

We performed image registration and estimated a liver template with average shape using an optimal normalisation technique by computing the Euclidean mean of the of the non-rigid transformations. This normalization technique has been shown to provide the optimal mapping and template shape [3]. Also, the computation of the Euclidean mean has been shown to provide a good estimation of the template construction [5]. We used rigid registration to align the subject segmentations to the template in order to account for orientation and position differences obtaining a transformation in a standard space for each subject. There are a few ways of fixing these difference of the subjects' meshes. Previous studies investigating the shape and motion of the left ventricle of the heart removed these differences by retaining the shape and size for analysis [5,19].

Statistical parametric mapping has been a useful technique in neuroimaging [20] and cardiac imaging [9], showing that such statistical methods may be utilised in modelling the relationship between phenotypic and genetic variation [18]. Our findings demonstrate that the 3D mesh-derived phenotypes of the liver in specific anatomical regions are associated with the anthropometric/phenotypic traits using the SPM framework. We also found that liver IDPs were significantly associated with higher S2S values, suggesting that increased liver fat and iron concentration may have an impact on the liver shape and structure. It is also notable that the liver S2S values increase with age and waist-to-hip ratio. These findings agree with previous studies that report an increase in liver fat and iron is associated with predictors of metabolic disease [15]. Interestingly, the pattern of changes in S2S values across the liver reflect differences that might be attributed to different lobes and segments of the liver described in the Couinaud system of classification. Further work to explore this in more detail may allow further mapping the associations between genetic variations and 3D phenotypes in specific anatomical regions.

Organ shape variations could become a powerful tool for assessing global changes associated with organ damage (liver fibrosis and cirrhosis), disease progression and remission (fatty liver, haemochromatosis, nonalcoholic steatohepatitis) and eventually treatment outcome. Moreover, this technique has the potential to be simultaneously applied in multi-organ approaches (e.g., liver, pancreas, kidneys, spleen) thus giving a more holistic overview of health and disease than what is currently available from single-organ measurements. Future work will apply this method on a larger cohort as well as on other organs in the abdominal cavity, such as the pancreas, spleen and kidneys.

In conclusion, we have constructed a surface mesh of the liver anatomy in a sample of subjects from the UK Biobank population. From the surface mesh, we presented a 3D mesh-derived phenotype and were able to quantify the anatomical relationships with the anthropometric/phenotypic traits in the liver using the mass univariate regression analysis. We believe that the mesh construction and statistical techniques will benefit future research in population-based cohort studied, in identifying associations between physiological, genetic and anthropometric effects on liver structure and function as well as in other abdominal organs.

Acknowledgements. This research has been conducted using the UK Biobank Resource under Application Number '44584' and was funded by Calico Life Sciences LLC.

References

1. Avants, B.B., Epstein, C., Grossman, M., Gee, J.C.: Symmetric diffeomorphic image registration with cross-correlation: evaluating automated labeling of elderly and neurodegenerative brain. Med. Image Anal. **12**, 1361–8415 (2008). https://doi.org/10.1016/j.media.2007.06.004

2. Avants, B.B., Tustison, N.J., Song, G., Cook, P.A., Klein, A., Gee, J.C.: A reproducible evaluation of ANTs similarity metric performance in brain image registration. NeuroImage **54**, 2033–2044 (2011). https://doi.org/10.1016/j.neuroimage.2010.09.025

3. Avants, B.B., et al.: The optimal template effect in hippocampus studies of diseased populations. Neuroimage **49**, 2457–2466 (2010). https://doi.org/10.1016/j.neuroimage.2009.09.062

4. Bagur, A.T., Ridgway, G., McGonigle, J., Brady, S.M., Bulte, D.: Pancreas Segmentation-Derived Biomarkers: Volume and Shape Metrics in the UK Biobank Imaging Study. In: Papież, B.W., Namburete, A.I.L., Yaqub, M., Noble, J.A. (eds.) MIUA 2020. CCIS, vol. 1248, pp. 131–142. Springer, Cham (2020). https://doi.org/10.1007/978-3-030-52791-4_11

5. Bai, W., et al.: A bi-ventricular cardiac atlas built from 1000+ high resolution MR images of healthy subjects and an analysis of shape and motion. Med. Image Anal. **26**(1), 133–145 (2015). https://doi.org/10.1016/j.media.2015.08.009

6. Benjamini, Y., Hochberg, Y.: Controlling the false discovery rate: a practical and powerful approach to multiple testing. Methodological **57**, 289–300 (1995). https://doi.org/10.1111/j.2517-6161.1995.tb02031.x

7. Beygelzimer, A., Kakadet, S., Langford, J., Arya, S., Moun, D., Li, S.: Fast Nearest Neighbor Search Algorithms and Applications (2019). https://rdrr.io/cran/FNN. R package version 1.1.3

8. Biffi, C.: An introduction to mass univariate analysis of three-dimensional phenotypes (2017). https://github.com/UK-Digital-Heart-Project/mutools3D. R package version 1.0

9. Biffi, C., et al.: Three-dimensional cardiovascular imaging-genetics: a mass univariate framework. Bioinformatics **34**, 97–103 (2018). https://doi.org/10.1093/bioinformatics/btx552

10. Bycroft, C., et al.: The UK Biobank resource with deep phenotyping and genomic data. Nature **562**(7726), 203–209 (2018). https://doi.org/10.1038/s41586-018-0579-z

11. Couinaud, C.: Le Foie: Études Anatomiques et Chirurgicales. Masson, Paris (1957)

12. Dixon, W.T.: Simple proton spectroscopic imaging. Radiology **153**(1), 189–194 (1984). https://doi.org/10.1148/radiology.153.1.6089263
13. Freedman, D., Lane, D.: A nonstochastic interpretation of reported significance levels. J. Bus. Econ. Stat. **1**, 292–298 (1983). https://doi.org/10.1080/07350015.1983.10509354
14. Guillaume, B., et al.: Improving mass-univariate analysis of neuroimaging data by modelling important unknown covariates: application to epigenome-wide association studies. NeuroImage **173**, 57–71 (2018). https://doi.org/10.1016/j.neuroimage.2018.01.073
15. Kühn, J.P., et al.: Prevalence of fatty liver disease and hepatic iron overload in a Northeastern German population by using quantitative MR imaging. Radiology **284**, 706–716 (2017). https://doi.org/10.1148/radiol.2017161228
16. Littlejohns, T.J., et al.: The UK Biobank imaging enhancement of 100,000 participants: rationale, data collection, management and future directions. Nat. Commun. **11**(1), 1–12 (2020). https://doi.org/10.1038/s41467-020-15948-9
17. Liu, Y., et al.: Genetic architecture of 11 abdominal organ traits derived from abdominal MRI using deep learning. eLife **10**, e65554 (2021)
18. de Marvao, A., et al.: Outcomes and phenotypic expression of rare variants in hypertrophic cardiomyopathy genes amongst UK Biobank participants. medRxiv (2021). https://doi.org/10.1101/2021.01.21.21249470
19. Medrano-Gracia, P., et al.: Left ventricular shape variation in asymptomatic populations: the multi-ethnic study of atherosclerosis. J. Cardiovasc. Magn. Reson. **16**, 56 (2014). https://doi.org/10.1186/s12968-014-0056-2
20. Penny, W., Friston, K., Ashburner, J., Kiebel, S., Nichols, T.: Statistical Parametric Mapping: The Analysis of Functional Brain Images. Elsevier/Academic Press, Amsterdam, Boston (2007). https://doi.org/10.1016/B978-0-12-372560-8.X5000-1
21. R Core Team: R: A Language and Environment for Statistical Computing. R Foundation for Statistical Computing, Vienna, Austria (2020). https://www.R-project.org
22. Ronneberger, O., Fischer, P., Brox, T.: U-Net: convolutional networks for biomedical image segmentation. In: Navab, N., Hornegger, J., Wells, W.M., Frangi, A.F. (eds.) MICCAI 2015, Part III. LNCS, vol. 9351, pp. 234–241. Springer, Cham (2015). https://doi.org/10.1007/978-3-319-24574-4_28
23. Schlager, S., Francois, G.: Manipulations of Triangular Meshes Based on the 'VCGLIB' API (2021). https://github.com/zarquon42b/Rvcg. R package version 0.19.2
24. Smith, S.M., Nichols, T.E.: Threshold-free cluster enhancement: addressing problems of smoothing, threshold dependence and localisation in cluster inference. NeuroImage **4**, 83–98 (2009). https://doi.org/10.1016/j.neuroimage.2008.03.061
25. Thomas, E.L., Fitzpatrick, J., Frost, G.S., Bell, J.D.: Metabolic syndrome, overweight and fatty liver. In: Berdanier, C., Dwyer, J., Heber, D. (eds.) Handbook of Nutrition and Food, pp. 763–768. CRC Press, Boca Raton, USA, 3rd edn. (2013). https://doi.org/10.1201/b15294
26. Winkler, A.M., Ridgway, G.R., Webster, M.A., Smith, S.M., Nichols, T.E.: Permutation inference for the general linear model. NeuroImage **92**, 381–397 (2014). https://doi.org/10.1016/j.neuroimage.2014.01.060
27. Younossi, Z.M.: Non-alcoholic fatty liver disease - a global public health perspective. J. Hepatol. **70**(3), 531–544 (2019). https://doi.org/10.1016/j.jhep.2018.10.033

Automatic Re-orientation of 3D Echocardiographic Images in Virtual Reality Using Deep Learning

Lindsay Munroe[1]([✉]), Gina Sajith[1], Ei Lin[1], Surjava Bhattacharya[1],
Kuberan Pushparajah[1,2], John Simpson[1,2], Julia A. Schnabel[1],
Gavin Wheeler[1], Alberto Gomez[1], and Shujie Deng[1]

[1] School of Biomedical Engineering and Imaging Sciences,
King's College London, London, UK
lindsay.munroe@kcl.ac.uk
[2] Department of Congenital Heart Disease, Evelina London Children's Hospital,
Guy's and St Thomas' National Health Service Foundation Trust, London, UK

Abstract. In 3D echocardiography (3D echo), the image orientation varies depending on the position and direction of the transducer during examination. As a result, when reviewing images the user must initially identify anatomical landmarks to understand image orientation – a potentially challenging and time-consuming task. We automated this initial step by training a deep residual neural network (ResNet) to predict the rotation required to re-orient an image to the standard apical four-chamber view). Three data pre-processing strategies were explored: 2D, 2.5D and 3D. Three different loss function strategies were investigated: classification of discrete integer angles, regression with mean absolute angle error loss, and regression with geodesic loss. We then integrated the model into a virtual reality application and aligned the re-oriented 3D echo images with a standard anatomical heart model. The deep learning strategy with the highest accuracy – 2.5D classification of discrete integer angles – achieved a mean absolute angle error on the test set of $9.0°$. This work demonstrates the potential of artificial intelligence to support visualisation and interaction in virtual reality.

Keywords: 3D echocardiography · Deep learning · Virtual reality

L. Munroe—This work is independent research funded by the National Institute for Health Research (NIHRi4i, 3D Heart Project, II-LA-0716-20001, https://www.3dheart.co.uk/). This work was also supported by the Wellcome/EPSRC Centre for Medical Engineering (WT203148/Z/16/Z). Lindsay Munroe and Suryava Bhattacharya would like to acknowledge funding from the EPSRC Centre for Doctoral Training in Smart Medical Imaging (EP/S022104/1). Authors also acknowledge financial support from the Department of Health via the National Institute for Health Research (NIHR) comprehensive Biomedical Research Centre award to Guy's and St Thomas' NHS Foundation Trust in partnership with King's College London and King's College Hospital NHS Foundation Trust.

ⓒ Springer Nature Switzerland AG 2021
B. W. Papież et al. (Eds.): MIUA 2021, LNCS 12722, pp. 177–188, 2021.
https://doi.org/10.1007/978-3-030-80432-9_14

1   Introduction

Congenital heart disease (CHD) is characterised by abnormalities of heart structures or associated blood vessels, and affects 8 out of every 1000 births in Europe [3]. Cardiovascular morphology is highly diverse across the CHD patient population, therefore understanding the spatial relationships between cardiac structures for each individual patient is crucial for surgical planning [18]. Magnetic resonance imaging (MR), computed tomography (CT) and echocardiography (echo) are the common imaging modalities that inform surgical planning and intervention guidance for CHD.

Compared to MR and CT, echo is safer, more affordable and portable, and can show real-time motion during the cardiac cycle – especially of fast moving structures such as the heart valves. 3D echo is particularly attractive for CHD treatment planning because it is beneficial in understanding the spatial relationships of cardiac structures, especially in more complex cases. For example, 3D echo captures a depth of field, enables arbitrary plane cropping, and is more accurate than 2D echo for evaluating left ventricular volumes and corresponding ejection fraction [2,14]. 3D echo can, however, be difficult to interpret for several reasons. Firstly, patients with CHD may have complex, atypical cardiac morphology that can be challenging to identify. Secondly, unlike CT or MR images that are usually displayed in a standard anatomical orientation, orientation of echo images is determined by the position and direction of the transducer during acquisition. The transducer is in turn limited by anatomy and acoustic access; typical acoustic windows are parasternal, apical and subcostal. Thirdly, not only can echo images be acquired from different orientations but they usually capture only a partial view of the anatomy, making it more difficult to place the image within the full-volume anatomy. Finally, 3D echo images are typically viewed on 2D screens that do not allow true depth perception [18].

Two possible solutions to improve understanding of the true three-dimensional nature of cardiac structures are 3D printing and Virtual Reality (VR). 3D printed models are an increasingly popular and effective tool to interpret cardiac morphology during CHD surgical planning [8,15]. However, 3D printed models are static and so do not reveal abnormalities that can only be detected with motion, such as valve abnormalities. Furthermore, segmentation and printing are time-consuming, and interrogation of internal cardiac structures is restricted since each model can only be cut a limited number of times. VR addresses the main limitations of 3D printing; for example, VR has retrospectively been demonstrated to improve surgeons' understanding of patient pathology of valve defects, which enabled then to fine-tune their surgical plans [11].

To further improve the interpretability of 3D echo images, displaying them in anatomic orientation consistent with other established imaging modalities is crucial [13]. Conventionally, the image orientation is manually identified from anatomical landmarks – a potentially challenging and time consuming task. To address this problem, we propose to automate the image orientation task using deep learning methods and apply it in a VR application. In summary, the contributions of this paper are as follows:

1. A deep residual neural network (ResNet) is trained to automatically rotate 3D echo images to a standard apical four-chamber view.
2. A comparative study of the model using three different preprocessing approaches and three different loss functions is conducted and the model with the best reorientation result on the test set is identified.
3. A VR application is implemented and demonstrated using the best reorientation model. A loaded 3D echo volume is automatically re-oriented and aligned to a 3D graphic model of an anatomical heart.

The best performance achieved from this study in re-orienting volumetric images to the reference view is a mean absolute angle error of 9.0°.

1.1   Related Work

Several recent studies have addressed the need to estimate object pose from images. RotNet predicts rotation for 2D images from the Google Streetview dataset [12]. A 50-layer ResNet that had been pre-trained on ImageNet was fine-tuned for the discrete angle classification task. The authors experimented with both classification and mean absolute angle error regression but the former achieved superior results. Gidari et al. predicted 2D image rotation as the surrogate task during self-supervised learning [12]. Four rotation angle classes (0°, 90°, 180° and 270°) were selected and four AlexNet models were simultaneously trained, one to predict each angle class. The loss function was defined as the mean cross entropy loss across the four models. Mahandren et al. predicted 3D axis-angle rotations as a regression task for 3D pose estimation of the Pascal 3D+ dataset [4]. Their multi-component model consisted of one common VGG feature network and a pose network for each object class. The VGG feature network weights are pre-trained on ImageNet data. The pose networks are trained from scratch and consist of three fully connected layers. The model was trained using mean squared error loss for the first 10 epochs followed by geodesic loss for the final 10 epochs. The overall median angle error was 15.38°.

Similar to Mahandren et al. [10], we estimated pose from 3D images using geodesic loss. We extended this previous study by exploring loss functions used in the other aforementioned works – cross entropy and mean absolute angle error regression. Furthermore, we trained a ResNet instead of a VGG network. ResNet achieved impressive gains in image classification accuracy over VGG in the 2015 ImageNet Large Scale Visual Recognition Challenge [6] and is less susceptible to shattered gradients [1].

2   Methodology

In this study, 47 echo datasets from 13 patients with hypoplastic left heart syndrome (HLHS) were manually annotated and a deep residual neural network was trained on these datasets [6]. Three data preprocessing strategies were explored: 1) using 3D images to estimate orientation, 2) using three orthogonal slices only (2.5D approach) and 3) using the central slice only (2D approach). Three different loss functions were investigated: 1) classification of orientation into discrete Euler angles, 2) regression with mean absolute angle error loss, and 3) regression with geodesic loss. In this section, we first describe the data and how we generated ground truth labels. We then present the methods used in this study: the network architecture, preprocessing and loss function strategies, training procedure and experiments, and finally how we evaluated model performance.

2.1 Data and Labelling

Data: Images ($n = 115$) had been acquired in a previous study at the Evelina Children's Hospital from 20 patients with hypoplastic left heart syndrome (HLHS), aged between 18 months and 3 years old. Images were acquired with the Philips iE33 system and a cardiac X5-1 3D transducer. Transthoracic volumes were acquired from the following transthoracic echocardiography windows: parasternal, parasternal oblique, apical and subcostal. The HLHS datasets acquired from the ultrasound examination were converted to Cartesian DICOM (Digital Imaging and Communications in Medicine) format using Philips QLAB cardiac analysis software (Philips Healthcare). The Cartesian DICOM files were converted to .mhd format using a python script in order to visualise images in the Medical Image Processing Toolbox [5]. The images are in 8-bit unsigned integer format.

Labelling approach: The images were visualised using the Medical Image Processing Toolbox [5] in MATLAB (The MathWorks Inc.). The ground truth rotation required to transform each image to the reference view was obtained as follows: we first selected one frame from the 4D dataset where the ventricular landmarks are most visible through the cardiac cycle. Since we assume the ground truth rotation is the same for all frames in a 4D dataset, the label obtained from the selected frame is valid for all frames in the 4D dataset. We then manually aligned markers with ventricular landmarks consistent with the reference view in each of the mutually orthogonal planes for the selected frame (Fig. 1). The reference view was chosen akin to the standard apical four chamber view in healthy hearts, showing the atrioventricular valves, the right atrium, the left atrium and the hypoplastic ventricle. The software computes the rigid transformation from the original image to the view defined by the markers. The output is a (4×4) matrix representing both rotation and translation. The matrices are converted into Euler angles and saved as a text file. We discarded translation parameters as these were not required for our application. We manually labelled 47 out of 115 images. We were not able to successfully label the remaining images due to shadow artefacts.

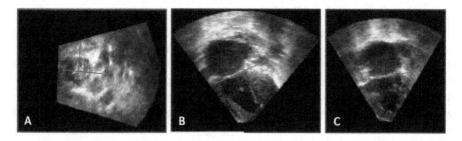

Fig. 1. Three orthogonal slices of a 3D echo image viewed in the Medical Image Processing Toolbox in MATLAB. Each plane is represented as a line of its bounding box colour in its two orthogonal planes e.g. plane A is represented as a red marker in plane B and C respectively. The markers are editable and their corresponding plane will be updated accordingly. To label the data the blue, green and red markers have been manually aligned to ventricular landmarks to correspond to the four chamber view. (Color figure online)

2.2 Methods

Network architecture: All experiments employed an 18-layer ResNet consisting of eight residual blocks and a final fully connected layer (Fig. 2) [6]. For the 2D and 2.5D approaches, 2D convolutional layers were employed whereas 3D convolutional layers were used in the 3D approach. The activation function of the final layer depends on the loss function strategy (see below).

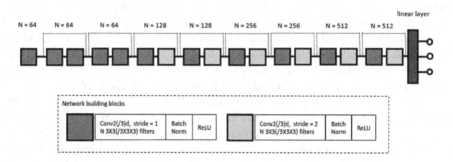

Fig. 2. The 18-layer residual neural network architecture trained for all experiments in this study.

Preprocessing: From each 4D image, end-systole (ES) and end-diastole (ED) 3D images were extracted. We discarded the other frames as they are highly correlated to the ES or ED frames. Each 3D image was re-sampled to $1mm^3$ voxel size and normalised to the intensity range $[0, 1]$ by dividing voxel intensities by 255. Three pre-processing approaches were explored for this project. The first is a 2D approach where only the mid-slice of the main orthogonal plane for each image is extracted for training. The second 2.5D approach extends the first by extracting mid-slices from all three planes then padding and stacking them to form a 3-channel 2D image. The third approach retains the data as a 3D volume. The 3D images were converted to 2D, 2.5D or retained as 3D depending on the experiment. Finally, images were resized to the maximum that were compatible with the available hardware: 2D, 2.5D and 3D images were resized to $1 \times 96 \times 96$, $3 \times 96 \times 96$ and $96 \times 96 \times 96$ respectively (original image size ranged from $87 \times 80 \times 70$ to $162 \times 148 \times 141$).

Loss functions: Three different strategies were implemented for the choice of loss function for the 2.5D approach. The first method represented the labels and predictions as Euler angles $[\theta_x, \theta_y, \theta_z]$ and the mean absolute angle error loss ($MAAE$) was applied (Eq. 1). In the second method, the labels and predictions were transformed to the axis-angle representation $[v_1, v_2, v_3]$ and the residual axis-angle computed. The loss function is the mean of the residual geodesic distances (d_{geo} in Eq. 2) [10]. Finally, classification of discrete integer angles was also explored using cross entropy loss.

$$MAAE(\theta_{label}, \theta_{pred}) = |180° - ||\theta_{label} - \theta_{pred}| - 180°|| \tag{1}$$

$$d_{geo} = \theta = \sqrt{v_1^2 + v_2^2 + v_3^2} \tag{2}$$

The activation function of the final fully connected layer in the network varied depending upon the loss function selected. The sigmoid function was applied for the first method, $\pi \tanh$ in the case of geodesic loss regression, and the softmax function was applied for classification.

Training: Prior to training, the data was split into training, validation and test sets with a $80\% : 10\% : 10\%$ split at the patient level. Images were augmented during training by applying random rotations in the range $[-10°, 10°]$ for each orthogonal axis and updating labels accordingly. The model was trained with a batch size of 10, 10 and 4 for 2D, 2.5D and 3D respectively. The Adam optimisation algorithm was used with parameters fixed as follows: the learning rate was set to $1e-5$, β_1 and β_2 were set to 0.9 and 0.999 respectively. The models were trained for 150 epochs with validation loss assessed and checkpoints saved every two epochs. The epoch where validation loss started to plateau was selected as the best model. The corresponding checkpoint was loaded and results assessed for the training and test set. Models were trained locally using a 6 GB Nvidia Quadro P3200 GPU. All code was implemented in the Python-based deep learning framework PyTorch.

Experiments: We ran five deep learning experiments for this study. First, we trained deep learning models using three different loss function strategies on 2.5D data. Then, two additional experiments explored 2D and 3D preprocessing strategies combined with the best loss function.

Performance evaluation: To assess model performance the mean absolute angle error (MAAE) defined in Eq. 1 between labels and predictions was computed for the training and test set. The labels and predictions were transformed, where necessary, to the Euler angle representation so the MAAE is interpretable. The MAAE was computed separately for the x, y and z axes to understand if the model training is focused more heavily towards one axis.

3 Results

Results are shown in Table 1. The most accurate approach is classification with 2.5D preprocessing, with MAAE of 9.0° on the test set. A visualisation of a random test data batch rotated by both the ground truth and predicted angles is shown in Fig. 3.

Table 1. Model mean absolute angle error (MAAE) for training and test set across five different experiments. MAAE is reported separately for rotation about the x,y and z axis. θ_{mean} is the mean of θ_x, θ_y and θ_z.

Experiment		Train MAAE (degrees)				Test MAAE (degrees)			
Dimension	Loss function	θ_x	θ_y	θ_z	θ_{mean}	θ_x	θ_y	θ_z	θ_{mean}
2.5D	Classification	13.9	15.5	23.6	17.7	**2.5**	18.1	**6.3**	**9.0**
2.5D	Regression	14.8	19.4	19.0	17.7	11.4	**5.8**	15.3	10.8
2.5D	Geodesic	18.1	29.9	25.8	24.6	26.4	18.6	32.7	25.9
2D	Classification	10.3	11.7	23.2	15.1	13.4	46.9	14.2	24.8
3D	Classification	11.5	15.9	23.6	17.0	12.7	21.6	33.5	22.6

The MAAE depends upon the extent of ground truth rotation, with larger rotations having larger MAAE results, as shown in Table 2. The ground truth rotations of our dataset are primarily smaller rotations: ground truth rotations ranged from 0.0° to 93.1° with mean rotation of 21.8° and std. dev. of 20.6° (mean of 13.8°, 26.2°, 25.4° and std. dev. of 10.8°, 22.0°, 24.0° for x, y, z axis ground truth rotations respectively). The training images have been augmented with additional random rotations, which is why the training MAAE can be larger than the test MAAE.

Table 2. Mean absolute angle error (MAAE) by ground truth angle range. The ground truth labels are from both the training and test set.

Ground truth angle range	Proportion of all ground truth labels	MAAE (degrees)
0–40°	82.3%	10.4
40–90°	16.3%	40.9
90+ °	1.4%	96.0

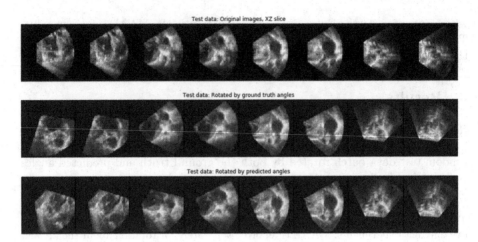

Fig. 3. A visual comparison of ground truth vs. predicted rotations of the test set from the best model. Row 1) is a random batch of images from the test set, row 2) shows the images rotated by the ground truth angles, row 3) shows the images rotated by the predicted angles. In all images, the mid-XZ slice has been selected from the 3D volume.

4   Applications

A VR application was implemented to align an echo image to a 3D anatomical cardiac model by integrating the best trained model obtained from the training process with a plug-in that enables viewing of the echo images in VR [16]. This section elaborates our approach of how the trained model is integrated and used in the Unity application.

4.1   Scene Setup

The application is set up within a Unity scene that consists of two objects: an anatomical model of a normal heart and a rendered volumetric echo image. The anatomical heart model was prefabricated and serves as an existing object in the scene, which does not need to be dynamically loaded during runtime. The echo volume, however, is loaded when the application starts running. The anatomical heart model was fabricated based on a stereolithography (STL) model[1] which was sliced in half approximately along the coronal plane to reveal its four-chamber apical view using Meshmixer (Autodesk). The heart model was placed in the scene with the four-chamber view facing the camera, as displayed on the right in Fig. 4.

The volume loader is already implemented by the plugin [16] but in a default orientation. Our trained model predicts rotation after the volume is loaded into the scene. The model input is the echo volume and the output is the three Euler

---

[1] Available at https://www.thingiverse.com/thing:852939.

angles, which are used to rotate the echo volume. The volume's four-chamber view is thus aligned with the anatomical heart model's four-chamber view. Model inference is fast, taking less than one second on a standard laptop with a 6GB GPU. Manually re-orienting the echo volume may take well over a minute, or be impossible, depending on the user's experience. An example of the resulting calibration is shown in Fig. 4. Once the calibration is complete, the anatomical heart model will automatically move in unison with the echo volume when a user interacts with the volume.

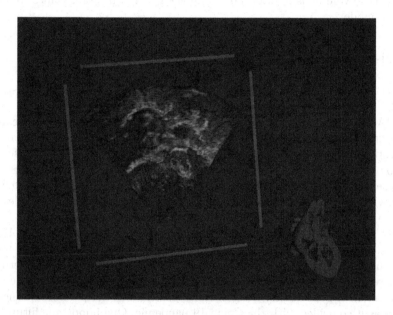

Fig. 4. The user interface (UI) layout that automatically informs users of the four chamber apical view. The network reorients the volume to the heart model in the bottom right corner, which is initialised in the four chamber apical view. The red edges are the edges of the cropping plane. (Color figure online)

4.2   Integration

The integration of the trained PyTorch model in Unity is implemented using sockets communication because the PyTorch model (built by Python) cannot be directly invoked by the native languages supported by Unity (C# and Javascript), but both C# and Python support sending and receiving data through sockets. An external server was scripted using Python for invoking the trained network model and sending its output to a client, and Unity acts as the client which is scripted in C# and added to the anatomical heart model.

However, for Python to run a server and then Unity to connect a client to it, Python and Unity must run in parallel but cannot be invoked simultaneously. To achieve this, parallel processing was used on the Unity side (in the form of a `Task[]` array object) where two threads were used to complete the two tasks in parallel. In the first thread, Unity calls the Python executable using the `ProcessStartInfo()` and instantiates the server. At the same time, the path of the echo volume data is passed to the server. Once the server is established the trained model can be invoked, the data path is then passed to the trained model and an inference is made by this model. In the other thread, a client is instantiated as a socket object and repeatedly tries to connect to the server. Once the connection is established, the network inference is received by Unity and the echo volume is re-orientated with the inferred rotation values to match the STL model.

5   Discussion

We successfully trained a ResNet to predict 3D rotations required to re-orient 3D echo images to the standard four chamber view. The best approach achieved 9.0° MAAE on the test set, even with the heterogeneous nature of the cardiac morphology present in our dataset and small sample size. Our method predicts rotations to the level of discrete angles as opposed to the less precise, four rotation classes proposed by Gadari et al. We achieve competitive performance to Mahendran et al. (our method achieved 9.0° mean error vs. 15.4° median error) with substantially fewer images in our training dataset, acknowledging the datasets and ground truth label range are very different in nature.

This study has three key limitations. Firstly, the labelled data is biased towards images of higher quality since the data was labeled by non-experts and some images could not be labeled due to shadow artefacts. Furthermore, labels were not reviewed by an expert. Ideally, cardiologists would review the labels and label poor quality images. However, this was not possible during this project as it coincided with the Covid-19 pandemic. One important future task is to improve the size and reliability of the training dataset by involving cardiologists in the labelling task. We could also further expand the training dataset by including additional CHD sub-types.

Secondly, the MAAE increases with the magnitude of ground truth rotation. Ideally, MAAE would be invariant to rotation magnitude. The model works reasonably well for small rotations within ±40° and thus is adequate for the majority of images in our dataset. However, the model is not well-suited for images that require re-orientation larger than ±40°. We hypothesise that a deeper network (such as a 50-layer ResNet) – combined with allowing any rotation angle during data augmentation – would be capable of learning more complex features required for re-orientations greater than ±40°. Wright et al. achieved good results on a similar task using a long-short term memory network (LSTM) [17]. The LSTM model applies small, successive rotations to the image until it reaches the reference view. A deeper network and LSTMs are both promising avenues to explore in future.

Thirdly, Hardware limitations in this study prevented training deeper, more sophisticated networks as mentioned above. Furthermore, the 3D preprocessing approach was hindered by computational capacity; it was necessary to decrease batch size, image size and number of network layers. Future access to a server with more computational capacity would enable exploration of more complex architectures and improve results for 3D approaches.

Several papers report the best accuracy with deep networks (such as a 50-layer ResNet) that are pre-trained and then fine-tuned for the re-orientation task [10,12]. Although there exists a large public dataset for 2D echo (CAMUS), the equivalent for 3D echo (CETUS) was temporarily unavailable due to database issues [7,9]. In future, if the CETUS or other similar public datasets become available, then pre-training a deeper network for this task would be a promising strategy to explore. Another possible alternative to explore is to modify model weights pre-trained using 2D images for use with 3D images.

In conclusion, a deep learning network to re-orient 3D echo images to a standard apical four chamber view was successfully trained and then integrated into virtual reality. The re-oriented echo volumes were then matched to a standard anatomical model in VR. This work demonstrates the potential of artificial intelligence to support visualisation and interaction in virtual reality, although a further user study is required to evaluate its clinical impact.

References

1. Balduzzi, D., et al.: The shattered gradients problem: if resnets are the answer, then what is the question? In: International Conference on Machine Learning. PMLR, pp. 342–350 (2017)
2. Cheng, K., et al.: 3D echocardiography: benefits and steps to wider implementation. Br. J. Cardiol. **25**, 63–68 (2018)
3. Dolk, H., et al.: Congenital heart defects in Europe: prevalence and perinatal mortality, 2000 to 2005. Circulation **123**(8), 841–849 (2011)
4. Gidaris, S., Singh, P., Komodakis, N.: Unsupervised rep- resentation learning by predicting image rotations. In: arXiv preprint arXiv:1803.07728 (2018)
5. Gomez, A.: MIPROT: A Medical Image Processing Toolbox for MAT-LAB. In: arXiv preprint arXiv:2104.04771 (2021)
6. He, K., et al.: Deep residual learning for image recognition. In: Proceedings of the IEEE Conference on Computer Vision and Pattern Recognition, pp. 770–778 (2016)
7. CREATIS Lab.-France. CETUS datasets. https://www.creatis.insa-lyon.fr/Challenge/CETUS/index.html. https://www.creatis.insa-lyon.fr/Challenge/CETUS/databases.html
8. Lau, I., Sun, Z.: Three-dimensional printing in congenital heart disease: a systematic review. J. Med. Radiat. Sci. **65**(3), 226–236 (2018)
9. Leclerc, S., et al.: Deep learning for segmentation using an open large-scale dataset in 2D echocardiography. IEEE Trans. Med. Imaging **38**(9), 2198–2210 (2019)
10. Mahendran, S., Ali, H., Vidal, R.: 3D pose regression using convolutional neural networks. In: Proceedings of the IEEE International Conference on Computer Vision Workshops, pp. 2174–2182 (2017)
11. Pushparajah, K., et al.: Virtual reality three dimensional echocardiographic imaging for planning surgical atrioventricular valve repair. JTCVS Tech. **7**, 269–277 (2021)

12. Saez, D.: Correcting Image Orientation Using Convolutional Neural Networks. http://d4nst.github.io/2017/01/12/image-orientation
13. Simpson, J., et al.: Image orientation for three-dimensional echocardiography of congenital heart disease. The Int. J. Cardiovasc. Imaging **28**(4), 743–753 (2012)
14. Simpson, J.M., Miller, O.: Three-dimensional echocardiography in congenital heart disease. Arch. Cardiovasc. Dis. **104**(1), 45–56 (2011)
15. Valverde, I., et al.: Three-dimensional printed models for surgical planning of complex congenital heart defects: an international multicentre study. Eur. J. Cardio-thorac. Surg. **52**(6), 1139–1148 (2017)
16. Wheeler, G., et al.: Virtual interaction and visualisation of 3D medical imaging data with VTK and unity. Healthc. Technol. Lett. **5**(5), 148–153 (2018)
17. Wright, R., et al.: LSTM spatial co-transformer networks for registration of 3D fetal US and MR brain images. In: Melbourne, A., et al. (eds.) PIPPI/DATRA -2018. LNCS, vol. 11076, pp. 149–159. Springer, Cham (2018). https://doi.org/10.1007/978-3-030-00807-9_15
18. Jia-Jun, X., et al.: Patient-specific three-dimensional printed heart models benefit preoperative planning for complex congenital heart disease. World J. Pediatr. **15**(3), 246–254 (2019)

A Simulation Study to Estimate Optimum LOR Angular Acceptance for the Image Reconstruction with the Total-Body J-PET

Meysam Dadgar[1,2]([✉]), Szymon Parzych[1,2], and Faranak Tayefi Ardebili[1,2]

[1] Marian Smoluchowski Institute of Physics, Jagiellonian University, Kraków, Poland
meysam.dadgar@doctoral.uj.edu.pl
[2] Total-Body Jagiellonian-PET Laboratory, Jagiellonian University, Kraków, Poland

Abstract. One of the directions in today's development of PET scanners is to increase their axial field of view (AFOV). Currently limited to several centimeters, AFOV of the clinically available PET tomographs results in a very low sensitivity ($\sim$1%) and requires an extended time for a scan of a whole human body. While these drawbacks are addressed in the so-called, Total Body PET concept (scanner with a significantly elongated field of view), it creates new challenges not only in the mechanical construction but also in the image reconstruction and event selection. The possibility of taking into account of large angle variety of lines of responses (LORs) contributes positively to the sensitivity of the tomograph. However, at the same time, the most oblique LORs have an unfavorable influence on the spatial resolution due to the parallax error and large contribution to the scatter fraction. This forces to determine a new factor - acceptance angle - which is a maximum azimuthal angle for which the LORs are still taken into image reconstruction. Correct determination of such factor is imperative to maximize the performance of a Total Body PET system since it introduces a trade-off between the two main characteristics of scanners: sensitivity and spatial resolution.

This work has been dedicated to the estimation of the optimal acceptance angle for the proposed by the Jagiellonian PET (J-PET) Collaboration Total Body tomograph. J-PET Collaboration introduces a novel, cost-effective approach to PET systems development with the use of organic scintillators. This simulation study provides evidence that the 45° acceptance angle cut can be an appropriate choice for the investigated scanner.

Keywords: Acceptance angle · Total Body J-PET · Sensitivity · Spatial resolution

1 Introduction

Positron Emission Tomography (PET) is an advanced diagnostic method that allows for non-invasive imaging of ongoing physiological processes in the human

© Springer Nature Switzerland AG 2021
B. W. Papież et al. (Eds.): MIUA 2021, LNCS 12722, pp. 189–200, 2021.
https://doi.org/10.1007/978-3-030-80432-9_15

body. Detection of malignant lesions, is one of the main clinical tasks of PET scan. The fact that the principle of operation of positron tomography is at the molecular level, opens a possibility to detect malignant tissues in very early stages. Furthermore, it is also used to monitor the treatment of patients. For that, the standardized uptake value (SUV) is being considered as an index based on the concentration of the radiopharmaceutical in the formerly detected lesion. Any change of this parameter between successive imaging sessions can be used to assess patient response to therapy [1].

Most of the current clinically available PET scanners have an axial field of view (AFOV) of 15–26 cm [2]. In order to image the entire body the iterative or continuous bed movement is applied allowing for part-by-part imaging of human and finally, after combination, for the whole body image. However, this method has a very low sensitivity of ∼1% which comes from the fact, that at any one time most of the body is outside the FOV and only a small fraction of emitted photons can be detected due to isotropic radiation [2,3]. In order to address those factors the Total Body PET concept, characterised by coverage of the entire human with the detector rings has been proposed. There are currently at least four groups/projects concerning such systems: the UC Davis EXPLORER Consortium [4,5], the Ghent University "PET 2020" [6], the Siemens Healthineers "Biograph Vision Quadra" [7] and the J-PET Collaboration [8–11] from the Jagiellonian University in Cracow, Poland. The benefits of TB PETs are not limited to just single bed position imaging or no need for motion correction but also, thanks to a huge gain in sensitivity, the possibility of an increase in the signal-to-noise ratio, reducement in time of the scan, or decrements in activity dose according to the ALARA principle [12]. Moreover, when it comes to lesion detectability, they can be used to detect even sub-centimeter specimens [2,3].

Nevertheless, long AFOV creates new challenges in image reconstruction and event selection. While the possibility of detection of additional LORs contributes positively to the sensitivity of the tomograph, it has an unfavorable influence on spatial resolution. Moreover, strong attenuation of oblique LORs in the body results in the increase of unwanted scattered coincidences and enlarges the parallax error. This forces to determine a new parameter such as an acceptance angle as a cut over all registered LORs. Acceptance angle is a maximum azimuthal angle for which the line of responses are still taken into image reconstruction (see Fig. 1) [13]. For different geometries of PET scanners, it can be defined either as an angle or as a ring difference. However, such an acceptance cut is creating a tradeoff between two main parameters of PET scanners: sensitivity and spatial resolution.

The presented study is focused on the Total Body system proposed by the Jagiellonian PET Collaboration. The J-PET Collaboration presents an innovative approach to the design of PET systems. In oppose to the common tomographs equipped with inorganic, radially arranged scintillator crystals, J-PET uses axially arranged plastic scintillator strips read out on both ends with silicon photomultipliers. With a geometry which allows to significantly reduce the

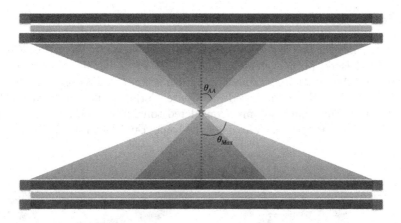

Fig. 1. Schematic cross section view of TB J-PET scanner simulated in this work. TB J-PET detector is composed of long strips of plastic scintillator (dark green) read out at two ends by photomultipliers (red). Two layers of scintillator strips are interleaved with the WLS layer (light green). It has 2 m long axial field of view and 78.6 cm of inner diameter. The structure of the detector is described more detailed in Fig. 2. The pink cone (wider one) created by the maximal angle θ_{Max} which for this geometry is equal to 69°, contains all possible LORs originating from the centrally located point source (marked as a star) which can be detected by the tomograph. The θ_{AA} denotes an exemplary acceptance angle for which only the LORs located within the blue cone (narrower one) are taken into account during further analysis. (Color figure online)

amount of needed electronics and smaller number of scintillators, J-PET strives to be the cost-effective competitor for PET imaging [9,11,14–19].

The effect of the oblique LORs has already been researched for the previously investigated Total Body J-PET system (TB J-PET) composed of a single layer with 384 plastic scintillators. The study was performed using the NEMA IEC-Body phantom based on the Filtered Back Projection (FBP) algorithm with STIR package [20].

The presented here simulation-based study was carried out in order to determine the proper acceptance angle for a newly proposed Total Body J-PET scanner with a multi-layer arrangement of plastic scintillators. Instead of the FBP, it was based on the Ordered Subset Expectation Maximization (OSEM) iterative image reconstruction algorithm. Recently, the iterative image reconstruction-based algorithms have been widely developed due to their superior performance in comparison with the traditionally used ones. They are significantly reducing noise and improving the image quality. OSEM is one of the main examples of such widely used iterative methods, which allows for a more precise model of PET acquisition procedure [21].

2   Methods

This study concerns one of the proposed and investigated by the J-PET Collaboration Total Body (TB) systems. The TB J-PET scanner has been simulated using Geant4 Application for Tomographic Emission (GATE) software [22–24]. GATE is a validated toolkit based on Monte Carlo simulations developed to research nuclear medicine systems. Considered tomograph consists of 24 panels which are parallel to the central axis of the tomograph (see Fig. 2).

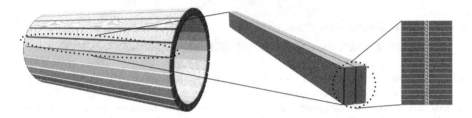

Fig. 2. (Left) Visualization of the Total Body J-PET scanner which consists of 24 axially arranged panels. (Middle) The panel is composed of 2 modules, each with 16 EJ230 plastic scintillation strips. (Right) In order to enhance axial resolution, modules are separated by an array of perpendicularly arranged WLS strips marked with green color (hash line pattern). (Color figure online)

Each panel is made from 2 modules separated by an array of wavelength-shifting (WLS) strips [25]. The module consists of 16 EJ230 "Eljen Technology" plastic scintillation strips with a dimension of $6 \times 30 \times 2000 \, \text{mm}^3$ located next to each other with 0.5 mm intervals between them. Each strip is coupled on both ends with a silicon photomultipliers [11].

For this study, two types of simulations have been performed using described geometry. The first simulation type included a 183 cm long linear source with a diameter of 1 mm and 1 MBq of total activity. In order to evaluate the contribution of phantom scatter coincidence events in all registered types of events, the study was also carried out with a centrally located, cylindrical phantom. The water filled phantom with 10 cm radius and 183 cm length[1] was simulated once without background activity and once with background activity of 10:1 target background ratio (TBR) to imitate the real, non-uniform activity distribution in the human body. The second group consisted of a situated in the center 1 MBq point like source placed inside a 20 cm long cylindrical air phantom with 20 cm radius and with 10:1 TBR. The hit-based result of the GATE simulation was analyzed using GOJA software. Gate Output J-PET Analyzer (GOJA) is a developed by J-PET Collaboration specialized software used for analyzing and construction of coincidence events based on the GATE hits output for J-PET-like tomographs [26]. For the case of this study, the time window

---

[1] The length of the source and phantom was set to 183 cm which represents the average man height in the tallest country in the world [2].

has been set to 5 ns, while lower energy threshold to 200 keV in order to minimize the number of detector scatter coincidences [27]. Due to the innovative geometric design of Total Body J-PET, most of the commonly used for image reconstruction softwares are not valid for the multi-layer PET scanner. One of the exceptions from that, is the Quantitative Emission Tomography Iterative Reconstruction (QETIR) software developed by Medisip group [28], which was chosen for the case of this study. Alongside image reconstruction application based on the OSEM algorithm, QETIR is also able to generate all of the needed requirements like sensitivity map and attenuation map. The generation of the sensitivity map was done with $3.125 \times 3.125 \times 3.125$ mm^3 voxel size and the reconstruction performed by 4 iterations with 25 subsets in each of them.

3   Results

In order to estimate a proper Acceptance Angle (from now on referred to as θ_{AA} angle) for the investigated Total Body J-PET scanner, four types of studies have been done based on the described simulations. Firstly, the effect of various θ_{AA} angles on the percentage share of each type of coincidence events in the total number of registered events (when $\theta_{AA} \equiv \theta_{Max} = 69°$) has been estimated for a group of simulations of a line source (see Fig. 3). The range of the tested angles covers the region from $10°$ to $65°$ angle. The influence of the $18°$ angle, which is a maximal achievable angle for traditional clinical PET tomographs was also inspected [29]. However, in this study, we are particularly emphasizing the $45°$ and $57°$ acceptance angles, which were already determined for different Total Body PET systems [30, 31].

While the number of each type of coincidences increases together with the widening of θ_{AA} angles, this relation is not linear especially for higher angles. Based on the simulation with line source (see Fig. 3a) it can be estimated, as anticipated in [13], that there is no significant increment in the number of each type of registered events for acceptance angle larger than $57°$. In case of having a phantom (see Figs. 3b and 3c) there is an addition of new type of coincidences which is a Phantom Scatter. For angles higher than $45°$ there is no meaningful change (only increase from 31.9% to 35.9% in case of simulation with background activity and from 30.9% to 34.3% for simulation without background activity) in the percentage share of True Coincidences. However, the $45°$ angle gives 14% reduction of undesirable Phantom Scatters, which is almost 5 times better than for $57°$ angle.

The second study concerned the determination of the influence of θ_{AA} angle on the sensitivity of TB J-PET tomograph. Sensitivity is one of the main parameters taken into account in lesion detectability of PET scanners [3]. Two parameters: Total Sensitivity and Sensitivity @ Center, were determined for each acceptance angle. The Total Sensitivity was calculated as a mean of sensitivity of every slice, while the sensitivity of a slice was estimated as the rate of registered events divided by the fraction of activity per slice. The Sensitivity @ Center is the sensitivity of the central slice (at 0 cm). Moreover, the same calculation

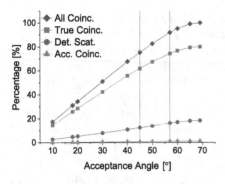

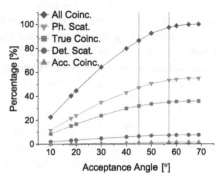

(a) Fraction of coincidences as a function of the acceptance angle determined based on the simulation of 1 MBq line source.

(b) Fraction of coincidences as a function of the acceptance angle determined based on the simulation of 1 MBq line source and 183 cm long water phantom with background activity.

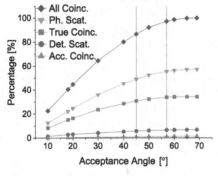

(c) Fraction of coincidences as a function of the acceptance angle determined based on the simulation of 1 MBq line source and 183 cm long water phantom without background activity.

Fig. 3. Plots represent the percentage share of all and each type of coincidence events for a given acceptance angle in the total number of registered events for 69° acceptance angle (which is the maximal possible angle which LOR can have in presented geometry). Abbreviations used in the legend: 'All Coinc.', 'Ph. Scat.', 'True Coinc.', 'Det. Scat.' and 'Acc. Coinc.' denote All Coincidences, Phantom Scatter, True Coincidences, Detector Scatter and Accidental Coincidences, respectively. Especially researched 45° and 57° angles are marked with vertical lines.

was performed also for only True Coincidences with a change in the definition of sensitivity in each slice from the rate of registered events to the rate of registered True Coincidence events. Figures 4a and 4b present the values of the first sensitivity parameter as a function of θ_{AA} from the simulation of line source without and with phantom, respectively. The results of the latter parameter are shown

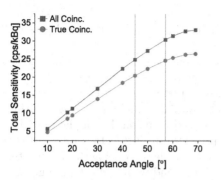

(a) Dependence of Total Sensitivity on acceptance angle determined based on the simulation of 1MBq line source.

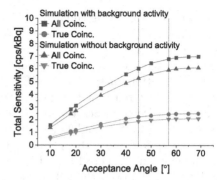

(b) Dependence of Total Sensitivity on acceptance angle determined based on the simulation of 1 MBq line source and 183 cm long water phantom.

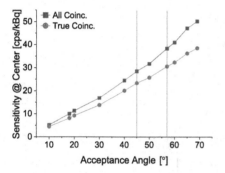

(c) Dependence of Sensitivity @ Center on acceptance angle determined based on the simulation of 1MBq line source.

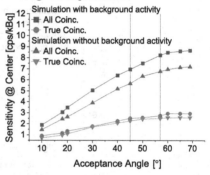

(d) Dependence of Sensitivity @ Center on acceptance angle determined based on the simulation of 1 MBq line source and 183 cm long water phantom.

Fig. 4. Results of Total Sensitivity and Sensitivity @ Center obtained with different acceptance angles for true (all) coincidences. The Total Sensitivity was calculated as a sum of sensitivities of each slice divided by the number of them, where the sensitivity of each slice is described as the rate of registered events resulting in true coincidences (any type of coincidences) divided by according fraction of activity. The Sensitivity @ Center is equal to the sensitivity of the central slice. Especially researched 45° and 57° angles are marked with vertical lines.

in the Figs. 4c and 4d. Based on the Fig. 4a, the acceptance angle has a huge impact on the Total Sensitivity. Considered 45° and 57° angles correspond to losses of ~24.3% (~22.7%) and ~7.7% (~7.1%) respectively for All (True) types of coincidence events. However, for the more realistic simulation with a phantom with warm background (see Fig. 4b), the influence of θ_{AA} angle is smaller and the 45° angle is reducing the Total Sensitivity only by ~13.4% in case of All events and ~11.2% in case of True Coincidences. The corresponding Sensitivity

@ Center (see Fig. 4d) is equal to 6.95 cps/kBq for All and 2.50 cps/kBq for True events, which is accordingly ∼80.3% and ∼85.3% of the maximal possible central sensitivity.

Furthermore, the effect of acceptance angle on a scatter fraction was investigated. The scatter fraction of the PET scanner quantifies the sensitivity of the detector to scattered radiation [32]. It was estimated as a ratio of the sum of Phantom and Detector Scatter to the number of All Coincidences. The relation between scatter fraction and θ_{AA} angle determined based on the simulations with 1 MBq line source and water phantom is presented in Fig. 5. In both cases the maximum minimization to 10° of the acceptance angle is reducing this parameter only by ∼2%. However, the noticeable growing trend with wider θ_{AA} suggests, that in case of higher activities the acceptance angle factor can contribute positively to the scatter fraction reduction.

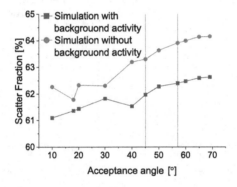

Fig. 5. The plot represents the scatter fraction as a function of the acceptance angle determined based on the simulations of 1 MBq line source and 183 cm long water phantom. Especially researched 45° and 57° angles are marked with vertical lines.

Spatial resolution is one of the most important characteristics of PET scanners, which determines the possible size of detectable lesions [32–34]. One of the classic approaches to investigate the quality of spatial resolution utilizes a Point Spread Function (PSF). PSF is defined as a full width at half maximum of the either transverse or axial one-dimensional projection of the slice of reconstructed image, which contains the radioactive source. In order to estimate the impact of the acceptance angle on the spatial resolution of Total Body J-PET, a point like source has been simulated inside a cylindrical air phantom with 10:1 target background ratio. Figure 6a presents values of both PSF parameters for 6 different acceptance angles, calculated based on the first iteration with 25 subsets of image reconstruction. The results show that increase of θ_{AA} has much worse influence on the axial PSF than on transverse PSF. In case of the latter, one can even observe an improvement of resolution for 57° in oppose to 45° angle. However, the effect of number of iterations in image reconstruction on the transverse resolution (see Fig. 6b) turns out to not only improve the results but also

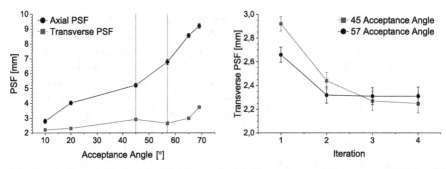

(a) Dependence of Axial (black) and Transverse (red) PSF on the acceptance angle for the image reconstructed with 1 iteration with 25 subsets.

(b) Dependence of Transverse PSF for 45° (red) and 57° (black) acceptance angle on the number of iterations in image reconstruction.

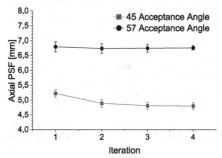

(c) Dependence of Axial PSF for 45° (red) and 57° (black) acceptance angle on the number of iterations in image reconstruction.

Fig. 6. Plots represent study of the effect of acceptance angle and image reconstruction iterations number on the spatial resolution. (Color figure online)

reverse the ratio between PSF for 45° and 57° angle. Nonetheless, the percentage difference between each calculated point for transverse PSF is much smaller and even this improvement is negligible in comparison to the deterioration in the axial resolution. Moreover, Fig. 6c shows that the influence of iterations number on the axial resolution is almost negligible for both angles. In case of the third and fourth iteration for which the ratio in transverse PSF is reversed, the 45° acceptance angle is almost 29% better than 57° in terms of axial resolution and results in axial PSF equal to 4.80(11) mm.

4 Conclusions

The determination of the proper acceptance angle is a mandatory requirement to maximize the performance of Total Body PET systems. The aim of this study was to estimate such optimal angle for the proposed and investigated by the J-PET Collaboration Total Body PET tomograph, based on the simulations performed using GATE software. The presented results show that 45° acceptance angle gives almost 5 times better reduction of undesirable Phantom Scatters than 57° angle. In case of the Total Sensitivity of investigated scanner, the same angle gives a ~13.4% (~11.2%) loss of maximal possible Total Sensitivity for All Coincidences (True Coincidences). No significant difference in the level of scatter fraction was observed. For both 45° and 57° angles it was estimated to ~62.0% (~63.3%) and ~62.4% (~63.9%), respectively for simulation with (without) background activity. However, different θ_{AA} angles have a major influence on the spatial resolution, especially on axial PSF. Discussed local improvement in transverse resolution for 57° angle in oppose to 45° disappears with higher numbers of iterations in image reconstruction, while there is no meaningful change in axial resolution. Based on provided evidence, the 45° acceptance angle seems to be an optimal choice for the Total Body J-PET tomograph.

Acknowledgment. This work was supported by Foundation for Polish Science through TEAM POIR.04.04. 00-00-4204/17, the National Science Centre, Poland (NCN) through grant No. 2019/35/B/ST2/03562 and grant PRELUDIUM 19, agreement No. UMO-2020/37/N/NZ7/04106.

The publication also has been supported by a grant from the SciMat Priority Research Area under the Strategic Programme Excellence Initiative at the Jagiellonian University.

References

1. Schaefferkoetter, J., Townsend, D.: A human-trained numerical observer model for PET lesion detection tasks. In: IEEE Nuclear Science Symposium and Medical Imaging Conference 2015, vol. 9999, pp. 1–3. San Diego, C (2015). https://doi.org/10.1109/NSSMIC.2015.7582063
2. Vandenberghe, S., Moskal, P., Karp, J.S.: State of the art in total body PET. EJNMMI Phys. **7**(1), 1–33 (2020). https://doi.org/10.1186/s40658-020-00290-2
3. Cherry, S.: Total-body PET: maximizing sensitivity to create new opportunities for clinical research and patient care. J. Nucl. Med. **59**(1), 3–12 (2018)
4. uEXPLORER Homepage. http://www.explorer.ucdavis.edu. Accessed 13 Mar 2021
5. Badawi, R., Shi, H., Hu, P., et al.: First human imaging studies with the EXPLORER total-body pet scanner. J. Nucl. Med. **60**(3), 299–303 (2019)
6. Vandenberghe, S., Geagan, M., Efthimiou, N.: PET2020 HRS: maximization of sensitivity and resolution using axial extension and patient adaptive rings in a high resolution long axial FOV scanner. In: European Journal of Nuclear Medicine and Molecular Imaging, vol. 46, p. 274 (2019)
7. Siemens Healthineers Homepage. http://www.siemens-healthineers.com/molecular-imaging/pet-ct/biograph-vision-quadra. Accessed 13 Mar 2021
8. J-PET Homepage. http://www.koza.if.uj.edu.pl/pet/. Accessed 13 Mar 2021
9. Moskal, P., Stępień, E.: Prospects and clinical perspectives of total-body PET imaging using plastic scintillators. PET Clin. **15**(4), 439–452 (2020)

10. Moskal, P., Jasińska, B., Stępień, E.: Positronium in medicine and biology. Nat. Rev. Phys. **1**, 527–529 (2019)
11. Moskal, P., Rundel, O., Alfs, D., et al.: Time resolution of the plastic scintillator strips with matrix photomultiplier readout for J-PET tomograph. Phys. Med. Biol. **61**(5), 2025–2047 (2016)
12. Karakatsanis, N., Fokou, E., Tsoumpas, C.: Dosage optimization in positron emission tomography: state-of-the-art methods and future prospects. Am. J. Nucl. Med. Mol. Imag. **5**(5), 527–547 (2015)
13. Efthimiou, N.: New challenges for PET image reconstruction for total-body imaging. PET Clin. **15**(4), 453–461 (2020)
14. Moskal, P., Niedźwiecki, S., Bednarski, T., et al.: Test of a single module of the J-PET scanner based on plastic scintillators. Nucl. Instr. Meth. A. **764**, 317–321 (2014)
15. Moskal, P., Kisielewska, D., et al.: Feasibility study of the positronium imaging with the J-PET tomograph. Phys. Med. Biol. **64** (2019)
16. Moskal, P., Kisielewska, D., et al.: Performance assessment of the 2 γ positronium imaging with the total-body PET scanners. EJNMMI Phys. **7**, 44 (2020)
17. Sharma, S., Chhokar, J., et al.: Estimating relationship between the Time Over Threshold and energy loss by photons in plastic scintillators used in the J-PET scanner. EJNMMI Phys. **7**, 39 (2020)
18. Moskal, P., Krawczyk, N., Hiesmayr, B.C., et al.: Feasibility studies of the polarization of photons beyond the optical wavelength regime with the J-PET detector. Eur. Phys. J. C. **78**(11), 970–980 (2018)
19. Moskal, P., Salabura, P., Silarski, M., et al.: Novel detector systems for the positron emission tomography. Bio-Algo. Med. Syst. **7**(2), 73–78 (2011)
20. Kopka, P., Klimaszewski, K.: Reconstruction of the NEAM IEX body phantom from J-PET total-body scanner simulation using STIR. Acta Phys. Pol. B. **51**, 357–360 (2020)
21. Zhu, Y.M.: Ordered subset expectation maximization algorithm for positron emission tomographic image reconstruction using belief kernels. J. Med. Imag. (Bellingham, Wash.) **5**(4), 044005 (2018)
22. Jan, S., Santin, G., Strul, D., et al.: GATE: a simulation toolkit for PET and SPECT. Phys. Med. Biol. **49**(19), 4543–4561 (2004)
23. Santin, J., Benoit, D., Becheva, E., et al.: GATE V6: a major enhancement of the GATE simulation platform enabling modelling of CT and radiotherapy. Phys. Med. Biol. **56**(4), 881–901 (2016)
24. Sarrut, D., BardiÃİs, M., Boussion, N., et al.: A review of the use and potential of the GATE Monte Carlo simulation code for radiation therapy and dosimetry applications. Med. Phys. **41**(6), 064301 (2014)
25. Smyrski, J., Alfs, D., Bednarski, T., et al.: Measurement of gamma quantum interaction point in plastic scintillator with WLS strips. Nucl. Instrum. Methods Phys. Res. A **851**(5), 39–42 (2017)
26. Dadgar, M., Kowalski, P.: GATE simulation study of the 24-module J-PET scanner: data analysis and image reconstruction. Acta Phys. Pol. B **51**, 309–320 (2020). Poland
27. Kowalski, P., Wiślicki, W., Raczyński, L., et al.: Scatter fraction of the J-PET tomography scanner. Acta Phys. Pol. B **47**(2), 549–560 (2016)
28. Medisip Homepage. https://www.ugent.be/ea/ibitech/en/research/medisip. Accessed 13 Mar 2021
29. Siemens Homepage. https://www.siemens-healthineers.com/molecular-imaging/pet-ct/biograph-vision. Accessed 13 Mar 2021

30. Cherry, S., et al.: Abstracts of the SPET conference 2018, pp. 1–2. EJNMMI, Ghent, Belgium (2018)
31. Zhang, X., Xie, Z., Berg, E., et al.: Total-body dynamic reconstruction and parametric imaging on the uEXPLORER. J. Nucl. Med. **61**(2), 285–291 (2019)
32. Kowalski, P., Wiślicki, W., Shopa, R., et al.: Estimating the NEMA characteristics of the J-PET tomograph using the GATE package. Phys. Med. Biol. **63**(16), 99–110 (2018)
33. NEMA Homepage. https://www.nema.org/standards/view/Performance-Measurements-of-Positron-Emission-Tomographs
34. Pawlik-Niedźwiecki, M., Niedźwiecki, S., Alfs, D., et al.: Preliminary studies of J-PET detector spatial resolution. Acta Phys. Pol. A **132**(5), 1645–1648 (2017)

Optimised Misalignment Correction from Cine MR Slices Using Statistical Shape Model

Abhirup Banerjee[1,2](✉) ⓘ, Ernesto Zacur[2], Robin P. Choudhury[1,3], and Vicente Grau[2] ⓘ

[1] Division of Cardiovascular Medicine, Radcliffe Department of Medicine, University of Oxford, Oxford, UK
{abhirup.banerjee,robin.choudhury}@cardiov.ox.ac.uk
[2] Institute of Biomedical Engineering, Department of Engineering Science, University of Oxford, Oxford, UK
vicente.grau@eng.ox.ac.uk
[3] Oxford Acute Vascular Imaging Centre, Oxford, UK

Abstract. Cardiac magnetic resonance (CMR) imaging is a valuable imaging technique for the diagnosis and characterisation of cardiovascular diseases. In clinical practice, it is commonly acquired as a collection of separated and independent 2D image planes, limiting its accuracy in 3D analysis. One of the major issues for 3D reconstruction of human heart surfaces from CMR slices is the misalignment between heart slices, often arising from breathing or subject motion. In this regard, the objective of this work is to develop a method for optimal correction of slice misalignments using a statistical shape model (SSM), for accurate 3D modelling of the heart. After extracting the heart contours from 2D cine slices, we perform initial misalignment corrections using the image intensities and the heart contours. Next, our proposed misalignment correction is performed by first optimally fitting an SSM to the sparse heart contours in 3D space and then optimally aligning the heart slices on the SSM, accounting for both in-plane and out-of-plane misalignments. The performance of the proposed approach is evaluated on a cohort of 20 subjects selected from the UK Biobank study, demonstrating an average reduction of misalignment artifacts from 1.14 ± 0.23 mm to 0.72 ± 0.11 mm, in terms of distance from the final reconstructed 3D mesh.

Keywords: Cardiac mesh reconstruction · Cine MRI · Misalignment correction · Statistical shape model

1 Introduction

Cardiac magnetic resonance (CMR) imaging is a noninvasive imaging modality for obtaining functional and anatomical information of the heart. Because of its ability to characterise soft tissues, CMR is increasingly used to evaluate

© Springer Nature Switzerland AG 2021
B. W. Papież et al. (Eds.): MIUA 2021, LNCS 12722, pp. 201–209, 2021.
https://doi.org/10.1007/978-3-030-80432-9_16

the myocardium, providing accurate assessments of left ventricular (LV) function, myocardial perfusion, oedema, and scar [3]. Clinically used cine MR studies acquire only a small number of image planes with good contrast between soft tissues at a reasonable temporal resolution, with an about 1.5×1.5 mm^2 in-plane resolution and 8–10 mm out-of-plane resolution. Although 2D CMR images can exhibit the presence of scar or oedema, or myocardial perfusion on heart structures, their visualisation on the 3D heart shape can help in the understanding of their locations and shapes. Heart shape has been shown to improve diagnosis and prediction of patient outcome, which necessitate the development of patient-specific 3D anatomical models of the human heart from CMR images.

The 3D heart mesh reconstruction from cine MR slices is usually affected by several artifacts, specifically the breathing artifact from each plane being acquired at separate breath-holds, and the data sparsity from limited number of image slices. Although significant research has been performed for reconstructing 3D surfaces from 2D images [4], a limited number of methods has attempted this problem for 3D heart meshes, specifically from sparse 2D cine MR slices. In order to compensate for the slice misalignment, some methods used slice-to-volume registration [5,12], while a geometry-based approach for smoothing epicardial shape using iterative application of in-plane translations was developed in [9]. One of the most common approaches for misalignment correction in heart slices is slice-to-slice registration. Villard et al. [11] optimised the similarity between intensities at the intersecting line between cine MR slices to achieve optimal consistency between cross sectional intensity profiles. Some methods preferred to use a fixed slice, usually long axis (LAX) slice, as a reference plane to align other slices [7].

Although there exist few research works that addressed the problem of misalignment correction from cine MR slices, most of them relied upon the consistency between sparse heart slices in 3D space and hence, could reliably account for only in-plane slice misalignments. In this regard, the objective of the current work is develop a method to optimally correct for slice misalignment, both in-plane and out-of-plane, for accurate patient-specific modelling of the heart surfaces. After extracting the heart contours, namely, LV and right ventricle (RV) endocardium and LV epicardium, we have first performed initial misalignment corrections, first using the image intensities and next using the heart contours, when both LAX and short axis (SAX) cine MR slices are available. In order to perform the optimised misalignment correction, we have first optimally fitted a statistical shape model (SSM) on the sparse heart contours in 3D space and then optimally aligned the heart slices on the SSM, accounting for both in-plane and out-of-plane movements. The performance of the proposed approach is evaluated by reconstructing biventricular heart surfaces from the optimally aligned cine MR slices using a surface generating algorithm [10]. On a cohort of 20 subjects randomly selected from the UK Biobank study [8], the proposed approach has been able to reduce the average misalignment from 1.14 ± 0.23 mm to 0.72 ± 0.11 mm, over the final reconstructed surface meshes.

2   Preprocessing and Initial Misalignment Corrections

2.1   Preprocessing

The MR slices for the standard cine MRI protocol are horizontal long axis (HLA) or 4 chamber view (Fig. 1, column 1), vertical long axis (VLA) or 2 chamber view (Fig. 1, column 2), left ventricular outflow tract (LVOT) or 3 chamber view (Fig. 1, column 3), and the stack of SAX views (Fig. 1, columns 4–5). In order to identify the epicardial contours and LV and RV endocardial contours from all cine MR slices, we have automatically segmented the heart slices in 4 classes, namely, LV cavity, LV myocardium, RV cavity, and background. We have employed the deep learning based method proposed by Bai et al. [2] for this step, since it has been shown to segment heart structures from cine MR slices with human-level accuracy. Since at this time we do not have any pre-trained network for the VLA and LVOT slices, we have manually contoured the LV and RV endocardial and LV epicardial surfaces on the LVOT slices and LV endocardial and epicardial surfaces on the VLA slices. Following segmentation, the heart contours are identified as the boundaries of segmented regions. The septal wall is identified as the intersection between LV epicardium and RV endocardium. An example of the input cine LAX and SAX slices with misalignment artifact is presented in the first row of Fig. 1. In Fig. 1, the HLA, VLA, and LVOT slices in first three columns have been annotated with red, green, and cyan colours, respectively; while the SAX slices in last two columns are annotated with yellow. The relative position of each slice is provided on all of the other slices. The HLA, VLA, and LVOT slices are presented with red, green, and cyan colours, respectively, on both SAX slices, while both SAX slices are presented as yellow lines over all three LAX slices.

2.2   Intensity and Contours Based Misalignment Corrections

In case both LAX and SAX cine MR slices are available, we have performed an initial misalignment correction based on intensity profiles [11]. The objective of this step is to provide spatial consistency to the slices in 3D space by comparing the intensity profiles at the line formed by the intersection between two slices. We have minimised the global motion discrepancy, defined as the sum of dissimilarity measures (here, normalised cross correlation) between a pair of intersecting slices, by estimating the optimal rigid transformation for each slice. We have adopted an iterative minimisation of partial terms, where, in each iteration, the rigid transformation for each slice is sequentially optimised, keeping others fixed.

Using the extracted heart contours, an iterative misalignment correction algorithm is next applied, in order to achieve in-plane alignment of LAX and SAX contours. The Euclidean distance between LAX and SAX contours is used as the dissimilarity measure in global motion discrepancy. A qualitative result of contours based misalignment correction on 2D cine MR slices is presented in second row of Fig. 1, while the sparse 3D representation of the same heart contours after misalignment correction is shown in first column of Fig. 2.

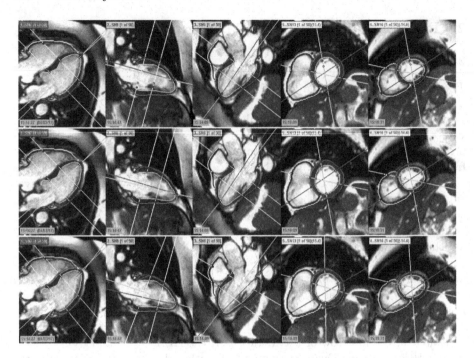

Fig. 1. From left to right: HLA view, VLA view, LVOT view, and two SAX slices. From top to bottom: the misaligned cine MR slices; after contours based misalignment correction; and after SSM based misalignment correction. The green, blue, red, yellow, and pink contours respectively denote the LV and RV endocardial contours, LV and RV epicardial contours, and the septal wall on 2D cine MR slices. The coloured dots represent the position of corresponding contours on the intersecting slice.

3   Proposed Misalignment Correction Using Statistical Shape Model

Our proposed approach of misalignment correction consists of two steps: fitting a statistical shape model (SSM) to the sparse heart contours in 3D space and then optimally aligning the heart slices to the fitted SSM.

3.1   Fitting the Statistical Shape Model

The optimum misalignment correction based only on the intensity profiles or the intersections between sparse LAX and SAX slices is an ill-posed problem and hence, can produce undesirable solutions if out-of-plane transformations are considered. Hence, the global motion discrepancy for both intensity and contours based misalignment corrections is optimised over only the in-plane rigid transformations. Now, although it can significantly reduce the misalignments, one major cause of misalignment, the motion artifacts between image acquisitions, cause out-of-plane movements. Here, the in-plane rigid transformation is defined as the

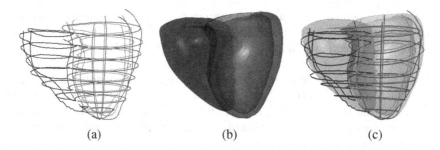

(a)   (b)   (c)

Fig. 2. (a) Sparse 3D contours after contours based misalignment correction; (b) the
fitted SSM; and (c) the optimal misalignment corrected heart contours on fitted SSM.

rigid transformation (translation and rotation) of the cine MR slice along the
original acquisition plane; while the out-of-plane rigid transformation can consider rigid transformation outside the acquisition plane for optimal correction of
the misalignment artifacts.

In order to estimate the optimal rigid transformations of the cine slices in
3D space for accurate surface reconstruction, we have employed a statistical
shape model of human heart ventricles by Bai et al. [1,6]. The SSM was created
by registering 1093 hearts, generated by high-resolution 3D cine MR images,
to the template space using rigid registration, thus removing the position and
orientation differences, and then by applying the principal component analysis
on the surface meshes of LV, RV, and both ventricles. The first 100 principal
components, accounting for 99.9% of the total shape variation in the dataset,
and the mean SSM are publicly available at http://wp.doc.ic.ac.uk/wbai/data/.

For our proposed misalignment correction, we have first extracted the LV and
RV endocardial and LV epicardial surfaces from the shapes of LV, RV, and both
ventricles of the model. The surface meshes from the model are then fitted to the
sparse heart contours in 3D space by the optimal estimation of the 100 principal
components of the SSM, followed by rigid transformation. The optimally fitted
SSM over the sparse heart contours of Fig. 2(a) is presented in Fig. 2(b). The
green, blue, and red surfaces in Fig. 2(b) respectively denote the LV and RV
endocardial and LV epicardial surfaces in the fitted SSM.

3.2   Misalignment Correction Using the SSM

Since the SSM was generated from high-resolution 3D cine MR images, there
is no misalignment artifact present in this model. Hence, the main objective of
obtaining the fitted SSM is to provide a reference for the alignment of heart
contours from the cine slices. In order to optimally remove the misalignment
artifacts, here we have applied both in-plane and out-of-plane 3D rigid transformations, i.e. translation t_i and rotation R_i for each slice i, to optimally align the
heart contours on cine MR slices to the fitted SSM. Along with accounting for
out-of-plane misalignments, another major advantage in the proposed SSM based
misalignment correction approach is that it can be applied even in the absence of

any LAX slices, which would make both intensity based and contours based mis-alignment corrections inoperable. An example of the performance of proposed misalignment correction using the fitted SSM on 2D cine MR slices is presented in the third row of Fig. 1, while the same misalignment corrected heart contours on fitted SSM in 3D space is presented in Fig. 2(c). In this regard, it should be noted that the sole objective of the SSM is for guiding a robust misalignment correction of the heart contours. Hence, after the misalignment correction step, the fitted shape model is discarded, and only the aligned heart contours are used for 3D heart surface reconstruction.

After the final alignment of heart contours, the 3D surface meshes are gener-ated, for both qualitative and quantitative evaluations, from each of the LV and RV endocardiums and the epicardium. We have applied the method proposed by Villard et al. [10], that can generate 3D surface mesh from sparse, hetero-geneous, non-parallel, cross-sectional, non-coincidental contours. The algorithm applies a composition of smooth deformations aiming at maximising fitting to image contours, thus ensuring a good matching of the reconstructed 3D biven-tricular surface to the input data, as well as optimal interpolation characteristics.

4 Experimental Analysis

In order to evaluate the reconstruction performance of the proposed misalign-ment correction approach, we have randomly selected a cohort of 20 subjects from the UK Biobank study [8]. The dataset includes an HLA, a VLA, an LVOT, and a stack of SAX cine MR slices for each subject. The number of SAX cardiac cine MR slices varies between 8 to 13, while the number of acquired frames over a cardiac cycle is 50. The images have in-plane resolution of 1.8269 mm and the distance between SAX slices is 10 mm. For each subject, the frame number 1 is selected as the end-diastolic frame in our analysis.

The qualitative performance of the 3D biventricular surface mesh reconstruc-tion over misaligned slices and after intensity based, contours based, and SSM based misalignment corrections are respectively presented in columns of Fig. 3. Over each reconstructed surface, the distances between sparse heart contours and the reconstructed surface are shown using a continuous color scale varying between 0–5 mm. The quantitative analysis is presented in Fig. 4 using box-plots, by computing the distances of LV endocardial, RV endocardial, and epicardial contours from the reconstructed LV endocardium, RV endocardium, and epi-cardium, respectively.

In our dataset of 20 subjects, the initial misalignment between the heart contours, measured with respect to the reconstructed heart surfaces, was average 0.98 ± 0.30 mm for the LV endocardial contours, average 1.14 ± 0.22 mm for the RV endocardial contours, and average 1.22 ± 0.26 mm for the epicardial contours. From the quantitative results presented in Fig. 4, it can be observed that the intensity based misalignment correction method has been able to significantly reduce the misalignment of heart contours to average 0.72 ± 0.14 mm for LV endocardial contours, average 1.04 ± 0.17 mm for RV endocardial contours, and

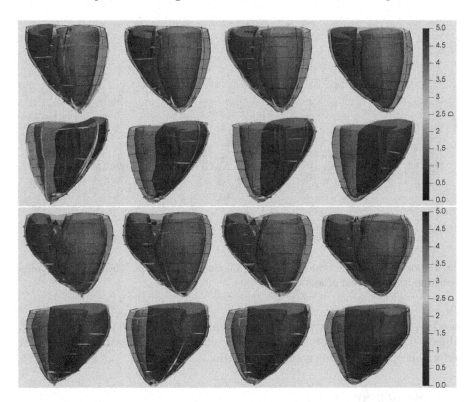

Fig. 3. From left to right: the reconstructed 3D heart surfaces from misaligned heart slices, after intensity based misalignment correction, after contours based misalignment correction, and after SSM based misalignment correction. Top: the anterior view; bottom: the posterior view. Over each reconstructed surface, the distances between heart contours and the surface are shown using a continuous color scale varying between 0–5 mm (defined at the right margin). The green, blue, and red surfaces respectively denote the LV and RV endocardium and the epicardium. (Color figure online)

average 1.04 ± 0.19 mm for epicardial contours. In most cases, the results are also statistically significant with respect to the parametric paired-t test, where 0.05 is the desired level of significance. On the other hand, the contours based misalignment correction method has been able to reduce the misalignment error to average 0.64 ± 0.11 mm for LV endocardial contours, average 0.82 ± 0.12 mm for RV endocardial contours, and average 0.94 ± 0.17 mm for epicardial contours.

From the quantitative results presented in Fig. 4 and in Fig. 3, it is clearly observed that the proposed SSM based misalignment correction method has been able to significantly reduce the misalignment artifacts from cine MR slices for all subjects below sub-voxel accuracy. After the proposed misalignment correction, the average misalignment errors for LV endocardial, RV endocardial, and epicardial contours are measured as 0.52 ± 0.11 mm, 0.66 ± 0.06 mm, and 0.89 ± 0.19 mm, respectively, for 20 subjects. In few subjects where the contours

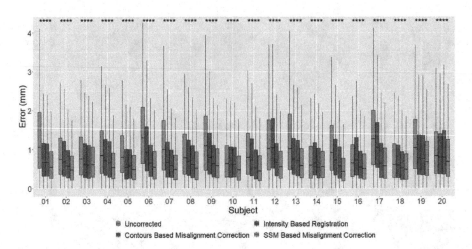

Fig. 4. Box-plots for the evaluation of misalignment correction. The '****' symbol is used to indicate p-value of less than 0.0001.

based misalignment correction is already achieving the optimal performance (e.g. subjects 03 and 14), the SSM based misalignment correction attains the similar performance, but with less standard deviation.

5   Conclusion

The accurate patient-specific 3D modelling of heart meshes from 2D cine MR slices is important for several reasons. On one hand, it is useful for visualising heart structures and functions, including pathologies, on 3D heart shape for improved diagnosis and prediction of patient outcome. On the other hand, it is necessary in modelling and simulation-based applications, such as in-silico clinical trials. In this regard, our main contribution in this work is to develop a completely automated statistical shape model based approach for misalignment correction of 2D cine MR slices for accurate modelling of the biventricular heart meshes. Not only the proposed method can account for both in-plane and out-of-plane misalignments, it is applicable in the presence of only SAX slices as well as on both LAX and SAX slices. The performance of the proposed approach has been evaluated on a cohort of 20 subjects, randomly selected from the UK Biobank study. Our SSM based method has been able to reduce the misalignment between heart slices from average 1.14±0.23 mm to average 0.72±0.11 mm, and has presented statistically significant misalignment correction performance as compared to the intensity based as well as contours based misalignment correction approaches.

Acknowledgments. This research has been conducted using the UK Biobank Resource under Application Number '40161'. The authors express no conflict of interest. The work was supported by the British Heart Foundation Project under Grant

HSR01230 and the CompBioMed 2 Centre of Excellence in Computational Biomedicine (European Commission Horizon 2020 research and innovation programme, grant agreement No. 823712). The authors acknowledge the use of services and facilities of the Institute of Biomedical Engineering, University of Oxford, UK and the Oxford Acute Vascular Imaging Centre, UK.

References

1. Bai, W., Shi, W., de Marvao, A., Dawes, T.J.W., et al.: A bi-ventricular cardiac atlas built from 1000+ high resolution MR images of healthy subjects and an analysis of shape and motion. Med. Image Anal. **26**(1), 133–145 (2015)
2. Bai, W., Sinclair, M., Tarroni, G., Oktay, O., et al.: Automated cardiovascular magnetic resonance image analysis with fully convolutional networks. J. Cardiovasc. Magn. Reson. **20**(65), 1–12 (2018)
3. Dall'Armellina, E., Karamitsos, T.D., Neubauer, S., Choudhury, R.P.: CMR for characterization of the myocardium in acute coronary syndromes. Nat. Rev. Cardiol. **7**(11), 624–636 (2010)
4. Khatamian, A., Arabnia, H.R.: Survey on 3D surface reconstruction. J. Inf. Process. Syst. **12**(3), 338–357 (2016)
5. Lötjönen, J., Pollari, M., Kivistö, S., Lauerma, K.: Correction of movement artifacts from 4-D cardiac short- and long-axis MR data. In: Barillot, C., Haynor, D.R., Hellier, P. (eds.) MICCAI 2004, Part II. LNCS, vol. 3217, pp. 405–412. Springer, Heidelberg (2004). https://doi.org/10.1007/978-3-540-30136-3_50
6. de Marvao, A., Dawes, T.J.W., Shi, W., Minas, C., et al.: Population-based studies of myocardial hypertrophy: high resolution cardiovascular magnetic resonance atlases improve statistical power. J. Cardiovasc. Magn. Reson. **16**(16), 1–10 (2014)
7. McLeish, K., Hill, D.L.G., Atkinson, D., Blackall, J.M., Razavi, R.: A study of the motion and deformation of the heart due to respiration. IEEE Trans. Med. Imaging **21**(9), 1142–1150 (2002)
8. Petersen, S.E., Matthews, P.M., Bamberg, F., Bluemke, D.A., et al.: Imaging in population science: cardiovascular magnetic resonance in 100,000 participants of UK Biobank-rationale, challenges and approaches. J. Cardiovasc. Magn. Reson. **15**(1), 46 (2013)
9. Su, Y., Tan, M., Lim, C., Teo, S., et al.: Automatic correction of motion artifacts in 4D left ventricle model reconstructed from MRI. In: Computing in Cardiology, pp. 705–708 (2014)
10. Villard, B., Grau, V., Zacur, E.: Surface mesh reconstruction from cardiac MRI contours. J. Imaging **4**(1), 1–21 (2018)
11. Villard, B., Zacur, E., Dall'Armellina, E., Grau, V.: Correction of slice misalignment in multi-breath-hold cardiac MRI scans. In: Mansi, T., McLeod, K., Pop, M., Rhode, K., Sermesant, M., Young, A. (eds.) STACOM 2016. LNCS, vol. 10124, pp. 30–38. Springer, Cham (2017). https://doi.org/10.1007/978-3-319-52718-5_4
12. Zakkaroff, C., Radjenovic, A., Greenwood, J., Magee, D.: Stack alignment transform for misalignment correction in cardiac MR cine series. University of Leeds, Technical report (2012)

Slice-to-Volume Registration Enables Automated Pancreas MRI Quantification in UK Biobank

Alexandre Triay Bagur[1,2]([⊠]) [iD], Paul Aljabar[2] [iD], Zobair Arya[2],
John McGonigle[2], Sir Michael Brady[2,3] [iD], and Daniel Bulte[1] [iD]

[1] Department of Engineering Science, University of Oxford, Oxford, UK
alexandre.triaybagur@eng.ox.ac.uk
[2] Perspectum Ltd., Oxford, UK
[3] Department of Oncology, University of Oxford, Oxford, UK

Abstract. Multiparametric MRI of the pancreas can potentially benefit from the fusion of multiple acquisitions. However, its small, irregular structure often results in poor organ alignment between acquisitions, potentially leading to inaccurate quantification. Recent studies using UK Biobank data have proposed using pancreas segmentation from a 3D volumetric scan to extract a region of interest in 2D quantitative maps. A limitation of these studies is that potential misalignment between the volumetric and single-slice scans has not been considered. In this paper, we report a slice-to-volume registration (SVR) method with multimodal similarity that aligns the UK Biobank pancreatic 3D acquisitions with the 2D acquisitions, leading to more accurate downstream quantification of an individual's pancreas T1. We validate the SVR method on a challenging UK Biobank subset of N = 50, using both direct and indirect performance metrics.

Keywords: Pancreas · Multiparametric · Magnetic resonance imaging · Volume · T1 · Slice-to-volume registration · UK biobank

1 Introduction

The incidence of chronic pancreas disease, particularly non-alcoholic fatty pancreas disease (NAFPD), is rising rapidly, reflecting the increasing worldwide prevalence of metabolic disease and obesity [1,2]. Fat infiltration due to obesity in the pancreas triggers an inflammatory response that can lead to chronic pancreatitis and ultimately to pancreatic cancer. While multiparametric magnetic resonance imaging (MRI) of the liver has become the gold standard tool for early detection, diagnosis and monitoring of chronic disease, this technique has been understudied in the pancreas. Multiparametric MRI provides the advantages of soft tissue contrast, lack of radiation, high accuracy and high precision, even in the most obese patients. This has resulted in the development of liver imaging biomarkers such as corrected T1 (cT1) [3] and proton density fat fraction

© Springer Nature Switzerland AG 2021
B. W. Papież et al. (Eds.): MIUA 2021, LNCS 12722, pp. 210–223, 2021.
https://doi.org/10.1007/978-3-030-80432-9_17

(PDFF) [4]. This motivates the development of quantitative imaging biomarkers for assessing the pancreas, which is increasingly important as it is fundamentally implicated in obesity-related conditions such as type 2 diabetes (T2D).

Volume [5], morphology [6], T1 [7] or PDFF [8] have been proposed as multi-parametric MRI biomarkers for the pancreas. The UK Biobank is a rich resource, currently acquiring dedicated pancreas volumetric and quantitative images from 100,000+ volunteers, alongside other non-imaging data [9], which enables the assessment of the aforementioned biomarkers. UK Biobank imaging data has been used in the past for establishing *inter alia* reference ranges [10], validation of novel processing methods [11], as well as automated processing and quality control [12].

Since quantitative parametric maps are primarily 2D, recent methods for analysing UK Biobank pancreas MRI data [13,14] have proposed using the 3D segmentation from a volumetric acquisition to extract a region of interest (ROI) from the 2D quantitative parametric maps. The method proposed in these studies uses the pancreas-specific volumetric scan and the 2D quantitative map derived from the multiecho gradient-recalled echo scan. The DICOM header coordinates are used to intersect the 3D segmentation from the volumetric scan with the 2D quantitative map in the same coordinate space. This approach is appealing over the alternative of segmenting the pancreas on the 2D quantitative maps directly, for several reasons: (1) only one segmentation method may be used for all quantitative map types, as opposed to training (and validating) map-specific methods; (2) fewer annotated subjects are needed in order to obtain a robust segmentation method: a model trained on 2D maps needs to have been exposed to many possible orientations and positions of slices through the pancreas observed in practice, and (3) the segmentation on the volumetric scan may be used for pancreas volumetry and morphometry, which provide insight into the pathophysiology of T2D.

However, the volumetric and slice acquisitions are acquired in separate breath-holds (usually 4 to 5 min apart in the case of T1), which may lead to misalignment between the scans due to breathing motion or scan re-positioning. Larger, more regular structures than the pancreas, such as the liver, may not be affected by misalignment to the same extent, and using the DICOM header position may be a reasonable approach. However, the pancreas is a small, irregularly-shaped organ that can vary significantly in appearance with changes in the viewing slice. Even slight motions cause the pancreas to move outside reference planes. In this context, using only the DICOM header information for alignment may be insufficient (Fig. 1, top); we will refer to this method as *No Registration* in this paper.

No considerations of inter-scan misalignment have previously been made for pancreas MRI analysis in the UK Biobank. We conjecture that accurate alignment of the volumetric and the slice acquisitions may improve quantitative pancreas analysis downstream, for example T1 and PDFF quantification. In this study, we propose using Slice-to-Volume Registration prior to ROI extraction and quantitative reporting. Slice-to-Volume Registration (SVR) aims to align data from more than one scan, one consisting of a planar acquisition (slice), and the other consisting of a 3D volume. SVR has been used in multiple applications, from real-time surgical navigation to volume reconstruction, with rigid registration and iconic matching criteria being the most commonly used SVR strategy [15]. Recent advances have shown the potential of deep learning within SVR, for instance used as feature extraction mechanisms towards robust matching criteria, or to drive SVR optimisation as a whole [16].

In this paper, we show an example implementation of automated SVR with a multimodal similarity criterion. We apply the method to pancreatic T1 scans, and demonstrate more accurate T1 quantification in the UK Biobank imaging substudy.

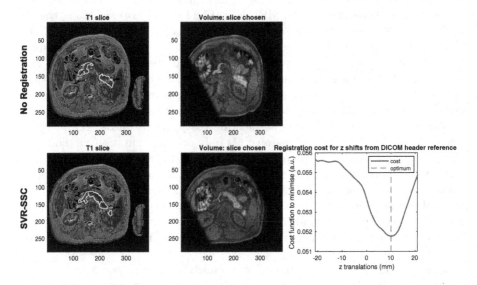

Fig. 1. An example of the need for alignment. (Top) Using No Registration (i.e. only using the DICOM header position). The segmentation resampled from the pancreas-specific volume (PSV) to the T1 slice is not aligned with the pancreas boundary, so T1 quantification is inaccurate. (Bottom) Using SVR-SSC. The output segmentation is aligned with the pancreas on the T1 slice. The method's optimal z translation was +10 mm from the DICOM header position. The cost evaluation mask used to evaluate image similarity for all PSV slices is shown as the red contour. The pancreas subcomponents head (blue), body (green), and tail (yellow), are shown on the PSV candidates. (Color figure online)

2   Materials and Methods

2.1   UK Biobank Data

In this work, we present an exemplary implementation of SVR for UK Biobank where the 3D volume used was the pancreas-specific volumetric scan (named

"Pancreas fat - DICOM", Data-Field 20202 on the UK Biobank Showcase[1]), that we refer to in this work as PSV. As the 2D image, we used the T1 map derived from the Shortened Modified Look-Locker Inversion recovery (ShMoLLI) scan (named "Pancreas Images - ShMoLLI", Data-Field 20259 on the UK Biobank Showcase). The T1 map was computed using a proprietary algorithm from Perspectum Ltd, previously used to compute T1 for UK Biobank ShMoLLI images of the liver [3].

Both scans were acquired using a Siemens Aera 1.5T (Siemens Healthineers AG, Erlangen, Germany). The PSV imaging data was acquired using the FLASH-3D acquisition, echo time (TE)/repetition time (TR) = 1.15/3.11 ms, voxel size = $1.1875 \times 1.1875 \times 1.6$ mm, with 10° flip angle and fat suppression. The PSV scans were resampled to 2 mm isotropic using linear interpolation. The ShMoLLI imaging data was collected using the same parameters than for the liver ShMoLLI, with voxel size = $1.146 \times 1.146 \times 8$ mm, TE/TR = 1.93/480.6 ms, 35° flip angle [9], and often had oblique orientation to better capture the pancreas. Only data from the first imaging visit (Instance 2) were used.

3D pancreas segmentations were predicted on the PSVs using the implementation of U-Net described in [17]. 14,439 subjects with both T1 and PSV scans were processed using (1) No Registration and (2) the proposed SVR method.

2.2   Slice-to-Volume Registration Method

We have observed that in practice translations in Z are the most prominent source of misalignment in the data. The SVR method we used is based on initial affine alignment in the XY plane, and posterior exhaustive search along the Z direction, above and below the DICOM header reference position ($z = 0$). At each z, the method evaluates image similarity between the resampled 3D volume (moving image) at z and the quantitative 2D slice (fixed image). The method then chooses the z that gives the highest image similarity between images, as illustrated in Fig. 2.

Resample Volume into 'Candidate' Slices. We evaluated similarity over a Z range that was computed from the volume total height (mm) as follows: (voxel Z resolution in mm × number of voxels in Z)/4. The Z range used was the same for all subjects. We incremented Z in steps of 1 mm. In this work, using the PSV as our volume rendered the SVR method's search range over Z = (2 mm × 42)/4 = $[-21, 21]$ mm.

At each z, the 3D PSV was resampled by considering the slice acquisition profile, since the T1 slice thickness is greater than the PSV slice thickness (8 mm and 2 mm, respectively). Resampled PSV slices across the T1 8 mm slice profile were weighted by a gaussian function (standard deviation of 1.4 mm) before being merged together. The corresponding 3D segmentation was resampled using nearest neighbour interpolation.

---

[1] https://biobank.ndph.ox.ac.uk/showcase/browse.cgi.

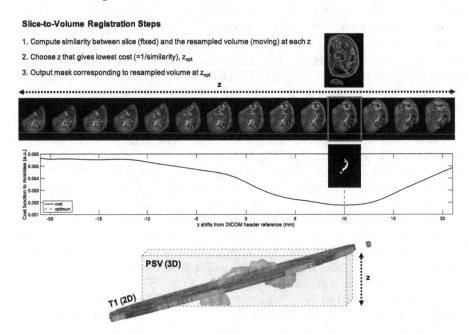

Fig. 2. (Top) Illustration of the Slice-to-Volume Registration procedure. The resampled pancreas-specific volume (PSV) candidates generated using resampling along Z are shown. In yellow is the chosen PSV candidate slice that gives highest image similarity with the quantitative slice (lowest registration cost). The corresponding resampled segmentation is chosen as the output segmentation for the quantitative slice. (Bottom) Alignment of the 3D PSV with the T1 slice and intersection of the segmentation from the 3D volume scan. The oblique T1 slice may extend beyond the PSV bounding box, which may cause missing data in the resampled candidates. (Color figure online)

Initial Within-Plane Alignment. An initial within plane registration step was carried out to align the the body contour in the T1 slice and PSV data. This allows to better compare candidate slices in the subsequent Z alignment step. The initial affine registration in XY was performed using the 'multimodal' configuration in Matlab R2019b *imregister* function, which uses Mattes Mutual Information as similarity metric. For this initial XY alignment, we used the resampled PSV at $z = 0$ mm and the T1 slice as the moving and fixed images, respectively. The resulting affine transformation in XY was subsequently applied to each resampled PSV 'candidate' along Z.

Evaluate Similarity over Z Range. For the exhaustive search along Z, image similarity at each z was evaluated only within a predefined cost evaluation mask, common to all 'candidates' (resampled PSV images). The cost evaluation mask was computed as the intersection of the body masks from all candidates, which were computed using a simple thresholding operation (>10). We evaluated similarity over the computed Z range, and the inverse of similarity was used as the cost function for alignment (see Fig. 2).

We explored 2 different similarity metrics: normalised mutual information (NMI) and Self-Similarity Context descriptors (SSC) based similarity from Heinrich et al. [18]. The SSC-based similarity computes the inverse of the squared differences between the SSC descriptors of image 1 and image 2 within the cost evaluation mask. This rendered two methods that we will refer to as *SVR-NMI* and *SVR-SSC* in the text.

Following the above steps, the z position for which the resampled 3D PSV gave the highest similarity (i.e., lowest cost) was selected, and the corresponding resampled 3D segmentation on the 2D slice was calculated. This output segmentation may be used for subsequent pancreas quantification on the 2D slice, for instance extracting global descriptive statistics (such as mean or median) or local measurements.

Figure 1 shows an example case after our method has been applied, including comparisons between using No Registration and using the SVR-SSC approach. Note the entire body of the pancreas (green) is missing from the No Registration candidate alignment, while it is present on the T1 map.

2.3 SVR Implementation and Inference at Scale

The implementation of the SVR method is available on GitHub[2]. The Matlab R2019b Compiler tool was used to package our SVR implementation into a Matlab application. The application was moved into a Docker image containing the compatible version of Matlab Runtime[3]. This enabled us to run the SVR method using Docker containers, in parallel and at scale on Amazon Web Services (AWS) EC2 instances, in order to process the large UK Biobank data set.

2.4 Automated Quality Control

For T1 quantification results in this paper, we opted for the SVR-SSC method due to its higher robustness in validation (see Sect. 3.2). The 14,439 subjects were processed by a quality control (QC) pipeline prior to T1 quantification. We first excluded those subjects where the resulting 2D pancreas mask was empty in either of the 2 methods. This was mostly due to the initial automated segmentation prediction being empty, for instance in the presence of imaging artefacts. For the remaining $N = 13,845$ subjects, we quality controlled the data by excluding those that met the following QC exclusion criteria: output 2D mask size <1000 pixels using the No Registration method ($N = 3,014$ excluded), percent overlap between the PSV at $z = 0$ and the T1 slice <50% ($N = 2,150$ excluded), or non-axial T1 acquisitions (see examples in Fig. 3) ($N = 106$ excluded). We also

---

[2] https://github.com/alexbagur/slice2vol.
[3] https://hub.docker.com/r/flywheel/matlab-mcr.

excluded those subjects where the SVR-SSC method gave an output z >10 mm (N = 317), in order to ensure there was enough overlap between the volume and the slice scans to compute image similarity. Figure 4 shows the histogram of output optimum z displacements for all subjects prior to this QC step. N = 8,829 subjects remained for analysis. Note that a given subject may be excluded using more than 1 of these QC exclusion criteria.

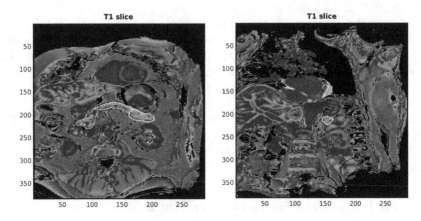

Fig. 3. Example of T1 slices from 2 different subjects where the primary view was coronal. These subjects were excluded from analysis.

2.5   SVR Validation

We performed direct validation of SVR using the alignment error from manually annotated reference positions. We also performed indirect validation by assessing quality of the output 2D segmentation, compared to manual delineations performed on the 2D slice directly.

Direct Validation Using Alignment Error. For alignment error, we selected the N = 50 most outlying cases from our T1 quantification experiment (see Sect. 3.1). We then manually annotated the reference z position that corresponded to the best alignment between the resampled PSV and T1 slice. Note the reference z is expressed relative to the DICOM header position. We then computed the alignment error of each method compared to the manually annotated z position. We did not consider translations in X or Y for this experiment.

Indirect Validation Using Segmentations. We had an available dataset of N = 157 from UK Biobank of healthy and diabetic subjects where manual delineations of pancreas had been performed on the T1 maps directly. These data had been quality-checked also (see Sect. 2.4). We evaluated output segmentation quality via overlap and surface distance measures, namely Dice

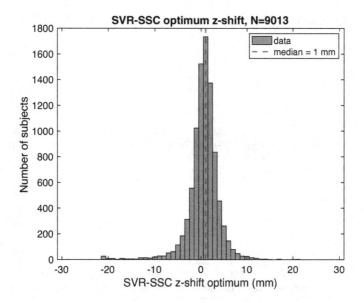

Fig. 4. Histogram of output optimum z-displacement by the SVR-SSC algorithm, showing a median displacement of $+1$ mm.

Similarity Coefficient (DSC) overlap and 95th percentile Hausdorff Distance (HD), respectively. The output 2D pancreas masks of all methods (No Registration, SVR-NMI, SVR-SSC) were compared to the reference manual delineations.

3   Results

3.1   T1 Quantification: No Registration vs SVR-SSC

We compared the T1 quantification results of the SVR-SSC method vs No Registration. Figure 5 shows the Bland-Altman density plot for $N = 8,829$ subjects, where T1 was quantified for each method using the median of the output segmentation. While the bias was small at 1.4 ms, the observed variability between No Registration T1 and SVR-SSC T1 of 53.3 ms renders the two approaches not equivalent for pancreatic T1 quantification at the subject level. We further explored these differences in our validation experiments (see Sect. 3.2), in order to determine which method had performed best.

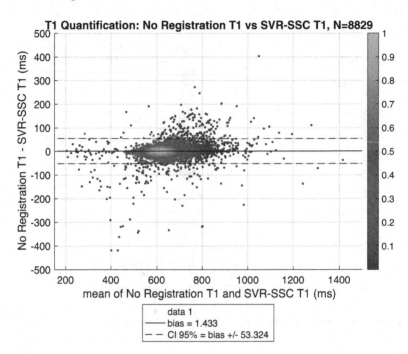

Fig. 5. Bland-Altman density plot showing T1 quantification using No Registration and the proposed SVR-SSC method. T1 is reported as the median of the output 2D pancreas segmentation for each method.

3.2 SVR Validation

Direct Validation Using Alignment Error. Figure 6 shows the alignment error in mm of each method in the selected $N = 50$ subset of the most outlying cases from our T1 quantification experiment (see Sect. 3.1). The median $\pm$ std for each method were: No Registration (8 ± 4.0) mm, SVR-NMI (3 ± 5.5) mm, SVR-SSC (3 ± 4.2) mm. The alignment error was substantially reduced using any of the two SVR implementations compared to No Registration. The differences between any of the two implementations vs No Registration were statistically significant (paired t-test, $p<1e-3$ for SVR-NMI, $p<1e-3$ for SVR-SSC). The differences between the two SVR methods, using 2 different similarity metrics, were not statistically significant ($p = 0.73$), though the SVR-SSC method appeared more robust to challenging examples.

Indirect Validation Using Segmentations. Figure 7 shows the segmentation quality of the three methods for the selected $N = 50$ subset of the most outlying cases from our T1 quantification experiment (see Sect. 3.1). We used DSC overlap and 95% HD metrics. The DSC median $\pm$ std for each method were: No Registration 0.799 ± 0.114, SVR-NMI 0.819 ± 0.118, SVR-SSC 0.822 ± 0.098. The differences between any of the two implementations vs No Registration were statistically significant (paired t-test, $p = 0.0069$ for SVR-NMI,

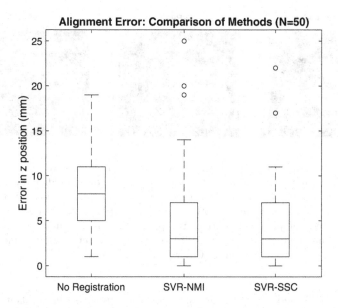

Fig. 6. Validation of SVR using alignment error in mm relative to the manually obtained reference positions. The alignment error was substantially reduced using any of the two SVR implementations compared to No Registration.

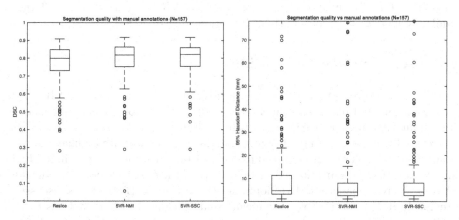

Fig. 7. Indirect validation of SVR was performed by assessing the quality of the output segmentations, in terms of Dice Similarity Coefficient (DSC) overlap (left) and 95th percentile Hausdorff surface distance (HD, right). The boxplots show summarised metrics for all three methods considered: No Registration, SVR-NMI and SVR-SSC.

p<1e-3 for SVR-SSC). The differences between the two SVR methods, using 2 different similarity metrics, were not statistically significant ($p = 0.1614$), though the SVR-SSC method appeared more robust to challenging examples, based on Fig. 8.

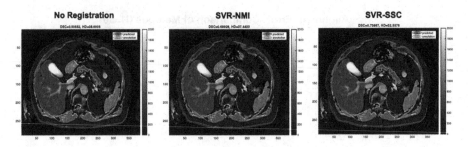

Fig. 8. Output segmentation quality compared to a manual annotation: a challenging example. The manual annotation is shown in yellow, while the predicted segmentation for each method is shown in red. Measured DSC for No Registration, SVR-NMI and SVR-SSC were 0.506, 0.689 and 0.739, respectively. Upon closer inspection, arguably the disconnected component is part of the pancreas, but the annotator might have excluded it from the quantifiable mask. T1 colormap was set to gray to increase visibility of the segmentation contours. (Color figure online)

The 95% HD median ± std for each method were: No Registration (4.123 ± 11.875) mm, SVR-NMI (3.606 ± 12.026) mm, SVR-SSC (3.606 ± 11.154) mm. The differences between any of the two implementations vs No Registration were statistically significant (paired t-test, $p = 0.0122$ for SVR-NMI, $p = 0.0054$ for SVR-SSC). The differences between the two SVR methods, using 2 different similarity metrics, were not statistically significant ($p = 0.83$).

4   Discussion and Conclusions

In this work, we report a method for Slice-to-Volume Registration (SVR) that enables automated pancreas segmentation and accurate downstream quantification in multiparametric MRI protocols like UK Biobank. We showed that SVR improved T1 segmentation quality (evaluated using overlap with manual annotations (Sect. 3.2)) as well as improved pancreas T1 quantification for an individual in UK Biobank. To our knowledge, this study is the first to report on utilising SVR for deriving pancreas MRI biomarkers from the UK Biobank. As discussed previously in [13], such a pipeline means that a segmentation model has to be built only for the 3D volumetric scan, which removes the need for generating new annotated data sets in order to segment each quantitative slice type.

In this study, we chose to use the 3D PSV scan for initial segmentation, and the 2D T1 map as our target quantitative slice to segment. However, the method is compatible with other 3D and 2D image types in UK Biobank, for instance the stitched 3D Whole-Body scans (named "Dixon technique for internal fat - DICOM", Data-Field 20201 on the UK Biobank Showcase) or the 2D quantitative PDFF maps derived from the multiecho gradient-recalled echo scan (Data-Field 20260). The choice of a feature-based similarity metric originally proposed in the context of multimodal image registration, the Self-Similarity Context descriptors (SSC) [18], was shown to work robustly on non-quantitative

(3D PSV data) and quantitative (2D T1 data) scans, and may transfer well to other data types.

The limitation of choosing the dedicated pancreas volume scan as our moving image was its limited total Z coverage of 84 mm. In practice, inter-scan motion may exceed this range. Furthermore, obliquely placed T1 slices (acquired to traverse the pancreas longitudinally) extended beyond the PSV bounding box at the Z range extrema (see Fig. 2, bottom), causing the PSV candidates at those extrema to have missing data (see Fig. 2, top), which in turn raised a set of challenges we have sought to address.

First, the cost evaluation mask, computed as the intersection of the body masks from all slices, was small (see Fig. 1, red contour). In those cases, the cost evaluation mask does not optimally include all the potentially useful image features for registration. This could have led to more noisy image similarity measurements and have introduced spurious local minima into the cost function. The alternative of considering a different cost evaluation mask for each z independently could have led to unfair comparisons of cost when selecting the global minimum. Second, we initially had considered an independent affine XY transformation for each z. However, candidates with missing data at the Z range extrema misled the registration procedure by producing a more noisy cost function. Using the same affine XY transformation on all candidates, obtained from the resampled candidate at z = 0 (avoiding missing data), addressed this problem, but could have introduced error.

Future work will use the Whole-Body scans from UK Biobank as our moving images for SVR. This will extend the SVR Z range and could lead to more robust estimates of similarity, as well as improved alignment in XY. However, obtaining robust 3D pancreas segmentations from the Whole-Body scans is challenging, as discussed in [14], since they are lower resolution compared to the dedicated pancreas volumes used in this work.

The proposed SVR methodology did not consider local deformations or deformations through the image plane, which can be expected in the abdominal region when the images are taken during different breath-holds. Our method's performance using axial affine registration and rigid alignment in Z encourages more advanced deformable image registration approaches, for instance those focusing on the organ surroundings rather than a global whole-body cost mask.

The exclusion criteria that we used during QC caused that nearly 39% of cases were excluded from reporting, which could have biased the comparison between the methods. This was due in part by upstream method failure, for instance where the segmentation model produced empty predictions or masks with a low number of voxels. We observed this effect mainly on images that contained image artefacts, such as wrap-around. Furthermore, we expect that the future work described above will increase our method's throughput, notably when using volumes with higher coverage since they will lead to full pancreas segmentations as well as to an extended Z range for SVR optimisation.

The small bias at 1.4 ms when comparing T1 quantification differences between No Registration and SVR-SSC indicates that we may be able to use

No Registration for T1 quantification at the population level, for instance for median pancreatic T1 in the UK Biobank. However, for individual subjects, or even for comparisons between relatively small groups (such as type 2 diabetics in UK Biobank), the clinically significant variability between methods renders the No Registration approach insufficient.

Moreover, note we quantified T1 for each method as the median of the output 2D segmentation, a relatively robust metric to outlier values. However, researchers often make distinctions in quantification between pancreatic head, body and tail [19,20]. When comparing medians of pancreas subsegments, we expect that small misalignments will amplify quantification differences between methods. This will further increase the need for registration in order to obtain accurate pancreas quantification in UK Biobank for given individuals. We have recently developed the first fully automated method for pancreas subsegmentation into head, body and tail [21] that, combined with SVR, will enable accurate MRI biomarker quantification regionally.

References

1. Mathur, A., et al.: Nonalcoholic fatty pancreas disease. **Hpb9**(4), 312–318 (2007). https://doi.org/10.1080/13651820701504157
2. Tariq, H., Nayudu, S., Akella, S., Glandt, M., Chilimuri, S.: Non-alcoholic fatty pancreatic disease: a review of literature. Gastroenterol. Res. **9**(6), 87–91 (2016). http://www.gastrores.org/index.php/Gastrores/article/view/731
3. Mojtahed, A., et al.: Reference range of liver corrected T1 values in a population at low risk for fatty liver disease–a UK Biobank sub-study, with an appendix of interesting cases. Abdom. Radiol. **44**(1), 72–84 (2019). http://link.springer.com/10.1007/s00261-018-1701-2
4. Reeder, S.B., Hu, H.H., Sirlin, C.B.: Proton density fat-fraction: a standardized MR-based biomarker of tissue fat concentration. J. Magn. Reson. Imaging **36**(5), 1011–1014 (2012). https://doi.org/10.1002/jmri.23741
5. Saisho, Y., et al.: Pancreas volumes in humans from birth to age one hundred taking into account sex, obesity, and presence of type-2 diabetes. Clin. Anat. **20**(8), 933–942 (2007). https://onlinelibrary.wiley.com/doi/10.1002/ca.20543
6. Al-Mrabeh, A., Hollingsworth, K.G., Steven, S., Taylor, R.: Morphology of the pancreas in type 2 diabetes: effect of weight loss with or without normalisation of insulin secretory capacity. Diabetologia **59**(8), 1753–1759 (2016). https://doi.org/10.1007/s00125-016-3984-6
7. Tirkes, T., Lin, C., Fogel, E.L., Sherman, S.S., Wang, Q., Sandrasegaran, K.: T 1 mapping for diagnosis of mild chronic pancreatitis. J. Magn. Reson. Imaging **45**(4), 1171–1176 (2017). https://doi.org/10.1002/jmri.25428
8. Kühn, J.P., et al.: Pancreatic steatosis demonstrated at mr imaging in the general population: clinical relevance. Radiology **276**(1), 129–136 (2015). http://pubs.rsna.org/doi/10.1148/radiol.15140446
9. Littlejohns, T.J., et al.: The UK Biobank imaging enhancement of 100,000 participants: rationale, data collection, management and future directions. Nat. Commun. **11**(1), 2624 (2020). https://doi.org/10.1038/s41467-020-15948-9, www.nature.com/articles/s41467-020-15948-9

10. Wilman, H.R., et al.: Characterisation of liver fat in the UK Biobank cohort. PLoS ONE **12**(2), 1–14 (2017). http://dx.doi.org/10.1371/journal.pone.0172921

11. Hutton, C., Gyngell, M.L., Milanesi, M., Bagur, A., Brady, M.: Validation of a standardized MRI method for liver fat and T2* quantification. PLOS ONE **13**(9), e0204175 (2018). https://dx.plos.org/10.1371/journal.pone.0204175

12. Tarroni, G., et al.: Large-scale quality control of cardiac imaging in population studies: application to UK Biobank. Sci. Rep. **10**(1), 2408 (2020). http://www.nature.com/articles/s41598-020-58212-2

13. Basty, N., Liu, Y., Cule, M., Thomas, E.L., Bell, J.D., Whitcher, B.: Automated measurement of pancreatic fat and iron concentration using multi-echo and T1-Weighted MRI data. In: 2020 IEEE 17th International Symposium on Biomedical Imaging (ISBI), vol. 2020-April, pp. 345–348. IEEE (2020). https://ieeexplore.ieee.org/document/9098650/

14. Liu, Y., et al.: Genetic architecture of 11 abdominal organ traits derived from abdominal MRI using deep learning, pp. 1–66 (2020)

15. Ferrante, E., Paragios, N.: Slice-to-volume medical image registration: a survey. Med. Image Anal. **39**, 101–123 (2017). https://linkinghub.elsevier.com/retrieve/pii/S1361841517300701

16. Hou, B., et al.: Predicting slice-to-volume transformation in presence of arbitrary subject motion. Lecture Notes in Computer Science (including subseries Lecture Notes in Artificial Intelligence and Lecture Notes in Bioinformatics), vol. 10434. LNCS, pp. 296–304 (2017)

17. Bagur, A.T., Ridgway, G., McGonigle, J., Brady, S.M., Bulte, D.: Pancreas segmentation-derived biomarkers: volume and shape metrics in the UK Biobank imaging study. In: Papież, B.W., Namburete, A.I.L., Yaqub, M., Noble, J.A. (eds.) MIUA 2020. CCIS, vol. 1248, pp. 131–142. Springer, Cham (2020). https://doi.org/10.1007/978-3-030-52791-4_11

18. Heinrich, M.P., Jenkinson, M., Papież, B.W., Brady, S.M., Schnabel, J.A.: Towards realtime multimodal fusion for image-guided interventions using self-similarities. In: Mori, K., Sakuma, I., Sato, Y., Barillot, C., Navab, N. (eds.) MICCAI 2013. LNCS, vol. 8149, pp. 187–194. Springer, Heidelberg (2013). https://doi.org/10.1007/978-3-642-40811-3_24

19. Nadarajah, C., et al.: Association of pancreatic fat content with type II diabetes mellitus. Clin. Radiol. **75**(1), 51–56 (2020). https://doi.org/10.1016/j.crad.2019.05.027

20. Sakai, N.S., Taylor, S.A., Chouhan, M.D.: Obesity, metabolic disease and the pancreas-Quantitative imaging of pancreatic fat. Br. J. Radiol. **91**(1089), 20180267 (2018). https://www.birpublications.org/doi/10.1259/bjr.20180267

21. Bagur, A.T., Ridgway, G., Brady, M., Bulte, D.: (Abstract accepted for presentation) Automated pancreas parts segmentation by groupwise registration and minimal annotation enables regional assessment of disease. In: ISMRM Annual Meeting (2021)

Image Segmentation

Deep Learning-Based Landmark Localisation in the Liver for Couinaud Segmentation

Zobair Arya$^{(\boxtimes)}$, Ged Ridgway, Arun Jandor, and Paul Aljabar

Perspectum Ltd., Oxford, UK
zobair.arya@perspectum.com

Abstract. Couinaud segmenation, which divides the liver into functional regions, is the most widely used functional anatomy of the liver and is important for surgical planning and lesion monitoring. Couinaud segmentation can be a time-consuming task, thereby necessitating automated methods. In this study, we propose a deep learning approach for automatically defining Couinaud segments 2 to 8, based on a novel application of automatic landmark localisation. We utilise a heatmap regression CNN to predict landmark locations in the liver, which can subsequently be used to derive the planes that divide the liver into Couinaud segments. A novel postprocessing step for reducing false-positive peaks in heatmaps and/or aiding quality control is also presented. We apply our approach to non-contrast T1-weighted MRI data and compare the accuracy of the derived segments to those obtained directly from a semantic segmentation network. We show that the approach we propose can match and potentially outperform the direct segmentation approach, and thus can be a good alternative option for automatic Couinaud segmentation.

Keywords: Deep learning · Landmark localisation · Heatmap regression · Couinaud segmentation · Liver segmenation · CNN

1 Introduction

Couinaud segmentation is a system for dividing the liver into nine functional regions based on vasculature [1] (the regions are numbered 1 to 8, but region 4 can be further divided into 4a and 4b). It is the most commonly used functional anatomy of the liver due to its usefulness in surgical planning and lesion monitoring [2]. Couinaud segmentation in practice can be time-consuming, meaning there is a need for automated methods of obtaining Couinaud segments. Before the increase in the populariy of deep learning, semi-automated methods such as those by Huang et al. [3] and Oliveira et al. [6] were proposed, based on techniques such as region-growing and mixture modelling. The primary disadvantage of these methods is that manual intervention is mandatory. More recently, a deep learning-based method utilising a 2D semantic segmentation U-Net has

© Springer Nature Switzerland AG 2021
B. W. Papież et al. (Eds.): MIUA 2021, LNCS 12722, pp. 227–237, 2021.
https://doi.org/10.1007/978-3-030-80432-9_18

been reported for use with CT data by Tian et al. [9], where the network is trained directly on the Couinaud segmentation masks.

In practice, Couinaud segments obtained from manual Couinaud segmentation are only an approximation, with the vasculature used as rough guide [2]. One pragmatic approach involves the identification of certain landmarks in the liver, which are subsequently used to define 3D planes that divide the liver into Couinaud segments. Therefore, an alternative to direct segmentation is to instead train a deep learning-based localisation network to locate these landmarks, and use the predicted landmark locations along with the set of plane creation rules to derive the Couinaud segments. The main benefit of this is related to the fact that deep-learning models for medical imaging data are not perfect, mainly due to the limited size of training datasets and the relative homogeneity of training datasets compared to the heterogeneity of data encountered when the model is applied in routine practice. Consequently, when deploying these models in a tool for real-world use, it is likely that some model predictions will need editing by an operator. It is quicker and easier to manually correct inaccurate landmarks than to manually reassign voxel labels to different regions, especially if multiple regions have been incorrectly segmented.

In medical imaging tasks, one deep learning-based landmark localisation approach has been to use CNN's with fully connected layers to directly regress the landmark coordinates. This is the basis for the method reported by Zhang et al. [11]. Another approach is to instead use U-Net-like models to regress heatmaps with peaks on the target landmark locations. Examples of this have been reported by Payer et al. [8] and Yang et al. [10]. For the relatively small datasets in medical imaging, heatmap regression can be more suitable. This is due to the models being fully convolutional, which avoids the need for fully connected layers and reduces the number of trainable parameters.

In this study, we make use of deep learning-based heatmap regression to build a landmark localisation model for predicting landmarks in the liver. We derive Couinaud segments 2 to 8 from the predicted landmarks and compare them with the segments obtained from a direct segmentation model. An issue that can arise in heatmap regression is that multiple candidate locations for a landmark can appear in predicted heatmaps [8,10]. Therefore, we also present a novel liver-specific postprocessing technique that attempts to address this issue and can additionally be used as a quality-control aid.

To our knowledge, this study is the first to report on deep learning-based landmark localisation applied to the liver and the first to use deep learning-based landmark localisation for automated Couinaud segmentation.

2   Methodology

2.1   Dataset

The bulk of the dataset used in this study consisted of 103 subjects from the HepaT1ca liver cancer study [5], with subjects scanned on either a 1.5T or 3T Siemens scanner. To increase the variability and size of this dataset, we further

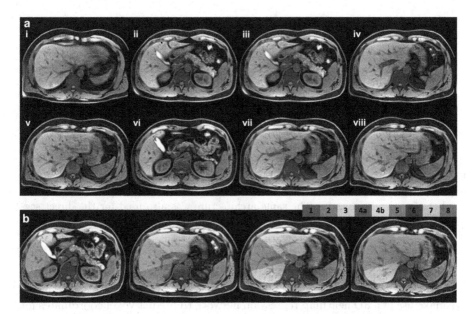

Fig. 1. a) The landmarks used for deriving Couinaud segments in this study: **i**) Inferior vena cava, inferior part (IVCi). **ii**) Inferior vena cava, superior part (IVCs). **iii**) Umbilical fissure (UF). **iv**) Right portal vein (RPV). **v**) Middle hepatic vein (MHV). **vi**) Gallbladder fossa (GBF). **vii**) Left portal vein (LPV). **viii**) Right hepatic vein (RHV). **b**) Couinaud segments derived from the landmarks. Note that the these images are different axial slices from the same subject.

added 8 subjects having liver disease and/or tumours scanned on a 3T GE scanner, in addition to 11 healthy volunteers scanned on either a Siemens, Philips or GE (at 1.5T or 3T) scanner. This resulted in a challenging dataset comprised of 122 subjects, which were randomly split into 85 training, 14 validation and 23 test cases.

To obtain ground truth landmark locations and Couinaud segmentation masks, the 8 landmarks shown in Fig. 1a were first manually located. Using a predefined set of rules based mainly on the guide by Germain et al. [2], planes were subsequently derived from combinations of these landmarks to divide a manually obtained liver mask into Couinaud segments 2 to 8. We delineated segment 1 manually since a simple planar surface does not always separate it well. An example Couinaud segmentation is shown in Fig. 1b. In the dataset we used, not all the subjects had a delineation for segment 1, with some subjects even having their segment 1 resected. However, in the way we derive Couinaud segments, we only use landmarking for deriving segments 2 to 8 (the implications of this are discussed in Sect. 4), and landmarking accuracy does not have any effect on segment 1. Therefore, we do not believe this to be an issue and, in this study, focus on how deep learning-based landmark localisation transfers to the prediction of segments 2 to 8.

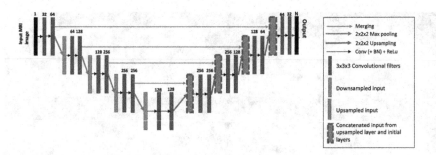

Fig. 2. The 3D U-Net network architecture used for landmark localisation based on heatmap regression, where there is a linear activation on the output and N is the number of target landmarks. The same architecture was also used for the direct segmentation task, where there is a softmax activation on the output and N is the number of class labels.

2.2   Landmark Localisation Model

We make use of a 3D heatmap regression network to predict voxel coordinates for a landmark. The architecture for this network is shown in Fig. 2. The architecture is the same as a typical U-Net used for semantic segmentation, except that each output channel is a heatmap with a peak on the landmark to be predicted, as opposed to softmax predictions for a label. The number of output channels is then equal to the number of target landmarks. The output layer is obtained through a $1 \times 1 \times 1$ convolution followed by a linear activation. To train such a model, each ground truth landmark needs to be converted to a heatmap image. We use the same heatmap generation approach as Payer et al. [8], although with some minor differences. Given N target landmarks, the value for a voxel with coordinates $\mathbf{x}_j$ in the heatmap for landmark L_i ($i = [1 \dots N]$ and $j = [1 \dots V]$, where V is total number of voxels) is given by the following Gaussian function:

$$g_i(\mathbf{x}_j; \sigma, \boldsymbol{\mu}_i) = \frac{\gamma}{(2\pi)^{3/2}\sigma^3} \exp\left(-\frac{\|\mathbf{x}_j - \boldsymbol{\mu}_i\|_2^2}{2\sigma^2}\right) \tag{1}$$

Here, $\boldsymbol{\mu}_i$ contains the coordinates for target landmark L_i, γ is a scaling factor (which we set to 1000) and σ governs the width of the peak in the heatmap. Unlike Payer et al., we do not make σ a trainable parameter and keep it fixed at 5 for all heatmaps. The mean squared error loss function is then used to train the network as follows:

$$\min_{\boldsymbol{\theta}} \frac{1}{NV} \sum_{i=1}^{N} \sum_{j=1}^{V} \|h_i(\mathbf{x}_j; \boldsymbol{\theta}) - g_i(\mathbf{x}_j; \sigma, \boldsymbol{\mu}_i)\|_2^2 \tag{2}$$

Network parameters are denoted by $\boldsymbol{\theta}$ and the heatmaps generated by the network are represented by $h_i(\mathbf{x}_j; \boldsymbol{\theta})$.

We explored two different approaches to inputting data into the model. In the first approach, we isotropically resized the input image and padded it with zeros as required so that the whole image fits into a model input size of 224, 192, 64 for the x, y and z axes respectively. This means that at inference time, only one run of the model is needed to obtain predictions. We refer to this as the 'full input' approach. In the second approach, we resampled the input image to a $2 \times 2 \times 4$ mm resolution, and extracted patches of size 144, 128, 64 from the image as input into the model. Patches were extracted randomly at training time and in a strided fashion at inference time. This means that at inference time, as many runs of model as needed are required to cover the entire image. We refer to this as the 'patch-based' approach. Other preprocessing steps, which were common to all approaches, included clamping the upper range of intensity values to the 99th percentile and afterwards normalising the intensity values to lie in between 0 and 100.

2.3 Direct Segmentation Model

In the landmark localisation approach, the landmarks predicted by the network are used to derive the Couinaud segmentation masks. Another possibility, as implemented by Tian et al. [9], is to use the segmentation masks (which in our case were derived from the ground truth landmarks) to train a semantic segmentation U-Net, thus directly estimating the Couinaud segments. We implemented this to enable a comparison to the landmark localisation approach. To ensure comparability, we used the same network architecture shown in Fig. 2. In a segmentation context, we obtained the output layer through a $1 \times 1 \times 1$ convolution followed by a softmax activation. The number of output channels was then equal to the number of class labels, which in our case was 9 (background plus segments 2 to 8). In addition, the categorical cross entropy loss function was used to train the network. As with the landmark localisation approach, we implemented both full input and patch-based versions of direct segmentation along with the same preprocessing steps.

2.4 Spatial Configuration Post-processing

As described above, the heatmap regression networks are trained on heatmaps that contain only one peak on a target landmark. However, a commonly reported issue with heatmap regression is that multiple peaks can sometimes be predicted in a heatmap for a target landmark at test time, with a local peak as opposed to the global peak corresponding to the correct landmark location [8]. To try and overcome this potential issue, we used a computationally inexpensive and novel postprocessing step specific to the liver, which is illustrated in Fig. 3. This method is motivated by the notion that, although the liver can be a highly variable organ across a population, we expect that the relative positioning of the landmarks we estimate in this study would still exhibit some degree of consistency.

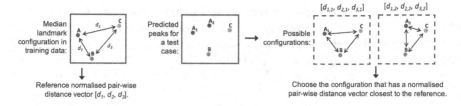

Fig. 3. An illustration of the postprocessing step we use to prune multiple peaks in predicted heatmaps.

In our postprocessing step, median pairwise distances between the landmarks in the training set are first calculated and stored as a vector, which we refer to as the reference vector. This is shown in the left of Fig. 3. Subsequently, if multiple peaks exist in the heatmap of one or more landmarks at test time, pairwise distance vectors are calculated for all possible combinations of landmarks, as shown in the right of Fig. 3. Finally, the set of landmark coordinates with the pairwise distance vector that is closest to the reference vector is chosen as the prediction. To take the overall size of the liver into account, every time a pairwise distance is calculated, it is normalised to the distance between IVCs and UF.

The actual peak finding step is as follows. We first smooth the heatmaps using a 3D Gaussian filter with a standard deviation of 5 voxels. We next apply a 3D maximum filter with a kernel size of 15 voxels. Voxels that then have the same value in both the Gaussian filtered and the maximum filtered images give the location of the peaks. To remove peaks due to noise, only peaks that have a value greater than 0.5 times the value of the global peak are kept. Additionally, to limit the number of possible landmark configurations, a maximum of 5 highest value peaks per heatmap are retained.

2.5   Training and Evaluation

The models were trained on an Nvidia Titan Xp using TensorFlow 2.1.0 along with the Adam optimiser, setting an initial learning rate of 5×10^{-5} based on previous work by Owler et al. [7]. We used a batch size of 1 to allow all the models to fit into GPU memory. Models were trained until the loss on the validation set stopped improving for 15 epochs. We additionally applied on-the-fly augmentation to every training batch. This consisted of random translations along each axis in the range $[-10\%, +10\%]$ of each axis's size, rotation around the z axis in the range $[-10°, +10°]$ and scaling by a factor in the range $[0.9, 1.1]$.

One of the ways the landmark localisation models were evaluated was to measure the mean distance across landmarks, in mm, between the predicted landmark locations and ground truth landmark locations in the test set:

$$\text{Landmark prediction error} = \frac{1}{N} \sum_{i=1}^{N} \|\hat{\mathbf{x}}_i' - \mathbf{x}_i'\|_2 \tag{3}$$

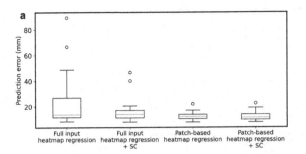
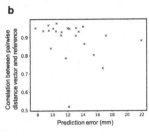

Fig. 4. a) The landmark prediction errors for the landmark localisation models. 'SC' denotes the case when the spatial configuration postprocessing step was applied to the heatmaps. **b)** A plot of the prediction error against the correlation between the corresponding and the reference pairwise distance vectors.

Table 1. The mean prediction error across subjects corresponding to the boxplots in Fig. 4a. Standard deviations are in parentheses.

	Full input	Full input + SC	Patch-based	Patch-based + SC
Prediction error (mm)	23.3 (19.9)	16.1 (9.1)	12.4 (3.1)	12.4 (3.5)

where $\mathbf{x}'_i$ is the ground truth location for landmark L_i and $\hat{\mathbf{x}}'_i$ is the location predicted by the model.

There is not 'one true' correct location for many of the landmarks, e.g. multiple places along the RHV can give rise to the same Couinaud segmentation. Since the end goal in this study is to be able to use the predicted landmarks to derive Couinaud segmentations, we also evaluated the landmark localisation models based on the Dice score (averaged across segments) between the landmark-derived and ground truth Couinaud segmentations. Moreover, this allowed the performance to be compared with the direct segmentation models, which we also evaluated using the mean Dice score. All the statistical tests in this study were carried out using the Wilcoxon signed-rank test.

3   Results

3.1   Landmarking Accuracy

Figure 4a shows the landmark prediction error for the two localisation models we implemented, showing results both with and without the spatial configuration (SC) postprocessing. The mean values across subjects for the models are shown in Table 1. In the cases where SC postprocessing was not used, we chose the global peak in the heatmaps as the landmark location.

In the heatmaps for the full input model, the multiple peak issue described earlier was evident for many cases, and this is reflected in the relatively large

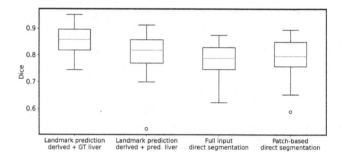

Fig. 5. Couinaud segmentation accuracy for the various approaches. GT - ground truth; Pred. - predicted.

Table 2. The mean segmentation accuracy across subjects corresponding to the box-plots in Fig. 5. LPD - Landmark prediction derived; DS - direct segmentation; GT - ground truth; Pred. - predicted. Standard deviations are in parentheses.

	LPD + GT liver	LPD + pred. liver	Full input DS	Patch-based DS
Dice	0.854 (0.054)	0.802 (0.08)	0.777 (0.069)	0.791 (0.075)

prediction errors for 'full input' in the figure. However, 'full input + SC' shows that our postprocessing step can be an effective way of reducing the influence of this issue. The patch-based model, on the other hand, did not seem to exhibit this multiple peak issue to the same extent, which is why there was no difference in performance when using the SC postprocessing. In addition, the figure shows that the patch-based method performed much better than the full input method and this difference was statistically significant ($P = .002$ when comparing 'patch-based' to 'full input + SC'). For these reasons, we chose to derive Couinaud segmentations from the landmarks predicted by the patch-based model without any SC postprocessing.

Although the patch-based model did not need the SC postprocessing, SC can still potentially be useful as a form of quality control. For each test case, we calculated the Pearson correlation coefficient between the pairwise distance vector from the patch-based model and the reference vector, and plotted this against the prediction error for the test case. This plot is shown in Fig. 4b. There is a clear relationship between this correlation value and the prediction error, meaning that in an automated workflow a correlation threshold could be chosen to aid in the decision of whether to manually check the results.

3.2   Couinaud Segmentation Accuracy

Figure 5 shows the Dice scores for the approaches we explored. Table 2 shows the corresponding mean values across subjects. Landmark prediction derived (LPD) refers to the Couinaud segments derived from the landmark locations predicted by the patch-based landmark localisation model. In this context, a separate mask

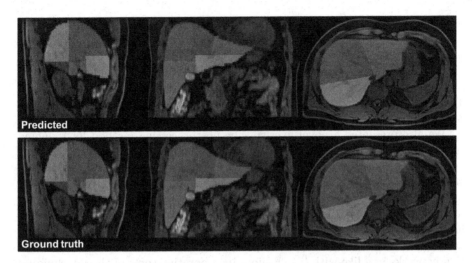

Fig. 6. Example Couinaud segmentation for a representative case, derived from the landmarks predicted by the patch-based heatmap regression model. Top - predicted, bottom - ground truth. Note that segment 1 is not labelled in these images.

for the liver is also needed. Therefore, we evaluated two scenarios: one where we used the ground truth liver mask (denoted by '+ GT liver') and one where we used a liver mask predicted by a separate liver delineation model (denoted by '+ pred. liver').

For direct segmentation (DS), the patch-based approach again performed better than the full input method, although the difference is not as large this time. Since direct segmentation does not require a separate liver mask, it would be fairest to compare the DS approaches with 'LPD + pred. liver'. The results suggest that the LPD approach performed better than both the DS approaches, although the difference was only statistically significant when comparing 'LPD + pred. liver' with 'Full input DS' ($P = .03$).

Dice scores in the region of 0.8 at first glance may not seem that high. However, typical Couinaud segmentations are only an approximation of the functional vasculature-based liver regions and can at times be very subjective. Furthermore, Fig. 6 shows the ground truth and predicted segments from 'LPD + pred. liver' for a representative case where the obtained Dice score was 0.85. The predicted segmentation shown would not be considered erroneous. Therefore, it would be fair to say that deep learning-based landmark localisation can be an effective way to automatically generate Couinaud segmentations.

4   Discussion and Conclusion

We have presented a novel application of deep learning-based landmark localisation to the liver to enable the estimation of Couinaud segments. We showed that a patch-based heatmap regression model was superior to the full input version in terms of landmark prediction error. One possible reason for this may be related to the different field-of-views (FOVs) that can be present in MRI scans. In the full input version, the liver in a large FOV scan would appear much smaller to the model compared to a scan with a tight FOV around the liver. This inconsistent scale may make it more likely to invoke false positive responses in the model. The input to the patch-based model, on the other hand, always has consistent scales across the different axes.

We also presented a novel liver-specific postprocessing method for reducing the effect of the multiple peak issue and/or obtaining a QC metric. It should be noted that we are not proposing this as a method for landmark localisation in general, since that was not the aim of this study. However, other more general techniques for addressing the multiple peak issue do exist. We did not choose to implement them since some require computationally expensive postprocessing [4], while others require changes to the network architecture itself [8], which would have made it difficult to compare the localisation and direct segmentation networks in a like-with-like fashion. Incorporating some of these techniques could comprise some of the future work.

We showed that the accuracy of the Couinaud segments derived from predicted landmarks was at least as good as, if not better than, those obtained from a direct segmentation model. This is a promising result that shows our proposed method for automatically obtaining Couinaud segments is a viable alternative to a direct semantic segmenation approach. As explained in the introduction, one scenario where our approach maybe more suitable could be when manual reviewing and editing by an operator is expected.

One limitation with the landmark localisation approach as we have presented it is that, due to Couinaud segment 1 not being derived from landmarks, a separate semantic segmentation model would have to be trained for predicting segment 1 to enable the automatic segmentation of all the segments. This increases the computational costs compared to using one direct segmentation model for all the segments. Nevertheless, benefits such as quicker manual editing and potentially more accurate segments still remain for segments 2 to 8.

In conclusion, we have shown that a deep learning heatmap regression-based landmark localisation model can be a viable option for automatically obtaining Couinaud segments from non-contrast T1-weighted MRI data. A patch-based version peformed better than a full input version, and the segments derived from the predicted landmark locations were slightly more accurate than those from a direct segmentation model.

References

1. Couinaud, C.: Le foie; études anatomiques et chirurgicales. Masson, Paris, France (1957)
2. Germain, T., Favelier, S., Cercueil, J.P., Denys, A., Krausé, D., Guiu, B.: Liver segmentation: practical tips. Diagn. Interv. Imaging **95**(11), 1003–1016 (2014)

3. Huang, S., Wang, B., Cheng, M., Wu, W., Huang, X., Ju, Y.: A fast method to segment the liver according to Couinaud's classification. In: Gao, X., Müller, H., Loomes, M.J., Comley, R., Luo, S. (eds.) MIMI 2007. LNCS, vol. 4987, pp. 270–276. Springer, Heidelberg (2008). https://doi.org/10.1007/978-3-540-79490-5_33

4. Liao, H., Mesfin, A., Luo, J.: Joint vertebrae identification and localization in spinal CT images by combining short- and long-range contextual information. IEEE Trans. Med. Imaging **37**(5), 1266–1275 (2018)

5. Mole, D.J., et al.: Study protocol: HepaT1ca - an observational clinical cohort study to quantify liver health in surgical candidates for liver malignancies. BMC Cancer **18**(1), 890 (2018)

6. Oliveira, D.A., Feitosa, R.Q., Correia, M.M.: Segmentation of liver, its vessels and lesions from CT images for surgical planning. BioMed. Eng. Online **10**, 30 (2011)

7. Owler, J., Irving, B., Ridgeway, G., Wojciechowska, M., McGonigle, J., Brady, S.M.: Comparison of multi-atlas segmentation and U-Net approaches for automated 3D liver delineation in MRI. In: Zheng, Y., Williams, B.M., Chen, K. (eds.) MIUA 2019. CCIS, vol. 1065, pp. 478–488. Springer, Cham (2020). https://doi.org/10.1007/978-3-030-39343-4_41

8. Payer, C., Štern, D., Bischof, H., Urschler, M.: Integrating spatial configuration into heatmap regression based CNNs for landmark localization. Med. Image Anal. **54**, 207–219 (2019)

9. Tian, J., Liu, L., Shi, Z., Xu, F.: Automatic couinaud segmentation from CT volumes on liver using GLC-UNet. In: Suk, H.-I., Liu, M., Yan, P., Lian, C. (eds.) MLMI 2019. LNCS, vol. 11861, pp. 274–282. Springer, Cham (2019). https://doi.org/10.1007/978-3-030-32692-0_32

10. Yang, D., et al.: Automatic vertebra labeling in large-scale 3D CT using deep Image-to-Image network with message passing and sparsity regularization. In: Niethammer, M., et al. (eds.) IPMI 2017. LNCS, vol. 10265, pp. 633–644. Springer, Cham (2017). https://doi.org/10.1007/978-3-319-59050-9_50

11. Zhang, J., Liu, M., Shen, D.: Detecting anatomical landmarks from limited medical imaging data using two-stage task-oriented deep neural networks. IEEE Trans. Image Process. **26**(10), 4753–4764 (2017)

Reproducibility of Retinal Vascular Phenotypes Obtained with Optical Coherence Tomography Angiography: Importance of Vessel Segmentation

Darwon Rashid[1]($\boxtimes$), Sophie Cai[4], Ylenia Giarratano[1], Calum Gray[2],
Charlene Hamid[2], Dilraj S. Grewal[4], Tom MacGillivray[2,3], Sharon Fekrat[4],
Cason B. Robbins[4], Srinath Soundararajan[4], Justin P. Ma[4],
and Miguel O. Bernabeu[1]

[1] Centre for Medical Informatics, Usher Institute, The University of Edinburgh,
Edinburgh, Scotland
d.rashid@sms.ed.ac.uk, miguel.bernabeu@ed.ac.uk
[2] Edinburgh Clinical Research Facility and Edinburgh Imaging,
University of Edinburgh, Edinburgh, Scotland
[3] Centre for Clinical Brain Sciences, University of Edinburgh, Edinburgh, Scotland
[4] Department of Ophthalmology, Duke University School of Medicine,
Durham, NC, USA

Abstract. Optical coherence tomography angiography (OCTA) is a
non-invasive imaging method that can visualize the finest vascular networks in the human retina. OCTA image analysis has been successfully
applied to the investigation of retinal vascular diseases of the eye and
other systemic conditions that may manifest in the eye. To characterize
and distinguish OCTA images from different pathologies, it is important
to identify quantitative metrics and phenotypes that have high reproducibility and are not overly susceptible to the effects of imaging artifacts. This paper demonstrates the reproducibility of several recently
demonstrated candidate OCTA quantitative metrics: mean curvature
and tortuosity of the whole, foveal, superior, nasal, inferior, and temporal regions; foveal and parafoveal vessel skeleton density; and finally,
foveal avascular zone area and acircularity index. This paper also highlights the importance of vessel segmentation choice on reproducibility
using two different segmentation methods: optimally oriented flux and
Frangi filter.

Keywords: OCTA imaging · Retinal vascular phenotype ·
Reproducibility

1 Introduction

Optical coherence tomography angiography (OCTA) is an effective imaging
modality that captures the finest vasculature within the retina with high resolution. The non-invasive nature of OCTA has consolidated this technology as

© Springer Nature Switzerland AG 2021
B. W. Papież et al. (Eds.): MIUA 2021, LNCS 12722, pp. 238–249, 2021.
https://doi.org/10.1007/978-3-030-80432-9_19

an investigative imaging modality for the detection of retinal vascular changes associated with retinal diseases in the eye as well as systemic conditions that manifest in the eye, what has been recently termed as *oculomics* [23]. Among these applications, OCTA has shown considerable potential in providing insight into diseases of the brain that are known to have a microvascular component. Recent studies have highlighted retinal microvascular biomarkers that may mirror the cerebrovascular changes in Alzheimer's disease (AD) [1,22,24]. Current research is focusing on the investigation of whether this could lead to non-invasive method for detecting brain diseases before clinical symptoms manifest. Furthermore, lower vessel density (VD) has been found in patients with diabetic retinopathy, and has been found to be useful for monitoring vascular pathologies [6,11]. Patients that have glacomatous eyes with central visual field defects were shown to have significantly larger foveal avascular zone (FAZ) area, lower FAZ circularity, and lower macular ganglion cell-inner plexiform layer thickness than patients with healthy eyes [9]. In addition, FAZ area has been shown to be larger in diabetic eyes with and without retinopathy than in healthy eyes [21]. The interested reader can refer to Hagag *et al.* for a survey of common quantitative phenotypes [6].

A currently recognized challenge in OCTA image analysis is the identification of vascular phenotypes that are reproducible across repeated images of the same eye regardless of imaging device and noise/artifacts affecting the image. Lei *et al.* explored the reproducibility of peripapillary retinal VD phenotypes on OCTA images of the same eye obtained with four separate devices [13]. Good reproducibility of the phenotypes vessel length density and perfused density across the four separate devices was reported as well as good repeatibility within devices. La *et al.* investigated the reproducibility of FAZ measurements [10], which they found to be reproducible across consecutive scans of the same eye. Corvi *et al.* found that another important factor to keep in consideration when investigating the reproducibility of OCTA phenotypes, in particular VD, fractal segmentation, and FAZ, across different OCTA imaging devices, is the lack of standardization on how retinal layers are segmented on OCT volumes [2]. Sampson *et al.* reported inconsistencies in measured phenotypes following a software update of a single device. The software update resulted in an increase in retinal thickness and reduction in VD in repeated measurements, which makes new images captured not backward compatible. Finally, an area that has received less attention up until recently is the evaluation of the reproducibility of OCTA phenotypes in eyes with pathology. Lee *et al.* explored reproducibility of VD measurements in diabetic macular oedema and retinal vein occlusion. The study concludes that VD measurements have good reproducibility for the retinal diseases considered. A key step in phenotype discovery is vessel segmentation. An increasing number of analysis pipelines have been proposed for the automated segmentation of retinal OCTA images. These range from traditional vessel enhancement and thresholding approaches to recent deep learning models [15,16]. However, recent investigations have shown that widely used retinal phenotypes are sensitive to the choice of the segmentation method [4]. For example, it has been found up to a 25% difference in VD accuracy depending on the segmentation method chosen.

The current literature has focused on retinal vascular phenotypes related to VD and FAZ morphology. However, less is known about the reproducibility of phenotypes capturing finer geometrical and topological properties of the networks. In a recent study, Giarratano *et al.* proposed a novel approach to phenotyping fovea-centred OCTA images based on expanding the set of commonly used phenotypes to include information about the geometry and topology of different foveal/parafoveal subregions. Furthermore, they demonstrated that these phenotypes were useful for classifying images of diabetic retinopathy and chronic kidney disease patients against controls [5]. However, the reproducibility of this approach has not been established. In this study, we aim to address this limitation.

The main contributions of this study are two-fold. First, we investigate reproducibility of a broad range of phenotypes including VD, FAZ area and acircularity index, vessel tortuosity and curvature in different foveal/parafoveal subregions over repeated imaging of the same eye using the same device. We find excellent reproduciblity, which demonstrates that this approach to deeply phenotyping retinal vascular networks is robust. Second, we investigate how the choice of vessel segmentation method affects the reproducibility of phenotypes for a single retinal image. Contrary to the first part of the study, we found that only a small subset of phenotypes (FAZ acircularity index and area) were reproducible between two different vessel segmentation pipelines. These findings demonstrate that contrary to previous studies, reproducibility of phenotypes should be investigated against ground truth values measured by expert graders rather than repeated imaging since stable measurements could still be affected by errors introduced by the vessel segmentation process. Further research should be devoted to establishing robust and reproducible vessel segmentation methods.

2   Methods

2.1   Participant Demographics and Imaging Protocol

The images included in this study were obtained between July 2017 – May 2018 and September 2019 – February 2020 from cognitively healthy patients at least 50 years of age imaged as part of an ongoing prospective study (Clinicaltrials.gov identifier, NCT 03233646). The study was approved by the Duke University School of Medicine (Durham, NC) Institutional Review Board for Human Research and conducted in accordance with the tenets of the Declaration of Helsinki. Written informed consent was obtained from all study subjects.

The method of patient recruitment has been previously described [24]. Control subjects were determined to be cognitively healthy based on the Montreal Cognitive Assessment, Trail Making Test, and Delayed Recall from the Consortium to Establish a Registry for Alzheimer's disease Word-List. Exclusion criteria included a history of non-AD dementia, poorly controlled hypertension, diabetes mellitus, demyelinating disorder, Early Treatment Diabetic Retinopathy Study (ETDRS) corrected visual acuity worse than 20/40 on date of image acquisition, self-reported refractive error outside the range of −6D to 6D, age-related macular degeneration, glaucoma, or vitreomacular pathology or surgery that could affect OCTA interpretation.

Study subjects underwent OCT imaging with the Zeiss Cirrus HD-5000 Spectral-Domain OCT with AngioPlex OCT Angiography (Carl Zeiss Meditec, Dublin, CA) with a wavelength of 840 nm, acquisition speed of 68,000 A-scans per second, internal motion tracking capacity, and an integrated OCTA analysis algorithm [18]. 3 × 3 mm scans centered on the fovea were obtained. This study focuses on the superficial capillary plexus due to the relatively improved visualization of microvasculature compared to the deep retinal plexus. Superficial capillary plexus segmentation was automatically determined by the OCTA software (Carl Zeiss Meditec, version 10.0.0.14618), with the inner boundary delineated by the inner limiting membrane and the outer boundary delineated by the inner plexiform layer, which was calculated as 70% of the distance from the internal limiting membrane to a line 110 mum above the retinal pigment epithelium, representing the estimated border of the outer plexiform layer. Segmentation boundaries for each scan were manually reviewed for accuracy. OCTA scans with visible vitreomacular pathology based on corresponding OCT imaging were excluded from the study. Superficial capillary plexus OCTA scans were exported in Tagged Image File Format (TIFF) for further image processing.

As part of the study protocol, patients had multiple OCTA images obtained from one or both eyes on the same day, offering the opportunity to assess the reproducibility of quantitative parameters from OCTA images obtained from the same study eye during the same imaging encounter. As determined by a trained grader (S.C.), pairs of images with at least one image having limited motion artifact and adequate vessel visualization were included in this study. Figure 1 displays an OCTA image of an eye (left) and the repeated OCTA image of the same eye (right) with mild motion artifacts. A total of 44 pairs of macular superficial capillary plexus OCTA images from 22 eyes of 22 unique patients (mean age 64.86+/−7.30 years, 77% female) were included for this study.

2.2 Image Analysis

The field of retinal image analysis has seen very substantial advances in the development of machine learning algorithms for the segmentation of blood vessels in fundus camera images (see [17] for a survey). These approaches rely on the availability of repositories of images and associated manual segmentations (e.g. Messidor [3], DRIVE [20]). Similar repositories are only starting to emerge for OCTA imaging [4,15,16], and therefore contemporary studies rely on classic vessel enhancement and thresholding approaches [8,25]. In the first part of the study, vessel segmentation was performed using the optimally oriented flux (OOF) algorithm [12] with double thresholding as described in [14] which was found to outperform other vessel enhancement filters and segmentation methods in the absence of manual segmentations to train deep learning methods [4].

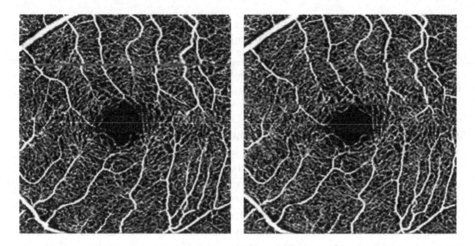

Fig. 1. OCTA image (left) and repeated OCTA image of the same eye (right).

Binary images were skeletonized using MATLAB R2020b software and converted into a graph object where each white pixel represented a node in the skeleton and touching pixels were connected with an edge. Each node stored information about its position in the Euclidean space and vessel radius in that point. Retinal phenotypes previously introduced in [5] were computed in the whole vascular network and in the five regions of interest in the retina: foveal, superior, nasal, inferior, temporal. Briefly, FAZ was computed using the doubly connected edge list (DCEL) data structure and the algorithm described in [19] to detect faces in a planar graph. Based on the detection of the nodes and edges which describe the FAZ boundary, we computed the FAZ area, and the acircularity index defined as the ratio between the FAZ perimeter over the perimeter of a circle with the same area as the FAZ. Vessel tortuosity was measured as the arc length over the chord ratio between branching points in the network. Since this metric does not take into account changes in sign of the curve, we also reported vessel curvature as introduced in [7]. In case of distributions, the mean was reported as summary statistic. In the second part of the study, we considered an alternative approach to vessel segmentation to compare to the aforementioned OOF method. Firstly, the vessels were enhanced using a Hessian-based Frangi filter. Next, the filtered image was thresholded using the Otsu and local thresholding methods. The binary maps from both methods were then combined into a final binary map in which the vessel structures from both maps align. Figure 2 displays two different skeletons obtained after using two different vessel enhancement filters (OOF and Frangi) on the same OCTA image.

2.3 Statistical Analysis

We compared repeated measurements with a paired t-test under the null hypothesis of equal means ($p < 0.05$ being the threshold for significance). In addition,

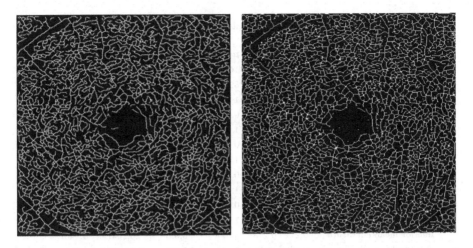

Fig. 2. Skeleton of OCTA image using the OOF method (left) and skeleton of the same OCTA image using the Frangi method (right).

we computed z-scores to estimate differences between means in a standardized way. Briefly, we define the z-score for any given phenotype as

$$z = \frac{\mu_1 - \mu_2}{\sigma_1}, \tag{1}$$

where $\mu_{1,2}$ are the means for the phenotype calculated over each set of repeated images (or over the two segmentation approaches considered) and $\sigma_{1,2}$ are the associated standard deviations. In the following section, statistical tests and z-scores are calculated for each phenotype over collections of a) two repeated images of the same eye or b) two different segmentations of the same eye image. To explore the results visually, Bland–Altman plots were used. Briefly, the plot identifies if there is a systematic bias in any of the measurements by displaying the mean difference between them and how differences are distributed around the mean. If the differences are normally distributed, then 95% of the data points should fall within the $\pm 1.96\sigma$ limits displayed.

3 Results

3.1 Microvascular Phenotype Reproducibility over Repeated OCTA Imaging

First, we investigate the reproducibility of the phenotypes over repeated imaging of the same eye. Table 1 summarizes the data. No statistically significant differences are observed for any of the phenotypes ($p > 0.05$ with z-scores ranging $[-0.64, 0.28]$). Figure 3 displays histograms and Bland-Altman plots for the phenotypes with the best z-score (left) and the worst z-score (right). We observe how differences are centred roughly around 0 and normally distributed around

Table 1. Summary of data for first experiment.

Phenotype	p	z	μ_1	μ_2	σ_1	σ_2
Parafoveal nodes	0.476	0.186	16281.681	15966.409	1654.868	2026.067
Foveal nodes	0.791	−0.026	899.636	906.909	286.878	269.276
Acircularity index	0.634	−0.103	1.555	1.569	0.135	0.178
FAZ area	0.518	−0.106	0.464	0.478	0.123	0.165
Whole mean tortuosity	0.597	−0.143	1.116	1.117	0.007	0.005
Whole mean curvature	0.172	−0.315	0.027	0.027	4.33E-4	3.77E-4
Fovea mean tortuosity	0.980	−0.009	1.133	1.134	0.040	0.045
Fovea mean curvature	0.784	0.095	0.021	0.021	0.002	0.002
Nasal mean tortuosity	0.292	−0.645	1.113	1.119	0.010	0.023
Nasal mean curvature	0.511	−0.210	0.021	0.021	5.496E-4	6.882E-4
Inferior mean tortuosity	0.623	0.132	1.111	1.109	0.013	0.010
Inferior mean curvature	0.271	0.283	0.021	0.021	5.264E-4	4.846E-4
Temporal mean tortuosity	0.760	−0.082	1.114	1.115	0.011	0.013
Temporal mean curvature	0.522	0.220	0.021	0.021	4.476E-4	6.123E-4
Superior mean tortuosity	0.618	−0.207	1.107	1.108	0.008	0.011
Superior mean curvature	0.344	−0.231	0.021	0.021	7.810E-4	7.147E-4

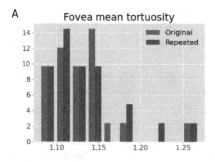
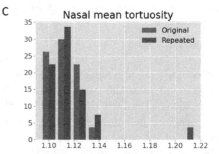
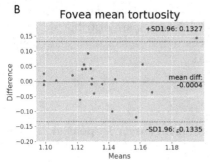
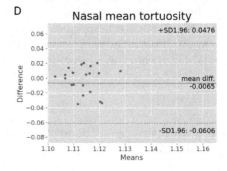

Fig. 3. Histogram for whole image mean tortuosity phenotype of original and repeated images (A), Bland-Altman plot for whole image mean tortuosity phenotype of original and repeated images (B), histogram for nasal mean tortuosity phenotype of original and repated images (C), and Bland-Altman plot for nasal mean tortuosity phenotype of original and repeated images (D).

it. These results confirm the reproducibility of the phenotypes over repeated imaging of the same eye.

3.2 Dependence of Microvascular Phenotypes on the Choice of Segmentation/Skeletonization Algorithm

Second, we investigate how the choice of vessel segmentation method affects the reproducibility of phenotypes calculated on the same retinal image. Table 2 summarizes the data. We found that only a small subset of phenotypes (FAZ acircularity index and area) were reproducible. The remaining phenotypes display differences as large as 9.4 standard deviations (*e.g.* mean curvature in the whole image). Figure 4 displays histograms and Bland-Altman plots for the phenotypes with the best z-score (left) and the worst z-score (right). We observe important biases in the calculation of curvature and tortuosity phenotypes across all regions of interest associated with the choice of segmentation algorithm.

Table 2. Summary of data for second experiment

Phenotype	p	z	μ_1	μ_2	σ_1	σ_2
Parafoveal nodes	<0.001	−0.424	16124.045	16920.704	1856.507	2018.106
Foveal nodes	<0.001	−0.782	903.273	1143.818	278.240	304.113
Acircularity index	0.324	−0.156	1.562	1.587	0.158	0.145
FAZ area	0.051	−0.175	0.471	0.497	0.146	0.167
Whole mean tortuosity	<0.001	5.921	1.117	1.078	0.006	0.003
Whole mean curvature	<0.001	9.492	0.027	0.023	4.120E-4	4.012E-4
Fovea mean tortuosity	<0.001	1.249	1.134	1.080	0.042	0.017
Fovea mean curvature	<0.001	1.441	0.021	0.017	0.003	0.002
Nasal mean tortuosity	<0.001	2.127	1.116	1.077	0.018	0.006
Nasal mean curvature	<0.001	5.477	0.021	0.0176	6.256-4	6.607E-4
Inferior mean tortuosity	<0.001	2.633	1.111	1.080	0.011	0.007
Inferior mean curvature	<0.001	5.691	0.021	0.018	5.116E-4	5.581E-4
Temporal mean tortuosity	<0.001	3.029	1.114	1.077	0.012	0.007
Temporal mean curvature	<0.001	5.740	0.021	0.018	5.387E-4	6.982E-4
Superior mean tortuosity	<0.001	3.077	1.107	1.078	0.010	0.006
Superior mean curvature	<0.001	3.942	0.021	0.018	7.542E-4	6.531E-4

4   Discussion

OCTA allows imaging of the finest vasculature in the foveal region of the retina. An increasing number of studies have investigated how changes to these vessels associate with retinal disease. Furthermore, several intriguing associations have been established between changes in foveal microvasculature and vascular complications of organs as diverse as the brain, heart, and kidney. As a consequence, much effort is devoted to identifying reliable retinal vascular biomarkers for the diagnosis of eye and systemic diseases.

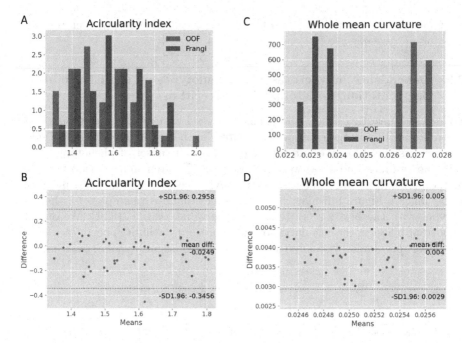

Fig. 4. Histogram for FAZ acircularity index phenotype for both vessel segmentation methods (A), Bland-Altman plot for FAZ acircularity index phenotype for both vessel segmentation methods (B), histogram for whole image mean curvature between both vessel segmentation methods (C), and Bland-Altman plot for whole image mean curvature between both vessel segmentation methods (D).

Despite promising results in this direction, a standard for retinal vascular phenotyping based on OCTA imaging is currently lacking. Recent studies have started to address the issue of reproducibility in OCTA phenotypes. As surveyed earlier, reproducibility has been investigated in terms of repeated imaging of the same eye on a single device, on multiple instances of the same device, on multiple devices by different manufacturers, or across multiple versions of the software running on a single device. Results so far have demonstrated good reproducibility of VD and FAZ metrics over repeated imaging of the same eye with the same device. This emphasizes the need of developing vendor-independent OCTA image analysis pipelines and evaluating their reproducibility across a broader range of vascular phenotypes in order to be able to consistently capture more subtle changes.

In the current study, we evaluated the reproducibility of OCTA vascular metrics describing FAZ shape and area, curvature of blood vessels, and vascular tortuosity with an emphasize on calculations over whole images and subregions of interest (foveal/parafoveal and foveal/superior/nasal/inferior/temporal). First, we investigated the reproducibility of OCTA phenotypes by comparing results over repeated images of the same eye using a standardized pipeline. We found that reproducibility was excellent when the same pipeline is applied to repeated images of the same eye. Next, we investigated how the step of image segmentation/skeletonization of raw images affects reproducibility by comparing the phenotypes calculated on the skeletons obtained after vessel enhancement with two popular algorithms in the literature. We found that FAZ shape phenotypes were reproducible between both approaches to segmentations but there was poor reproducibility for VD, tortuosity, and curvature. We attribute this finding to significant differences in vessel segmentation. This is in agreement with previous investigations of reproducibility of the VD phenotype [4]. Future work should address how the differences observed impact image classification and potential clinical diagnosis capabilities.

To conclude, our study highlights the need to develop standardized OCTA image analysis frameworks and to validate OCTA retinal vascular phenotypes in order to ensure reproducibility of clinical findings. Contrary to previous studies, reproducibility of phenotypes should be investigated against ground truth values measured by expert graders rather than repeated imaging since stable measurements could still be affected by errors introduced by the vessel segmentation process. Further research should be devoted to establishing robust and reproducible vessel segmentation methods. Future work should also investigate the reproducibility of phenotypes in eyes from patients with various pathological conditions and different demographics. In addition, with larger datasets, it will be important to examine more rigorously the effects of different degrees and types of imaging artifact on the reproducibility of OCTA phenotypes.

Acknowledgements. DR and YG were supported by two Medical Research Council Precision Medicine Doctoral Training Programme scholarships (MR/N013166/1). MOB is supported by grants from EPSRC (EP/R029598/1, EP/T008806/1), Fondation Leducq (17 CVD 03), the European Union's Horizon 2020 research and innovation programme under grant agreement No. 801423, and British Heart Foundation/The Alan Turing Institute under a Cardiovascular Data Science Award. TJM and MOB acknowledge the funders of the SCONe project (https://www.ed.ac.uk/ophthalmology/scone). This project was supported in part by the Alzheimer's Drug Discovery Foundation and the Heed Foundation (SC) and the VitreoRetinal Surgery Foundation (SC).

References

1. Bulut, M., et al.: Evaluation of optical coherence tomography angiographic findings in Alzheimer's type dementia. Br. J. Ophthalmol. **102**(2), 233–237 (2018). https://doi.org/10.1136/bjophthalmol-2017-310476
2. Corvi, F., Pellegrini, M., Erba, S., Cozzi, M., Staurenghi, G., Giani, A.: Reproducibility of vessel density, fractal dimension, and foveal avascular zone using 7 different optical coherence tomography angiography devices. Br. J. Ophthalmol. **186**, 25–31 (2018)
3. Decencière, E., et al.: Feedback on a publicly distributed image database: the messidor database. Image Anal. Stereology **33**(3), 231–234 (2014)
4. Giarratano, Y., et al.: Automated segmentation of optical coherence tomography angiography images: benchmark data and clinically relevant metrics. Transl. Vis. Sci. Technol. **9**(13), 5 (2020)

5. Giarratano, Y., et al.: A framework for the discovery of retinal biomarkers in optical coherence tomography angiography (OCTA). In: Fu, H., Garvin, M.K., MacGillivray, T., Xu, Y., Zheng, Y. (eds.) OMIA 2020. LNCS, vol. 12069, pp. 155–164. Springer, Cham (2020). https://doi.org/10.1007/978-3-030-63419-3_16

6. Hagag, A.M., Gao, S.S., Jia, Y., Huang, D.: Optical coherence tomography angiography: technical principles and clinical applications in ophthalmology. Taiwan J. Ophthalmol. **7**(3), 115 (2017)

7. Hart, W.E., Goldbaum, M., Côté, B., Kube, P., Nelson, M.R.: Measurement and classification of retinal vascular tortuosity. Int. J. Med. Informatics **53**(2–3), 239–252 (1999). https://doi.org/10.1016/S1386-5056(98)00163-4

8. Kim, A.Y., Chu, Z., Shahidzadeh, A., Wang, R.K., Puliafito, C.A., Kashani, A.H.: Quantifying microvascular density and morphology in diabetic retinopathy using spectral-domain optical coherence tomography angiography. Invest. Ophthalmol. Vis. Sci. **57**(9), OCT362-OCT370 (2016). https://doi.org/10.1167/iovs.15-18904

9. Kwon, J., Choi, J., Shin, J.W., Lee, J., Kook, M.S.: Alterations of the foveal avascular zone measured by optical coherence tomography angiography in glaucoma patients with central visual field defects. Invest. Ophthalmol. Vis. Sci. **58**(3), 1637–1645 (2017)

10. La Spina, C., Carnevali, A., Marchese, A., Querques, G., Bandello, F.: Reproducibility and reliability of optical coherence tomography angiography for foveal avascular zone evaluation and measurement in different settings. Retina **37**(9), 1636–1641 (2017)

11. Lavia, C., Couturier, A., Erginay, A., Dupas, B., Tadayoni, R., Gaudric, A.: Reduced vessel density in the superficial and deep plexuses in diabetic retinopathy is associated with structural changes in corresponding retinal layers. PloS ONE **14**(7), e0219164 (2019)

12. Law, M.W.K., Chung, A.C.S.: Three dimensional curvilinear structure detection using optimally oriented flux. In: Forsyth, D., Torr, P., Zisserman, A. (eds.) ECCV 2008. LNCS, vol. 5305, pp. 368–382. Springer, Heidelberg (2008). https://doi.org/10.1007/978-3-540-88693-8_27

13. Lei, J., et al.: Repeatability and reproducibility of superficial macular retinal vessel density measurements using optical coherence tomography angiography en face images. JAMA Ophthalmol. **135**(10), 1092–1098 (2017). https://doi.org/10.1001/jamaophthalmol.2017.3431

14. Li, A., You, J., Du, C., Pan, Y.: Automated segmentation and quantification of OCT angiography for tracking angiogenesis progression. Biomed. Opt. Express **8**(12), 5604 (2017). https://doi.org/10.1364/boe.8.005604

15. Li, M., et al.: Ipn-v2 and octa-500: Methodology and dataset for retinal image segmentation. arXiv preprint arXiv:2012.07261 (2020)

16. Ma, Y., et al.: Rose: a retinal oct-angiography vessel segmentation dataset and new model. IEEE Trans. Med. Imaging **40**(3), 928–939 (2021). https://doi.org/10.1109/TMI.2020.3042802

17. Mookiah, M.R.K., et al.: A review of machine learning methods for retinal blood vessel segmentation and artery/vein classification. Med. Image Anal. **68**, 101905 (2020)

18. Rosenfeld, P.J., et al.: Zeiss angioplex^{TM} spectral domain optical coherence tomography angiography: technical aspects. OCT Angiography Retinal Macular Dis. **56**, 18–29 (2016)

19. Schneider, S., Sbalzarini, I.F.: Finding faces in a planar embedding of a graph. Technical Report, MOSAIC Group, MPI-CBG (2015)

20. Staal, J., Abràmoff, M.D., Niemeijer, M., Viergever, M.A., Van Ginneken, B.: Ridge-based vessel segmentation in color images of the retina. IEEE Trans. Med. Imaging **23**(4), 501–509 (2004)
21. Takase, N., Nozaki, M., Kato, A., Ozeki, H., Yoshida, M., Ogura, Y.: Enlargement of foveal avascular zone in diabetic eyes evaluated by en face optical coherence tomography angiography. Retina **35**(11), 2377–2383 (2015)
22. Van De Kreeke, J.A., et al.: Optical coherence tomography angiography in preclinical Alzheimer's disease. Br. J. Ophthalmol. 157–161 (2019). https://doi.org/10.1136/bjophthalmol-2019-314127
23. Wagner, S.K., et al.: Insights into systemic disease through retinal imaging-based oculomics. Transl. Vis. Sci. Technol. **9**(2), 6 (2020) https://doi.org/10.1167/tvst.9.2.6, https://tvst.arvojournals.org/article.aspx?articleid=2761238
24. Yoon, S.P., et al.: Retinal microvascular and neurodegenerative changes in alzheimer's disease and mild cognitive impairment compared with control participants. Ophthalmol. Retina **3**(6), 489–499 (2019)
25. Zhang, M., Hwang, T.S., Dongye, C., Wilson, D.J., Huang, D., Jia, Y.: Automated quantification of nonperfusion in three retinal plexuses using projection-resolved optical coherence tomography angiography in diabetic retinopathy. Invest. Ophthalmol. Vis. Sci. **57**(13), 5101–5106 (2016). https://doi.org/10.1167/iovs.16-19776

Fast Automatic Bone Surface Segmentation in Ultrasound Images Without Machine Learning

Shihfan Jack Tu[1]([✉])[ID], Jules Morel[1], Minsi Chen[2][ID], and Stephen J. Mellon[1][ID]

[1] Oxford Orthopaedic Engineering Centre, Nuffield Department of Orthopaedics, Rheumatology and Musculoskeletal Sciences, University of Oxford, Oxford, UK
{jack.tu,stephen.mellon}@ndorms.ox.ac.uk
[2] Department of Computer Science, University of Huddersfield, Huddersfield, UK
M.Chen@hud.ac.uk
https://www.ndorms.ox.ac.uk/research-groups/oxford-orthopaedic-engineering-centre

Abstract. Reconstructing 3D bone images with 2D clinical ultrasound image is one of the primary developmental trends of computer-assisted orthopaedic surgery procedures, and real-time bone segmentation is required for such development. We previously presented a dynamic programming method with local phase tensor extraction for bone structure segmentation that could process one ultrasound frame with a true positive ratio of 71% in approximately 1 s. The present study aimed to reduce the segmentation time to enable real-time computational capacity for clinical application developments. A simplified bone probability algorithm was optimised by systematically identifying and removing the components which cost most computing resources. The segmentation results produced by the bone probability method were compared to the local phase method, and manual segmentation carried out by clinical experts. The proposed method had higher recall metric (0.67) than the local phase method (0.61), while the computational time is reduced to 0.02 s per image. However, the bone probability method did not perform as well as the local phase method in specificity and precision metrics. In conclusion, the simplified version of the segmentation algorithm improved computational speed and promised an advantage in further real time application developments, but additional functions that can improve accuracy and further extensive validations are still required before further clinical application developments.

Keywords: Bone segmentation · Ultrasound imaging · Bone probability map

This work was supported by Wellcome Trust ISSF/University of Oxford MLSTF. The authors would like to thank Prof Irina Voiculescu and Mr Ziyang Wang for supporting and discussing the machine learning aspect. The authors also acknowledge the reviewers for the constructive feedback.

B. W. Papież et al. (Eds.): MIUA 2021, LNCS 12722, pp. 250–264, 2021.
https://doi.org/10.1007/978-3-030-80432-9_20

1   Introduction

Reliable automatic segmentation of bone in ultrasound images has a variety of potential applications within orthopaedics. Paired with a robust registration algorithm [7] and an external localiser, auto-segmentation can be used to accurately locate and track bony anatomy in 3D space.

The motion analysis with ultrasound system (CAT&MAUS) used in our work is composed of 2D B-mode ultrasound (US) scanners and an optoelectronic motion analysis system to directly locate and track under-skin bony landmarks for the purposes of measuring kinematics at joints. It has been previously applied to track patella motion during knee flexion [10] and the greater trochanter during gait [6,8].

Although several methods exist, reliably and quickly segmenting the bone structure in US images remains a crucial challenge for the development of clinical systems. For example, US bone scans do not provide complete anatomical boundaries and shape, and the bone structures give a strong acoustic response and provide a shadowing effect below them [4]. An automatic bone segmentation algorithm using local phase features and dynamic programming [5] was developed to resolve the issue of strong acoustic response anatomical structures, such as fascia and tendon, being incorrectly recognized as a bone. This method reduced the need for manual segmentation and can be integrated with a motion capture system [9]. However, some limitations still need to be conquered to achieve the ultimate goal, which is to combine automatically segmented bone contours from 2D freehand US scans with positional information obtained from the probe transducer and to transform the data into a 3D bone model at a real-time pace. Although the automatic algorithm's outcome quality has the potential to enable this goal, the most critical issue to resolve is to reduce the long processing time.

Ultrasound image processing methods based on machine learning have been proposed for bone surfaces segmentation from ultrasound data. For example, deep learning network architecture, termed U-net and convolutional neural network [1,11,12,14,15]. A well-trained machine learning model could help with the bone model reconstruction's processing speed and promise wider applications for clinical usage. A previous study reported that after 6 h of training, the machine learning programme based on the convolutional neural network (CNN) has the potential to segment one image in 52 milliseconds [1]. Although the speed seems promising, this method still requires a long local phase extraction procedure. This is a limitation that our previous method shares [5]. The local phase extraction was the most time-consuming processing step in the procedure. Considering the success of the machine learning methods is dependent on a number of training images, anatomical variation present in the training data, and the quality of the image. Manual labelling is still required in training data preparation. The proposed method in this paper can be used to reduce or remove this requirement.

The current study proposes methods to reduce the computation time to segment and register bone surface from specific ultrasound images, which contains a high-intensity bone response profile, corresponding to the bone surface, followed by shadow region. Our target was to perform segmentation for one frame within 0.05 s in order to treat more than 20 frames per second, which is the common acquisition sampling rate of most US probes. Our ultimate developmental focus is to improve segmentation speed for use as a step within rapid reconstruction of bone position in 3D via registration. The secondary aim of the current study was to explore a fast non-machine-learning segmentation method that could offer acceptable accuracy and precision results with minimum human input.

2   Methods

2.1   Simplified Segmentation Method with Bone Probability Map

The simplified *bone probability method* is based on the previous method [5], but some functions have been sacrificed to improve processing efficiency. For example, the local phase extraction involving multidimensional inverse fast Fourier transform was removed from *bone probability method* and saved $\sim$ 190 milliseconds. Additionally, some new functions were added to compensate for the accuracy lost due to the deletion. To maximise the performance efficiency, only six amended steps were applied in the present method. Details of each function are described as below (Fig. 1):

Bone Shadow. Due to the high acoustic impedance difference between bone and soft tissue, most US energy is reflected by the bone surface. The region below the bone surface usually appears to be dark in a US image. In the original algorithm [5], the shadow value of the $pixel(a, b)$ was calculated by a Gaussian weighted accumulation of the pixels below. However, considering the fact that the variance ($\sigma^2 < 1$) has been reduced by using a hardware pre-set, the calculation can be simplified as follows:

$$\exp\left(-\frac{(x-1)^2}{2.\sigma^2}\right) \simeq \begin{cases} 1 \text{ si } x = 1 \\ 0 \text{ sinon.} \end{cases} \tag{1}$$

and the *Bone_shadow* can then be generated:

$$Sum(r, c) = \sum_{i=r}^{R} I(i, c) \tag{2}$$

$$Mask(r, c) = 1 - \frac{Sum(r, c) - min_{r,c}(Sum)}{max_{r,c}(Sum) - min_{r,c}(Sum)} \tag{3}$$

$$Bone\_shadow = Mask^{Power} \tag{4}$$

where r and c are the rows and column indexes in the US image, R is the number of rows in the US image, and I is the pixel intensity matrix. The parameter *Power* needs to be adjusted and introduced to manage the *Mask* before generating Bone shadow. Figure 2b demonstrates a bone shadow mask.

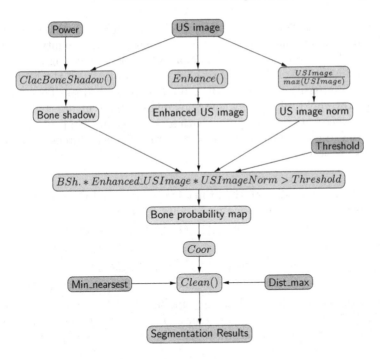

Fig. 1. Flowchart of the method developed in this article

Enhance Function. An additional function *Enhance()* has been used to enhance the US image before generating a bone probability map. This function calculated the derivative of the echo intensity, which measured the sensitivity to the echo intensity change. The output image has enhanced edges of the high intensity area, where the soft tissue features were suppressed, and the intensity of bone edges remains at the same level. Figure 2b is an example of the enhanced ultrasound image, and Fig. 2e demonstrates a selected column of the echo signal beam (*columnindex* = 222). The peak of the signal (*rowindex* = 50) remains the same level, and the density of reflective artefact (*rowindex* = 118) was reduced. $\forall\, c$, if $r = 1$ or $r = R$,

$$Enhanced\_US\_image(r, c) = 0 \tag{5}$$

else,

$$Enhanced\_US\_image(r, c) = 1 - (2 * I_{(r,c)} - I_{(r+1,c)} - I_{(r-1,c)}) \tag{6}$$

where c is the column index, R is the number of rows in the US image, and I is the pixel intensity matrix.

Bone Probability. The final bone probability map was generated by multiplying the bone shadow map, the enhanced US image and the normalised US image. A *threshold* value was applied to eliminate pixels with low probabilities. This parameter is adjustable to improve segmentation quality, and it may need to be adjusted for specific acquisition protocols (depending on the probe and acquisition mode). A higher threshold value can decrease the likelihood of labelling high intensity features, such as tendon, as bony areas; Therefore, with a very high threshold value, the whole bone might remain undetected; and noisy labels will be generated when applied with a very low threshold value. The ideal way of setting the threshold value was to extract the echo intensity value from a "boneless" region.

Segmentation. According to the principle of ultrasound beam reflection, only one pixel per column on the bone probability map (column c, row s_c) is recognised to initiate the segmentation (Eq. 7). The recognition was performed as follows:

$$\forall c, \exists s_c \in [0, R], s = \sum_{i=r}^{R} i * Bness(i, c) \tag{7}$$

where c is the column index, R is the number of rows in the US image, and *Bness* is the bone probability map described above. After generating the bone probability map, the segmentation can then be obtained as follows:

$$Segmentation'(r, c) = \begin{cases} 1 \text{ if } & r = s_c \\ 0 \text{ else} \end{cases} \tag{8}$$

Clean Function. A *Clean()* function has been created to improve the segmentation and remove parasite pixels. This step was based on user observation of the preliminary segmentation label (Figure 2f). Mislabelled pixels were usually isolated from the main cluster of correct labels. Typically, the pixel density around the false positives area was lower, and it can be examined by using the process below:

$$Sum'(r, c) = \sum_{|j-c|=0}^{2.dist\_max} \left(\sum_{|i-r|=0}^{dist\_max} Segmentation'(i, j) \right) \tag{9}$$

$$Segmentation(r, c) = \begin{cases} 1 \text{ if } & Sum'(r, c) > min \quad nearest \\ 0 \text{ else} \end{cases} \tag{10}$$

The new method gave up the principle that was applied in the original method, which is connecting up nearest neighbour segmented pixels into a continuous line, and the *Clean()* function considered the labels as a cluster of the point cloud. Clusters with continuous shape were included, and the random clusters were excluded. The user can monitor the performance and adjust two parameters to control the level of cleaning. The result of the clean function is demonstrated in Fig. 2g.

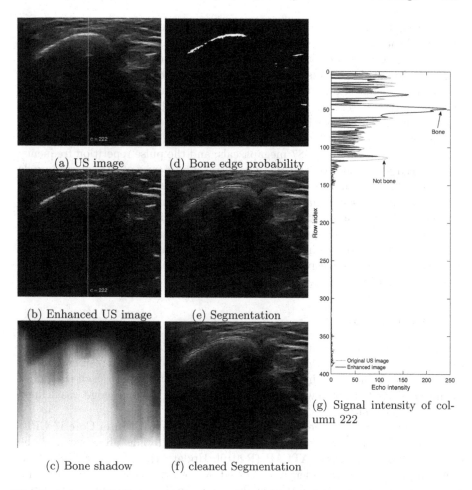

(a) US image   (d) Bone edge probability

(b) Enhanced US image   (e) Segmentation

(c) Bone shadow   (f) cleaned Segmentation

(g) Signal intensity of column 222

Fig. 2. An example of processing results. (a) US image; (b) Enhanced US image; (c) Bone shadow; (d) Bone edge probability map; (e) Segmentation label before clean; (f)Cleaned segmentation label; (g) An example for the *Enhanced*() function, where the y axis is depth of the US image (row index) and the y axis is the intensity of selected column ($c = 222$). the grey line represents the original image and the black line represents the enhanced US image

2.2   Image Acquisition and Hardware Pre-sets

A total of 1393 B-mode US images were collected using a hand-held portable linear US scanner (L7, Clarius Mobile Health CO., Canada) with probe frequency of 4–13 MHz from two young volunteers by sweeping the greater trochanter, the medial femur epicondyle, the medial tibial surface and the radial tuberosity with two hardware pre-sets of the scanner. Each US image was 458×480 pixels, and the depth settings and image resolutions were 4 cm and 0.086 mm. Collected images were independently reviewed, and two experienced practitioners manually segmented the bone surfaces for validation purposes.

To maximise the performance for automatic bone edge recognition, several manufacturer's pre-sets were tested, and one set of parameters originally designed for ocular examination was selected for the further experiments to test the automatic segmentation method. A few mechanical indexes have been adjusted in comparison with the common b-mode settings (Table 1). For example, the derated peak rarefactional pressure associated with the transmit pattern was reduced from 2.66 to 0.487 megapascals; and the pulse repetition frequency was reduced from 9600 to 4800 Hz. The ultrasound power of the selected mode was also reduced from 9.33 to 0.153 milliwatts. Figure 3 demonstrates an example of scanning the same anatomical feature with different hardware parameters. Although scanning muscle and bones with the adjusted mode designated for ocular examinations would suppress soft-tissue resolutions, it saved 50 milliseconds and produced a better accuracy of bone edge labelling in our preliminary testing. Therefore, comprehensive quantitative comparisons between scanning modes were performed in the current study.

2.3   Algorithm Testing

The performance of the simplified *bone probability method* was compared against the original *local phase method* [5] and the manual segmentation results. In order to reduce the influence of human error, the average position in each column of the image where both reviewers had labelled a pixel as the bone edge was defined as ground truth data.

Experiments were carried out on a Mac computer with Intel Core i7 CPU, 16 gigabyte RAM, and AMD Radeon Pro 5300M graphic processor and the testing environment was MATLAB (R2019b Update 4). A total of 1393 ultrasound images were independently processed by both the *local phase method* and *bone probability method* segmentation algorithms on the same computer to determine the segmentation speed. The processing time of each image was timed and recorded with *MATLAB function tic, toc*. Two-way ANOVA tests were applied to examine the difference in mean processing times across total images between methods and acquisition modes. The results of both methods were compared with a manually labelled data set using sensitive and specific tests to evaluate the performance. The Accuracy, Recall, Precision, Specificity and F1 score were calculated to exam the performance. Sørensen–Dice coefficient between each automatic and manual label was computed to evaluate the similarity of the shape. All performance metrics between methods and acquisition modes were compared with two-way ANOVA.

Table 1. Acoustic output parameters comparison between pre-sets

Parameters	Unit	B-mode	Adjusted mode
Mechanical Index			
Pr3[a]	MPa	2.66	0.487
Axial distance	Cm	1.90	0.90
Centre frequency	MHz	7.04	6.79
Pulse duration	μsec	0.180	0.183
Pulse repetition frequency	Hz	9600	4800
Pr at PIImax[b]	MPa	4.21	0.601
The soft tissue thermal Index			
Ultrasonic power	mW	9.33	0.153
Active aperture dimensions	cm	1.34	0.499
Centre frequency	MHz	7.04	6.79
Focal length (parallel)	cm	2	0.9
Focal length (vertical)	cm	4	4

[a] *peak rarefactional pressure (in MPa) that has been derated by 0.3 dB/MHz-cm*
[b] *Peak rarefaction pressure at maximum pulse intensity integral*

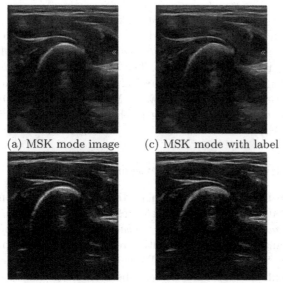

(a) MSK mode image   (c) MSK mode with label

(b) Adjusted mode image (d) Adjusted mode with label

Fig. 3. An example of comparison between hardware pre-sets. a)MSK mode, b)Adjusted mode, c)labelled MSK image, d)labelled adjusted image (segmentation labels were generated with the local phase method [5])

2.4   Performance Testing Against a Machine Learning Model

To explore the performance comparison against machine learning frameworks for image segmentation, we compared the bone probability method with a 2D U-Net model [11] trained with a separate set of 235 images. Among these images, 200 was used for training, and 35 was used for validation. Images were normalised and resized to 256 × 256 pixels. The code was run in Python using TensorFlow with an Ubuntu 20.04 Operating System on an Nvidia GeForce RTX3090 GPU with 24 GB memory and Intel i9-10900K CPU. After training, an independent set of 15 images was used for the performance test. The results of *bone probability method* and 2D U-Net were compared with the contour masks used in the training process. The same set of evaluation scores for assessing the quality of the segmentation between methods were computed, and paired t-tests were performed to exam the systematic difference. Segmentation times were not compared because two methods were performed on different machines, and the image size was adjusted to a smaller scale in this sub-experiment.

3   Results

3.1   Processing Time

On average, the *bone probability method* took 17.6 ± 5.5 and 19.9 ± 8.1 milliseconds to segment one US image acquired with standard MSK mode and adjusted mode, while the original *local phase method* needed 327.2 ± 14.1 and 335.4 ± 14.2 milliseconds to process the image. The *bone probability method* was significantly quicker than the *local phase method* ($p < 0.01$). Both methods needed less time to process images acquired with MSK mode ($p < 0.01$). The interaction between two factors (processing methods and acquisition methods) was significant ($p < 0.01$)

3.2   Quantitative Comparison Between Methods

The results from both automatic segmentation methods were compared to the manual segmentation. Table 2 shows the results of the comparison with both the *local phase method* and the *bone probability method*. All included images contained bony anatomy; the segmentation results labelled no bone in the image were considered "detection failure". There was no detection failure reported from this particular set of data. As can be seen from Table 1, despite the difference being small, *local phase method* was statistically better than *bone probability method* in Accuracy, Specificity, Precision, Dice and F_1 Score ($p < 0.01$). Notably, *bone probability method* has higher recall metrics when labelling images acquired with adjusted mode ($p < 0.01$).

The comparison between common musculoskeletal mode and adjusted mode can also be seen in Table 2. The ANOVA showed that both methods had better performance in processing images acquired with the adjusted acquisition mode.

Table 2. Classification table comparing *Local Phase method* and *Bone Probability method* against the manual segmentation (with the best performance scores hightlighted in red).

AcqMode	Local Phase method		Bone Probability method	
	MSK	Adjusted	MSK	Adjusted
Accuracy	0.9993(0.0005)	0.9993(0.0005)	0.9984(0.0007)	0.9985(0.0006)
Recall	0.6459(0.2297)	0.6139(0.2550)	0.4520(0.2591)	0.6663(0.2441)
Specificity	0.9998(0.0003)	0.9998(0.0003)	0.9991(0.0005)	0.9989(0.0004)
Precision	0.7885(0.2384)	0.8110(0.2758)	0.5539(0.2937)	0.6688(0.2533)
Dice	0.6675(0.2142)	0.6754(0.2510)	0.4205(0.2181)	0.5722(0.1960)
F_1	0.7025(0.1868)	0.7180(0.2104)	0.5324(0.2171)	0.6580(0.2309)

values: mean (standard deviation)

3.3 Qualitative Comparison Between Methods

The qualitative results of two tested methods are shown in Fig. 4, where the red pixels indicate bone labels produced by *local phase method*, while magenta pixels indicate the labels by the *bone probability method* for the segmentation. Bone segmentation results against experienced manual labels are presented side by side in the figure. Two acquisition settings for B-mode US data were also presented for comparison. First and second columns were the same image captured with standard MSK mode, and the third and fourth columns were images captured with the adjusted acquisition mode.

Investigating the segmentation results, we can infer that when the images were captured with the standard MSK settings, the proposed *bone probability method* did not perform as well as the *local phase method*:

- the small gap from the manual labels
- missing bone boundaries
- false bone labels when strong reflection occurred on the sharp bony surface

Although the performance increased when the images were captured with the adjusted mode, some false positive segmentation is still visible. On the other hand, the original method using *local phase feature categorisation* performed better with the adjusted acquisition mode. No false-positive segmentation was observed, despite some false negative being observed. Qualitatively, the *bone probability method* achieves similar performance for the data set that captured with adjusted mode.

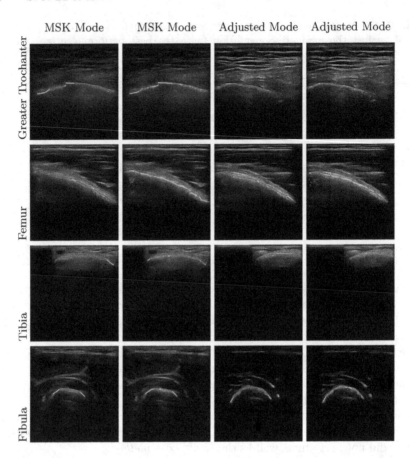

Fig. 4. Bone labels obtained from the *local phase method* (red) and proposed *bone probability method* (magenta) to manual segmentation (green). (Color figure online)

3.4   Performance Comparison with U-Net

The evaluation results of segmentation methods with and without machine learning were compared to the manual segmentation. Table 3 shows the numerical results of the comparisons. The *Bone probability method* achieved a same level of specificity($p = 0.65$) with the 2D U-NET model, and a better recall metric ($p < 0.01$). However, the performance of proposed method was slightly weaker in Accuracy ($p < 0.01$), Precision ($p < 0.01$), Dice ($p < 0.01$) and F_1 Score ($p = 0.02$), despite the difference being small.

Table 3. Classification table comparing machine learning (2D U-NET) method and *Bone Probability method* against the manual segmentation (with the better score hightlighted in red).

Methods	Accuracy	Recall	Specificity	Precision	Dice	F1
BP method	0.9987	0.9843	0.9987	0.7846	0.7232	0.9161
2D U-NET	0.9995	0.9596	0.9996	0.9643	0.8843	0.9725
p value	< 0.01	< 0.01	0.65	< 0.01	< 0.01	0.02

4   Discussion and Conclusion

In this study, we attempted to use a simplified *bone probability method* for bone segmentation from US data. Our previous method [5] that incorporated local phase images in conjunction with B-mode US data achieved an acceptable level in quality bone edge localisation. However, the processing time is a significant barrier to the clinical application that we are currently developing. We have investigated how to achieve a similar quality of bone labelling without combining information from local phase images and B-mode US data since it is the most time-consuming step. Our result demonstrates that the proposed *bone probability method* successfully reduced the processing time from $\sim$ 335 ms to $\sim$ 19 ms per image in the size of 458×480 pixels. This is due to the use of bone probability map for effectively preventing the bone shadow region from being processed, while the original *local phase method* had to process the entire image. This processing speed enables us to perform $\sim$ 50 bone edge segmentation per second, which promises real-time data processing while capturing the B-mode US data. This was an essential improvement for the proposed kinematic assessment system with ultrasound.

Although the computational time performance has been significantly improved by giving up the local phase extraction, the qualitative performance assessment was unsatisfactory. For our primary purpose of building an automatic 3D bone surface reconstruction method for doing kinematic assessments, the most crucial performance metrics are specificity and precision. High specificity means fewer false-negative labels that create ghost points when transforming 2D points into the 3D coordinate system and require extra cleaning work. Similarly, high precision means a greater number of true-positive labels, reducing the total time required for scanning and building the completed surface. The original *local phase method* has achieved an adequate level of specificity regardless of the acquisition mode (mean 0.9998, SD:0.0003), and the *bone probability method* was able to achieve a similar level at 0.9989 (SD: 0.0004) with the data set acquired with adjusted mode. However, the *bone probability method* could not improve Precision, although this method did have the best Recall metric across all the conditions. Overall, the *bone probability method* did not perform better than the *local phase method* quantitatively or qualitatively. These outcomes suggest that local phase extraction may still be required to obtain the same level of accuracy, but a faster alternative method is desired.

The high computational time required for the local phase image features extraction is a noticeable drawback for real-time bone edge segmentation. Alsinan et al. [1] reported that it takes 1 s on average to perform local phase extraction. Our original method was also bottlenecked by the same process, although its execution time can be improved by using more powerful computing hardware. Minimising computational bottleneck can also be achieved by parallelisation. In particular, GPGPUs (general-purpose graphics processing units) are highly suited to the computations commonly found in segmentation algorithms [13]. For operations with low data dependency, such as bone shadow estimation, US image enhancement and normalisation, the throughput may be increased by a factor of 10. This can significantly decrease the time consumed by these bottlenecks.

Although applying a learning free method has a few advantages, the robustness of this type of algorithm remains a challenge. When comparing with the 2D U-net model, the *bone probability method* was able to achieve a similar level of specificity at 0.9987 (2D U-Net: 0.9996) and a better recall score at 0.9843 (2D U-Net: 0.9596). However, the overall scores of the learning free method were not better. Currently, researchers are still attempting different frameworks to achieve a better performance, for example, deploying a two-stage design for U-Net [3], or a novel generative adversarial network architecture [2]. The basic image preset and adjustment mentioned in the present work may also be considered when designing a new machine learning framework. We are open to exploring the use of a machine learning approach if it can operate at a sufficient speed and without the need for expensive hardware. However, creating a suitable training set by relying on expert observers to label images manually is also a relative drawback of machine learning approaches.

The present study attempted to use the simplest steps to achieve a similar level of segmentation as a previous method. A fast and robust method that can be deployed on a standard machine is required. Our future work will involve (1) extensive validation of bone labelling against known geometry; (2) improving the accuracy of the *bone probability method* by adding more functions or attempting to reduce the computational cost of local phase feature extraction, (3) replacing the local phase extraction part by using the mask produced by our *bone probability method* and combine with fusion network models proposed by Ronneberger et al. [11] and Alsinan et al. [1].

References

1. Alsinan, A.Z., Patel, V.M., Hacihaliloglu, I.: Automatic segmentation of bone surfaces from ultrasound using a filter-layer-guided CNN. Int. J. Comput. Assist. Radiol. Surg. **14**(5), 775–783 (2019). https://doi.org/10.1007/s11548-019-01934-0
2. Alsinan, A.Z., Patel, V.M., Hacihaliloglu, I.: Bone shadow segmentation from ultrasound data for orthopedic surgery using GAN. Int. J. Comput. Assist. Radiol. Surg. **15**(9), 1477–1485 (2020). https://doi.org/10.1007/s11548-020-02221-z

3. Amiri, M., Brooks, R., Behboodi, B., Rivaz, H.: Two-stage ultrasound image segmentation using U-Net and test time augmentation. Int. J. Comput. Assist. Radiol. Surg. **15**(6), 981–988 (2020). https://doi.org/10.1007/s11548-020-02158-3
4. Foroughi, P., Boctor, E., Swartz, M.J., Taylor, R.H., Fichtinger, G.: P6d–2 ultrasound bone segmentation using dynamic programming. In: 2007 IEEE Ultra Sonics Symposium Proceedings, pp. 2523–2526. IEEE (2007). https://doi.org/10.1109/ULTSYM.2007.635
5. Jia, R., Mellon, S.J., Hansjee, S., Monk, A.P., Murray, D.W., Noble, J.A.: Automatic bone segmentation in ultrasound images using local phase features and dynamic programming. In: 2016 IEEE 13th International Symposium on Biomedical Imaging (ISBI), pp. 1005–1008 (2016). https://doi.org/10.1109/ISBI.2016.7493435
6. Jia, R., Monk, A.P., Murray, D.W., Mellon, S.J., Noble, J.A.: Greater trochanter tracking in ultrasound imaging during gait. In: 2015 IEEE 12th International Symposium on Biomedical Imaging (ISBI), pp. 260–263 (2015). https://doi.org/10.1109/ISBI.2015.7163863
7. Jia, R., Mellon, S., Monk, P., Murray, D., Noble, A.: Globally optimal registration for describing joint kinematics. Procedia Comput. Sci. **90**, 188–193 (2016). https://doi.org/10.1016/j.procs.2016.07.016, 20th Conference on Medical Image Understanding and Analysis (MIUA 2016)
8. Jia, R., Mellon, S., Monk, P., Murray, D., Noble, J.A.: A computer-aided tracking and motion analysis with ultrasound (CAT & MAUS) system for the description of hip joint kinematics. Int. J. Comput. Assist. Radiol. Surg. **11**(11), 1965–1977 (2016). https://doi.org/10.1007/s11548-016-1443-y
9. Jia, R., Monk, P., Murray, D., Noble, J.A., Mellon, S.: Cat & maus: a novel system for true dynamic motion measurement of underlying bony structures with compensation for soft tissue movement. J. Biomech. **62**, 156–164 (2017). https://doi.org/10.1016/j.jbiomech.2017.04.015
10. Monk, A.P., et al.: Measurement of in-vivo patella kinematics using motion analysis and ultrasound (maus). In: 2013 IEEE International Symposium on Medical Measurements and Applications (MeMeA), pp. 257–260 (2013). https://doi.org/10.1109/MeMeA.2013.6549747
11. Ronneberger, O., Fischer, P., Brox, T.: U-net: convolutional networks for biomedical image segmentation. In: Navab, N., Hornegger, J., Wells, W.M., Frangi, A.F. (eds.) MICCAI 2015. LNCS, vol. 9351, pp. 234–241. Springer, Cham (2015). https://doi.org/10.1007/978-3-319-24574-4_28
12. Salehi, M., Prevost, R., Moctezuma, J.-L., Navab, N., Wein, W.: Precise ultrasound bone registration with learning-based segmentation and speed of sound calibration. In: Descoteaux, M., Maier-Hein, L., Franz, A., Jannin, P., Collins, D.L., Duchesne, S. (eds.) MICCAI 2017. LNCS, vol. 10434, pp. 682–690. Springer, Cham (2017). https://doi.org/10.1007/978-3-319-66185-8_77
13. Smistad, E., Falch, T.L., Bozorgi, M., Elster, A.C., Lindseth, F.: Medical image segmentation on GPUs: A comprehensive review. Med. Image Anal. **20**(1), 1 – 18 (2015). https://doi.org/10.1016/j.media.2014.10.012, https://www.sciencedirect.com/science/article/pii/S1361841514001819
14. Villa, M., Dardenne, G., Nasan, M., Letissier, H., Hamitouche, C., Stindel, E.: FCN-based approach for the automatic segmentation of bone surfaces in ultrasound images. Int. J. Comput. Assist. Radiol. Surg. **13**(11), 1707–1716 (2018). https://doi.org/10.1007/s11548-018-1856-x

15. Wang, P., Patel, V.M., Hacihaliloglu, I.: Simultaneous segmentation and classification of bone surfaces from ultrasound using a multi-feature guided CNN. In: Frangi, A.F., Schnabel, J.A., Davatzikos, C., Alberola-López, C., Fichtinger, G. (eds.) MICCAI 2018. LNCS, vol. 11073, pp. 134–142. Springer, Cham (2018). https://doi.org/10.1007/978-3-030-00937-3_16

Pancreas Volumetry in UK Biobank: Comparison of Models and Inference at Scale

James Owler[1(✉)], Alexandre Triay Bagur[1,2], Scott Marriage[1], Zobair Arya[1], Paul Aljabar[1], John McGonigle[1], Sir Michael Brady[1,3], and Daniel Bulte[2]

[1] Perspectum Ltd., Oxford, UK
james.owler@perspectum.com
[2] Department of Engineering Science, University of Oxford, Oxford, UK
[3] Department of Oncology, University of Oxford, Oxford, UK

Abstract. The UK Biobank imaging sub-study enables large-scale measurement of pancreas volume, an important biomarker in metabolic disease, including diabetes. Previous methods utilised a pancreas-specific (PS) 3D MRI UK Biobank acquisition to automatically measure pancreas volume. This may lead to a clinically significant underestimation of volume, due to partial coverage of the pancreas in these acquisitions. To address this, we propose a pipeline for the accurate measurement of pancreas volume using stitched whole-body (WB) 3D MRI UK Biobank acquisitions and deep learning-based segmentation. We implement and compare the performance of six different U-Net-like model architectures, leveraging attention layers, recurrent layers, and residual blocks. Furthermore, we investigate pancreas volumetry in 42,313 subjects, separated by sex, and present novel results concerning the change in pancreas volume throughout the course of a day (diurnal variation). To the best of our knowledge, this is the largest pancreas volumetry study to date and the first to propose a pipeline using the whole-body UK Biobank MRI acquisitions to measure pancreas volume.

Keywords: Pancreas segmentation · Deep learning · UK Biobank

1 Introduction

Pancreas volume has been shown to change with age and in diseases such as pancreatitis, type 1 diabetes, and type 2 diabetes [1–5]. Pancreas volume is typically measured following segmentation. Manual pancreas segmentation is labour-intensive and the delineation of a three-dimensional shape, with ill-defined boundaries extending across multiple views, is prone to substantial inter-rater and intra-rater variability. Automating pancreas segmentation can alleviate these problems by increasing reproducibility and decreasing subjectivity. Automation can also allow for large-scale model deployment, leading to practical implementations for clinical decision-making and population health research.

© Springer Nature Switzerland AG 2021
B. W. Papież et al. (Eds.): MIUA 2021, LNCS 12722, pp. 265–279, 2021.
https://doi.org/10.1007/978-3-030-80432-9_21

Deep learning-based pancreas segmentation methods have been proposed to automate pancreas segmentation [6,7].

UK Biobank is one of the largest resources of imaging and non-imaging medical data in the world. This resource opens up the possibility of exploratory research into the realm of precision medicine [8]. The UK Biobank imaging sub-study aims to scan a total of 100,000 volunteers at multiple timepoints [9]; part of this sub-study includes numerous MRI imaging acquisitions. One such acquisition is a dedicated pancreas volumetric interpolated breath-hold examination (VIBE). Previous works have used this pancreas-specific scan for pancreas segmentation, volumetry estimation, shape measurement and downstream pancreatic quantification [10,11]. The high resolution nature of the pancreas-specific scan allows models to more easily learn useful representations of the pancreas and to better quantify biomarkers like surface lobularity [5,12]. However, in many cases, there is only partial coverage of the pancreas, often missing parts of the pancreas head region (Fig. 1). The longitudinal nature of UK Biobank motivates the need for accurate and precise pancreas volume measurement, in order to detect small (clinically) meaningful changes caused by aging and pathological processes. The partial coverage effect observed in pancreas-specific images could lead to significant inaccuracies in volume measurements, hindering the ability to detect such small changes. For reference, a study by Saisho et al. [2] reported a $74.9\,cm^3$ total pancreas volume in type 2 diabetics and $70.0\,cm^3$ volume in age-, sex-, and BMI-matched controls.

The UK Biobank imaging protocol also contains a whole-body (neck-to-knee) 2-point Dixon acquisition, acquired sequentially in multiple breath-hold volumes. Although they have been acquired at slightly lower resolution than the pancreas-specific scan, they have the advantage that they contain full coverage of the pancreas. To the best of our knowledge, no study investigating pancreas volumetry in UK Biobank has used these whole-body acquisitions to measure pancreas volume. These whole-body scans can provide accurate and consistent volume measurements across the UK Biobank population. In addition, they can provide insight into interrelations with other imaging (e.g. quantitative MRI, volumetry from other organs) and non-imaging (e.g. blood tests, diagnoses, genetics) biomarkers, in order to better understand and improve treatment of disease.

In this work, we propose a novel pipeline to study pancreas volumetry in UK Biobank. This pipeline can be easily extended to study other organs captured in the whole-body acquisition. The pipeline includes stitching, registration to a common coordinate space, cropping out the abdominal region, deep learning-based segmentation model training, testing and prediction. We leverage a previously presented pancreas segmentation model [11] to build a large annotated dataset of whole-body pancreas segmentations. We implement six different U-Net [13] based models that have been proposed in the literature, some with existing pancreas segmentation applications [14], and compared their performance. Finally, we estimate the pancreas volume under-estimation caused by the partial coverage in the pancreas-specific acquisitions and investigate pancreas volume population metrics, including diurnal variation of pancreas volume.

Pancreas-specific   **Whole-Body**

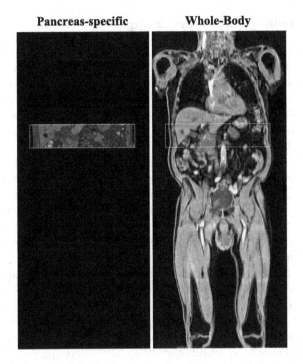

Fig. 1. A coronal slice through the 3D pancreas-specific scan (left), overlayed on a coronal slice through the 3D whole-body scan (right). The bottom of the pancreas has been cropped out in the PS scan which has led to the underestimation of pancreas volume. The measured pancreas volume from the PS scan was 78.6 cm^3, whereas the measured pancreas volume from the whole-body scan was 91.9 cm^3 – a 14.4 % underestimation.

2 Materials and Methods

2.1 Data Acquisition

Imaging data was acquired with a Siemens Aera 1.5 T (Siemens Healthineers AG, Erlangen, Germany). "Pancreas fat - DICOM" (Data-Field ID 20202 in the UK Biobank Showcase[1]) volumetric acquisition, which we are referring to here as the pancreas-specific (PS) scan, targets the abdominal location of the pancreas. The PS scan used the FLASH-3D acquisition (echo time (TE)/repetition time (TR) = 1.15/3.11 ms, voxel size = 1.1875 × 1.1875 × 1.6 mm), with 10° flip angle and fat suppression.

"Dixon technique for internal fat - DICOM" (Data-Field ID 20201) volumetric acquisition, which we are referring to here as the whole-body (WB) scan, involved multiple dual echo Dixon VIBE volumes, acquired over the course of 6 min to provide water/fat separated overlapping volumes from the neck to the knees. Further details about the acquisition can in found in [15].

---

[1] https://biobank.ndph.ox.ac.uk/showcase/browse.cgi.

2.2 Data Labelling and Preprocessing

First, overlapping volumes from the WB acquisition were automatically stitched together, following a round of N4 bias field correction [16] on each volume block. To deal with an MRI wrapping artefact present in the data, the top two and the bottom two slices from each block were removed prior to stitching. Linear normalisation was then applied across blocks, to correct for any cross-block intensity inhomogeneity. This resulted in a relatively 'clean' stitched image, with a homogeneous intensity throughout.

Second, in order to build an annotated dataset for model training, we leveraged a model previously trained on PS acquisition images that were labelled by an expert [11]. Labels, obtained from the PS model were propagated across to the WB images using DICOM header geometry information. 200 WB images, along with their propagated label counterpart, were selected by visual inspection. Subjects that had minimal movement between the two types of acquisitions and thus, whose propagated labels best aligned with the WB image were selected. Corrections, including the addition of the missing pancreas information, were made after the alignment. Corrections were performed using the 3D brush tool in ITK-SNAP [17]. This labelling process allowed us to quickly build an annotated dataset.

Third, a representative WB image was selected to provide a dedicated 'reference' coordinate space to which all other WB images were aligned via affine image registration. Here, we used the affine registration implementation from ANTS [18]. This enabled us to heuristically crop out the abdominal region from the WB image, resulting in a consistent image size and similar coverage of abdominal organs across all subjects. As we are only interested in the pancreas region, training a model using the full WB image would have been inefficient in terms of GPU memory useage, in addition to potentially degrading the performance of the model. The same registration and cropping was applied to the WB pancreas label. Segmentations were transformed back to their original coordinate space, via the inverse affine transformation, when calculating a final measurement of pancreas volume.

Lastly, we clipped image intensities at the 99th percentile value, normalised voxel values between 0 and 1, and randomly split the dataset as follows: 70% training, 20% testing, and 10% validation.

2.3 Model Architectures

We compared the performance of six different semantic segmentation networks based on previously reported architectures. The first of these was a conventional 3D U-Net with an encoder path, decoder path and skip connections [19]. The details of this network can be found in Figs. 2a and 3a.

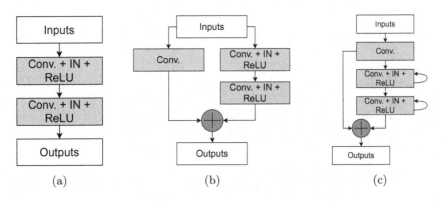

Fig. 2. Different variants of convolutional blocks. (a) Conventional convolutional block, (b) Residual convolutional block, (c) Recurrent residual convolutional block.

The second network was an attention U-Net (AU-Net) as reported by Oktay et al. [14]. The idea behind AU-Nets is to include a mechanism such that the network learns to focus only on the features of an image that are relevant to the task. Given that the pancreas encompasses only a small part of the cropped whole-body image, AU-Nets have potential to be suitable for our application. We implemented the AU-Net by, prior to the concatenation step in each skip connection, 'gating' the incoming feature maps from the encoder layer as shown in Fig. 3b. This gating step is as follows: the n_x feature maps from the encoder layer are denoted by x, and the corresponding n_g feature maps from the decoder layer that are typically concatenated with x in a skip connection are denoted by g. First, x is downsampled to the same spatial resolution as g. A set of attention weights, α, is then obtained through the following operation: $\alpha = \sigma_2(\psi^T(\sigma_1(W_x^T x + W_g^T g + b_1)) + b_2)$. Here, W_x and W_g represent $1 \times 1 \times 1$ convolution operations that output n_x features, σ_1 is a ReLu activation function, ψ is a $1 \times 1 \times 1$ convolution that outputs 1 feature map, σ_2 is a sigmoid activation function and the b vectors are bias terms. Finally, α is resampled to the same spatial resolution as x, and x is multiplied by α to give the attention gated features $\hat{x}$.

The third architecture we investigated was a residual U-Net (RU-Net) as reported by Alom et al. [20]. Figure 2a shows that our conventional U-Net architecture consists of blocks that have two convolutional + instance-normalisation + ReLu (Conv + IN + ReLu) layers. In the RU-Net, these blocks are altered so that output of the second Conv + IN + ReLu layer is summed with a feature set based on the input to the first Conv + IN + ReLu layer, as shown in Fig. 2b. Note that the 'Conv' layer in Fig. 2b is a $1 \times 1 \times 1$ convolutional layer in the residual path to ensure that the dimensions of the two feature sets to be summed match. The motivation behind RU-Nets is that the residual paths can improve optimisation during training.

The fourth architecture that we investigated, also reported by Alom et al. [20], was a recurrent residual U-Net (R2U-Net). It builds on top of

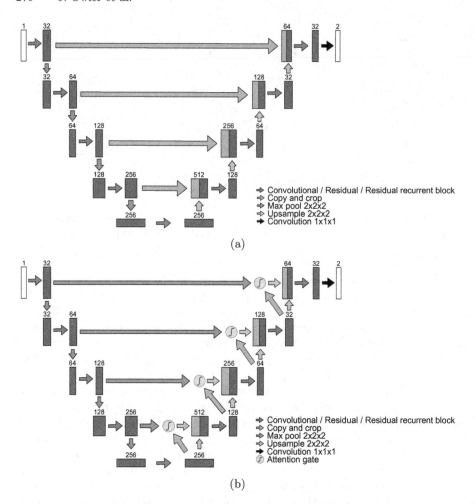

Fig. 3. The base U-Net model architectures for (a) a conventional U-Net and (b) an attention U-Net. Note that the blue arrows represent a block of operations, where the details of the block depend on the specific U-Net variant being implemented. The details for each type of block are shown in Fig. 2.

RU-Net by using recurrent convolutional layers (RCL), as opposed to conventional convolutional layers. R2U-Net has been shown to improve performance when compared to a conventional U-Net. In an RCL, features are 'accumulated' in a layer by forward propagating through the layer multiple times, and taking into account the feature maps from the previous propagations. For example, if x is the original input to the RCL, then the output of the RCL can be calculated by $o_t = W_f^T z_t + b$, where W_f represents a convolution, $z_1 = x$ for $t = 1$ and $z_t = o_{t-1} + z_{t-1}$ for $t > 1$. The output at each time-step is fed through a ReLu activation. We implemented the R2U-Net with two time-steps. The details of the

R2U-Net block is shown in Fig. 2c. As before, a 'Conv' layer is used to ensure that the dimensions match for any summation of features.

The U-Net variants above are modular, meaning they can be further combined with one another to utilise each of their potential benefits. Consequently, we investigated two further variants. The first was a residual U-Net with attention-gating (RAU-Net), the second was a recurrent residual U-Net with attention-gating (R2AU-Net).

2.4   Model Training and Testing

Before training, weights in each network were initialised using Glorot uniform initialisation [21]. Data augmentation and random shuffling were performed 'on-the-fly'. Augmentation included random rotations, within a range of -10° to +10° about the inferior-superior axis. We also randomly scaled the image size to between 95% and 105% of the original image dimensions. Zero-padding was used to keep to a consistent input dimension. Each network was trained for 100 epochs, with a batch size of 1 and a learning rate of 0.0005. Model weights were saved each time there was a decrease in validation loss. ADAM [22] optimisation was used, with a combined cross-entropy and soft dice loss function as shown in the following equation:

$$
L = \frac{1}{N}\frac{1}{K}\sum_{i}^{N}\sum_{c}^{K}\left(1 - \frac{2\sum_{j}^{M} p_{icj}y_{icj} + \delta}{\sum_{j}^{M}(p_{icj} + y_{icj}) + \delta} - \frac{1}{M}\sum_{j}^{M} y_{icj}\log p_{icj}\right) \tag{1}
$$

where $i = [1..N]$ is the subject index, $c = [1..K]$ is the class index, $j = [1..M]$ is the voxel index, p_{icj} is the predicted probability of voxel j for subject i belonging to class c, y_{icj} is a binary variable equal to 1 if the ground truth class for voxel j in subject i is c and 0 otherwise, and $\delta = 0.01$.

The models, training pipeline, and testing pipeline were all implemented using PyTorch[2]. Each model took approximately 5 h to train on an NVIDIA Tesla V100 GPU.

During inference, final pancreas segmentations were obtained by thresholding the label probabilities at 0.5, a threshold that selects the class label with the largest posterior probability under the assumption of equal misclassification costs. These segmentation labels were then resampled back to the original image size, using nearest-neighbour interpolation. Pancreatic volumes were calculated from the final segmentation labels.

To evaluate each model we used the commonly reported Dice Similarity Coefficient (DSC) and the 95th percentile of the Hausdorff Distance metric (HD95). Results from this evaluation can be seen in Sect. 3.1.

---

[2]   https://pytorch.org/.

2.5   Model Inference at Scale

After evaluating the performance of each model variant, we used AU-Net to automatically segment 42,313 pancreas volumes in UK Biobank. The stitching, cropping, and pre-processing of each WB image was the same as described in Sect. 2.2. By utilising Terraform[3] to orchestrate multiple Amazon Web Services EC2 instances, we were able to obtain all 42,313 pancreas segmentations, and their corresponding volumes, in less than 4 h.

Pancreas Volume Within the UK Biobank Population. Using these automatically obtained pancreas segmentations, we compared volume measurements from the proposed WB segmentation method with volume measurements derived from the previously proposed PS segmentation model [11]. We used this comparison to estimate the extent of pancreas volume under-estimation in the PS scans. We also calculated the average pancreas volume for males, females, and combined males and females. These results are presented in Sect. 3.2 and 3.3, respectively.

Pancreas Volume Diurnal Variation. As an applied use-case of automated pancreas volumetry, and to demonstrate the versatility of large-scale research resources such as UK Biobank, we further investigated the natural change in pancreas volume (if any) throughout the course of the day. In UK Biobank, each subject is scanned just once, at an imaging session typically between the times of 9 am and 7 pm. Here, we rounded the timepoint at which a subject was scanned to the nearest hour; subjects scanned after 7:30 pm and before 8:30am were excluded to keep the number of subjects in each sub-group to greater than 1000. This resulted in 11 unique groups of pancreas volumes (mean n = 3791, range n = 2796–4194). We then calculated the median pancreas volume at each timepoint and observed the change in median volume between those timepoints. The sheer scale of UK Biobank allows us to investigate average changes in the human body, with the noise present from individual measurements largely mitigated. Using the whole-body images to measure pancreas volume also mitigates added noise from partial coverage in the PS acquisitions. One could measure the volume of the pancreas for the same individual at multiple timepoints throughout the day; however, it can be argued that observing the same average phenomena in tens of thousands of people provides greater validity to the result. Pancreas volume diurnal variation results are presented in Sect. 3.4.

3   Results

3.1   Model Evaluation

Qualitative Results. Figure 4 shows a qualitative comparison of the different models. Although it is difficult to draw any firm conclusions from a qualitative

---

[3] https://www.terraform.io/.

evaluation of predicted segmentations, one noticeable observation is that each of the automated models in Fig. 4 struggle to segment the pancreas towards the lowest extent of the organ. See Sect. 4 for further discussion.

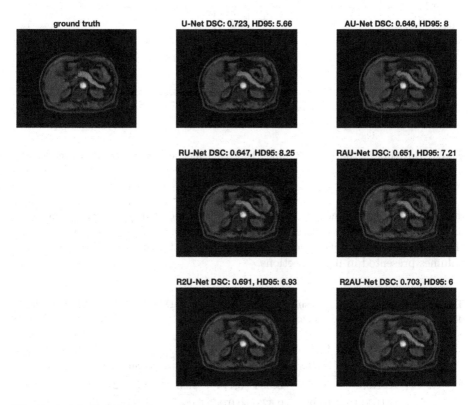

Fig. 4. Axial slice through a cropped WB image, showing automated pancreas segmentations and their respective performance metrics, for each type of model being compared.

Quantitative Results. Table 1 shows the DSC and HD95 metrics for the various models we investigated. Figures 5a and 5b also show boxplots for the these metrics. The conventional U-Net and R2U-Net models both resulted in the highest mean DSC, while the conventional U-Net resulted in the highest median DSC. When considering the standard deviation of DSC, it was R2U-Net that appeared to be least susceptible to outliers, with AU-Net in close second. In terms of HD95, AU-Net resulted in the lowest mean score and tightest standard

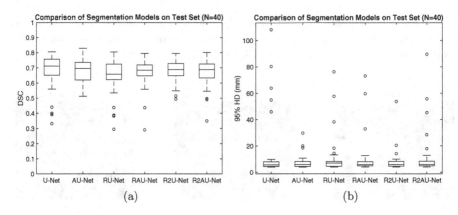

Fig. 5. Boxplots of the (a) DSC and (b) HD95 metrics for each type of model we investigated.

deviation. In general, no one model appeared to be clearly superior. The performances were similar across the board. Based on the fact that it was one of the models least susceptible to outliers, we decided to use AU-Net to derive the volumes presented in future sections.

Table 1. Model evaluation results. DSC - dice score; HD 95 - 95 percentile Hausdorff distance; SD - standard deviation.

Model	DSC			HD95		
	Median	Mean	SD	Median	Mean	SD
U-Net	0.712	**0.681**	0.114	5.66	13.7	23.2
AU-Net	0.696	0.675	0.078	6.00	**7.54**	**4.94**
RU-Net	0.658	0.652	0.112	7.07	10.1	14.4
RAU-Net	0.685	0.668	0.090	5.83	10.2	13.9
R2U-Net	0.689	**0.681**	**0.070**	5.83	7.87	8.04
R2AU-Net	0.689	0.670	0.088	6.00	11.29	16.51

3.2   Comparison with Volumetry from Pancreas-Specific Scan

Figure 6 shows the histogram of differences between volumes derived from the PS and the WB scans. We observed a mean difference of $11.7\,\mathrm{cm}^3$, or 14.25%, ($p = 1.4 \times 10^{-288}$) between the two approaches.

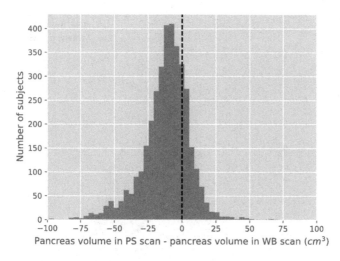

Fig. 6. Comparison of measured pancreas volume from the pancreas-specific scan segmentation model and the proposed whole-body scan segmentation model. A mean volume difference of $11.7\,\mathrm{cm}^3$ (14.25%) was observed. n = 3672.

3.3 UK Biobank Population Volumetry

Table 2 shows median pancreas volume in UK Biobank for both males, females, and combined males and females. Figure 7 shows histograms of pancreas volumes for males and females. These results show a 13% difference between the average volume of the pancreas in males when compared with females.

Table 2. Average pancreas volumes in UK Biobank for males, females, and combined males and females.

	Number Quantified	Median (cm^3)	SD (cm^3)
Male	20395	70.0	19.8
Female	21918	62.0	17.1
Combined	42313	65.5	18.9

3.4 Pancreas Volume Diurnal Variation.

Fig. 8 shows that there is a marked variation in the volume of the pancreas throughout the course of a day. The largest change in total pancreatic volume of a 5.73% reduction ($p = 6.80 \times 10^{-20}$) was observed between the hours of 9am and 3pm. The largest change over the course of an hour, observed between 10 am and 11 am, was a reduction of 2.57% ($p=1.01\times10^{-5}$).

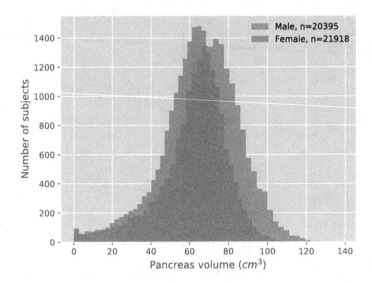

Fig. 7. Histograms of pancreas volume in UK Biobank for males and females. n = 42313.

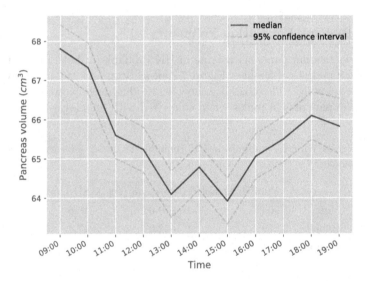

Fig. 8. Pancreas volume diurnal variation in UK Biobank (combined male and female). n = 41704.

4   Discussion and Conclusion

Pancreas volumetry in UK Biobank measured using the proposed pipeline, notably segmenting the pancreas on the WB acquisitions, agrees with reported values for nominally healthy populations [2] (N = 1,721). On the other hand, pancreas volumetry performed using the PS acquisitions [10,11], underestimated volume by an average of $11.7\,cm^3$ compared to our method. To the best of our knowledge, this work is the largest attempt at accurate pancreas volume measurement in a nominally healthy population, which may be used for reference in future studies of age and disease.

Using the PS segmentation model allowed us to exploit the prior expert knowledge distilled in the annotations used for training the original model. It also enabled cheap, fast generation of good-quality starting estimates of pancreas labels for the WB images. One limitation of this approach is that, while these starting estimates were manually corrected when necessary, the selection of a subset of 'good' starting candidates from a larger dataset could have biased our dataset for WB model training. For instance, if we consider the extreme case of selecting a subset where no annotations are needed, the dataset might become limited towards those small pancreata that already fit in the PS scan volume, though that could be partly addressed using data augmentation.

We chose AU-Net to run at scale as it was less susceptible to outliers in both DSC and HD95. The differences in performance between the segmentation models were not found to be significant (when using a paired t-test); however, the test set was a relatively small sample. The 'effect size' of differences, in which it is difficult to gain insight to with a paired t-test alone, could be investigated further with more sophisticated Bayesian testing [23].

Although the performance of all the models presented here are on-par with other state-of-the-art pancreas segmentation methods [24], there is scope for improvement. Due to all of the cropped UK Biobank whole-body images being the same resolution, a patch-based segmentation approach could improve segmentation performance. This would mean that neither the input image or the output label would require any resampling, thereby avoiding resampling errors at object boundaries. This type of resampling error can be particularly detrimental in smaller organs, such as the pancreas, when using methods like nearest neighbour interpolation. This resampling error could also lead to inaccuracies in pancreas surface lobularity measures.

In terms of the diurnal variation of pancreas volume, we are not aware of the biological mechanism that causes this change; however, a similar pattern has been observed in other organs in the body [25]. It is important to note that there is a marked change in pancreas volume throughout the day, which should be considered when making clinical decisions based on volume assessments. This change could be corrected for via normalisation, although more experimentation is needed to tease out any unforeseen biases in the data before presenting any correction methods.

In conclusion, we have highlighted clinically significant underestimation of pancreas volume in UK Biobank, caused by partial coverage in the pancreas-specific acquisition. We presented a comparison of 6 different variants of U-Net models for pancreas segmentation in whole-body MRI. We also proposed a pipeline for efficient data labelling, using a previously trained PS model, and deployment of a trained model on a large scale. We believe the culmination of large data sources, such as UK Biobank, with deep learning methods, and cloud computing has exciting potential to provide a better insight into population health, allow for the exploration of novel biomarkers, and improve patient care.

Acknowledgements. We would like to acknowledge Perspectum Ltd and the Engineering and Physical Sciences Research Council (EPSRC) for funding and support. This research has been conducted using the UK Biobank resource under application 9914.

References

1. Schrader, H., et al.: Reduced pancreatic volume and β-cell area in patients with chronic pancreatitis. Gastroenterology **136**(2), 513–522 (2009). http://dx.doi.org/10.1053/j.gastro.2008.10.083
2. Saisho, Y., et al.: Pancreas volumes in humans from birth to age one hundred taking into account sex, obesity, and presence of type-2 diabetes. Clin. Anat. **20**(8), 933–942 (2007)
3. Saisho, Y.: Pancreas volume and fat deposition in diabetes and normal physiology: consideration of the interplay between endocrine and exocrine pancreas. Rev. Diabet. Stud. **13**(2–3), 132–147 (2016)
4. Macauley, M., Percival, K., Thelwall, P.E., Hollingsworth, K.G., Taylor, R.: Altered volume, morphology and composition of the pancreas in type 2 diabetes. PLoS ONE **10**(5), 1–14 (2015)
5. Al-Mrabeh, A., et al.: 2-year remission of type 2 diabetes and pancreas morphology: a post-hoc analysis of the DiRECT open-label, cluster-randomised trial. Lancet Diabetes Endocrinol. **8**(12), 939–948 (2020). http://dx.doi.org/10.1016/S2213-8587(20)30303-X
6. Cai, J., Lu, L., Xing, F., Yang, L.: Pancreas segmentation in CT and MRI via task-specific network design and recurrent neural contextual learning. In: Lu, L., Wang, X., Carneiro, G., Yang, L. (eds.) Deep Learning and Convolutional Neural Networks for Medical Imaging and Clinical Informatics. ACVPR, pp. 3–21. Springer, Cham (2019). https://doi.org/10.1007/978-3-030-13969-8_1
7. Isensee, F., Jaeger, P.F., Kohl, S.A., Petersen, J., Maier-Hein, K.H.: nnU-Net: a self-configuring method for deep learning-based biomedical image segmentation. Nat. Methods **18**(2), 203–211 (2021). http://dx.doi.org/10.1038/s41592-020-01008-z
8. Sudlow, C., et al.: Uk biobank: an open access resource for identifying the causes of a wide range of complex diseases of middle and old age. Plos Med **12**(3), e1001779 (2015)
9. Littlejohns, T.J., et al.: The UK Biobank imaging enhancement of 100,000 participants: rationale, data collection, management and future directions. Nature Commun. **11**(1), 1–12 (2020). http://dx.doi.org/10.1038/s41467-020-15948-9
10. Liu, Y., et al.: Genetic architecture of 11 abdominal organ traits derived from abdominal MRI using deep learning, pp. 1–66 (2020)
11. Bagur, A.T., Ridgway, G., McGonigle, J., Brady, S.M., Bulte, D.: Pancreas segmentation-derived biomarkers: volume and shape metrics in the UK biobank imaging study. In: Papież, B.W., Namburete, A.I.L., Yaqub, M., Noble, J.A. (eds.) MIUA 2020. CCIS, vol. 1248, pp. 131–142. Springer, Cham (2020). https://doi.org/10.1007/978-3-030-52791-4_11

12. Calandra, A., Sartoris, R., Lee, K.J., Gauss, T., Vilgrain, V., Ronot, M.: Quantification of pancreas surface Lobularity on CT: a feasibility study in the normal pancreas (2020)

13. Ronneberger, O., Fischer, P., Brox, T.: U-net: convolutional networks for biomedical image segmentation. In: Navab, N., Hornegger, J., Wells, W.M., Frangi, A.F. (eds.) MICCAI 2015. LNCS, vol. 9351, pp. 234–241. Springer, Cham (2015). https://doi.org/10.1007/978-3-319-24574-4_28

14. Oktay, O., et al.: Attention u-net: Learning where to look for the pancreas. arXiv preprint arXiv:1804.03999 (2018)

15. Linge, J., et al.: Body composition profiling in the UK biobank imaging study. Obesity **26**(11), 1785–1795 (2018)

16. Tustison, N.J., et al.: N4ITK: improved N3 bias correction. IEEE Trans. Med. Imaging **29**(6), 1310–1320 (2010). https://www.ncbi.nlm.nih.gov/pubmed/20378467, www.ncbi.nlm.nih.gov/pmc/PMC3071855/

17. Yushkevich, P.A., et al.: User-guided 3D active contour segmentation of anatomical structures: significantly improved efficiency and reliability. NeuroImage **31**(3), 1116–1128 (2006)

18. Avants, B.B., Tustison, N., Song, G.: Advanced normalization tools (ants). Insight J. **2**(365), 1–35 (2009)

19. Milletari, F., Navab, N., Ahmadi, S.A.: V-net: fully convolutional neural networks for volumetric medical image segmentation. In: Proceedings - 2016 4th International Conference on 3D Vision, 3DV 2016, pp. 565–571 (2016)

20. Alom, M.Z., Yakopcic, C., Hasan, M., Taha, T.M., Asari, V.K.: Recurrent residual U-Net for medical image segmentation. J. Med. Imaging **6**(1), 014006 (2019)

21. Glorot, X., Bengio, Y.: Understanding the difficulty of training deep feedforward neural networks. J. Mach. Learn. Res. **9**, 249–256 (2010)

22. Kingma, D.P., Ba, J.: Adam: A method for stochastic optimization. arXiv preprint arXiv:1412.6980 (2014)

23. Benavoli, A., Corani, G., Demšar, J., Zaffalon, M.: Time for a change: a tutorial for comparing multiple classifiers through Bayesian analysis. J. Mach. Learn. Res. **18**, 1–36 (2017)

24. Heinrich, M.P., Oktay, O., Bouteldja, N.: OBELISK-Net: Fewer layers to solve 3D multi-organ segmentation with sparse deformable convolutions. Med. Image Anal. **54**, 1–9 (2019)

25. Owler, J., McGonigle, J., Robson, M., Brady, M., Banerjee, R.: Liver volume diurnal variation in UK biobank. In: The Liver Meeting Digital Experience^{TM}. AASLD (2020)

Ensemble of Deep Convolutional Neural Networks with Monte Carlo Dropout Sampling for Automated Image Segmentation Quality Control and Robust Deep Learning Using Small Datasets

Evan Hann$^{(\boxtimes)}$ ⓘ, Ricardo A. Gonzales ⓘ, Iulia A. Popescu ⓘ, Qiang Zhang ⓘ, Vanessa M. Ferreira ⓘ, and Stefan K. Piechnik ⓘ

Oxford Centre for Clinical Magnetic Resonance Research, Division of Cardiovascular Medicine, Radcliffe Department of Medicine, University of Oxford, Oxford, UK
evan.hann@cardiov.ox.ac.uk

Abstract. Recent progress on deep learning (DL)-based medical image segmentation can enable fast extraction of clinical parameters for efficient clinical workflows. However, current DL methods can still fail and require manual visual inspection of outputs, which is time-consuming and diminishes the advantages of automation. For clinical applications, it is essential to develop DL approaches that can not only perform accurate segmentation, but also predict the segmentation quality and flag poor-quality results to avoid errors in diagnosis. To achieve robust performance, DL-based methods often require large datasets, which are not always readily available. It would be highly desirable to be able to train DL models using only small datasets, but this requires a quality prediction method to ensure reliability. We present a novel segmentation framework utilizing an ensemble of deep convolutional neural networks with Monte Carlo sampling. The proposed framework merges the advantages of both state-of-the-art deep ensembles and Bayesian approaches, to provide robust segmentation with inherent quality control. We successfully developed and tested this framework using just a small MRI dataset of 45 subjects. The framework obtained high mean Dice similarity coefficients (DSC) for segmentation of the endocardium (0.922) and the epicardium (0.942); importantly, segmentation DSC can be accurately predicted with low mean absolute errors (≤ 0.035), in the absence of the manual ground truth. Furthermore, binary classification of segmentation quality achieved a near-perfect accuracy of 99%. The proposed framework can enable fast and reliable medical image analysis with accurate quality control, and training of DL-based methods using even small datasets.

Keywords: Automated quality assessment · Segmentation · Ensemble learning · Monte Carlo sampling

© Springer Nature Switzerland AG 2021
B. W. Papież et al. (Eds.): MIUA 2021, LNCS 12722, pp. 280–293, 2021.
https://doi.org/10.1007/978-3-030-80432-9_22

1 Introduction

Cardiovascular diseases (CVD) are a leading cause of mortality worldwide [1]. Cardiac magnetic resonance (CMR) imaging is a powerful tool in the diagnosis and treatment of CVD, providing comprehensive analysis of cardiac structure and function, especially the left ventricle (LV). Accurate segmentation of the LV is an essential step for the quantification of clinically important parameters, such as volumes, ejection fraction and mass. Despite advances of automated segmentation methods, manual delineations and quality assurance are still the current clinical standard for performing and validating automated segmentation.

Automated LV segmentation has been extensively studied over the past decade, with progress ranging from classical machine learning to advanced deep learning (DL) approaches. The latter was recently enabled by data availability and hardware development. There have been a number of international challenges and collective efforts to benchmark state-of-the-art segmentation accuracy, providing valuable CMR cine SSFP images of the LV, such as the Sunnybrook Cardiac Dataset [2], the Automatic Cardiac Diagnosis Challenge [3], and the UK Biobank [4].

Given the time-consuming task of manual annotation of CMR images in typical clinical workflow, there is significant interest in fully automatic segmentation. Initial efforts required manual extraction of relevant image features with prior knowledge to achieve satisfactory accuracy. A series of LV segmentation methods have been proposed using the publicly available Sunnybrook Cardiac Dataset of 45 subjects [2]. Among others, the proposed approaches use deformable models [5, 6], image-based [7–9] and model-based [10, 11] methods. However, the hand-crafted approaches can fall short in generalizability when dealing with unfamiliar new data. Furthermore, they often require manual adjustments, which limit implementation of fully-automatic tools in modern clinical practice.

With recent advancements of DL, data-driven neural networks can learn end-to-end for image segmentation, reducing the need for hand-crafted approaches. Nevertheless, even state-of-the-art DL methods can still fail on unfamiliar testing data [3]. Case-by-case visual inspection of segmentation quality is still necessary, which is laborious, time-consuming, and defies the benefits of fully-automated methods. Moreover, to achieve robust performance, end-to-end deep learning-based methods require larger and more representative datasets [3, 12], which can be time-consuming to curate and not always readily available. Training of DL models requiring only small datasets would be desirable, but demands a quality prediction method in real-world applications, to flag poor-quality results. We therefore present a DL approach, with automated quality prediction, which holds the DL models accountable, even when trained on small datasets. We validated this novel framework on the Sunnybrook Cardiac Dataset for LV segmentation.

1.1 Related Work

There is increasing interest in developing accountable DL-based segmentation methods with inherent quality control. Bayesian approaches have been proposed to provide means of uncertainty estimation for prediction. In particular, Monte Carlo sampling-based neural networks have been used to perform medical image segmentation, as well as quality control [13, 14]. To implement the Monte Carlo sampling approach, a deep convolutional neural network can be modified by adding dropout units, which randomly "turn off" some internal connections within the neural network [13, 14]. While dropout units are activated only for training in standard DL, they can be activated for testing or deployment to generate many different segmentation samples. The agreement among the samples can be exploited to predict segmentation evaluation metrics, such as Dice similarity coefficient (DSC), without the need of a reference manual segmentation. [13] has successfully demonstrated the capability of the Monte Carlo dropout (MCD) approach for whole brain segmentation.

Alternatively, deep ensembles have also been used to estimate uncertainty and predict segmentation quality [15–17]. Successful applications include segmentation of the brain, prostate, and cardiovascular structures [15, 17, 18]. Similar to Monte Carlo sampling, deep ensembles also generate multiple candidates, then exploit the agreement among candidates to predict output quality or uncertainty. The difference is that a single trained neural network with Monte Carlo dropout can theoretically generate unlimited number of segmentation candidates, whereas the number of candidates generated by deep ensembles is limited by the number of independent neural networks trained. For example, an ensemble of 50 independently-trained neural networks can generate up to 50 different segmentation candidates [18]. This makes deep ensembles more computationally expensive to train and deploy than the Monte Carlo dropout approach. Despite this disadvantage, deep ensembles tend to generate more diverse prediction samples, offering higher accuracy and robustness in uncertainty estimation compared to Bayesian approaches [18, 19]. In addition, selecting the segmentation candidate with the best predicted quality as the final output for deep ensembles can improve the overall accuracy and robustness [15, 17]. The same mechanism has not been applied to Bayesian approaches. Therefore, deep ensembles and Bayesian approaches have their own merits and pitfalls.

It has been shown that using an ensemble of multiple MC-dropout models can improve classification accuracy for handwritten digit and character recognition tasks [20]. In this work, we further explore the idea of combining novel deep ensemble frameworks such as [15, 17] and Bayesian approaches for reliable medical image segmentation and quality control.

1.2 Contributions

The contributions of this work are as follows: (1) we propose a novel ensemble of deep convolutional neural networks with Monte-Carlo dropout to merge the advantages of both deep ensembles and Bayesian approaches for reliable medical image segmentation and quality control; (2) we show that deep ensembles can generate diverse segmentation candidates for reliable quality prediction; (3) we add Monte Carlo dropout in the individual neural networks to efficiently generate a large number of segmentation samples; (4) the proposed framework adopts a novel automatic selection of the final optimal segmentation from multiple candidates [15, 17], and we demonstrate that the proposed framework can produce more accurate segmentation; (5) the proposed approach predicts the quality of segmentation accurately even when trained with a highly-limited (small) dataset.

2 Methodology

2.1 Data

The Sunnybrook Cardiac Dataset [2] comprises 45 subjects divided into normal controls and 3 different pathological groups: heart failure with ischemia, heart failure without ischemia, and hypertrophic cardiomyopathy. The dataset was randomly split into 38 training subjects (355 images) and 7 testing subjects (65 images). The testing data comprised of two subjects from each of the 3 pathological groups, and one subject from the normal control. For each subject, the short-axis cine SSFP CMR images were provided with manually drawn contours on both endocardial and epicardial borders at end-diastole, which were considered the ground truth for training and testing in this work. Images at end-systole were not used for the development of this work, as only the endocardial contours were provided, without the epicardial contours. The training data were augmented by randomly rotating within $\pm 10°$ to prevent overfitting. In total, 85200 augmented training images were generated.

2.2 Overview of the Ensemble Framework

The proposed ensemble framework (Fig. 1A) involved multiple independently-trained U-nets [21] implemented with MCD and their combined segmentation models generated via a label voting scheme [22], with a quality control pipeline to predict the segmentation

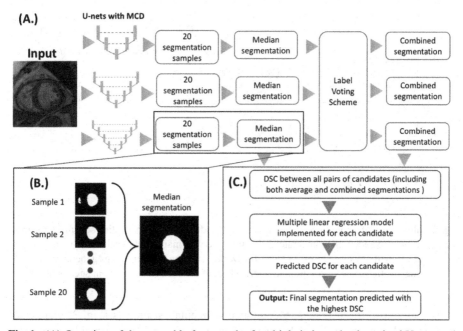

Fig. 1. (A) Overview of the ensemble framework of multiple independently-trained U-nets and combined segmentation models; (B) illustration of generating the median segmentation from 20 samples by each U-net, and (C) the segmentation quality control pipeline.

accuracy and to select the optimal result [15, 17]. The MCD approach (Fig. 1B) used the median of 20 generated segmentation samples for each MCD U-net. The quality control pipeline (Fig. 1C) calculated inter-candidate DSC for quality prediction via multiple linear regression, and selected the final optimal segmentation.

2.3 U-Nets with Monte Carlo Dropout

In the proposed ensemble framework, 6 U-nets [21] with different numbers of convolutional layers (7, 11, 15, 19, 23, 27) were implemented based on [15, 17] to perform segmentation of the LV endocardium and epicardium. By varying the number of convolutional layers across individual U-nets, it was expected to increase prediction diversity of the ensemble for robust quality control. The U-nets were modified by adding MCD units similar to [14]. The dropout units were activated during both training and testing with a dropout rate of 0.5. In this work, each U-net was set to generate 20 different segmentation samples for each anatomical structure (the endocardium or the epicardium) in a given input, as shown in Fig. 1B. The median segmentation candidate was calculated as the mean of the 20 Monte Carlo samples, with thresholding at 0.5, to obtain a binary mask. In other words, 6 median segmentations were produced from a total of 120 samples by the 6 U-nets.

2.4 Combined Segmentation Models

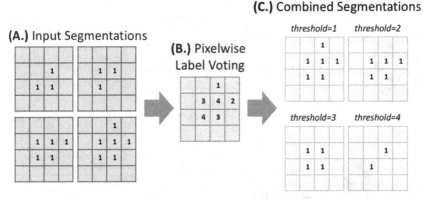

Fig. 2. Illustration of a combined segmentation of 4 models using a label voting scheme: (A) input segmentations are added up to generate (B) a pixelwise vote map, used to calculate (C) combined segmentations with different thresholds. In this work, 6 median segmentations generated by the 6 U-net models with Monte Carlo dropout approach were considered as the input segmentations.

In addition to the 6 U-nets, 6 combined segmentation models (Fig. 2) were also implemented via a pixelwise label voting scheme [15, 17, 22] to provide additional segmentation candidates for the ensemble. Figure 2 exemplifies the process of generating combined segmentations from 4 models. The input (Fig. 2A) is the median segmentations independently generated by the multiple MCD U-nets. The input segmentations are added up pixel-by-pixel (Fig. 2B) to produce multiple combined segmentations with different thresholds (Fig. 2C). In this work, 6 combined segmentations were generated for each input medical image.

2.5 Prediction of Segmentation Quality

For the quality control component (Fig. 1C), a multiple linear regression model was implemented for each of the 12 candidate segmentation models (including both U-nets with MCD and combined models) based on [15, 17] to predict the ground truth DSC, calculated between the candidate segmentation and the manual ground truth segmentation. The independent variables of the regression model were inter-candidate DSCs calculated between all possible pairs of the 12 candidate segmentations. Via the regression model, the inter-candidate DSCs can associate to the ground truth DSC. The regression parameters have been established using the same ground truth data used for training each individual neural network. Once trained, the regression models can predict DSC of the test segmentation on a per-case basis and in the absence of a manual ground truth segmentation. In this work, the proposed framework adopted a novel mechanism to choose the best final output, with the highest predicted DSC, from multiple candidate segmentations [15, 17].

We also implemented another segmentation quality prediction method based on [13] for comparison. This DSC prediction was calculated by averaging over the DSCs of all possible pairs of Monte Carlo segmentation samples, available only in the MCD models, excluding the combined models in the evaluation.

2.6 Evaluation

Each of the 12 candidate models implemented in the ensemble framework was evaluated for its segmentation performance, measured in terms of mean DSC (and standard deviation), independently for the endocardium and the epicardium. For the U-nets with MCD, only the median segmentations, not the Monte Carlo segmentation samples, were evaluated.

For the quality control component, the regression-based DSC prediction was evaluated independently for each candidate model for both the endocardium and the epicardium. The mean absolute error (MAE) and the Pearson correlation coefficient (r) were calculated between the predicted DSC and the observed ground-truth DSC derived from the manual segmentation. In addition, evaluation of the Monte Carlo-based DSC prediction was also reported for comparison.

3 Experimental Results

The methods were implemented in Python using TensorFlow, Keras and Scipy modules. The neural networks were trained for 240 epochs each, taking 6 h and 48 min in total, with an additional 6 min for the DSC regression models, on a desktop computer equipped with a NVIDIA Titan X GPU. The testing on 7 subjects (65 images) took 12 min and 7 s (i.e. 11 s per image).

3.1 Segmentation Performance

The mean DSC results for all the candidate segmentation models and the proposed ensemble framework are shown for both the endocardium and the epicardium (Table 1). The best mean DSC obtained by a single U-net model was 0.916 (U-net 15) in segmenting the endocardium, and 0.939 (U-net 23) in segmenting the epicardium. The best combined model (Combined Model 3) achieved a mean DSC of 0.920 and 0.941 for the endocardium and the epicardium, respectively. In comparison, the proposed framework outperformed all single and combined models, with a mean DSC of 0.922 and 0.942 for the endocardium and the epicardium, respectively. Furthermore, the framework, comprising of U-nets with Monte Carlo dropout (MCD) in this work, also achieved better performance than the reported results in [17], which implemented U-nets without MCD for the ensemble using the same training and testing datasets. This demonstrates the potential improvement on robustness and accuracy brought forth by integrating the deep ensemble framework with the Bayesian approach, subject to further cross-validation to mitigate the limitation of having a small testing dataset.

Table 1. Mean Dice similarity coefficients (DSCs) for U-nets with Monte-Carlo Dropout (MCD), Combined Models, and the proposed ensemble framework. Standard deviations shown in brackets.

Model	Endocardium DSC	Epicardium DSC
U-net 7 with MCD	0.486 (0.270)	0.569 (0.266)
U-net 11 with MCD	0.878 (0.172)	0.895 (0.166)
U-net 15 with MCD	0.916 (0.127)	0.938 (0.107)
U-net 19 with MCD	0.913 (0.128)	0.936 (0.124)
U-net 23 with MCD	0.915 (0.130)	0.939 (0.124)
U-net 27 with MCD	0.913 (0.128)	0.934 (0.127)
Combined model 1	0.810 (0.161)	0.856 (0.131)
Combined model 2	0.913 (0.127)	0.935 (0.123)
Combined model 3	0.920 (0.126)	0.941 (0.122)
Combined model 4	0.916 (0.127)	0.936 (0.126)
Combined model 5	0.887 (0.177)	0.904 (0.175)
Combined model 6	0.550 (0.297)	0.617 (0.286)
Proposed framework	**0.922 (0.125)**	**0.942 (0.122)**

Figure 3 shows an example of an apical slice image in the testing dataset (Fig. 3A), with the corresponding manual segmentation of the epicardium (Fig. 3B), and the segmentations by U-net 23 (Fig. 3C), Combined Model 3 (Fig. 3D), and Combined Model 2 chosen by the ensemble framework (Fig. 3E). Despite U-net 23 and Combined Model 3 respectively being the best among the U-nets and the Combined Models, they were outperformed by the proposed ensemble framework when compared to the ground truth manual segmentation. The framework chose the segmentation generated by Combined

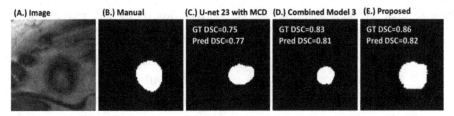

Fig. 3. Example of (A) an input image with (B) its corresponding manual segmentation of the epicardium, (C) segmentation generated by the best single neural network – U-net 23 with Monte Carlos dropout (MCD), (D) segmentation generated by the best Combined Model – Combined Model 3, and (E) final optimal segmentation chosen by the proposed ensemble framework – Combined Model 2, for the epicardium. The corresponding ground truth (GT) Dice similarity coefficients (DSCs) and the predicted (Pred) DSCs are shown.

Model 2, as its predicted DSC (0.82) was higher than the predicted DSCs for U-net 23 (0.77) and Combined Model 3 (0.81). This demonstrates that the on-the-fly selection of segmentation can improve overall segmentation quality by choosing the most-optimal candidate.

3.2 Regression-Based DSC Prediction Accuracy

For the evaluation of the DSC prediction via multiple linear regression, the mean absolute errors (MAE) and Pearson correlation coefficients (r) are reported in Table 2 for both the endocardium and the epicardium. All the regression models achieved excellent performance in predicting the ground truth DSC, with very low MAE (from 0.011 to 0.035) and very high Pearson r (0.90 to 1.00).

The scatter plots (Fig. 4) also reflect the high agreement between the DSC prediction (x-axis) and the ground truth (y-axis) for both the endocardium (Fig. 4A) and the epicardium (Fig. 4B). Most cases clustered closely along the identity line, indicating very accurate DSC predictions. Using a binary threshold at 0.7, the segmentations were classified into good (≥ 0.7) or poor quality (< 0.7) with an excellent accuracy of 98% and 99% for the endocardium and the epicardium, respectively, consistent with the performance reported in [15]. This demonstrates the accuracy and practicality of the proposed quality predictions to flag potentially problematic segmentations to human attention for clinical applications.

3.3 Comparison with Monte Carlo-Based DSC Prediction

The Monte Carlo (MC)-based DSC prediction [13] was also evaluated for comparison. The MC-based prediction achieved generally good performance (Table 3), but with higher MAE (from 0.52 to 0.177) and lower Pearson r (0.54 to 0.98) when compared to the regression-based prediction (Table 2). Moreover, the scatter plots (Fig. 5) show that the data points deviate farther from the identity line, with a lower classification accuracy (95%), compared to the regression-based prediction (Fig. 4). Thus, regression-based DSC prediction demonstrated the expected advantages over the intrinsic MC-based agreement measures.

Table 2. Mean absolute error (MAE) and Pearson coefficient (r) for DSC prediction using regression described in [15, 17]. All r had p < 0.0005.

Model	Endocardium MAE	r	Epicardium MAE	r
U-net 7 with MCD	0.016	1.00	0.011	1.00
U-net 11 with MCD	0.026	0.97	0.018	0.97
U-net 15 with MCD	0.030	0.92	0.021	0.97
U-net 19 with MCD	0.030	0.93	0.023	0.97
U-net 23 with MCD	0.032	0.92	0.020	0.97
U-net 27 with MCD	0.028	0.93	0.024	0.96
Combined model 1	0.035	0.94	0.021	0.97
Combined model 2	0.032	0.90	0.023	0.97
Combined model 3	0.032	0.91	0.022	0.97
Combined model 4	0.030	0.94	0.021	0.96
Combined model 5	0.027	0.97	0.023	0.97
Combined model 6	0.019	1.00	0.014	1.00
Proposed framework	0.034	0.90	0.023	0.97

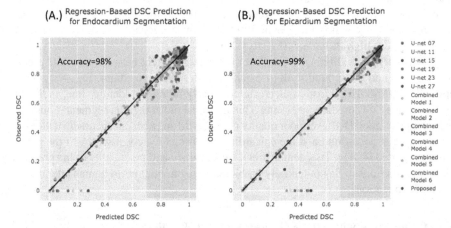

Fig. 4. Scatter plots of the regression-based predicted Dice similarity coefficient (DSC) (x-axis) versus the observed ground-truth DSC (y-axis) for (A) the endocardium and (B) the epicardium. With the quality prediction dichotomized by a binary threshold of 0.7, the DSC prediction achieved a very high classification accuracy of 98% and 99% for the endocardium and the epicardium, respectively. The black diagonal line is the identity line.

An example is shown in Fig. 6 showing an input image (Fig. 6A), the corresponding manual segmentation (Fig. 6B), and the automatic epicardium segmentation (Fig. 6C), with a table detailing the DSC prediction results (Fig. 6D). The automatic segmentation was derived from the median of the 20 segmentation samples generated by the U-net 15 with MCD. The MC-based quality control method falsely predicted a high DSC of 0.917 (Fig. 6D top row) with an incorrectly predicted label of "good quality" for the automatic segmentation, while the regression-based method predicted a low DSC of 0.301 (Fig. 6D middle row). The regression-based method achieved a result closer to the ground truth DSC of 0.145 (Fig. 6D bottom row), and also correctly flagged the poor-quality segmentation.

Table 3. Mean absolute error (MAE) and Pearson coefficient (r) for DSC prediction using average DSC over all possible pairs of Monte Carlo samples based on [13], available to U-nets with Monte-Carlo dropout (MCD) only. All r had $p < 0.0005$.

Model	Endocardium		Epicardium	
	MAE	r	MAE	r
U-net 7 with MCD	0.177	0.86	0.150	0.92
U-net 11 with MCD	0.052	0.88	0.033	0.94
U-net 15 with MCD	0.062	0.54	0.045	0.74
U-net 19 with MCD	0.068	0.89	0.048	0.96
U-net 23 with MCD	0.064	0.93	0.047	0.98
U-net 27 with MCD	0.066	0.91	0.059	0.95

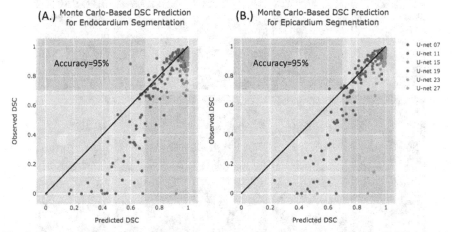

Fig. 5. Scatter plots of the Monte Carlo-based predicted Dice similarity coefficient (DSC) (x-axis) versus the observed ground-truth DSC (y-axis) for (A) the endocardium and (B) the epicardium for U-nets 7 to 27. With a binary threshold of 0.7, the DSC prediction achieved a segmentation quality classification accuracy of 95% for both the endocardium and the epicardium. The black diagonal line is the identity line.

(A.) Image   (B.) Manual   (C.) Automatic

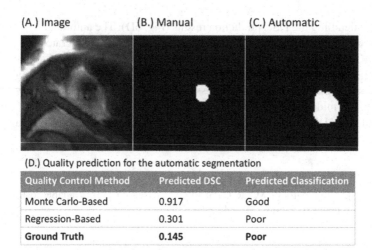

(D.) Quality prediction for the automatic segmentation

Quality Control Method	Predicted DSC	Predicted Classification
Monte Carlo-Based	0.917	Good
Regression-Based	0.301	Poor
Ground Truth	**0.145**	**Poor**

Fig. 6. Example of an (A) input image with (B) its manual segmentation and (C) a poor-quality automatic segmentation, obtained by averaging 20 samples generated by U-net 15 with Monte Carlo dropout (MCD). Table (D) shows quality prediction of the automatic segmentation by the Monte Carlo-based method (top row), the regression-based method (middle row), with the ground truth (bottom row).

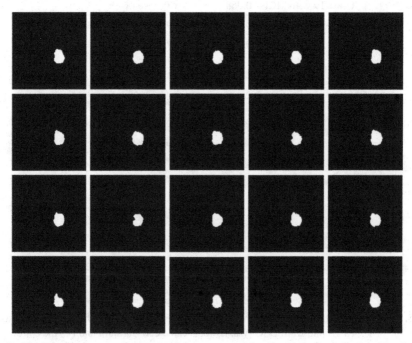

Fig. 7. 20 Monte Carlo segmentation samples generated for the median segmentation in Fig. 6C are shown. The samples lacked diversity in prediction as they highly resemble each other, causing a high Monte Carlo-based predicted Dice similarity coefficient (0.917) despite low agreement with the ground-truth Dice similarity coefficient (0.145).

Figure 7 and 8 are shown for further insights into the differences in prediction performance by the two quality control methods. The MC segmentation samples for the automatic segmentation (Fig. 6C) are shown in Fig. 7. Despite having 20 segmentation samples, the MC samples lacked diversity in prediction and were prone to making the same segmentation mistake – falsely locating the epicardium. This led to an undesirable consequence of predicting a high DSC while the actual ground truth DSC was low. Figure 8 shows the 12 candidate segmentations, which were utilized for the DSC prediction via multiple linear regression in the proposed ensemble framework. Compared with MC samples, the segmentations show more prediction diversity, consistent with the observed advantage of deep ensembles reported in [19].

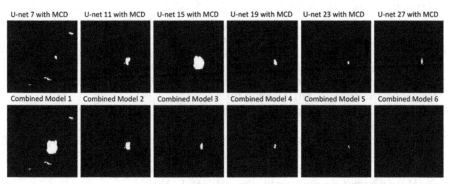

Fig. 8. The proposed framework provided 12 candidate segmentations, with high prediction diversity, are shown for the same input image in Fig. 6A. The segmentation generated by U-net 15 was compared with other candidates to predict a Dice similarity coefficient (0.301), correctly classifying the segmentation as bad quality.

4   Conclusion

In this work, we validated a novel deep ensemble segmentation framework integrated with Bayesian Monte Carlo sampling. The proposed framework can delineate the left ventricular endocardium and epicardium with a high mean DSC of 0.922 and 0.942, respectively. It has inherent quality control, which can predict the segmentation quality in terms of expected DSC with excellent accuracy. We have shown that the regression-based DSC prediction integrated in the framework outperformed the conventional Monte Carlo-based approach, which lacked prediction diversity. This framework successfully merged the advantages of deep neural network ensembles and Bayesian approximation, enabling reliable automatic image segmentation, even for deep learning models trained on small datasets. This can potentially accelerate the advancement of deep learning approaches for diagnostic imaging by reducing requirements of large training datasets.

References

1. Roth, G.A., et al.: Global, regional, and national age-sex-specific mortality for 282 causes of death in 195 countries and territories, 1980–2017: a systematic analysis for the Global Burden of Disease Study 2017. The Lancet **392**(10159), 1736–1788 (2018)
2. Radau, P., et al.: Evaluation framework for algorithms segmenting short axis cardiac MRI. MIDAS J. **49** (2009)
3. Bernard, O., et al.: Deep learning techniques for automatic MRI cardiac multi-structures segmentation and diagnosis: is the problem solved? IEEE Trans. Med. Imaging **37**(11), 2514–2525 (2018)
4. Petersen, S.E., et al.: Imaging in population science: cardiovascular magnetic resonance in 100,000 participants of UK Biobank - rationale, challenges and approaches. J. Cardiovasc. Magn. Reson. **15**(1), 46 (2013)
5. Constantinides, C., et al.: Semi-automated cardiac segmentation on cine magnetic resonance images using GVF-Snake deformable models. MIDAS J. **77** (2009)
6. Casta, C., et al.: Evaluation of the dynamic deformable elastic template model for the segmentation of the heart in MRI sequences. MIDAS J. (2009)
7. Huang, S., et al.: Segmentation of the left ventricle from cine MR images using a comprehensive approach. MIDAS J. (2009)
8. Lu, Y., et al.: Automatic image-driven segmentation of left ventricle in cardiac cine MRI. MIDAS J. (2009)
9. Jolly, M.: Fully automatic left ventricle segmentation in cardiac cine MR images using registration and minimum surfaces. MIDAS J. (2009)
10. O'Brien, S., Ghita, O., Whelan, P.F.: Segmenting the left ventricle in 3D using a coupled ASM and a learned non-rigid spatial model. MIDAS J. (2009)
11. Wijnhout, J., et al.: LV challenge LKEB contribution: fully automated myocardial contour detection. MIDAS J. (2009)
12. Chen, C., et al.: Deep learning for cardiac image segmentation: a review. Front. Cardiovasc. Med. **7** (2020)
13. Roy, A., Conjeti, S., Navab, N., Wachinger, C.: Inherent brain segmentation quality control from fully convnet monte carlo sampling. In: Frangi, A.F., Schnabel, J.A., Davatzikos, C., Alberola-López, C., Fichtinger, G. (eds.) Medical Image Computing and Computer Assisted Intervention – MICCAI 2018: 21st International Conference, Granada, Spain, September 16-20, 2018, Proceedings, Part I, pp. 664–672. Springer International Publishing, Cham (2018). https://doi.org/10.1007/978-3-030-00928-1_75
14. DeVries, T., Graham, T.: Leveraging Uncertainty Estimates for Predicting Segmentation Quality. arXiv pre-print server (2018)
15. Hann, E., et al.: Quality control-driven image segmentation towards reliable automatic image analysis in large-scale cardiovascular magnetic resonance aortic cine imaging. In: Shen, D., et al. (eds.) MICCAI 2019. LNCS, vol. 11765, pp. 750–758. Springer, Cham (2019). https://doi.org/10.1007/978-3-030-32245-8_83
16. Lakshminarayanan, B., Pritzel, V., Blundell, C.: Simple and scalable predictive uncertainty estimation using deep ensembles. arXiv pre-print server (2017)
17. Hann, E., et al.: Deep neural network ensemble for on-the-fly quality control-driven segmentation of cardiac MRI T1 mapping. Med. Image Anal. **71**, 102 (2021)
18. Mehrtash, A., et al.: Confidence calibration and predictive uncertainty estimation for deep medical image segmentation. IEEE Trans. Med. Imaging **39**(12), 3868–3878 (2020)
19. Fort, S., Hu, H., Lakshminarayanan, B.: Deep ensembles: a loss landscape perspective. arXiv pre-print server (2020)

20. Pop, R., Fulop, P.: Deep ensemble bayesian active learning : addressing the mode collapse issue in monte carlo dropout via ensembles. arXiv pre-print server (2018)

21. Ronneberger, O., Fischer, P., Brox, T.: U-Net: convolutional networks for biomedical image segmentation. In: Navab, Nassir, Hornegger, Joachim, Wells, William M., Frangi, Alejandro F. (eds.) Medical Image Computing and Computer-Assisted Intervention – MICCAI 2015: 18th International Conference, Munich, Germany, October 5-9, 2015, Proceedings, Part III, pp. 234–241. Springer International Publishing, Cham (2015). https://doi.org/10.1007/978-3-319-24574-4_28

22. Li, X., et al.: Estimating the ground truth from multiple individual segmentations incorporating prior pattern analysis with application to skin lesion segmentation. IEEE (2011)

Reducing Textural Bias Improves Robustness of Deep Segmentation Models

Seoin Chai, Daniel Rueckert, and Ahmed E. Fetit[(✉)]

Imperial College London, London, UK
a.fetit@imperial.ac.uk

Abstract. Despite advances in deep learning, robustness under domain shift remains a major bottleneck in medical imaging settings. Findings on natural images suggest that deep neural models can show a strong textural bias when carrying out image classification tasks. In this thorough empirical study, we draw inspiration from findings on natural images and investigate ways in which addressing the textural bias phenomenon could bring up the robustness of deep segmentation models when applied to three-dimensional (3D) medical data. To achieve this, publicly available MRI scans from the Developing Human Connectome Project are used to study ways in which simulating textural noise can help train robust models in a complex semantic segmentation task. We contribute an extensive empirical investigation consisting of 176 experiments and illustrate how applying specific types of simulated textural noise prior to training can lead to *texture invariant* models, resulting in improved robustness when segmenting scans corrupted by previously unseen noise types and levels.

Keywords: Textural bias · Domain shift · Robustness · Segmentation

1 Introduction

In medical imaging research, using convolutional neural networks (CNNs) is a popular way of classifying and segmenting scans. However, medical scans can contain subtle visual noise and low-frequency textural patterns that are inherent to acquisition protocols or hardware. As a result, a network may be optimised to reduce empirical risk on data from one domain, but its performance can degrade when applied to a different domain, such as a different hospital or imaging centre [4,8]. In a study on natural images by Geirhos et al. [3], separate CNNs were trained on standard ImageNet [2] data as well as a stylised version of the same data, introducing conflicting textures to the input images. In doing so, the authors showed that training networks on stylised images has led to a reduction of the textural bias phenomenon as the models focused on more robust shape features. In this study, we draw inspiration from the work reported in [3] and hypothesise that addressing this phenomenon in medical imaging can lead to CNNs that are *texture invariant* and hence more resilient to changes in data

© Springer Nature Switzerland AG 2021
B. W. Papież et al. (Eds.): MIUA 2021, LNCS 12722, pp. 294–304, 2021.
https://doi.org/10.1007/978-3-030-80432-9_23

distribution. Specifically, our motivation is to find out whether training segmentation CNNs under certain textural noise settings could help improve model robustness. Long term benefits of this include flexibility to variations in acquisition protocols, scanner hardware, hospital infrastructure, or image resolution often faced in realistic healthcare scenarios and routine clinical practice.

In this regard, we carried out an empirical investigation using neuroimaging data publicly available from the Developing Human Connectome Project (dHCP)[1] [1]. In doing so, we simulated different categories and levels of textural noise by applying several permutations of filtering techniques to the scans. In a series of 176 experiments, we trained 11 models and subsequently tested them on 16 different held-out sets in order to evaluate which settings can best generalise to previously unseen noise, thereby thoroughly simulating performance under various types and severities of domain shift.

Our contribution is an extensive empirical study which demonstrated that training a deep segmentation model on data corrupted with certain combinations of textural noise can in fact improve model robustness on new, previously unseen noise types and levels. We believe that this is due to the models being incentified to learn anatomical and tissue-specific features during training, as opposed to low-frequency textural patterns that may be brittle and domain specific.

The rest of the paper is structured as follows: in Sect. 2.1 we present a summary of the dataset used, followed by an explanation of the three main textural noise simulation techniques used in Sect. 2.2. Then, we give a summary of the CNN architecture used in Sect. 2.3, as well as a breakdown of the 176 experiments conducted in Sect. 2.4. We present and discuss details of the experimental results in Sect. 3, which includes tables of three insightful models used in our experiments, as well as visual examples of how the most robust model performs on previously unseen noise levels. Finally, we present conclusions and directions for future work in Sect. 4.

2 Materials and Methods

2.1 Dataset

The dataset consisted of 70 3D T2-weighted brain MRI scans publicly available from the dHCP neonatal cohort. The segmentation maps had 10 classes, corresponding to: zero-pixel background, cerebrospinal fluid (CSF), cortical grey matter (cGM), white matter (WM), background bordering brain tissues, ventricles, cerebellum, deep grey matter (dGM), brainstem, and hippocampus. The scans covered an age range of 24.3–42.2 weeks. The data was available in NIfTI format; Fig. 1 shows an example scan and corresponding tissue labels. We carried out a pre-processing step where each scan was independently normalised to zero-mean and unit-variance.

---

[1] http://www.developingconnectome.org/project/.

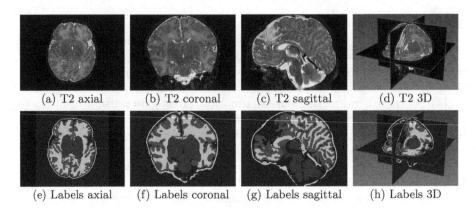

(a) T2 axial   (b) T2 coronal   (c) T2 sagittal   (d) T2 3D

(e) Labels axial   (f) Labels coronal   (g) Labels sagittal   (h) Labels 3D

Fig. 1. Example T2-weighted neonatal brain scan and corresponding segmentation labels from the publicly available dHCP dataset.

2.2   Textural Noise Simulation

Blur is a rather unique family of MRI artefacts because it can be introduced into a scan after the acquisition stage, such as during post-processing or in the manifestation of pathological conditions [7]; simulating blur textural artefacts was therefore important and relevant for this study. We used the well-established Gaussian blur, which can be produced by continuously applying a Gaussian filter to the image. Different degrees of blurring can be obtained by altering the σ parameter, where higher values of σ give blurrier transformations. We employed Gaussian filters from the library `scikit-image` [10] with $\sigma = \{1, 2, 3, 4, 5\}$ to give five different degrees of Gaussian blurred datasets named gaus01-05, respectively. Figure 2 shows examples of noisy images.

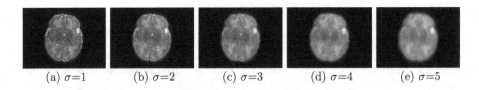

(a) $\sigma=1$ (b) $\sigma=2$ (c) $\sigma=3$ (d) $\sigma=4$ (e) $\sigma=5$

Fig. 2. Axial slices of Gaussian blurred brain images with different values of σ.

Second, we simulated further blur manifestations using median filters, which replace the pixel value with the median of the neighbouring pixels. We utilised filters from `scipy` [9], each with a parameter `size` that specifies the neighbourhood distance used to compute the median where higher values of `size` result in smoother filter output. Three different degrees of median filtered images with `size= {2, 5, 8}` were generated. Figure 3 shows examples of median-filtered images generated in this study.

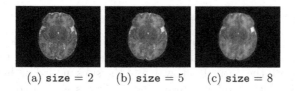

(a) `size = 2`   (b) `size = 5`   (c) `size = 8`

Fig. 3. Axial slices of median smoothed images with different values of `size`.

We also explored *impulse noise* corruptions, such as those introduced by noisy communications channels, faulty memory locations, or damage in channel decoders [6]. The impulse noise generating filter we used in this study is based on the salt-and-pepper (SNP) technique, which randomly generates black and white pixels on the image of interest. The function we used takes into account a parameter called `prob`, where $0 \leq$ `prob` ≤ 0.5. A random number is generated for each pixel; if it is less than `prob` then the function paints the pixel with black, if it is greater than `1-prob` then it paints the pixel with white, otherwise the pixel is left unchanged. In other words, the higher the value of `prob`, the noisier the output can become. We used different values for `prob`, in particular `prob` ={0.01, 0.03, 0.05, 0.07, 0.10, 0.15, 0.20}, to create seven different noisy datasets named snp_prob. Examples of axial slices are shown in Fig. 4.

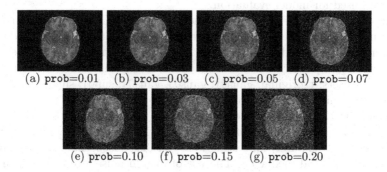

(a) `prob=0.01`   (b) `prob=0.03`   (c) `prob=0.05`   (d) `prob=0.07`

(e) `prob=0.10`   (f) `prob=0.15`   (g) `prob=0.20`

Fig. 4. Axial slices of images injected with salt-and-pepper noise with different `prob` values.

It is important to note that when we carried out the experiments, which are discussed in Sect. 2.4, we reserved some of the generated noisy datasets solely for use at test time. This was done in order to simulate various degrees of domain shift encountered in routine clinical practice (e.g. changes in magnetic field strength, scanner hardware, acquisition protocols, hospital's digital communications infrastucture) that may not be experienced during training.

2.3 CNN Architecture

We used DeepMedic, an open source 3D architecture with multiple convolutional pathways originally built for brain lesion segmentation tasks [5]. Our CNN had eight layers in both the normal pathway and the subsampled pathways with the kernel size of 3^3. Two parallel subsampled pathways were used, giving a total of three pathways. We also employed residual connections between layers 3 and 4, layers 5 and 6, and layers 7 and 8. The number of feature maps used in each fully connected layer was 250. We ran the experiment for 100 epochs where each epoch comprised of 20 subepochs. In every subepoch, images from five cases were loaded and the training samples were extracted. We used a batch size of 10, in addition to a predefined learning rate scheduler which schedules every 8 epochs starting from the 24th epoch. Finally, we used RMSProp as an optimiser, and L1 and L2 regularisation with values 10^{-5} and 10^{-4}.

2.4 Experiments

Of the 70 scans provided, we randomly selected 50 scans for training, 10 for validation, and 10 for testing. This is a fairly complex tissue segmentation task because visual characteristics of the scans can greatly vary throughout brain development, e.g. as the brain matures, the occurrence of dark intensities present in white matter regions gradually increases. The Dice similarity coefficient (DSC) metric was used for model evaluation.

Since the motivation of the experiments was to find out whether training models under a certain textural noise setting could potentially help with domain shift, some of the noisy datasets created above were intentionally not used during model training. The CNN was trained on 11 different combinations of the following datasets: t2original, gaus01, gaus03, gaus04, snp01, snp05, snp10 and median05, as summarised in Table 1.

Each of the 11 models was then evaluated on 16 held-out test sets. The total number of experiments carried out was therefore 176. Note that the unseen noise types/levels reserved purely for testing were gaus02, gaus05, snp03, snp07, snp15, snp20, median02, and median08. The use of unseen noise levels at test time is important to simulate realistic domain shifts introduced in healthcare settings, e.g. changes in magnetic field strength or scanner hardware. Our goal was to find out whether training on certain combinations of textural noise can lead to *textural invariant* models, and hence lead to robustness on new, previously unseen noise.

3 Results and Discussion

All 11 models were first tested on held out data that represented the settings they were trained on. The models were then tested on held out data that represented settings which are completely new and are assumed to simulate domain shift. For succinctness, we present in detail 3 of the 11 models, hence showing the most

Table 1. Model names and the corresponding combinations of training data.

Model name	Training data combination
M_T2ORIGINAL	t2original (baseline, noise free)
M_GAUS01	gaus01 ($\sigma = 1$)
M_GAUS03	gaus03 ($\sigma = 3$)
M_GAUS04	gaus04 ($\sigma = 4$)
M_GAUS010304	gaus01, gaus03, gaus04 ($\sigma = 1, 3, 4$)
M_SNP01	snp01 (prob $= 0.01$)
M_SNP05	snp05 (prob $= 0.05$)
M_SNP10	snp10 (prob $= 0.10$)
M_SNP010510	snp01, snp05, snp10 (prob $= 0.01, 0.05, 0.10$)
M_MEDIAN05	median05 (size $= 5$)
M_GAUS010304_SNP010510	gaus01, gaus03, gaus04 ($\sigma = 1, 3, 4$),
	snp01, snp05, snp10 (prob $= 0.01, 0.05, 0.10$)

insightful results in the tables below. The reported DSC values corresponded to the following tissue classes: *1)* zero-intensity background, *2)* CSF, *3)* cGM, *4)* WM, *5)* background bordering brain tissues, *6)* ventricles, *7)* cerebellum, *8)* dGM, *9)* brainstem, and *10)* hippocampus.

We observed that the model trained on noise-free scans achieved highest overall robustness on data similar to the data it was trained on and where textural transformation was only lightly applied; a rather expected finding (see Table 2). For instance, the noise-free model was able to achieve a DSC of over 86% for all 10 classes on gaus01, but its performance immensely dropped to 0–1% with 4 of those classes on gaus05. This shows that the baseline model trained using a conventional approach does not immediately generalise to unseen noise categories and levels, highlighting the problem of domain shift often experienced with a wide range of realistic scenarios, e.g. variations in scanner manufacturer, magnetic field strengths, or acquisition protocols.

A similar pattern emerged when we trained models on images smoothed out by the Gaussian filter only, we observed that they generalised on Gaussian filtered images that were generated using values of σ close to those used on training data. For example, a model which was trained only on Gaussian noise $\sigma = 3$ showed the highest overall DSC values on datasets gaus03 and gaus04, but not on other variations of the Gaussian images. Noteworthy, we also observed that when a model was trained on a Gaussian dataset created with a very large value for σ, it did not segment the noise-free data correctly. We believe this is because the use of such blur filters alone may have completely distorted useful anatomical features within the scans. Interestingly, however, when a model was trained on multiple Gaussian datasets generated with a variety of σ values, it generalised well on the noise-free data as well as on Gaussian filtered images, even for the cases where the σ values used were different from that used in

Table 2. DSCs achieved by the baseline, noise-free model (M_T2ORIGINAL) on the 16 versions of the held-out test set. Data highlighted in bold correspond to settings where the model achieves over 90% DSC for *all* classes, demonstrating particularly high levels of robustness. The model achieved highest overall robustness on test sets that are noise-free and where textural transformation was only lightly applied, reflecting the common problem of domain shift in imaging.

Data/Class	1	2	3	4	5	6	7	8	9	10
t2original	99.35	95.99	96.64	97.57	91.96	95.39	97.78	97.06	97.12	92.98
gaus01	98.87	90.21	91.97	94.30	86.43	90.79	95.97	95.28	95.67	89.90
gaus02	96.69	54.18	33.40	65.59	70.94	62.57	59.77	56.65	88.93	66.08
gaus03	94.98	15.67	03.84	51.65	60.83	25.27	44.42	42.43	69.00	13.47
gaus04	94.71	00.78	01.75	53.41	54.99	00.25	33.44	45.49	16.04	00.00
gaus05	93.98	00.19	01.61	47.00	45.44	00.08	35.81	36.73	00.33	00.05
snp01	99.22	87.14	85.02	84.45	89.62	88.95	94.02	86.11	92.51	83.83
snp03	93.96	45.62	48.08	14.73	63.21	19.28	16.85	05.85	47.76	10.85
snp05	55.33	26.53	06.63	01.60	23.90	00.32	00.01	00.33	01.79	00.17
snp07	16.99	21.49	00.03	00.08	12.38	00.10	00.00	00.02	00.01	00.01
snp10	03.03	07.80	00.00	00.00	04.05	00.00	00.00	00.00	00.00	00.00
snp15	00.15	00.53	00.00	00.00	00.24	00.00	00.00	00.00	00.00	00.00
snp20	00.02	00.12	00.00	00.00	00.01	00.00	00.00	00.00	00.00	00.00
median02	98.91	87.54	88.86	92.58	81.99	88.95	96.22	95.22	94.11	88.33
median05	99.18	85.15	86.57	91.43	88.50	86.17	89.59	91.57	94.81	86.82
median08	98.38	61.75	46.73	73.58	76.40	59.42	57.81	63.40	87.58	48.14

training. Moreover, segmentation performance on the median filtered data also increased. Nevertheless, the models did not result in high DSCs for the Gaussian datasets with the highest σ value or the salt-and-pepper noise. This behaviour was consistent with the performance of models trained solely on median filtered images, which showed low levels of robustness on heavier filtered median image and Gaussian filtered images, as well as a lack of generalisation when tested on salt-and-pepper filtered images.

Interestingly, and unlike the Gaussian image trained models, models trained on salt-and-pepper noise showed high levels of robustness on all the lighter noise levels, some heavier noise levels, as well as completely different noise categories. For instance, model M_SNP01 (fairly low levels of salt-and-pepper noise, `prob` = 0.01) showed high levels of robustness on data with heavier salt-and-pepper noise, on the original dHCP images, as well as on the lightly texturised Gaussian and median images. This was also consistent with the robustness observed with M_SNP10 (trained on very high levels of salt-and-pepper noise, `prob` = 0.10) which is detailed in Table 3.

Table 3. DSCs achieved by the model trained on high levels of salt-and-pepper noise (M_SNP10, `prob` = 0.01) on the 16 versions of the held-out test set. Data highlighted in bold correspond to settings where the model achieves over 90% DSC for *all* classes, demonstrating particularly high levels of robustness. A key observation is that this model demonstrated high levels of robustness on all 10 classes for noise-free data (t2original), data injected with lower levels of salt-and-pepper noise (e.g. snp01), as well as higher levels of salt-and-pepper noise (e.g. snp15).

Data/Class	1	2	3	4	5	6	7	8	9	10
t2original	99.30	95.17	95.84	96.80	91.15	94.20	97.25	96.39	96.38	91.41
gaus01	98.86	90.37	91.58	93.74	86.45	90.80	95.66	94.66	94.77	88.77
gaus02	97.32	50.32	55.60	69.55	71.64	73.49	87.67	80.20	89.31	81.33
gaus03	96.07	20.73	16.64	59.98	58.44	52.44	83.01	69.76	83.82	71.18
gaus04	95.55	08.15	07.77	57.44	48.09	25.70	77.99	62.70	71.66	57.42
gaus05	95.48	03.40	06.46	56.26	37.80	12.07	70.84	56.04	47.10	41.67
snp01	99.31	95.34	95.97	96.94	91.33	94.61	97.34	96.54	96.59	91.73
snp03	99.31	95.59	96.18	97.13	91.56	94.98	97.45	96.74	96.82	92.26
snp05	99.32	95.75	96.31	97.27	91.65	95.11	97.51	96.81	96.85	92.50
snp07	99.32	95.83	96.39	97.36	91.69	95.00	97.51	96.84	96.88	92.71
snp10	99.33	95.86	96.39	97.39	91.67	95.14	97.56	96.84	96.93	92.72
snp15	99.32	95.51	95.92	97.11	91.40	94.51	97.39	96.66	96.73	92.08
snp20	99.04	94.08	94.37	96.09	89.12	93.57	96.53	95.59	95.99	89.69
median02	98.87	87.03	89.04	92.68	82.00	87.89	95.69	94.81	93.65	87.74
median05	99.17	84.75	87.21	91.36	88.14	87.05	93.27	92.65	93.56	86.79
median08	98.63	62.46	59.60	76.26	78.12	68.34	84.49	83.03	87.76	78.60

Our findings therefore suggest that models trained on impulse noise corruptions, namely salt-and-pepper textured images, can generalise well on images similarly corrupted by salt-and-pepper noise as well as on blur smoothed images. This leads us to believe that the model developed invariance to low frequency textural patterns and hence used relatively useful information about the global anatomical structure of the brain when carrying out tissue segmentation. Models trained on blur smoothed images, however, only generalised to images smoothed to similar degrees but not on images smoothed using other techniques.

Finally, Table 4 shows that the model trained on data injected with different degrees of both Gaussian and salt-and-pepper noise (gaus01, gaus03, gaus04 with $\sigma = 1, 3, 4$; snp01, snp05, snp10 with `prob` = 0.01, 0.05, 0.10) achieves the best overall robustness across the 16 versions of the test set, where even the heaviest filtered images had very few segmentation inaccuracies. For instance, this model demonstrated high levels of robustness on all 10 classes for noise-free data, on data injected with a previously unseen degree of salt-and-pepper noise (e.g. DSCs of 88%–99% on snp20), as well as on data with a previously

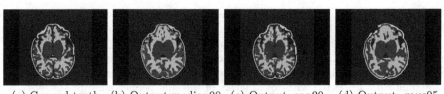

(a) Ground-truth   (b) Output:median08   (c) Output: snp20   (d) Output: gaus05

Fig. 5. Example performance of the model trained on several levels of Gaussian and salt-and-pepper noise on previously unseen textural noise levels.

Table 4. DSCs achieved by the model trained on several levels of Gaussian ($\sigma = 1$, 3, 4) and salt-and-pepper noise (`prob` = 0.01, 0.05, 0.10) on the 16 versions of the held-out test set. Data highlighted in bold correspond to settings where the model achieves over 0.9 DSC for *all* classes, demonstrating particularly high levels of robustness. This model exhibited the best overall robustness across all levels and types of noise tested, as well as on all 10 classes studied.

Data/Class	1	2	3	4	5	6	7	8	9	10
t2original	99.28	94.62	95.21	96.45	90.83	94.24	97.18	96.34	96.14	92.11
gaus01	99.27	94.52	95.00	96.34	90.64	94.47	97.24	96.36	96.19	92.06
gaus02	99.18	89.87	90.89	92.64	88.05	91.37	96.21	95.68	95.36	90.83
gaus03	99.10	89.26	89.68	92.25	86.41	91.74	96.15	95.49	95.07	90.00
gaus04	98.97	85.77	85.94	89.14	83.92	90.28	95.52	94.88	94.41	88.29
gaus05	98.56	72.98	73.12	79.46	75.91	82.91	92.77	92.60	91.74	83.86
snp01	99.28	94.87	95.42	96.62	90.92	94.43	97.26	96.38	96.19	92.09
snp03	99.28	95.06	95.56	96.73	91.00	94.56	97.30	96.44	96.20	92.01
snp05	99.28	95.10	95.57	96.71	90.99	94.50	97.32	96.39	96.22	91.78
snp07	99.27	95.11	95.54	96.66	90.92	94.43	97.21	96.36	96.27	91.68
snp10	99.26	95.06	95.42	96.55	90.78	94.41	97.19	96.21	96.11	91.64
snp15	99.23	94.56	94.63	95.86	90.25	92.84	96.80	95.48	95.49	90.03
snp20	99.07	92.99	91.95	93.74	88.33	88.39	94.86	92.96	94.40	86.02
median02	98.90	87.01	88.24	92.08	81.71	88.41	95.92	94.57	93.42	87.88
median05	99.27	88.30	91.02	94.30	89.18	88.50	96.10	95.36	95.04	90.19
median08	98.86	72.79	77.41	84.45	79.29	74.46	92.41	92.62	91.24	84.39

unseen degree of Gaussian noise (e.g. DSCs of 72%–98% on gaus05). To illustrate, Fig. 5 shows examples of the predicted segmentation of the model on the heaviest transformed images of the three filter categories used in these experiments. From the figure it is clear that the model achieved excellent mapping of the brain tissue regions. This shows a tremendous improvement from the baseline model which was trained using a conventional, noise-free approach and thus failed severely on the heavily corrupted images (e.g. DSCs of 0% for all 10 classes on snp20, see Table 2).

4   Conclusion

The medical domain contains all the factors that make machine learning algorithms challenging to translate to routine practice. Predictive models would have to adapt to unexpected circumstances when solving perceptual tasks or planning treatment strategies, due to the infinite variability in human nature and healthcare systems. In medical image segmentation, a major bottleneck that is often encountered when developing machine learning models is the issue of robustness under domain shift, where changes in hospital infrastructure, acquisition hardware, or image resolution cause seemingly accurate models to immensely fail at test time.

In this study, we hypothesised that addressing the textural bias phenomenon in medical imaging can lead to CNNs that are *texture invariant* and hence more resilient to changes in data distribution. Specifically, our motivation was to find out whether training deep segmentation models under the right textural noise settings could help improve model robustness. In this regard, we carried out an extensive empirical study consisting of 176 experiments for a complex brain tissue segmentation task, exploring the effect of training with several permutations of three main types of noise: Gaussian blur, median filter blur, and impulse salt-and-pepper noise. A key finding of our work is that training a deep segmentation model on neuroimaging data injected with certain combinations of textural noise can indeed improve model robustness on new, previously unseen noise levels. We believe that this is due to the models being incentified to learn anatomical and tissue-specific features, as opposed to low-frequency textural patterns that may be brittle and domain specific (e.g. inherent to a given scanner or acquisition protocol).

In terms of extending the investigation further, the next natural step would be to conduct a similar experiment on even more permutations of smoothing, salt-and-pepper generating, or even contrast changing filters to see if training models in these settings can yield even better segmentation results. Additionally, an investigation on whether this hypothesis is valid for non-simulated domain shift of neuroimaging data - ideally using data acquired across different sites - ought to be carried out. If findings are consistent with this study, generalisation to further modalities would be important to explore.

Acknowledgments. The research leading to these results has received funding from the European Research Council under the European Union's Seventh Framework Programme (FP/2007-2013)/ERC Grant Agreement no. 319456. We are grateful to the families who generously supported his trial.

References

1. Bastiani, M., et al.: Automated processing pipeline for neonatal diffusion MRI in the developing human connectome project. NeuroImage **185**, 750–763 (2019). https://doi.org/10.1016/j.neuroimage.2018.05.064

2. Deng, J., Dong, W., Socher, R., Li, L.J., Li, K., Fei-Fei, L.: Imagenet: a large-scale hierarchical image database. In: 2009 IEEE Conference on Computer Vision and Pattern Recognition, pp. 248–255. IEEE (2009)

3. Geirhos, R., Rubisch, P., Michaelis, C., Bethge, M., Wichmann, F.A., Brendel, W.: Imagenet-trained cnns are biased towards texture; increasing shape bias improves accuracy and robustness. In: 7th International Conference on Learning Representations, ICLR. OpenReview.net (2019). https://openreview.net/forum?id=Bygh9j09KX

4. Kamnitsas, K., et al.: Unsupervised domain adaptation in brain lesion segmentation with adversarial networks. In: Niethammer, M., et al. (eds.) IPMI 2017. LNCS, vol. 10265, pp. 597–609. Springer, Cham (2017). https://doi.org/10.1007/978-3-319-59050-9_47

5. Kamnitsas, K., et al.: Efficient multi-scale 3D CNN with fully connected CRF for accurate brain lesion segmentation. Med. Image Anal. **36**, 61–78 (2017). https://doi.org/10.1016/j.media.2016.10.004

6. Mousavi, S.M., Naghsh, A., Manaf, A.A., Abu-Bakar, S.A.R.: A robust medical image watermarking against salt and pepper noise for brain MRI images. Multimedia Tools Appl. **76**(7), 10313–10342 (2016). https://doi.org/10.1007/s11042-016-3622-9

7. Osadebey, M.E., Pedersen, M., Arnold, D.L., Wendel-Mitoraj, K.E.: Blind blur assessment of MRI images using parallel multiscale difference of gaussian filters. Biomed. Eng. Online **17**(1), 1–22 (2018)

8. Perone, C.S., Ballester, P.L., Barros, R.C., Cohen-Adad, J.: Unsupervised domain adaptation for medical imaging segmentation with self-ensembling. NeuroImage **194**, 1–11 (2019). https://doi.org/10.1016/j.neuroimage.2019.03.026

9. Virtanen, P., et al.: SciPy 1.0 contributors: SciPy 1.0: fundamental algorithms for scientific computing in python. Nat. Methods **17**, 261–272 (2020)

10. van der Walt, S., et al.: scikit-image: Image processing in python (2014). CoRR abs/1407.6245, http://arxiv.org/abs/1407.6245

Generative Models, Biomedical Simulation and Modelling

HDR-Like Image Generation to Mitigate Adverse Wound Illumination Using Deep Bi-directional Retinex and Exposure Fusion

Songlin Hou[1], Clifford Lindsay[2], Emmanuel Agu[1(✉)], Peder Pedersen[1], Bengisu Tulu[1], and Diane Strong[1]

[1] Department of Computer Science, Worcester Polytechnic Institute, Worcester, USA
emmanuel@wpi.edu

[2] Department of Radiology, University of Massachusetts Medical School, Worcester, USA

Abstract. Periodic assessment is necessary to evaluate the healing progress of chronic wounds. Image analyses using computer vision algorithms have recently emerged as a viable alternative that has been demonstrated by prior work. However, the performance of such image analysis methods degrade on captured in adverse illumination, which is common in many indoor environments. To mitigate these lighting problems, High Dynamic Range (HDR) image enhancement techniques can be used to mitigate over- and under-exposure issues and preserve the details of scenes captured in non-ideal illumination. In this paper, we address over- and under-exposure simultaneously using a deep learning-based bi-directional illumination enhancement network that is able to generate over- and under-exposed images that are then fused into a final image with enhanced illumination. In rigorous evaluations using metrics including structure similarity, peak signal-noise ratio and changes in segmentation accuracy, our proposed method outperformed the state-of-the-art (SSIM scores $0.76 \pm 0.04/0.69 \pm 0.08$ on bright/dark images, PSNR scores 28.60 ± 0.70 on dark images and DSC scores $0.76 \pm 0.09/0.74 \pm 0.09$ on bright/dark images).

Keywords: Image enhancement · Wound assessment · Deep learning · Image segmentation

1 Introduction

Chronic wounds are a major and growing public health burden [1] and can take up to 13 months to heal with a 60–70% probability of recurrence [2]. If not treated properly, chronic wounds may become infected resulting in amputation and even death. Wound care involves periodic assessment via visual inspections by wound experts, requiring frequent hospital visits. Smartphone imaging with algorithmic assessment using computer vision methods has recently emerged as

© Springer Nature Switzerland AG 2021
B. W. Papież et al. (Eds.): MIUA 2021, LNCS 12722, pp. 307–321, 2021.
https://doi.org/10.1007/978-3-030-80432-9_24

a viable method to standardize the care provided to patients in their homes. Such methods analyze the visual characteristics of wounds such as the wound size, depth and tissue composition in order to evaluate their healing progress [3,4]. However, smartphone wounds imaged in natural environments such as homes and clinics be underexposed (too dark), overexposed (too bright) or have negative-saturated pixels (values approaching 0). Such adverse illumination can cause errors in wound assessment using automated computer vision methods. Wang [5] found that slight deviations from optimal lighting conditions resulted in errors in the detected wound area wherein healthy skin was mis-detected as part of the wound. Consequently, methods for mitigating adverse lighting conditions in order to maintain clinically acceptable wound assessment accuracy are important.

The dynamic range of images refers to the ratio between the brightest and darkest part of an image [6]. Low Dynamic Range (LDR) imaging uses an 8-bit unsigned integer to capture only 255 levels of illumination per pixel. High Dynamic Range (HDR) imaging uses 16 or 32-bit floating-point numbers to capture an almost infinite number of color levels [7]. In medical analyses such as chronic wound assessment, HDR images are preferred to LDR images as they capture higher contrasts and finer texture details. Moreover, HDR techniques can be used to address exposure issues (images appearing too bright or too dark), improving the performance of a wide range of chronic wound imaging tasks including classification, segmentation and wound assessment.

However, while some high-end smartphones now come equipped with HDR modes, many low-end phones are still not equipped with HDR capabilities. Consequently, techniques such as exposure bracketing [8], which can convert LDR images to HDR, have emerged as viable methods to mitigate exposure issues s [9]. In these HDR recovery methods, images captured with different exposures are used to recover the irradiance map, which is then mapped to a narrower range (0–255) suitable for display using tone mapping. HDR inference from a single LDR image is more challenging as information is lost in ill-exposed regions. Recent work has proposed HDR recovery from a single LDR image directly [10,11], by synthesizing exposure-bracket images in order to simulate traditional HDR image generation [12,13]. Image fusion using multiple LDR images to infer an HDR image has been found to generate better images than tone mapping [14].

In this paper, we propose a method to recover an HDR image from a single LDR image to mitigate exposure issues. Our method is based on Retinex theory, which models an image S as an element-wise multiplication of the reflectance map R that describes the material and texture information and an illumination map I that contains the light intensity information ($S = R \otimes I$). First, a deep Retinex model was used to synthesize two images that have higher and lower exposures. To mitigate information loss in the synthesized images, we designed a detail enhancement pipeline to recover details lost in saturated and negatively-saturated regions. In the last step, the two synthesized detail-recovered images are merged with the original image, creating a single image with better illumination. Exposure fusion is used to merge images with different exposures. The outputs of our method are "HDR-like" because we produce an LDR image of higher visual quality but do not generate the radiance map (HDR image). Our proposed method outperforms state-of-art HDR generation methods on several metrics including structure similarity ($0.76 \pm 0.04/0.69 \pm 0.08$ on bright/dark images), peak signal-ratio score (28.60 ± 0.70 on dark images) and dice coefficient scores ($0.76 \pm 0.09/0.74 \pm 0.09$ on bright/dark images).

Most prior work to generate an HDR from LDR image(s) [11–13,15] has focused on improving visual quality by performing tasks such as predicting pixel information in completely saturated/negatively-saturated regions. However, the proposed image enhancements were not faithful as required for accurate medical image analyses such as chronic wound assessment. For instance, Endo et al. [13] showed that HDR images generated using an image-to-image translation network from [16] with direct mapping are usually inconsistent in quality with lots of noise due to the randomness introduced during model training. Moreover, prior work typically focused on mitigating either over- or under-exposure, but not both. To the best of our knowledge, our work is the first HDR generation method that enhances both over- and under-exposed LDR images faithfully for medical assessment. The limited availability of medical images also poses a significant challenge, which we overcome by using an indirect approach that is not limited by the availability of HDR ground-truth images. The exposure fusion [17] method that we utilize for merging up-exposed and down-exposed images has been found to produce more natural-looking than images generated by tone mapping [14]. Finally, to reduce noise artifacts, we also denoise images in a post-process that removes highlights in down-exposed images. Our contributions are listed as follows.

- We propose the Bi-directional Retinex Model network structure that generates up- and down-exposed images from an input image.
- We introduce detail-recovered exposure fusion as a method to enhance texture details that are hard to recover in highly saturated and negative saturated image regions.
- In the rigorous evaluation of our method on a dataset with wound images captured in systematically varied illumination, our method outperforms the state-of-the-art HDR generation methods in terms of image similarity and change in semantic segmentation performance metrics.

2   Related Work

Multi-exposure Fusion: Exposure fusion techniques utilize information from images captured or synthesized at multiple exposures to infer the HDR-like image with optimal exposure. Song et al. [18] use a probabilistic model that preserves the calculated image luminance levels and suppresses reversals in the image luminance gradients. Mertens et al. [17] compute a perceptual quality measure for each pixel in the multi-exposure sequence encoding desirable qualities such as saturation and contrast for fusion. Liu et al. [19] perform signal decomposition using independent component analysis to restore texture and color information during fusion. [20] introduce membership functions, which assign weights to every pixel in images based on exposure rates and pixel illumination values. In our work, we merge images using a membership function similar to Vonikakis et al. [20] to merge images. A key difference is that using exposure fusion to generate the final output, an intermediate fusion is performed to enhance the texture details in highly saturated and negatively saturated regions. Our method ensures that texture details are preserved maximally during exposure fusion.

Inverse Tone Mapping: Inverse tone mapping or reverse tone mapping are referred as inverse Tone Mapping Operator (iTMO) [21] or reverse Tone Mapping Operator (rTMO) in the literature. It is the dual of tone mapping and infers an HDR from LDR image, which is the opposite of tone mapping [21]. iTMO is similar to the conventional HDR imaging pipeline, except that only one LDR image taken with single exposure level is utilized. Endo et al. [13] propose a technique for synthesizing bracket images using Camera Response Functions (CRFs) selected from 201 curves in Grossberg and Nayar's Database of Response Function [22], They use k-means clustering and an encoder-decoder model to learn the mapping between the input LDR image and synthesized bracket images. Liu et al. [12] model the HDR-to-LDR image generation process with three steps (dynamic range clipping, non-linear mapping and quantization) and inverse each step by designing a dedicated neural network. [11] design a multi-scale CNN architecture named ExpandNet that learns the direct mapping from LDR image to its HDR counterpart. HDRCNN [10] predicts missing details in over-exposed regions during HDR generation. Apart from directly generating an HDR image, some works [23,24] generate tone-mapped images or HDR-like images instead of the original HDR images. Besides, some iTMOs [25,26] focus on reconstructing saturated or over-exposed regions to recover lost information in highlight areas. Similar to other iTMO methods, our proposed method only requires one LDR as input, however, we focus on the generating the tone-mapped image instead of the HDR image itself.

3   Our Approach: Algorithmic Pipeline

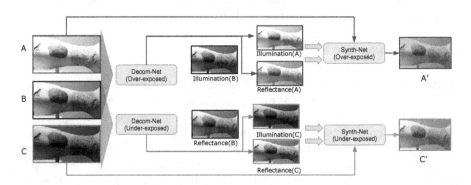

Fig. 1. Model for image bracketing to generate up/down exposed images

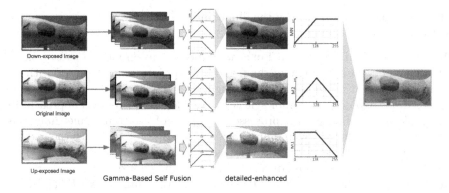

Fig. 2. Exposure fusion and HDR-like image generation

The overall flow of our pipeline has two phases, In phase one (Fig. 1), an ensemble bi-directional model generates an over-exposed and an under-exposed image. The model contains two identical sub-models that are used for image decomposition (Decom-Net) and bracket image synthesis (Synth-Net) respectively. In phase two, the original image is fused with the over- and under-exposed images in two steps that enhance details as well as global illumination, producing an HDR-like image. The details of these two phases are expounded on in the following sections.

Bidirectional Retinex Model: Our bi-directional Retinex model is an ensemble model consisting of Up and Down-exposure sub-models. These two sub-models have similar network structures but are trained with different data pairs. Under-exposed images (image C) are used in training the up-exposure model and Over-exposed images (image A) are used in training the down-exposure model. Both models utilize images labeled with the correct exposure as the ground truth. Image decomposition based on the Retinex theory without any prior information is a mathematically ill-posed problem. To admit a solution, we make assumptions that: 1) all three images $(A - C)$ have the same reflectance map. 2) the illumination map is smooth in regions where strong structures (such as boundaries) are not present.

For the Decom-Net, we adopt a similar network structure as in Wei et al. [27], which uses an optimization approach with the afore-mentioned assumptions as constraints. However, instead of minimizing the difference between the reflectance maps of under-exposed and target images, we consider three images. This reduces the influence of exposures and enables the sharing of network parameters between the two Decom-Nets in the Up and Down-exposure models for more stable training. We implement the Synth-Net with an encoder-decoder network with skip-connections inspired by the generator implementation in Pix2pix model [16]. However, instead of using 3 input channels (for RGB images), we used 4 input channels (3 channels for the reflectance map and 1 channel for the illumination map) in Synth-Net. Each channel of the output reflectance map is multiplied by the illumination map in a pixel-wise manner to synthesize an Up/Down-exposed image $(A' - C')$.

Detail-Recovered Exposure Fusion: Exposure fusion merges three images with different exposures (high, medium and low exposures) into an image without tone-mapping. Rather than directly taking the three images (the original image and up/down-exposed image) for synthesis, we adopted a two-level approach. In the first level, three images are processed using one of two transformer functions $(TF_-$ and $TF_+)$ followed by an illumination-based fusion to enhance the texture details. In the second level, the three images produced in the first level are used to synthesize the output again during illumination-based fusion. The pixel values are normalized in the range $(0-1)$. The two transformer functions TF_- and TF_+ are defined in 1 and 2 respectively.

$$TF_-(img, \lambda) = 1 - (1 - img)^{\frac{1}{\lambda}} \tag{1}$$

$$TF_+(img, \lambda) = img^{\frac{1}{\lambda}} \tag{2}$$

As gamma correction is commonly used to improve contrast in dark images [28], we design TF_- and TF_+ based on gamma correction with TF_- enhancing the details in over-exposed and TF_+ enhancing the under-exposed regions. Parameter img is a $H \times W \times 3$ matrix representing the input RGB image. And λ is a small custom parameter. As for $TF_+(img, \lambda)$, $lambda$ are the inverse of γ in gamma correction. $TF_-(img, \lambda)$ works by applying gamma correction on the color-inverted version of the input rather than the input itself. Then, we invert the color after gamma correction to generate the output. We used the illumination-based fusion method proposed in [20] for the 1st and 2nd level fusion due to its simplicity and minimal introduction of artifacts. The core concept behind this fusion method is assigning different weight functions for pixels in each image based on the exposures so that pixels in the bright region of under-exposed images, the dark region of over-exposed images and the middle range of well-exposed images can be emphasized more using higher weights. Each weight function is applied on all the channels and the fusion output is the weighted sum of all images (usually 3 images). The pseudo-code of the illumination-based fusion and the detail-recovered exposure fusion are shown in Algorithm 1 and 2 respectively. The illumination-based fusion is invoked in the detail-recovered exposure fusion.

Global Illumination Adjustment: To ensure optimal visual quality in terms of illumination, a global multiplier ψ is introduced to adjust the global illumination. We define ψ in.

$$\psi = \min(\frac{1}{\max(img^*)}, \frac{\max(\text{avg}(img), C)}{\text{avg}(img^*)}) \tag{3}$$

img is the original image and img^* is the fused image. C is a constant that controls the average pixel values of the fused image in case the original image is too dark or too bright. It can be considered as the "preferred" average pixel

---

Algorithm 1. Illumination-based Fusion

1: **procedure** FUSE(img_+, img, img_-)   ▷ RGB images with high/median/low exposures
2:    $rgb\_list \leftarrow \{img_+, img, img_-\}$
3:    $weight\_list \leftarrow \{\}$
4:    **for** rgb in rgb_list **do**
5:        $ycbcr \leftarrow$ convert rgb to YCbCr space.
6:        $Ych \leftarrow$ get the Y channel from $ycbcr$.
7:        $weight\_list$.add(get_weight_matrix(Ych))
8:    **end for**
                                    ▷ weight_list contains weights for rgb_list
9:    $img^* \leftarrow$ weighted sum of rgb_list
10:    $img^* \leftarrow$ normalized(img^*)
11:    **return** img^* ▷ Fused image. Clip the pixel values within the range of $(0-1)$
12: **end procedure**

Algorithm 2. Detail-recovered Exposure Fusion

1: **procedure** DR-FUSE(img_+, img, img_-)   ▷ Up/Original/Down exposed images
2:    $img\_list_+ \leftarrow \{\}$
3:    $img\_list_- \leftarrow \{\}$
4:    $img\_list \leftarrow \{\}$
5:    **for** λ in 1..3 **do**
6:        img_list_-.add($TF_-(img_-, \lambda)$)
7:        img_list_+.add($TF_+(img_+, \lambda)$)
8:    **end for**
9:    img_list.add($TF_+(img, 2)$)
10:    img_list.add(img)
11:    img_list.add($TF_-(img, 2)$)
                                    ▷ Perform first level fusion on 3 images
12:    $img_+^{(1)} \leftarrow$ FUSE(img_list_+)
13:    $img_-^{(1)} \leftarrow$ FUSE(reversed(img_list_-))
14:    $img^{(1)} \leftarrow$ FUSE(img_list)
15:    $img^* \leftarrow$ FUSE($img_+^{(1)}, img^{(1)}, img_-^{(1)}$)   ▷ Perform second level fusion
16:    **return** img^* ▷ Fused image.
17: **end procedure**

---

value of the fused image. We also limit pixels of the adjusted image to the $(0-1)$ range by not allowing the value of ψ exceeding $1/\mathbf{max}(img^*)$. We set $C = 0.6$ for all our experiments.

4   Evaluation and Results

Wound Illumination Varying Dataset (IVDS) Dataset: To train and evaluate our proposed methods we utilized the Illumination Varying Dataset (IVDS) [29]. The IVDS contains 55440 images (7 sub-datasets with 7920 images in each sub-dataset) of a wound moulage [30]. Captured under various illumination conditions/smartphone cameras/parameters. The moulage was molded from an actual patient and several conditions are presented including venous ulcers and cellulitis. The illumination included 33 different light sources, 16 light color temperatures ranging from 3200K to 5600K and 15 light intensities ranging from 10 to 255. To reduce bias caused by auto adjustments by smartphone cameras, we focus on the subset of wound images taken with fixed manual configurations (*Pixel-M1* and *Pixel-M2*). The ISO sensitivities of *Pixel-M1* and *Pixel-M2* are 50 and 72, while the exposure time are 62.00 ms and 31.25 ms. White balance is turned off for both two datasets.

Data Preparation: To train the bi-directional Retinex model, we created "over-exposed", "correct-exposed" and "under-exposed" images (images A, B and C in Fig. 1). In the IVDS dataset, there are 15 images which are different on their light source intensities for each (camera, parameters) pair, and we manually labelled images as "bright", "normal" and "dark" for training. For each (camera, parameters) pair, we sampled an equal number of "bright" and "dark" images in order to ensure a balanced dataset. We only select "normal" images that have the least over/under exposed regions. We then randomly selected 80% of the labelled data as training data and use the remaining 20% for testing.

Training the Bi-Directional Retinex Model: We generate training pairs using the labelled data with permutation. Each pair consists of one "bright" image, one "normal" image and one "dark" image. Except for the light intensities, all the configurations remain the same for the three images in each pair. The "bright"/"normal" images are used to train the down-exposure model while the "dark"/"normal" images are used to train the up-exposure model. In our experiment, we use 1267 pairs for training and 317 pairs for testing. All the images are re-scaled to $512 * 256$ for fast training and testing. We trained the bi-directional Retinex model for 100 epochs in total that took approximately 20 min. After hyper-parameter search, we utilized a batch size of 16, patch size of 48, the Adam optimizer and a learning rate of 0.001.

HDR Target Creation: To evaluate the performance of our proposed method, we generated the HDR wound image as the ground-truth and calculated its similarity with multiple LDR wound images by recovering the radiance map [8]. Four LDR images were used to synthesize the HDR target. In our experiment, the light source intensities of selected four LDR images are 10, 80, 140 and 250 and all of them are from the *Pixel-M1* dataset with the same light source position(index number 9) and light temperatures(index number 102). These four images are shown in Fig. 3. We compared the *Drago*, *Mantiuk* and *Reinhard* tone mappers for converting the HDR wound image into an HDR-like image. Among the three mappers, *Drago* performed the best, with minimal over/under exposed areas and it preserved most texture details.

(a) light intensity 10 (b) light intensity 80 (c) light intensity 140 (d) light intensity 250

Fig. 3. LDR images used for HDR target creation

4.1 Evaluation Experiments

Evaluation via Visual Comparison: We compared results generated with our proposed method with those generated from HDRCNN [10], ExpandNet [11] and SingleHDR [12]. Sample results are shown in Fig. 4. The 1st column in Fig. 4 shows the original images that are used as inputs (the first 3 images are over-exposed and the last 3 images are under-exposed). Each row shows the HDR-like images generated using different methods. HDRCNN mitigates the over-exposed issues at the cost of reducing the color contrast. SingleHDR is prone to saturated issues for both over and under-exposed images. The performance of ExpandNet is roughly equivalent to our method in over-exposed images. However, ExpandNet may introduce significant artifacts in saturated regions. Overall, the enhanced results generated by our method have the best visual quality.

Evaluations Using Image Similarity Metrics: We used image similarity metrics to measures differences between the 1) the original images and the HDR target image (**original** scores in Fig. 5), and 2) the generated HDR-like images and the HDR target image (**enhanced** scores in Fig. 5). The distributions of the two similarities are then compared. We also included the similarities between the images generated directly with illumination-based fusion without detail recovery and the HDR target image (**directly-fused** scores in Fig. 5). We used the Structural Similarity Index (SSIM) [31] and Peak Signal-Noise Ratio (PSNR) [32] as image similarity metrics. PSNR (5) estimates absolute errors while SSIM (4) takes into consideration inter-correlations between spatially-close pixels. SSIM is calculated on several windows ($N \times N$ region) of an image. For two $N \times N$ windows x and y, μ_x and μ_y denote the average values of pixels in x and y. σ_x^2, σ_y^2 and σ_{xy} denote variance of pixels in x, variance of pixels in y and covariance of pixels in x, y.

$$SSIM(x,y) = \frac{(2\mu_x\mu_y + c_1)(2\sigma_{xy} + c_2)}{(\mu_x^2 + \mu_y^2 + c_1)(\sigma_x^2 + \sigma y^2 + c_2)} \qquad (4)$$

PSNR is calculated based on the mean square error (MSE) between two images s and t.

$$PSNR(s,t) = 10log_{10}(\frac{255^2}{MSE(s,t)}) \qquad (5)$$

where

$$MSE(s,t) = \frac{1}{MN}\sum_{i=1}^{M}\sum_{j=1}^{N}(s_{ij} - t_{ij})^2 \qquad (6)$$

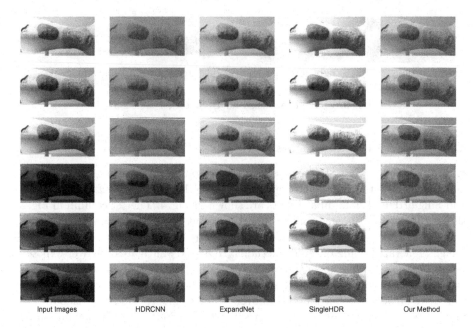

Input Images   HDRCNN   ExpandNet   SingleHDR   Our Method

Fig. 4. Visual comparison on enhanced images. The first column contains the input images (moulage) from the IVDS dataset. The first 3 rows include HDR results generated from 3 over-exposed images and the last 3 rows include HDR results with under-exposed images. Except for our method, all other enhanced images(column 2–4) are tone-mapped with *Drago* using the same settings (gamma = 1.6, saturation = 1.45).

To demonstrate the effectiveness of HDR generation from images with different exposure issues, we divided the test set into over and under-exposed image sets (both sets have 317 images) and calculated the similarity scores separately. The distribution of the SSIM scores is shown in Fig. 5(a–b) and the distribution of the PSNR scores is shown in Fig. 5(c–d). The HDR-like images generated achieved higher similarity scores, indicating a higher resemblance to the HDR target image. Also, compared with direct fusion, the detail-recovered exposure fusion method achieves significantly higher similarity scores for both over- and under- exposed images using both metrics. As the HDR target image is designed to contain rich texture information with minimal exposure issues, images generated using detail-recovered exposure fusion achieve higher similarity scores than direct fusion, indicating higher performance in recovering texture information. We also compared our method with the state-of-art HDR generation methods trained using an 80%:20% train: test split on the *Pixel-M1* and *Pixel-M2* datasets (See Table 1). **SSIM+** means the scores are only calculated using images that are labelled as "bright", while **SSIM-** means the scores are calculated using only images labelled as "dark". A similar notation is used for **PSNR+** and **PSNR-**. Average and standard deviations of each metric were recorded in this table. Our method achieves the highest scores of all methods compared against for **SSIM+**, **SSIM-** and **PSNR-** metrics.

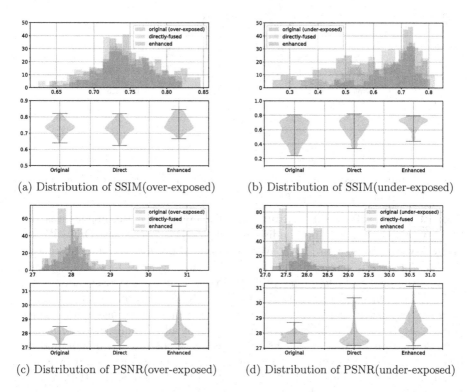

(a) Distribution of SSIM(over-exposed)   (b) Distribution of SSIM(under-exposed)

(c) Distribution of PSNR(over-exposed)   (d) Distribution of PSNR(under-exposed)

Fig. 5. Distribution of SSIM and PSNR. (a) and (b) show the distribution of SSIM scores on over- and under-exposed images respectively. (c) and (d) are similar with PSNR as the metric.

Evaluation Using Segmentation Accuracy: We also used change in semantic segmentation accuracy as an evaluation metric since semantic segmentation is an important task in medical image analyses including wounds. We used *Dice Similarity Coefficient*(DSC) [33] as the metric for measuring segmentation accuracy compared to ground truth.

In our experiments, we train models to differentiate the chronic wound regions(positive) and the rest(negative) on *Pixel-M1* and *Pixel-M2* datasets. 80% of the data is used in training and the remaining 20% is used for evaluation(same as other methods). The equation for DSC is $DSC = 2TP/(2TP + FP + FN)$. TP, FP and FN are abbreviations of true positive, false positive and false negative. The comparison of DSC scores is listed in Table 1. **DSC-** and **DSC+** are scores calculated on under and over-exposed images respectively. We can find our method achieves the highest DSC scores for both under and over-exposed

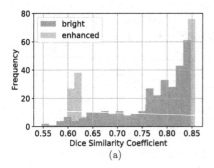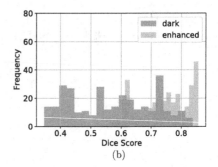

(a)   (b)

Fig. 6. .

among all methods compared against. And compared with the original images, the images enhanced with our method result in higher DSC scores.

Score Under Different Illumination: The enhancement of images was analyzed with different lighting conditions using SSIM, PSNR and DSC. We use image L-values to estimate their average illumination. The L-value is the average value of the luminance channel $L(S)$ of an image S in CIELAB or LAB space can be expressed as $Lvalue = avg(L(S))$. Figure 7 shows how SSIM, PSNR and DSC change with different L-values respectively. For clearer visualization, all data points plotted in Fig. 7 are smoothed using moving average with window size 10. As the L-value increases, SSIM scores of all methods increase in general. Beyond an L-value of 70, the SSIM scores of the original image, and the image generated by ExpandNet and our method start to decrease. SSIM scores of HDRCNN is the most steady among all methods and SingleHDR has the best SSIM scores for very bright images (L-value larger than 80). Our method achieves the highest PSNR scores for dark (L-value around 35) and bright (L-value around 65) images. And ExpandNet achieves better scores in the middle range of L-values. Our method generally achieves better DSC scores on dark images (L-value below 55), and ExpandNet has roughly equivalent performance with our method on bright images.

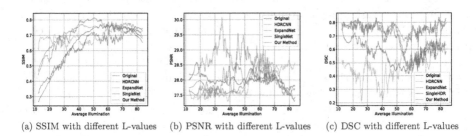

(a) SSIM with different L-values   (b) PSNR with different L-values   (c) DSC with different L-values

Fig. 7. SSIM/PSNR/DSC with Different L-values. Results are processed with moving average(window size equals to 10) for better visualization.

Table 1. Comparison with other methods

Methods	SSIM+	SSIM-	PSNR+	PSNR-	DSC+	DSC-
Original	0.74 ± 0.04	0.56 ± 0.14	27.97 ± 0.27	27.76 ± 0.26	0.75 ± 0.07	0.58 ± 0.14
HDRCNN	0.73 ± 0.02	0.66 ± 0.06	**28.46 ± 0.29**	27.53 ± 0.28	0.41 ± 0.18	0.55 ± 0.11
ExpandNet	0.74 ± 0.04	0.68 ± 0.11	27.97 ± 0.35	27.97 ± 0.84	0.75 ± 0.11	0.73 ± 0.11
SingleHDR	0.66 ± 0.06	0.60 ± 0.11	27.90 ± 0.21	28.02 ± 0.22	0.60 ± 0.12	0.55 ± 0.14
Our Method	**0.76 ± 0.04**	**0.69 ± 0.08**	28.25 ± 0.76	**28.60 ± 0.70**	**0.76 ± 0.09**	**0.74 ± 0.09**

5 Conclusion and Future Work

In this paper, we propose a method for generating HDR-like images using a deep Retinex model and exposure fusion. Up-exposure and down-exposure models are trained to create bracket images. To mitigate information loss in over and under-exposed image regions when fusing, we applied two transformer functions to recover the textural details. Exposure fusion is used in the 2-step fusion to generate the final output. Experiments show our method achieves higher scores in most of metrics (SSIM, PSNR, DSC) for both dark and bright images on chronic wound images in the IVDS dataset.

The work presented in this paper is a first step in developing our light mitigation method, which focused on algorithm development and initial validation. In future work, we will evaluate our method in a wider range of clinical settings and investigate whether wound image enhancement can improve the decision-making process of clinicians. We will also explore improvements in the performance of automated wound assessment using computer vision on images labeled with clinically validated wound assessment rubrics such as the Photographic Wound Assessment Tool (PWAT) and the Braden Scale.

References

1. Martinengo, L., et al.: Prevalence of chronic wounds in the general population: systematic review and meta-analysis of observational studies. Ann. Epidemiol. **29**, 8–15 (2019)
2. Richmond, N.A., Maderal, A.D., Vivas, A.C.: Evidence-based management of common chronic lower extremity ulcers. Dermatol. Ther. **26**(3), 187–196 (2013)
3. Wang, L., Pedersen, P.C., Strong, D.M., Tulu, B., Agu, E., Ignotz, R.: Smartphone-based wound assessment system for patients with diabetes. IEEE Trans. Biomed. Eng. **62**(2), 477–488 (2014)
4. Kuang, B., et al.: Assessment of a smartphone-based application for diabetic foot ulcer measurement. Wound Repair and Regeneration (2021)
5. Wang, L.: System designs for diabetic foot ulcer image assessment. System **2016**, 03–07 (2016)
6. Myszkowski, K., Mantiuk, R., Krawczyk, G.: High dynamic range video. Syn. Lect. Comput. Graph. Anim. **1**(1), 1–158 (2008)
7. Reinhard, E., Heidrich, W., Debevec, P., Pattanaik, S., Ward, G., Myszkowski, K.: High Dynamic Range Imaging: Acquisition, Display, and Image-Based Lighting. Morgan Kaufmann, Burlington (2010)

8. Debevec, P.E., Malik, J.: Recovering high dynamic range radiance maps from photographs. In: ACM SIGGRAPH 2008 Classes, pp. 1–10 (2008)
9. Licciardo, G.D., Cappetta, C., Di Benedetto, L.: Dynamic range enhancement for medical image processing. In: 2017 7th IEEE International Workshop on Advances in Sensors and Interfaces (IWASI), pp. 219–223. IEEE (2017)
10. Eilertsen, G., Kronander, J., Denes, G., Mantiuk, R.K., Unger, J.: HDR image reconstruction from a single exposure using deep CNNs. ACM Trans. Graph. (TOG) **36**(6), 1–15 (2017)
11. Marnerides, D., Bashford-Rogers, T., Hatchett, J., Debattista, K.: Expandnet: a deep convolutional neural network for high dynamic range expansion from low dynamic range content. In: Computer Graphics Forum, vol. 37, no. 2, pp. 37–49. Wiley Online Library (2018)
12. Y.-L. Liu, et al.: Single-image HDR reconstruction by learning to reverse the camera pipeline. In: Proceedings IEEE CVPR, pp. 1651–1660 (2020)
13. Endo, Y., Kanamori, Y., Mitani, J.: Deep reverse tone mapping. ACM Trans. Graph. **36**(6), 177–181 (2017)
14. Bachoo, A.K.: Real-time exposure fusion on a mobile computer (2009)
15. Kim, J.H., Lee, S., Jo, S., Kang, S.-J.: End-to-end differentiable learning to HDR image synth. for multi-exposure images. arXiv preprint arXiv:2006.15833 (2020)
16. Isola, P., Zhu, J.-Y., Zhou, T., Efros, A.A.: Image-to-image translation with conditional adversarial networks. In: Proceedings of the CVPR, pp. 1125–1134 (2017)
17. Mertens, T., Kautz, J., Van Reeth, F.: Exposure fusion: a simple and practical alternative to high dynamic range photography. Comput. Graph. Forum **28**(1), 161–171 (2009). Wiley Online Library
18. Song, M., Tao, D., Chen, C., Bu, J., Luo, J., Zhang, C.: Probabilistic exposure fusion. IEEE Trans. Image Process. **21**(1), 341–357 (2011)
19. Li, H., et al.: Denoising scanner effects from multimodal MRI data using linked independent component analysis. Neuroimage **208**, 116388 (2020)
20. Vonikakis, V., Bouzos, O., Andreadis, I.: Multi-exposure image fusion based on illumination estimation. In: Proceedings of IASTED SIPA, pp. 135–142 (2011)
21. Banterle, F., Ledda, P., Debattista, K., Chalmers, A.: Inverse tone mapping. In: Proceedings of the 4th International Conference on Computer Graphics and Interactive Techniques in Australasia and Southeast Asia, pp. 349–356 (2006)
22. Grossberg, M.D., Nayar, S.K.: What is the space of camera response functions? In: Proceedings of IEEE CVPR, vol. 2, pp. II-602. IEEE (2003)
23. Zhang, J., Lalonde, J.-F.: Learning high dynamic range from outdoor panoramas. In: Proceedings IEEE International Conference on Computer Vision, pp. 4519–4528 (2017)
24. Kalantari, N.K., Ramamoorthi, R.: Deep high dynamic range imaging of dynamic scenes. ACM Trans. Graph. **36**(4), 144–151 (2017)
25. Banterle, F., Ledda, P., Debattista, K., Bloj, M., Artusi, A., Chalmers, A.: A psychophysical evaluation of inverse tone mapping techniques. In: Computer Graphics Forum, vol. 28, no. 1, pp. 13–25 (2009). Wiley Online Library
26. Kovaleski, R.P., Oliveira, M.M.: High-quality reverse tone mapping for a wide range of exposures. In: Proceedings of SIBGRAPI, pp. 49–56. IEEE (2014)
27. Wei, C., Wang, W., Yang, W., Liu, J.: Deep retinex decomposition for low-light enhancement. arXiv preprint arXiv:1808.04560 (2018)
28. Khunteta, A., Ghosh, D., et al.: Fuzzy rule-based image exposure level estimation and adaptive gamma correction for contrast enhancement in dark images. In: IEEE International Conference on Signal Processing, vol. 1, pp. 667–672. IEEE (2012)

29. Iyer, A.B.: Let there be light...characterizing the effects of adverse lighting on semantic segmentation of wound images and mitigation using a deep retinex model. Masters thesis, Worcester Polytechnic Institute (2020)
30. Vinnie venous insufficiency leg model. https://vatainc.com/product/vinnie-venous-insufficiency-leg-model. Accessed 10 May 2021
31. Wang, Z., Bovik, A.C., Sheikh, H.R., Simoncelli, E.P.: Image quality assessment: from error visibility to structural similarity. IEEE Trans. Image Process. **13**(4), 600–612 (2004)
32. Hore, A., Ziou, D.: Image quality metrics: PSNR vs. SSIM. In: 2010 20th International Conference on Pattern Recognition, pp. 2366–2369. IEEE (2010)
33. Dice, L.R.: Measures of the amount of ecologic association between species. Ecology **26**(3), 297–302 (1945)

Deep Learning-Based Bias Transfer for Overcoming Laboratory Differences of Microscopic Images

Ann-Katrin Thebille[1], Esther Dietrich[1], Martin Klaus[1,2], Lukas Gernhold[2],
Maximilian Lennartz[3], Christoph Kuppe[4], Rafael Kramann[4],
Tobias B. Huber[2], Guido Sauter[3], Victor G. Puelles[2],
Marina Zimmermann[1,2(✉)], and Stefan Bonn[1(✉)]

[1] Institute of Medical Systems Biology, Center for Biomedical AI (bAIome),
University Medical Center Hamburg-Eppendorf, Hamburg, Germany
`marina.zimmermann@zmnh.uni-hamburg.de, sbonn@uke.de`
[2] III. Department of Medicine, University Medical Center Hamburg-Eppendorf,
Hamburg, Germany
[3] Institute of Pathology, University Medical Center Hamburg-Eppendorf,
Hamburg, Germany
[4] Institute of Experimental Medicine and Systems Biology, and Division of
Nephrology and Clinical Immunology, RWTH Aachen University, Aachen, Germany

Abstract. The automated analysis of medical images is currently limited by technical and biological noise and bias. The same source tissue can be represented by vastly different images if the image acquisition or processing protocols vary. For an image analysis pipeline, it is crucial to compensate such biases to avoid misinterpretations. Here, we evaluate, compare, and improve existing generative model architectures to overcome domain shifts for immunofluorescence (IF) and Hematoxylin and Eosin (H&E) stained microscopy images. To determine the performance of the generative models, the original and transformed images were segmented or classified by deep neural networks that were trained only on images of the target bias. In the scope of our analysis, U-Net cycleGANs trained with an additional identity and an MS-SSIM-based loss and Fixed-Point GANs trained with an additional structure loss led to the best results for the IF and H&E stained samples, respectively. Adapting the bias of the samples significantly improved the pixel-level segmentation for human kidney glomeruli and podocytes and improved the classification accuracy for human prostate biopsies by up to 14%.

Keywords: CycleGAN · Fixed-Point GAN · Domain adaptation · H&E staining · Unsupervised learning · Immunofluorescence microscopy

1 Introduction

Deep learning (DL) applications play an increasingly important role in medicine, yet, "everyone participating in medical image evaluation with machine learning

© Springer Nature Switzerland AG 2021
B. W. Papież et al. (Eds.): MIUA 2021, LNCS 12722, pp. 322–336, 2021.
https://doi.org/10.1007/978-3-030-80432-9_25

is data starved" [15]. When the same imaging technique is used, e.g. confocal microscopy of a specific tissue type with a matching staining protocol, image analysis networks should be applicable to datasets not seen during training without a substantial drop in performance. However, due to bias introduced during the acquisition or processing of datasets ('domain shift'), the generalizability of deep neural networks is negatively affected. If bias cannot be accounted for, models have to be re-trained every time a new dataset becomes available. In consequence, new data can only be used if a large cohort of labeled samples exists or can be created. Traditional techniques try to compensate domain shift through normalization or simple image transformations, like an increase of the contrast through histogram equalization [12]. However, as indicated by de Bel et al. [4], simple approaches are limited in how much bias they can capture and adjust, leading to insufficient transformations.

Recent evidence suggests that deep generative models could be utilized to modify samples, creating novel images that are close to the reference domain while not altering the original content. This way, images of a new dataset can be adjusted to the bias of the reference domain as a pre-processing step ('bias transfer'), in order to avoid re-training large image analysis networks. The goal is that new data of the same modality can be handled correctly and without a large drop in performance. Necessarily, content preservation is of the utmost importance since hallucination artifacts introduced by the generative models could lead to misdiagnosis. Reliable bias transfer approaches would also enable the usage of DL in settings that do not allow for frequent creation and retraining of models, such as decision support systems in hospitals.

In this paper, we aim to improve existing generative models to enable stable bias transfer for medical histopathology images. In addition, we propose guidelines for testing and evaluating bias transfer models in settings with similar transformation goals. To benchmark the quality of the bias transfer for three state-of-the-art generative models, cycle-consistent generative adversarial networks (cycleGANs) [24], U-Net cycleGANs [4], and Fixed-Point GANs [6], we measured the content preservation, target domain adaptation, and impact on image segmentation and classification performance. To increase the performance of the models, we tested three additional losses designed to improve the quality in terms of content ('MS-SSIM loss') [1], structure integrity ('structure loss') [16] and intensity of transformation ('additional identity loss') [4]. As a baseline, histogram matching [12] for color correction in a decorrelated color space [18] ('color transfer') is included in our evaluation, which utilizes a single, random image to represent the target domain.

2   Related Work

In a medical context, most datasets that require bias transfer are unpaired, meaning that a ground truth for the transformed images does not exist. A well-established approach for learning unpaired image-to-image translations is using CycleGANs [24]. CycleGANs have already been used for stain transforming renal tissue sections [4] or histological images of breast cancer [20]. Unfortunately, every paper introduces a new variation of cycleGAN. To the best of our knowledge, an extensive comparison of bias transfer algorithms for microscopy images does not exist yet. Selecting a fitting approach is a non-trivial task.

One common enhancement is using a U-Net structure for the generators [4, 17]. In a U-Net, the encoder and decoder structures of the neural network are connected via skip connections between layers of the same size [19]. Thus, the generators can pass information directly without transitioning the bottleneck, improving the level of detail of the images. In this paper, we refer to the modified cycleGAN approach as U-Net cycleGAN.

Besides architecture modifications or simple hyperparameter adaptations like changing the learning rate [4,17], or the number of images per batch [4,20], a frequent change is adding additional losses to incorporate demands the network has to fulfill. Armanious et al. used the multi-scale structural similarity index (MS-SSIM) [23] in [1] as an additional cycle loss between the original and the cycle-reconstructed images. Their goal was to penalize structural discrepancies between the images. In their experiments, the additional loss led to sharper results with better textural details. Another loss has been proposed by de Bel et al. in [4]. They included an additional identity loss that is decreased to zero over the first 20 epochs of training. The loss is an addition to the original identity loss of cycleGANs [24]. In their experiments, the loss stabilized the training process and led to faster convergence since it forces the generator to look for transformations close to identity first, effectively shrinking the solution space. Moreover, Ma et al. proposed a loss based on the sub-part of the structural similarity index (SSIM) that only evaluates local structure changes [16]. They proposed the loss to enhance the quality of endoscopy and confocal microscopy images that suffer from "intensity inhomogeneity, noticeable blur and poor contrast" [16]. The structure loss directly compares the original and transformed images.

Unfortunately, cycleGANs are designed for transforming between two domains only. Therefore, transforming multiple domains to the reference domain requires individual models and training runs. In [6], Choi et al. proposed Star-GANs, which transform between any number of domains with a single generator and discriminator. The generator produces samples based on the input image and the label of the target domain. Even if only two domains exist, StarGANs can potentially outperform cycleGAN-based approaches due to the benefit that all data samples are used to train a single generator. Siddiquee et al. proposed Fixed-Point GANs (FPGs), which extend StarGANs with a conditional identity loss, thus reducing hallucination artifacts [21]. If a new dataset could be adjusted by adding a new condition to FPG, bias transfer would be an exceptionally fast solution for the generalizability problem of image analysis networks.

3   Methodology

In the following, the overall workflow, as well as the generative models (Subsect. 3.1) and our datasets (Subsect. 3.2) are introduced.

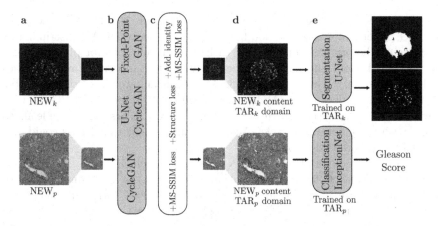

Fig. 1. Overview of the image analysis pipeline including bias transfer: First, the original input images are downsampled **a** and transformed by the generative models **b** trained with or without additional losses **c**. Afterwards, they are upsampled to the original size **d**. The transformed images are then used as inputs for the segmentation or classification networks **e**. NEW_k and TAR_k refer to the domains of the kidney samples, whereas NEW_p and TAR_p are the domains of the prostate samples.

As part of this work, several generative approaches are introduced into an existing image analysis pipeline. An overview of the complete workflow with bias transfer is shown in Fig. 1. As bias transfer is used as a pre-processing step here, the image size of the transformed images should be identical to the original size. However, generative models are usually not designed for images larger than 256×256 pixels. We used the Laplacian pyramid-based approach introduced in [11] to downsample the images to 256×256 pixels pre- (Fig. 1 a → b), and back to the original size post-transformation (Fig. 1 c → d), which is 1024×1024 pixels for the kidney (upper row) and 2048×2048 for the prostate images (lower row). The approach involves replacing the smallest layer of the pyramids with the respective transformed image for upsampling. Hence, the edges of the original image, which are typically lost at a low resolution, are added back into the image, boosting the quality of the transformed image.

3.1 Generative Approaches

For this paper, we implemented, tested, and modified three state-of-the-art generative models for image generation, as shown in Fig. 1b. For this purpose, we re-created the original implementation by Zhu et al. for cycleGANs [24] in Tensorflow 2.0 with a PatchGAN discriminator that judges 16×16 patches per image. The U-Net cycleGAN architecture has been constructed by adding skip connections to the generators.

For FPG, the generator input consists of an image and the one-hot encoded label of the target domain. Otherwise the generator architecture is identical to cycleGAN. Instead of using the discriminator structure described in the original FPG paper, we created a modified version of the cycleGAN discriminator by replacing the original output layer with two outputs, one for predicting the image's domain and one for the authenticity of the individual image patches (real vs. generated). The modified cycleGAN discriminator has fewer trainable parameters and judges more patches per image than the original FPG discriminator. In [14] it has been shown that judging more patches per image leads to images with more detail for cycleGANs. It seems evident that FPGs could also benefit from this modification.

The hyperparameters for cycleGAN and U-Net cycleGAN are mainly based on the original cycleGAN implementation. The networks are trained for up to 200 epochs with an initial steady learning rate of 0.0005 (original: 0.0002) for the first 100 epochs and a linearly decaying learning rate that reaches 0 after 200 epochs. For FPG, an initial learning rate of 0.0001 has been used, as proposed by Choi et al. [6]. We used a batch size of one.

To improve the generative performance and reduce hallucination artifacts of the three models, we included additional terms in the loss functions (see Fig. 1c) specifically the additional identity loss $\mathcal{L}_{+\mathrm{id}}$ [4], the MS-SSIM loss $\mathcal{L}_{+\mathrm{ms\text{-}ssim}}$ [1] and the structure loss $\mathcal{L}_{+\mathrm{struc}}$ [16]. The original losses of (U-Net) cycleGANs are the adversarial loss (adv), the cycle-reconstruction loss (cyc) and the identity loss (id). For FPGs, the original losses are the cycle-reconstruction loss (cyc), the domain-classification loss ($domain$), the gradient penalty loss (gp) and the conditional identity loss (id). For our experiments, the losses have been weighted with $\lambda_{\mathrm{adv}} = 1$, $\lambda_{\mathrm{cyc}} = 10$, and $\lambda_{\mathrm{id}} = 10$ (original: $\lambda_{\mathrm{id}} = 5$) for cycleGAN and U-Net cycleGAN and with $\lambda_{\mathrm{cyc}} = \lambda_{\mathrm{gp}} = \lambda_{\mathrm{id}} = 10$ and $\lambda_{\mathrm{domain}} = 1$ for FPG. All additional losses have been weighted with $\lambda = 5$. We added the structure loss and the MS-SSIM loss individually and, for the MS-SSIM loss, in combination with the additional identity loss ('combined losses'). The structure loss was not combined with the other additional losses since the direct comparison of the original and transformed images covers similar objectives as the 'combined losses'. Our implementation is publicly available on GitHub.[1]

3.2 Data

In this paper, we perform bias transfer for two modalities, IF images of kidney biopsies and H&E stained tissue microarray (TMA) spots of prostate biopsies. In Fig. 1, the upper row shows the workflow for the kidney and the lower row for the prostate biopsy samples. Each modality includes two sub-datasets (domains) originating from different hospitals and has a specified transformation direction. This is due to the fact that bias transfer is applied here to overcome the domain shift between a new and a target domain which has been used for training a segmentation or classification network, as outlined in Fig. 1e.

---

[1] https://github.com/imsb-uke/bias-transfer-microscopy.

Kidney Biopsies. The kidney dataset consists of 2D kidney IF images [25]. The images have been used to train a modified U-Net with dual output for the automatic segmentation of the glomeruli and their podocytes, an important cell type inside the glomeruli that is indicative of the health of the kidney [25]. To highlight the glomeruli and podocytes, three different biomarkers were used: DAPI for cell nuclei (blue), WT1 for podocyte cytoplasm (red), and DACH1 for podocyte nuclei (green). The images originate from two distinct hospitals (two domains: the target domain 'TAR$_k$' and the new domain 'NEW$_k$'), which were imaged by two different operators with differing confocal laser scanning microscopes (Nikon A1 Confocal and Zeiss Confocal LSM 800). In total, subset TAR$_k$ contains 269 images and subset NEW$_k$ contains 375 images. Training and evaluating the segmentation requires annotations. For this purpose, a subset of the images has been annotated by medical experts (TAR$_k$: 109, NEW$_k$: 90). Further details on the datasets and their biological meaning can be found in [25].

For bias transfer, the split into training, validation, and test sets was performed randomly, with the constraints that images originating from the same patient (up to 16) remain in the same set, that a ratio of approximately 70% for training and 15% each for validation and test sets is reached and that the validation and test sets only consist of annotated images. While bias transfer does not require a ground truth, masks are necessary to evaluate the segmentation quality achieved on the transformed images. Accordingly, the TAR$_k$ images were split into 180 for training, 44 for validation, and 45 for testing. The NEW$_k$ dataset was split into 285 images for training, 46 for validation, and 44 for testing.

Prostate Biopsies. The prostate dataset consists of circular biopsies (spots) originating from radical prostatectomies (RPE) that have been assembled with the help of TMAs. The images originate from two different hospitals ('TAR$_p$' and 'NEW$_p$'). The TAR$_p$ dataset consists of 2866 and the NEW$_p$ dataset of 886 images. Both datasets were created for staging prostate cancer with the help of Gleason patterns, which categorize the shape of the glands. Tissue can be classified as benign or as Gleason pattern 3, 4, or 5, whereas a higher number represents worse tumor tissue. The Gleason score (GS) is then calculated as the sum of the most prevalent and the worst occurring patterns in a sample [5]. Unfortunately, the inter-pathologist agreement on GSs is usually low [10], which is why automated and stable GS prediction is of high clinical value.

We trained an InceptionNet-V3-based classification network, pre-trained on ImageNet [8], on images of the TAR$_p$ dataset, reaching a test accuracy of 81.6%. To exclude image background only the innermost 2048×2048 pixels of the spots have been used. The classification was limited to single Gleason patterns only ('0' = benign, '3+3', '4+4', and '5+5') since TMA spots with two (or more) differing Gleason patterns could potentially contain patterns that occur solely outside of the innermost pixels. During training, all images are pre-processed with normalization, and the data is augmented with random rotations, shearing, shifting, and flipping. After pre-processing, the images are resized to 224×224 pixels, which is the input size of the InceptionNet.

The NEW$_p$ dataset is publicly available [2] and does not include Gleason score annotations. However, the images have been annotated via segmentation masks to identify areas containing a specific Gleason pattern. To evaluate the classification performance of our network, we used the segmentation masks to calculate the Gleason scores of the image centers. We then used the same split into training, validation, and test sets that was defined for the dataset in the original publication [3]. The training and validation sets have been annotated by one pathologist (pathologist 1) and the test set has been annotated independently by two pathologists (pathologist 1 and pathologist 2), resulting in 506 images (429 with a single Gleason pattern) for training, 133 (111) for validation, and 245 (199 – pathologist 1, 156 – pathologist 2) images for testing. For the TAR$_p$ dataset, the split was determined according to a random stratified sampling strategy, resulting in the same label distribution for the training, validation, and test sets. The resulting split consists of 2000 (1136 with a single Gleason pattern) images for training, 432 (246) images for validation, and 434 (245) images for testing. All images were annotated by the same pathologist (pathologist 3). Nonetheless, the label distribution is not uniform, TAR$_p$ overrepresents 3+3 samples. In contrast, NEW$_p$ mostly contains 4+4 samples.

4   Results and Discussion

We based the evaluation of the generative approaches on three factors: content preservation, target domain imitation, and impact on the segmentation or classification performance. Two metrics have been selected for a quantitative evaluation of the transformation quality: the structural similarity index (SSIM) [22] and the Fréchet Inception Distance (FID) [13]. The SSIM is a well-established metric that has been used here to calculate the degradation of structural information (content) between the original (Fig. 1a) and the transformed (Fig. 1d) input images. The FID measures the feature-wise distance between two domains. While bias transfer can be trained on all images, for the prostate dataset, the impact on the classification can only be evaluated on the single Gleason pattern images. Nonetheless, we decided to evaluate all images regarding the SSIM and FID since the differing annotations given by pathologist 1 and 2 create two non-identical subsets of test images. For our evaluation, the FID between the target domain and the transformed images is calculated and compared to the FID between the original domains. A lower FID implies that images are visually closer to the target domain after bias transfer. Since the FID highly depends on the image content, validation and test scores should be considered separately.

Finally, for the kidney data, the segmentation accuracy pre- and post-bias transfer is compared. The segmentation network predicts one segmentation mask for the glomerulus as a whole and one for the podocytes (see Fig. 1e). The quality of the segmentation is measured with three Dice scores [9], the pixel-wise segmentation of the glomeruli and podocytes and the object-wise segmentation of the podocytes. For the prostate data, the classification accuracy and the macro F1 scores pre- and post-bias transfer are compared. The macro F1 score can give insights into the classification performance for underrepresented classes. Considering class imbalance is especially important here since the label distributions of TAR_p and NEW_p do not match.

Every variation has been trained five times with different random seeds, including the baseline. The training epoch with the lowest generator validation loss for transformations from NEW to TAR has been used for the evaluation of the individual runs. For each approach, the run with the best results on the validation set has been evaluated on the test set.

4.1   Results

Kidney Biopsies. The SSIM and FID scores for the validation set are visualized in Fig. 2. They indicate that the transformations performed by the U-Net cycleGAN and FPG variations had the highest quality, with U-Net cycleGAN with structure loss leading to the overall best and most stable SSIM scores. Regarding the FID, U-Net cycleGAN with combined losses largely improved the distance to the target domain for the validation and test sets. The combination of additional identity and MS-SSIM loss (combined losses) had a stabilizing effect on the training process of cycleGAN and U-Net cycleGAN. The structure loss had a positive effect on U-Net cycleGAN and FPG, however, it led to mode collapse for cycleGAN, only producing a single output irrespective of the input.

The segmentation scores on the test set are shown in Table 1. Here, 'Test TAR_k' only includes images not used for training the segmentation U-Net (21 images). Color transfer worsened the segmentation scores, despite improving the FID. As for the SSIM and FID scores, the U-Net cycleGAN and Fixed-Point GAN variations led to the best results. For those approaches, three out of four variations significantly improved the pixel-level Dice score for the glomeruli and all four for the podocytes. However, unlike FPG, not all runs of U-Net cycleGAN worsened the object-level Dice score for the podocytes. Overall, U-Net cycleGANs with combined losses performed the best for the task at hand. The aforementioned approach produced hallucination artifact-free images that significantly improved the pixel-level segmentation scores of the glomeruli (0.909 to 0.923, $p = 0.005$), and podocytes (0.730 to 0.797, $p < 0.0001$), due to a strong adaption to the target domain. An example transformation performed by U-Net cycleGAN and its effect on the segmentation result can be found in [25].

Prostate Biopsies. For the prostate biopsies, the boxplots in Fig. 3 show that FPG trained with the structure loss outperformed the other approaches regarding content preservation. While the overall lowest FID score on the test set was achieved by cycleGAN with structure loss, this result is not reproducible for different random seeds – the training process is not stable enough.

When we look at the accuracies achieved on the validation set in Table 2, FPG with structure loss achieved the largest improvement (from 0.576 to 0.698). However, this was not reproduced on the test set or for the macro-weighted F1 scores. Regarding the test accuracies, cycleGAN with combined losses and

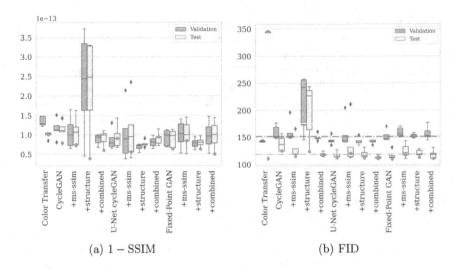

(a) $1 - $ SSIM   (b) FID

Fig. 2. Boxplots visualizing the transformation metrics $1-$SSIM (a) and FID (b) for the validation and test sets of the kidney biopsies. The dashed lines highlight the original FID scores and the red dots show the values achieved by the runs that performed best on the validation set for each variation. (Color figure online)

Table 1. Means (μ_{test}) and standard deviations (σ_{test}) of the Dice scores on the original test sets and the relative performance for the transformed images. 'Dice glom. pix.' evaluates the pixel-wise segmentation of the glomeruli, and 'Dice podo. pix.' of the podocytes. 'Dice podo. obj.' evaluates the object-wise segmentation of the podocytes. Significance ($p < 0.05$) is marked with $*$.

	Dice glom. pix.		Dice podo. obj.		Dice podo. pix.	
	μ_{test}	σ_{test}	μ_{test}	σ_{test}	μ_{test}	σ_{test}
Test TAR$_k$	0.953	0.02	0.933	0.06	0.929	0.01
Test NEW$_k$	0.909	0.05	0.877	0.05	0.730	0.07
Color transfer	−0.158*	0.32	−0.163*	0.27	−0.079	0.26
CycleGAN	−0.002	0.06	−0.053*	0.10	+0.020	0.05
+ MS-SSIM	+0.014	0.05	−0.013	0.10	+0.055*	0.04
+ structure	−0.020	0.10	−0.055*	0.09	−0.017	0.09
+ combined	−0.005*	0.05	−0.031*	0.09	+0.046	0.05
U-Net CycleGAN	+0.018*	0.04	**+0.015**	0.05	+0.061*	0.04
+ MS-SSIM	+0.020*	0.03	−0.002	0.07	+0.039*	0.04
+ structure	+0.017	0.04	−0.023	0.11	+0.064*	0.05
+ combined	+0.022*	0.03	+0.000	0.08	**+0.067***	0.03
Fixed-Point GAN	−0.002	0.06	−0.019	0.09	+0.060*	0.05
+ MS-SSIM	**+0.031***	0.02	−0.005	0.07	+0.040*	0.04
+ structure	+0.022*	0.04	−0.007	0.07	+0.055*	0.04
+ combined	+0.025*	0.03	−0.002	0.07	+0.048*	0.04

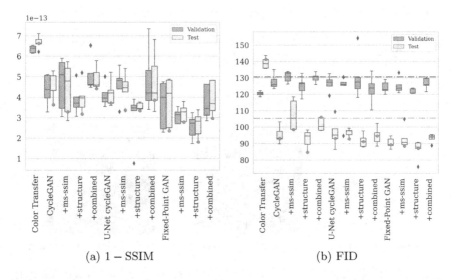

(a) $1 - $ SSIM   (b) FID

Fig. 3. Boxplots visualizing the transformation metrics $1 - $ SSIM (a) and FID (b) for the validation and test sets of the prostate biopsies. The dashed lines highlight the original FID scores and the red dots mark the results achieved by the runs that performed best on the validation set for each variation. (Color figure online)

Table 2. Means (μ_{val}) and standard deviations (σ_{val}) of the accuracy and macro-weighted F1 scores for the classification of the Gleason scores on the original validation and test sets and the relative performance for the transformed images. For the test set, the predicted Gleason scores are compared to the annotations of two medical professionals (μ_{test1} and μ_{test2}). The best results are bold.

	Accuracy				F1 score			
	μ_{val}	σ_{val}	μ_{test1}	μ_{test2}	μ_{val}	σ_{val}	μ_{test1}	μ_{test2}
Test TAR$_p$	0.763	0.00	0.816	0.816	0.665	0.00	0.742	0.742
Test NEW$_p$	0.576	0.00	0.256	0.217	0.454	0.00	0.249	0.242
Color transfer	−0.117	0.01	−0.105	−0.108	−0.178	0.03	−0.096	−0.107
CycleGAN	+0.073	0.04	+0.090	+0.038	+0.066	0.04	+0.069	+0.032
+ MS-SSIM	+0.063	0.06	+0.065	+0.012	+0.086	0.04	+0.050	+0.009
+ structure	+0.057	0.03	+0.100	+0.083	+0.046	0.03	+0.064	+0.060
+ combined	+0.111	0.00	+0.140	**+0.141**	**+0.120**	0.01	+0.108	**+0.128**
U-Net CycleGAN	+0.095	0.03	+0.095	+0.070	+0.086	0.02	+0.072	+0.066
+ MS-SSIM	+0.099	0.02	+0.135	**+0.141**	+0.102	0.02	+0.107	+0.120
+ structure	+0.097	0.04	+0.100	+0.096	+0.090	0.03	+0.075	+0.086
+ combined	+0.064	0.03	+0.120	+0.083	+0.065	0.03	+0.082	+0.067
Fixed-Point GAN	+0.109	0.00	**+0.145**	+0.128	+0.106	0.01	**+0.112**	+0.113
+ MS-SSIM	+0.117	0.00	+0.090	+0.051	+0.106	0.00	+0.068	+0.054
+ structure	**+0.122**	0.02	+0.125	+0.089	+0.107	0.02	+0.097	+0.082
+ combined	+0.117	0.01	+0.140	+0.096	+0.109	0.00	+0.111	+0.086

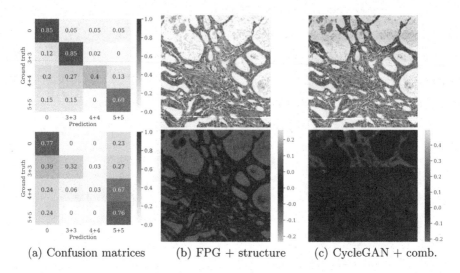

(a) Confusion matrices   (b) FPG + structure   (c) CycleGAN + comb.

Fig. 4. (a) Relative confusion matrices for the classification of the test sets for TAR_p (top) and NEW_p annotated by pathologist 1 (bottom), as well as transformed test images with and without hallucination artifacts and their corresponding transformation heatmaps in (b) and (c).

FPG resulted in the best scores. While FPG achieved the best accuracy and F1 scores for the images annotated by pathologist 1, cycleGAN with combined losses outperformed all other approaches for the images annotated by pathologist 2. Yet, the images transformed by cycleGAN with combined losses contained hallucination artifacts, as is reflected in the SSIM scores (Fig. 3a). Unfortunately, for the classification network at hand, a direct link between transformation and classification quality does not seem to exist. In Fig. 4c an image transformed by cycleGAN with combined losses can be seen. The transformation adds a purple streak across the upper third of the image, which is also reflected in the heatmap of the changes in pixel intensities through the transformation. In contrast, the change heatmap of the images transformed by FPG with structure loss indicates that the color intensity of the background was reduced and that small cell structures remained intact (see Fig. 4b). Hence, to avoid basing the classification on hallucination artifacts, the best bias-transfer approach cannot be judged based on classification accuracies only.

Another factor that could contribute to the mismatch of transformation and classification quality is the very low accuracy on the original NEW_p test images, which is 25.6% for the images annotated by pathologist 1 and 21.7% for pathologist 2, compared to 81.6% achieved on the TAR_p test set. The samples of the two domains were annotated by different pathologists, who might interpret structures disparately, resulting in different Gleason scores. As a result, a transformation to the target domain might not be sufficient to reach high accuracy on this dataset. This is amplified by the fact that the classification network has the lowest accuracy for $4 + 4$ samples (see Fig. 4a), which is the most frequent class in NEW_p. For the original images of NEW_p most of the $4 + 4$ samples were misclassified as $5 + 5$ or 0 (benign) (see Fig. 4a). As a consequence, an accidental improvement of the classification accuracy due to hallucination artifacts cannot be ruled out. All those reasons indicate that FPG with structure loss may still be considered the best performing bias transfer for the prostate biopsy samples, despite not resulting in the highest classification scores. It stably led to improved accuracies and F1 scores across all runs while keeping the balance between not adding hallucination artifacts to the images and imitating the target domain.

4.2 Discussion

Currently, bias can be overcome with hand-engineered methods [4] or with deep generative models, which are universal function approximators. While our datasets do not require a strong transformation, the simple baseline approach was not a sufficient solution. Like for many simple approaches, the transformation is applied uniformly across the whole image. Therefore, non-linear bias cannot be captured. The resulting, tinted images hindered the segmentation of the glomeruli and their podocytes, as well as the classification of prostate cancer. In contrast, nine generative approaches significantly improved the pixel-wise segmentation of the podocytes and six the segmentation of the glomeruli. The classification accuracies and F1 scores were improved by every variation. However, as indicated by the varying SSIM scores, not all approaches resulted in artifact-free images. In our experiments, the widely used cycleGAN architecture did not perform well for bias transfer without any modifications. The training process was unstable and often led to hallucination artifacts. This matches reports in previous studies. Manakov et al. described that, in their experiments, vanilla cycleGANs generated blurry images with checkerboard-like artifacts [17].

CycleGANs were created for general image-to-image transformation. While goals like identity-transforming images that belong to the target domain and reconstructing the original from a transformed image are already incorporated in the basic model, other objectives have to be added explicitly and balanced carefully to achieve maximum performance for the task at hand. Otherwise the instability of the cycleGAN training process can be amplified, leading to complete mode collapse, as we experienced with the kidney biopsies.

FPGs were invented to convert diseased to healthy samples, which requires removing structures from the images. For bias transfer, however, the content of the image has to stay the same. Since this is not explicitly safeguarded by FPG, it is not surprising that the additional losses had a positive impact on the transformation. U-Net cycleGANs, on the other hand, did not benefit as much from the losses as the other approaches. The structure loss (and the combined losses) did however further stabilize the training process since the transformation goals were explicitly incorporated into the loss function.

In our evaluation, the transformation quality was well reflected in the segmentation scores, but only partially for the classification. It is particularly important to note that if the best approach is selected solely based on the resulting classification or segmentation scores, hallucination artifacts might be missed. The artifacts can introduce a new type of bias to the images which could 'accidentally' improve the segmentation or classification scores. Finally, another factor that contributed to the mismatch of transformation quality and classification scores for the prostate biopsies is the inter-annotator variance.

Multiple factors play a role in deciding which of the three architectures should be used to perform bias transfer. When a large number of domains are available, the training time can become a limiting factor. Since U-Net cycleGANs only transform between two domains, every target domain requires a separate model and training. Another common problem is a lack of data. If few samples exist (e.g. <100 per domain), U-Net cycleGANs might have too many trainable parameters to learn an adequate transformation. Both of these issues would warrant selecting FPGs instead since all domains are used to train a single generator. The cycleGAN variations did not match the performance of U-Net cycleGAN and FPG. Therefore, we do not recommend using vanilla cycleGANs for bias transfer in medicine. Regarding the evaluation of generative approaches, no consensus on adequate evaluation metrics exists. Here, the complementary metrics SSIM and FID were good indicators of hallucination artifacts and therefore suitable for evaluating bias transfer.

4.3   Limitations

The selection of the models and the additional losses in this paper focused on bias transfer for domains with small differences between them, i.e., a shift in staining. Therefore, the results might not be reproducible for less similar domains. Additionally, bias transfer has to be performed with caution if the domains have an inherent content bias. A disease can become part of the 'style' if one domain contains more healthy and the other more malformed samples [7].

5   Conclusion and Outlook

The goal of this paper was to determine which deep generative approach results in the best bias transfer for medical images originating from IF confocal microscopy of kidney biopsies and H&E stained microscopy images of prostate biopsies. The performance on the test set mostly corroborated our findings obtained on the validation data. U-Net cycleGAN with combined losses and FPG with structure loss had a stable training process and created hallucination artifact-free images that imitated the target domain, improving the segmentation or classification performance. Our results show that bias transfer for histopathological images benefits from adding structure-based losses since they help with content preservation. The combination of MS-SSIM loss and additional identity loss is especially helpful if bias transfer is performed for domains that only require small changes. For medical image datasets with larger differences, e.g. a different staining type, the additional identity loss should only be used if it is weighted much lower than the MS-SSIM loss. Furthermore, additional losses are not a universal solution. The individual network architectures and approaches have to be adapted to the task at hand, as our differing results for the two modalities have shown. In future projects, similar datasets from additional sites could be investigated to get a firmer grasp on which transformation goals can be covered by the selected additional losses and which further limitations might prevent using deep learning-based bias transfer for medical datasets.

Acknowledgements. This work was supported by DFG (SFB 1192 projects B8, B9 and C3) and by BMBF (eMed Consortia 'Fibromap').

References

1. Armanious, K., Tanwar, A., Abdulatif, S., Küstner, T., Gatidis, S., Yang, B.: Unsupervised adversarial correction of rigid mr motion artifacts. In: 2020 IEEE 17th International Symposium on Biomedical Imaging (ISBI), pp. 1494–1498 (2020)
2. Arvaniti, E., et al.: Replication Data for: Automated Gleason grading of prostate cancer tissue microarrays via deep learning (2018)
3. Arvaniti, E., et al.: Automated Gleason grading of prostate cancer tissue microarrays via deep learning. Scientific Reports (2018)
4. de Bel, T., Hermsen, M., Jesper Kers, R., van der Laak, J., Litjens, G.: Stain-transforming cycle-consistent generative adversarial networks for improved segmentation of renal histopathology. In: Proceedings of The 2nd International Conference on Medical Imaging with Deep Learning, pp. 151–163 (2019)
5. Chen, N., Zhou, Q.: The evolving gleason grading system. Chin. J. Cancer Res. **28**(1), 58–64 (2016)
6. Choi, Y., Choi, M., Kim, M., Ha, J.W., Kim, S., Choo, J.: StarGAN: unified generative adversarial networks for multi-domain image-to-image translation. In: Proceedings of the IEEE Computer Society Conference on Computer Vision and Pattern Recognition, pp. 8789–8797 (2018)
7. Cohen, J.P., Luck, M., Honari, S.: Distribution matching losses can hallucinate features in medical image translation. In: Frangi, A.F., Schnabel, J.A., Davatzikos, C., Alberola-López, C., Fichtinger, G. (eds.) MICCAI 2018. LNCS, vol. 11070, pp. 529–536. Springer, Cham (2018). https://doi.org/10.1007/978-3-030-00928-1_60
8. Deng, J., Dong, W., Socher, R., Li, L.J., Li, K., Fei-Fei, L.: ImageNet: a large-scale hierarchical image database. In: 2009 IEEE Conference on Computer Vision and Pattern Recognition, pp. 248–255 (2009)
9. Dice, L.R.: Measures of the amount of ecologic association between species. Ecology **26**(3), 297–302 (1945). https://www.jstor.org/stable/1932409
10. Egevad, L., et al.: Standardization of Gleason grading among 337 European pathologists. Histopathology **62**(2), 247–256 (2013)
11. Engin, D., Genc, A., Ekenel, H.K.: Cycle-dehaze: enhanced cyclegan for single image dehazing. In: IEEE Computer Society Conference on Computer Vision and Pattern Recognition Workshops, pp. 938–946 (2018)
12. Gonzalez, R.C., Woods, R.E., Masters, B.R.: Digital Image Processing, 3rd edn. Pearson, London (2007)
13. Heusel, M., Ramsauer, H., Unterthiner, T., Nessler, B., Hochreiter, S.: GANs trained by a two time-scale update rule converge to a local Nash equilibrium. In: Proceedings of the 31st International Conference on Neural Information Processing Systems, pp. 6629–6640. NIPS'17 (2017)

14. Isola, P., Zhu, J.Y., Zhou, T., Efros, A.A.: Image-to-image translation with conditional adversarial networks. In: Proceedings - 30th IEEE Conference on Computer Vision and Pattern Recognition, CVPR 2017, pp. 5967–5976 (2017)
15. Kohli, M.D., Summers, R.M., Geis, J.R.: Medical image data and datasets in the era of machine learning - whitepaper from the 2016 c-mimi meeting dataset session. J. Digital Imag. **30**(4), 392–399 (2017)
16. Ma, Y., et al.: Cycle structure and illumination constrained GAN for medical image enhancement. In: Martel, A.L., et al. (eds.) MICCAI 2020. LNCS, vol. 12262, pp. 667–677. Springer, Cham (2020). https://doi.org/10.1007/978-3-030-59713-9_64
17. Manakov, I., Rohm, M., Kern, C., Schworm, B., Kortuem, K., Tresp, V.: Noise as domain shift: denoising medical images by unpaired image translation. In: Wang, Q., et al. (eds.) DART/MIL3ID -2019. LNCS, vol. 11795, pp. 3–10. Springer, Cham (2019). https://doi.org/10.1007/978-3-030-33391-1_1
18. Reinhard, E., Ashikhmin, M., Gooch, B., Shirley, P.: Color transfer between images. IEEE Comput. Graph. Appl. **21**(5), 34–41 (2001)
19. Ronneberger, O., Fischer, P., Brox, T.: U-Net: convolutional networks for biomedical image segmentation. In: Navab, N., Hornegger, J., Wells, W.M., Frangi, A.F. (eds.) MICCAI 2015. LNCS, vol. 9351, pp. 234–241. Springer, Cham (2015). https://doi.org/10.1007/978-3-319-24574-4_28
20. Shaban, M.T., Baur, C., Navab, N., Albarqouni, S.: Staingan: stain style transfer for digital histological images. In: 2019 IEEE 16th International Symposium on Biomedical Imaging (ISBI 2019), pp. 953–956 (2019)
21. Siddiquee, M.M.R., et al.: Learning fixed points in generative adversarial networks: from image-to-image translation to disease detection and localization. In: 2019 IEEE/CVF International Conference on Computer Vision (ICCV), pp. 191–200 (2019)
22. Wang, Z., Bovik, A.C., Sheikh, H.R., Simoncelli, E.P.: Image quality assessment: from error visibility to structural similarity. IEEE Trans. Image Process. **13**(4), 600–612 (2004)
23. Wang, Z., Simoncelli, E.P., Bovik, A.C.: Multi-scale structural similarity for image quality assessment. In: The Thirty-Seventh Asilomar Conference on Signals, Systems and Computers 2003, pp. 1398–1402 (2003)
24. Zhu, J., Park, T., Isola, P., Efros, A.A.: Unpaired image-to-image translation using cycle-consistent adversarial networks. In: 2017 IEEE International Conference on Computer Vision (ICCV), pp. 2242–2251 (2017)
25. Zimmermann, M., et al.: Deep learning-based molecular morphometrics for kidney biopsies. JCI Insight **6** (2021)

Dense Depth Estimation from Stereo Endoscopy Videos Using Unsupervised Optical Flow Methods

Zixin Yang[1]([✉]), Richard Simon[2], Yangming Li[3], and Cristian A. Linte[1,2]

[1] Center for Imaging Science, Rochester Institute of Technology,
Rochester, NY 14623, USA
yy8898@rit.edu
[2] Biomedical Engineering, Rochester Institute of Technology,
Rochester, NY 14623, USA
[3] Electrical Computer and Telecommunications Engineering Technology,
Rochester Institute of Technology, Rochester, NY 14623, USA

Abstract. In the context of Minimally Invasive Surgery, estimating depth from stereo endoscopy plays a crucial role in three-dimensional (3D) reconstruction, surgical navigation, and augmentation reality (AR) visualization. However, the challenges associated with this task are three-fold: 1) feature-less surface representations, often polluted by artifacts, pose difficulty in identifying correspondence; 2) ground truth depth is difficult to estimate; and 3) an endoscopy image acquisition accompanied by accurately calibrated camera parameters is rare, as the camera is often adjusted during an intervention. To address these difficulties, we propose an unsupervised depth estimation framework (END-flow) based on an unsupervised optical flow network trained on un-rectified binocular videos without calibrated camera parameters. The proposed END-flow architecture is compared with traditional stereo matching, self-supervised depth estimation, unsupervised optical flow, and supervised methods implemented on the Stereo Correspondence and Reconstruction of Endoscopic Data (SCARED) Challenge dataset. Experimental results show that our method outperforms several state-of-the-art techniques and achieves a close performance to that of supervised methods.

Keywords: Stereo endoscopy · Depth estimation · Self supervised learning · Stereo matching · Optical flow

1 Introduction

In the context of Minimally Invasive Surgery (MIS), dense depth perception from endoscopy is a prerequisite for surgical robotics Augmented Reality (AR) [12] and computer vision-based navigation systems [23], as such applications require registration of pre-operative data, such as CT/MRI to intra-operative video data. Dense depth perception is also a fundamental component of simultaneous

© Springer Nature Switzerland AG 2021
B. W. Papież et al. (Eds.): MIUA 2021, LNCS 12722, pp. 337–349, 2021.
https://doi.org/10.1007/978-3-030-80432-9_26

localization and mapping (SLAM) [31] and three-dimension (3D) reconstruction [16]. However, estimating depth from endoscopic images is very challenging due to wet and feature-less surfaces, the presence of imaging artifacts, the presence of surgical instruments, and varying lighting conditions.

Depth can be estimated from different types of endoscopic images [7], including structured light endoscopy [14], monocular endoscopy [24] and stereo endoscopy [38]. By analyzing the deformation between the known projected light pattern and received projected pattern, structure light endoscopy can sparsely and accurately reconstruct tissue with no limitations due to texture information. Thus, structured light endoscopy is often used to reconstruct the ground truth depth [1,30]. However, this technique requires specialized processing hardware and is sensitive to environment lighting, limiting its application *in vivo*.

Recovering depth from monocular endoscopic video can be achieved via SLAM [3,20], Shape from Shading [28], and Structure from Motion (SfM) [18,37], as well as machine learning [24] and its integration with SfM [16]. However, monocular depth estimation is the most challenging and least accurate technique: not only does it require the estimation of the camera pose, which is difficult to obtain due to the lack of photometric constancy cross frames, but scale ambiguity is a common, inherent problem in monocular dense depth estimation. To mitigate these limitations, most efforts have been shifted to estimating depth from stereo endoscopy [1,7].

Estimating depth in stereo endoscopy can be achieved via densely matching pixels from a pair of binocular images. Following intrinsic and extrinsic camera calibration, matched points can be triangulated to recover depth using both classical and deep learning methods.

Classical methods include dense optical flow [5,26] and stereo matching methods [6,11], with the latter being the most common method in estimating depth [29]. Several traditional stereo matching methods have been applied [2,31,40] in MIS. However, despite achieving accurate results in the feature-rich region, stereo matching methods often lead to holes and speckle in texture-less surfaces, occlusions, repetition patterns, non-Lambertian surfaces, and specularities, which are common in endoscopy images. As such, parameter tuning and post-processing are necessary.

With the rapid development of deep learning, methods based on the convolutional neural network (CNN) have surpassed traditional methods in several public benchmarks, such as SceneFlow [21] and KITTI platform [22]. However, using CNNs in a supervised fashion in endoscopic videos is challenging, as ground truth depth is difficult to obtain. Visentini-Scarzanella *et al.* [33] trained and validated CNNs on phantom bronchus data from CT data. Similarly, Mahmood *et al.*[19] trained CNNs on synthetic texture-free images generated from digital colon phantom and validated on real images using adversarial learning to transfer real images to synthetic images. Lastly, Wang *et al.* [35] trained and validated a stereo matching CNN on simulated binocular data. A disparity dataset from CT [4] with a limited number of frames was created from *ex vivo* small porcine full torso cadavers and was used to assess several publicly supervised stereo matching methods.

Recently, self-supervised methods [9,39,42] that utilize image reconstruction as supervision signals have achieved remarkable results in self-driving cars. Their common approach is to formulate a depth estimation problem as the minimization of a photometric reprojection loss at the training stage. Self/unsupervised methods include self-supervised stereo matching [34], self-supervised depth estimation [9,39,42], and unsupervised optical flow [15]. However, they have been rarely studied on stereo endoscopic images.

To our best knowledge, the only self-supervised stereo matching method implemented on a pair of stereo endoscopic images was reported by Ye *et al.* in [38]. Nevertheless, there have been several works reported that focus on estimating monocular endoscopic image depth. In [16], Liu *et al.* incorporated recomputed matched points and camera pose from the SfM to train a self-supervised monocular depth estimation network on sinus video. Similarly, Ozyoruk *et al.* [24] jointly estimate camera pose and depth on synthetically generated data.

In the training stage, a self-supervised depth estimation network requires calibrated stereo camera parameters, while a self-supervised stereo matching requires rectified images. Nevertheless, both limit their application in MIS. During an intervention, the surgeon adjusts focus to adapt to anatomical targets, therefore invalidating pre-calibrated parameters [27], while binocular images can be rectified through an uncalibrated stereo rectification approach, which uses matched points to estimate the fundamental matrix. However, on a pair of featureless frames, accurately matched points are limited, and rectification error is introduced in this process [17]. Unsupervised optical flow methods have the advantage that, during training, they do not require calibrated camera parameters or an un-calibrated stereo rectification process.

To estimate depth from stereo endoscopic videos, we present an unsupervised optical flow network (END-flow). Compared with other methods, it does not require any ground truth labels, calibrated camera parameters, or rectified images for training. This work represents the first effort to use an unsupervised optical flow network to estimate depth from a stereo endoscopic video to the best of our knowledge. In addition, we also introduce an auto-masking and a sparse flow loss function to improve further accuracy beyond that achieved via the techniques disseminated to date.

2 Methods

The goal of this work is to first learn the optical flow mapping without the need of ground truth depth or camera parameters and subsequently recover depth from the optical flow with camera parameters in inference. Previous works [9, 15,39,41] have established solid baseline in unsupervised learning. In this section, we first introduce the general image reconstruction objective loss functions for unsupervised optical flow learning, then introduce our proposed enhancements in training using END-flow.

2.1 Baseline Unsupervised Optical Flow Loss Functions

In the absence of ground truth, one alternative is to use image reconstruction as the supervisory signal. The common approach is to formulate a photometric loss between the original image and the warped image. For two images, target image I_t and source image I_s, I_s can be warped to I_t via predicted optical flow mapping transformation $F_{t,s}$ to create the synthesis view of I_t' using

$$I_t'(\boldsymbol{p}) = I_s(\boldsymbol{p} + F_{t,s}),\tag{1}$$

where $\boldsymbol{p}$ is the pixel coordinates on the target image. Following [9,39,41], the photometric loss can be established using

$$L_p = \frac{\alpha}{2}(1 - \text{SSIM}(I_t, I_t')) + (1 - \alpha)\|I_t - I_t'\|_1,\tag{2}$$

where $\alpha = 0.85$, $SSIM$ is the similarity structure index [36]. However, the photometric loss may not be valid in texture-less regions; instead, at these locations, an edge-aware smoothness [8] term is commonly coupled with L_p, taking the form shown below:

$$L_s = |\partial_x F_{t,s}|\, e^{-|\partial_x I_t|} + |\partial_y F_{t,s}|\, e^{-|\partial_y I_t|}.\tag{3}$$

Overall, the total loss function for training via unsupervised optical flow mapping is of the form:

$$L_{flow} = L_p + \beta L_s,\tag{4}$$

where β is commonly set to 0.1.

2.2 Proposed Method

Previous works in optical flow networks take sequential images as input; here, we extend the method to find stereo correspondences on a pair of binocular images. The pipeline associated with the training stage is shown in Fig. 1. We adopt PWC-net [32] as our backbone network to predict forward flow, left to right, and backward flow, right to the left. The basic unsupervised loss function for a pair of binocular images is therefore given by:

$$L_{flow} = \sum_{I_t \in (I_l, I_r))} (L_p + \beta L_s).\tag{5}$$

Auto-masking. To handle occlusions and feature-less regions where photometric consistency is not valid, we utilize the auto-masking method proposed in [9] to select a valid region for photometric loss calculation:

$$M_p = \big[\, \|I_t - I_t'\|_1 < \|I_t - I_{t'}\|_1 \,\big].\tag{6}$$

Here [] is the Iverson bracket, taking the value 1 if the statement inside the bracket is true and otherwise taking the value 0. $I_{t'}$ is the other image in the pair of images, and I_t' is the synthesis image.

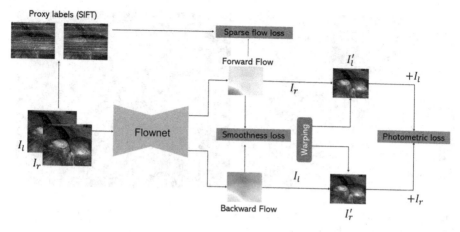

Fig. 1. Overview of proposed method. We adopt PWC-net [32] as our Flow-net. Proxy labels are generated from SIFT as a supervision signal. A sparse flow loss is calculated between proxy labels and predicted forward flow. Smoothness loss is calculated on forward flow and backward flow. The difference between the warped and input images forms the photometric loss.

Sparse Flow Loss. In addition to the basic unsupervised loss function, a sparse flow loss is included. We used an illumination-invariant feature descriptor, the Scale Invariant Feature Transform (SIFT), to find matched key-points within a pair of stereo images. Key-points are used to estimate the fundamental matrix, which is then used along with the RANSAC method to eliminate outliers. These matched points are further processed to generate a sparse flow map from the left image to the right image, to serve as a proxy label for supervision. The sparse flow F_{SIFT} loss is defined by:

$$L_{sf} = \frac{1}{|M_{SIFT}|} \sum_p M_{SIFT}(p)\|F_{SIFT}(p) - F_{l,r}(p_a)\|_1, \qquad (7)$$

where $F_{l,r}$ is sparse flow map generated from SIFT, M_{SIFT} is the mask where sparse keypoints exist, and $|M_{SIFT}|$ stands for the number of matched points.

The overall loss function L is formulated as

$$L = \sum_{I_t \in (I_l, I_r))} (M_p L_p + \beta L_s) + \gamma L_{sf}, \qquad (8)$$

where γ is the weight for sparse flow loss and is empirically set to 0.15.

To recover depth from optical flow in inference, following [10,41], we adopt the mid-point triangulation method using the stereo calibration parameters, which has a linear solution.

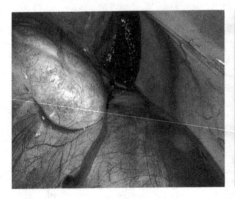

Fig. 2. An example of an endoscopic image, as well as its ground truth reconstructed depth from the SCARED dataset.

3   Dataset and Implementation

We conducted all experiments on the SCARED dataset[1] (the Stereo Correspondence and Reconstruction of Endoscopic Data). The dataset contains binocular images of abdominal anatomy from fresh porcine cadavers collected by a Da Vinci Xi endoscope, along with the associated camera parameters, camera poses, and ground truth depth maps generated using structure light. One sample is shown in Fig. 2. The data employed in this work consists of seven training datasets and two testing datasets.

All experiments are conducted on a GTX 2070 GPU, and all methods are implemented on Pytorch. We train the models [25] using the Adam optimizer [13] with a learning rate 10^{-4} and a batch size of 8. Images are enhanced with CLAHE (contrast limited adaptive histogram equalization) and resized to 256×320. Data augmentation only includes random flip, which also mimics the real scenario, especially in the event that the left - and right- images are flipped.

4   Results

4.1   Evaluation Metrics

We use the following metrics for evaluation: 1) the mean absolute distance (MAD (mm)), 2) the absolute relative error (AbsRel), and 3) the root mean squared error (RMSE (mm)), defined by the following equations:

$$MAD = \frac{1}{n}\sum_{i=1}^{n} |\hat{d}_i - d_i|, \tag{9}$$

$$AbsRel = \frac{1}{n}\sum_{i=1}^{n} \frac{|\hat{d}_i - d_i|}{d_i}, \tag{10}$$

---

[1] https://endovissub2019-scared.grand-challenge.org/.

$$RMSE = \sqrt{\frac{1}{n} \sum_{i=1}^{n} |\hat{d}_i - d_i|^2}, \tag{11}$$

where n denotes the number of pixels, $\hat{d}_i$ and d_i represent ground truth depth and predicted depth of the pixel i, respectively.

4.2 Comparison with State-of-the-Art Depth Reconstruction Methods

Table 1. Comparison between several state-of-the-art depth stereo reconstruction methods and our proposed method (END-flow) in terms of Mean Absolute Distance (MAD), Absolute Relative Error (AbsRel) and Root-Mean-Squared Error (RMSE) in mean ± std. The statistical significance of the END-flow results against other methods is identified by $*(p < 0.005)$.

Method	MAD (mm)	AbsRel (%)	RMSE (mm)
* SGM [11]	9.37 ± 2.95	7.44 ± 3.47	8.37
* PASM [34]	16.87 ± 4.83	22.12 ± 4.78	20.72
* Monodepth2 [9]	7.03 ± 3.83	9.373 ± 4.514	9.02
* AR-flow [15]	6.65 ± 3.50	8.509 ± 3.965	9.40
END-flow	**5.40 ± 3.92**	**7.17 ± 5.20**	**7.55**

We first compare the results achieved using our proposed method to those obtained using several state-of-art methods, including the traditional stereo matching method SGM [11], unsupervised stereo matching method PASM [34], self-supervised depth estimation method Monodepth2 [9], unsupervised optical flow method AR-flow. Both SGM and PASM require rectified images as input, while Monodepth2 requires camera parameters. Our method, END-flow, does not require stereo rectification or camera parameters for training, which is advantageous in the endoscopy application. These results are summarized in Table 1.

It has been noted that the SCARED dataset was reported to have a calibration error [1]. After close examination, datasets 1–3 featuring minor calibration errors are used for comparison. We use the shortest video in each dataset for testing, the remaining for training and validation. In total, there are 7092 image pairs used for training, 787 image pairs used for validation, and 613 image pairs used for testing. The results in Table 1 suggest that our method achieves the best performances. The differences between the MAD errors from END-flow and other methods are statistical significance, characterized by $p < 0.005$.

Qualitative results are presented in Fig. 3. In comparison with other methods, SGM [11] fails to find correspondences in the ambiguous region, presenting black holes, while PASM [34] designed for rectified natural images with high dependence on epipolar constraints shows bad performance. Despite the fact that

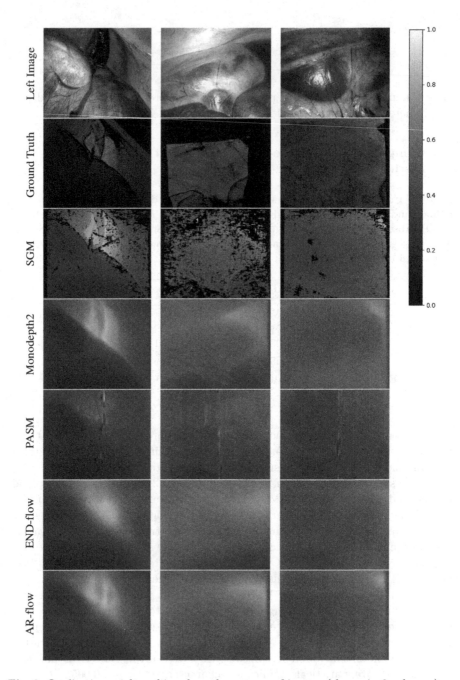

Fig. 3. Qualitative results achieved on three types of images (shown in 3 columns) as part of SCARED dataset using several techniques (illustrated in each row). Predicted depth maps are normalized by the maximum values of the ground truth depth map for enhanced visualization.

Monodepth2 predicts depth maps with sharp boundaries, our evaluation revealed that it tends to lose real scale. This may result from the fact that Monodepth2 takes one image to predict the depth, which is different from other methods that require two images. Moreover, over-enhanced edges on some images may result from texture changes, not necessarily depth changes. As such, the Monodepth2 technique may lead to over-enhanced edges when estimating depth from one image. Our method, END-flow, predicts depth with finer detail than AR-flow, while not lose real scale as Monodepth2.

With calibrated camera parameters, images can be rectified. When training or make predictions based on rectified images, optical flow can be constrained to the horizontal direction, the disparity, and this approach may help improve depth estimation. However, endoscopy image acquisition accompanied by accurately calibrated camera parameters is rare, as the camera is often adjusted during an intervention. To mitigate this inconvenience, our proposed method, END-flow, has the advantage of not requiring accurately calibrated camera parameters for training.

4.3   Ablation Study

To identify the contribution brought forth by each of the individual components integrated into our proposed pipeline, specifically auto-masking M_p and sparse flow loss L_{sf}, we conduct an ablation study to evaluate the performance of each pipeline component. This study is summarized in Table 2. Both M_p and L_{sf} alone, as well as their combination $M_p + L_{sf}$ yield statistically significant improvement in MAD compared to the baseline ($p < 0.005$).

Table 2. Ablation study showing the improvement in MAD (mm) (mean±std) in response to augmenting the baseline technique with the auto-masking M_p, sparse flow loss L_{sf} and their combination $M_p + L_{sf}$.

Method	MAD (mm)	AbsRel (%)	RMSE (mm)
Baseline	8.40 ± 4.08	11.02 ± 4.60	11.38
Baseline + M_p	7.46 ± 4.25	9.87 ± 4.95	10.20
Baseline + L_{sf}	5.59 ± 3.92	7.37 ± 5.17	7.76
Baseline + $M_p + L_{sf}$	5.40 ± 3.92	7.17 ± 5.20	7.55

4.4   Comparison with Top Methods in the SCARED Challenge

We further compare our method with winners' methods reported in the SCARED challenge [1], shown in Table 3. Winners were Trevor Zeffiro and Jean-Claude Rosenthal. We train our method on seven training sub-datasets and test on two testing sub-datasets. Note that these winners' methods utilized ground truth depth for training their networks, while our proposed architecture method achieves competitive results without using the ground truth depth labels.

Table 3. Comparison (in terms of mean MAD (mm)) between END-flow and best performing methods reported in the SCARED challenge.

	Trevor Zeffiro [1]	J.C. Rosenthal [1]	END-flow
testDataset1	3.60	3.44	4.77
testDatseta2	3.47	4.05	4.76

5   Conclusion

We have presented a dense depth estimation method based on an unsupervised optical flow network named END-flow. This method poses several advantages over previous techniques: 1) it can be trained on original videos without access to camera calibration parameters and stereo rectification or ground-truth labels; and 2) it integrates key-points matching to facilitates training. We deployed this method on several datasets available as part of the SCARED challenge; the results achieved using END-flow are comparable to those achieved using state-of-art methods, as well as the best-performing methods reported in the challenge. Specifically, we demonstrate that END-flow outperforms the state-of-the-art traditional and self/unsupervised methods and achieves comparatively performance against the best-performing supervised methods reported in the challenge. Future work will focus on estimating the confidence of unsupervised optical flow methods, which will benefit the down-stream analysis and integration of traditional depth estimation methods. Following additional work on the topic and further software improvement, we plan to release a link to a repository consisting of open-source code to the community.

References

1. Allan, M., et al.: Stereo correspondence and reconstruction of endoscopic data challenge. arXiv preprint arXiv:2101.01133 (2021)
2. Bernhardt, S., Abi-Nahed, J., Abugharbieh, R.: Robust dense endoscopic stereo reconstruction for minimally invasive surgery. In: Menze, B.H., Langs, G., Lu, L., Montillo, A., Tu, Z., Criminisi, A. (eds.) MCV 2012. LNCS, vol. 7766, pp. 254–262. Springer, Heidelberg (2013). https://doi.org/10.1007/978-3-642-36620-8_25
3. Chen, L., Tang, W., John, N.W., Wan, T.R., Zhang, J.J.: Slam-based dense surface reconstruction in monocular minimally invasive surgery and its application to augmented reality. Comput. Methods Prog. Biomed **158**, 135–146 (2018)
4. Eddie"Edwards, P., Psychogyios, D., Speidel, S., Maier-Hein, L., Stoyanov, D.: Serv-ct: a disparity dataset from ct for validation of endoscopic 3d reconstruction. arXiv e-prints pp. arXiv-2012 (2020)
5. Farnebäck, G.: Two-frame motion estimation based on polynomial expansion. In: Bigun, J., Gustavsson, T. (eds.) SCIA 2003. LNCS, vol. 2749, pp. 363–370. Springer, Heidelberg (2003). https://doi.org/10.1007/3-540-45103-X_50
6. Geiger, A., Roser, M., Urtasun, R.: Efficient large-scale stereo matching. In: Kimmel, R., Klette, R., Sugimoto, A. (eds.) ACCV 2010. LNCS, vol. 6492, pp. 25–38. Springer, Heidelberg (2011). https://doi.org/10.1007/978-3-642-19315-6_3

7. Geng, J., Xie, J.: Review of 3-d endoscopic surface imaging techniques. IEEE Sens. J. **14**(4), 945–960 (2013)
8. Godard, C., Mac Aodha, O., Brostow, G.J.: Unsupervised monocular depth estimation with left-right consistency. In: Proceedings of the IEEE Conference on Computer Vision and Pattern Recognition, pp. 270–279 (2017)
9. Godard, C., Mac Aodha, O., Firman, M., Brostow, G.J.: Digging into self-supervised monocular depth estimation. In: Proceedings of the IEEE/CVF International Conference on Computer Vision, pp. 3828–3838 (2019)
10. Hartley, R.I., Sturm, P.: Triangulation. Comput. Vision Image Underst. **68**(2), 146–157 (1997)
11. Hirschmuller, H.: Accurate and efficient stereo processing by semi-global matching and mutual information. In: 2005 IEEE Computer Society Conference on Computer Vision and Pattern Recognition (CVPR'05), vol. 2, pp. 807–814. IEEE (2005)
12. Kalia, M., Navab, N., Salcudean, T.: A real-time interactive augmented reality depth estimation technique for surgical robotics. In: 2019 International Conference on Robotics and Automation (ICRA), pp. 8291–8297. IEEE (2019)
13. Kingma, D.P., Ba, J.: Adam: a method for stochastic optimization. arXiv preprint arXiv:1412.6980 (2014)
14. Lin, J., et al.: Endoscopic depth measurement and super-spectral-resolution imaging. In: Descoteaux, M., Maier-Hein, L., Franz, A., Jannin, P., Collins, D.L., Duchesne, S. (eds.) MICCAI 2017. LNCS, vol. 10434, pp. 39–47. Springer, Cham (2017). https://doi.org/10.1007/978-3-319-66185-8_5
15. Liu, L., et al.: Learning by analogy: Reliable supervision from transformations for unsupervised optical flow estimation. In: Proceedings of the IEEE/CVF Conference on Computer Vision and Pattern Recognition, pp. 6489–6498 (2020)
16. Liu, X., et al.: Reconstructing sinus anatomy from endoscopic video – towards a radiation-free approach for quantitative longitudinal assessment. In: Martel, A.L., Martel, A.L., et al. (eds.) MICCAI 2020. LNCS, vol. 12263, pp. 3–13. Springer, Cham (2020). https://doi.org/10.1007/978-3-030-59716-0_1
17. Luo, X., Jayarathne, U.L., McLeod, A.J., Pautler, S.E., Schlacta, C.M., Peters, T.M.: Uncalibrated stereo rectification and disparity range stabilization: a comparison of different feature detectors. In: Medical Imaging 2016: Image-Guided Procedures, Robotic Interventions, and Modeling, vol. 9786, p. 97861C. International Society for Optics and Photonics (2016)
18. Lurie, K.L., Angst, R., Zlatev, D.V., Liao, J.C., Bowden, A.K.E.: 3d reconstruction of cystoscopy videos for comprehensive bladder records. Biomed. Opt. Exp. **8**(4), 2106–2123 (2017)
19. Mahmood, F., Durr, N.J.: Deep learning and conditional random fields-based depth estimation and topographical reconstruction from conventional endoscopy. Med. Image Anal. **48**, 230–243 (2018)
20. Mahmoud, N., Collins, T., Hostettler, A., Soler, L., Doignon, C., Montiel, J.M.M.: Live tracking and dense reconstruction for handheld monocular endoscopy. IEEE Trans. Medical Imag. **38**(1), 79–89 (2018)
21. Mayer, N., et al.: A large dataset to train convolutional networks for disparity, optical flow, and scene flow estimation. In: Proceedings of the IEEE Conference on Computer Vision and Pattern Recognition, pp. 4040–4048 (2016)
22. Menze, M., Geiger, A.: Object scene flow for autonomous vehicles. In: Proceedings of the IEEE Conference on Computer Vision and Pattern Recognition, pp. 3061–3070 (2015)
23. Mirota, D.J., Ishii, M., Hager, G.D.: Vision-based navigation in image-guided interventions. Ann. Rev. Biomed. Eng. **13** (2011)

24. Ozyoruk, K.B., et al.: Endoslam dataset and an unsupervised monocular visual odometry and depth estimation approach for endoscopic videos. Med. Image Anal., 102058 (2021)
25. Paszke, A., et al.: Automatic differentiation in pytorch (2017)
26. Phan, T.B., Trinh, D.H., Lamarque, D., Wolf, D., Daul, C.: Dense optical flow for the reconstruction of weakly textured and structured surfaces: Application to endoscopy. In: 2019 IEEE International Conference on Image Processing (ICIP), pp. 310–314. IEEE (2019)
27. Pratt, P., Bergeles, C., Darzi, A., Yang, G.Z.: Practical intraoperative stereo camera calibration. In: International Conference on Medical Image Computing and Computer-Assisted Intervention. pp. 667–675. Springer (2014)
28. Ren, Z., He, T., Peng, L., Liu, S., Zhu, S., Zeng, B.: Shape recovery of endoscopic videos by shape from shading using mesh regularization. In: Zhao, Y., Kong, X., Taubman, D. (eds.) ICIG 2017. LNCS, vol. 10668, pp. 204–213. Springer, Cham (2017). https://doi.org/10.1007/978-3-319-71598-8_19
29. Scharstein, D., Szeliski, R.: A taxonomy and evaluation of dense two-frame stereo correspondence algorithms. Int. J. Comput. Vision 47(1), 7–42 (2002)
30. Scharstein, D., Szeliski, R.: High-accuracy stereo depth maps using structured light. In: 2003 IEEE Computer Society Conference on Computer Vision and Pattern Recognition, 2003, Proceedings, vol. 1, pp. I-I. IEEE (2003)
31. Song, J., Wang, J., Zhao, L., Huang, S., Dissanayake, G.: Mis-slam: real-time large-scale dense deformable slam system in minimal invasive surgery based on heterogeneous computing. IEEE Rob. Autom. Lett. 3(4), 4068–4075 (2018)
32. Sun, D., Yang, X., Liu, M.Y., Kautz, J.: Pwc-net: CNNs for optical flow using pyramid, warping, and cost volume. In: Proceedings of the IEEE Conference on Computer Vision and Pattern Recognition, pp. 8934–8943 (2018)
33. Visentini-Scarzanella, M., Sugiura, T., Kaneko, T., Koto, S.: Deep monocular 3D reconstruction for assisted navigation in bronchoscopy. Int. J. Comput. Assist. Radiol. Surg. 12(7), 1089–1099 (2017)
34. Wang, L., et a.: Parallax attention for unsupervised stereo correspondence learning. IEEE Trans. Pattern Anal. Mach. Intell. (2020)
35. Wang, X.Z., Nie, Y., Lu, S.P., Zhang, J.: Deep convolutional network for stereo depth mapping in binocular endoscopy. IEEE Access 8, 73241–73249 (2020)
36. Wang, Z., Bovik, A.C., Sheikh, H.R., Simoncelli, E.P.: Image quality assessment: from error visibility to structural similarity. IEEE Trans. Image Process 13(4), 600–612 (2004)
37. Widya, A.R., Monno, Y., Okutomi, M., Suzuki, S., Gotoda, T., Miki, K.: Whole stomach 3D reconstruction and frame localization from monocular endoscope video. IEEE J. Transl. Eng. Health Med. 7, 1–10 (2019)
38. Ye, M., Johns, E., Handa, A., Zhang, L., Pratt, P., Yang, G.Z.: Self-supervised siamese learning on stereo image pairs for depth estimation in robotic surgery. In: Hamlyn Symposium on Medical Robotics (2017)
39. Yin, Z., Shi, J.: Geonet: Unsupervised learning of dense depth, optical flow and camera pose. In: Proceedings of the IEEE Conference on Computer Vision and Pattern Recognition, pp. 1983–1992 (2018)
40. Zampokas, G., Tsiolis, K., Peleka, G., Mariolis, I., Malasiotis, S., Tzovaras, D.: Real-time 3D reconstruction in minimally invasive surgery with quasi-dense matching. In: 2018 IEEE International Conference on Imaging Systems and Techniques (IST), pp. 1–6. IEEE (2018)

41. Zhao, W., Liu, S., Shu, Y., Liu, Y.J.: Towards better generalization: joint depth-pose learning without posenet. In: Proceedings of the IEEE/CVF Conference on Computer Vision and Pattern Recognition, pp. 9151–9161 (2020)
42. Zhou, T., Brown, M., Snavely, N., Lowe, D.G.: Unsupervised learning of depth and ego-motion from video. In: Proceedings of the IEEE Conference on Computer Vision and Pattern Recognition, pp. 1851–1858 (2017)

Image Augmentation Using a Task Guided Generative Adversarial Network for Age Estimation on Brain MRI

Ruizhe Li[1]($\boxtimes$), Matteo Bastiani[2], Dorothee Auer[2], Christian Wagner[1], and Xin Chen[1]

[1] IMA/LUCID Group, School of Computer Science, University of Nottingham, Nottingham, UK
Ruizhe.Li@nottingham.ac.uk
[2] School of Medicine, University of Nottingham, Nottingham, UK

Abstract. Brain age estimation based on magnetic resonance imaging (MRI) is an active research area in early diagnosis of some neurodegenerative diseases (e.g. Alzheimer, Parkinson, Huntington, etc.) for elderly people or brain underdevelopment for the young group. Deep learning methods have achieved the state-of-the-art performance in many medical image analysis tasks, including brain age estimation. However, the performance and generalisability of the deep learning model are highly dependent on the quantity and quality of the training data set. Both collecting and annotating brain MRI data are extremely time-consuming. In this paper, to overcome the data scarcity problem, we propose a generative adversarial network (GAN) based image synthesis method. Different from the existing GAN-based methods, we integrate a task-guided branch (a regression model for age estimation) to the end of the generator in GAN. By adding a task-guided loss to the conventional GAN loss, the learned low-dimensional latent space and the synthesised images are more task-specific. It helps to boost the performance of the down-stream task by combining the synthesised images and real images for model training. The proposed method was evaluated on a public brain MRI data set for age estimation. Our proposed method outperformed (statistically significant) a deep convolutional neural network based regression model and the GAN-based image synthesis method without the task-guided branch. More importantly, it enables the identification of age-related brain regions in the image space. The code is available on GitHub (https://github.com/ruizhe-l/tgb-gan).

Keywords: Generative adversarial network · Brain age regression · Data augmentation

1 Introduction

Data-driven deep learning methods have achieved the state-of-the-art performance in various medical image analysis tasks, which normally require a large amount of data for model training. However, collecting a large quantity of data with high-quality annotation is a big challenge in medical imaging. Data augmentation is one of the widely used methods to address this problem. Classical image augmentation techniques (e.g.

© Springer Nature Switzerland AG 2021
B. W. Papież et al. (Eds.): MIUA 2021, LNCS 12722, pp. 350–360, 2021.
https://doi.org/10.1007/978-3-030-80432-9_27

geometric transformation, colour space augmentation, random erasing, etc.) are not effective enough to simulate different imaging variations realistically. Therefore, more sophisticated methods based on generative modelling were proposed by researchers [1–3].

The basic idea of generative modelling based image augmentation is to learn the underlying data distribution from a set of real images, then new images can be synthesised by sampling the learned distribution. In 2014, generative adversarial network (GAN), proposed by Goodfellow et al. [4], was utilised for generating high-quality images from a random latent vector. Subsequently, many image reconstruction and image-to-image translation methods were proposed based on GAN. In 2016, Larsen et al. proposed VAE-GAN which combined variational autoencoder (VAE) [5] and GAN by integrating the decoder of VAE and the generator of GAN. VAE-GAN learns to minimise the dissimilarity of the latent features of the reconstructed image and the input image, as well as a classical GAN-based adversarial loss. Similarly, another image to image translation method was proposed by Isola et al. in 2017 [6]. The authors replaced the generator in GAN with an encoder-decoder based deep convolutional neural network (CNN). A pixel-wise loss was also added to measure the similarity of the original images and the reconstructed images. Both methods have achieved superior performance in generating realistic images. However, this type of methods can only synthesise images from the same distribution of the input dataset without the capability of specifying certain characteristics in the image (e.g. synthesise face images for the same person but at a different age).

To overcome this problem, several methods were proposed in recent years. In 2017, a conditional generative adversarial network (CGAN) [7] based face generation model was proposed by Antipov et al. [8]. It firstly trains an encoder on a dataset that is generated from a pre-trained age-condition GAN. Then a face recognition neural network is used to minimise the distance between real images and synthesised images to preserve the person's identity. Subsequently, an image of certain age of the same person can be generated by conditioning on a given age. Another face ageing method was proposed by Wang et al. in 2018 [9]. It also contains an identity-preserved control loss which minimises the distance between the feature maps of the real and synthesised images that were obtained from a pre-trained feature extraction model. Additionally, a pre-trained age classification model is used to generate an age classification loss for forcing the generated face being in the specified age group.

In medical imaging applications, Xia et. al [10] proposed to learn a joint distribution of brain images and ages using GAN, and also added an identity-preserve loss to generate more age-related images. Besides, several image augmentation methods proposed recently that applied feature interpolation in latent feature space to generate specific target images [11, 12]. However, most of these studies focus on improving the quality of the synthesised images and rarely simultaneously investigate the effects of using the synthesised images for down-stream tasks, such as classification of a disease. Arguably, a globally realistically-looking synthesised image does not necessarily contain useful image features for classifying a disease, as most of the information are in a local image region and are concealed by superposed and more distinguishable features (e.g. age and gender variations).

In this paper, we propose a GAN based image synthesis method which aims to generate images that can be used to improve the performance of an age estimation task. The main contributions of this paper are summarised as follows. (1) We integrate a task guided branch to the end of the generator in GAN, and a task-specific loss is added to the conventional GAN loss for guiding the learned low-dimensional latent space to be more task-specific. Hence the synthesised images are also more task-specific. (2) We evaluated our method on a public brain MRI dataset for the task of age estimation. The task-guided branch in this case is an age regression model. We have demonstrated the effectiveness of the proposed method by ablation studies. The results have shown that our method is able to generate age-specific MRI images and therefore boosting the performance of the age estimation model.

2   Methodology

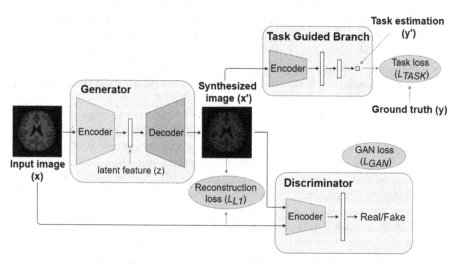

Fig. 1. Overview of the proposed method.

The overview of our method is depicted in Fig. 1. The proposed method mainly consists of three network blocks: a generator, a discriminator and a task-guided branch. The generator and discriminator form an image reconstruction network, which is similar to the GAN architecture used in [6]. The task-guided branch is concatenated to the end of the generator, which is a regression network for age estimation. The whole network is trained by optimising an objective function that consists of three terms: a pixel-wise reconstruction loss L_{L1}, a GAN loss L_{GAN} and a task-guided loss L_{TASK}. The details of each components are described in the following subsections.

2.1 GAN Based Image Synthesis

To achieve image synthesis, we use a modified 3D image to image GAN [6]. To overcome the mode collapse problem in conventional GAN, we use the loss proposed in WGAN-GP [13]. The generator of our GAN model consists of an encoder and a decoder. Both of them have 5 layers of feature maps by applying one $3 \times 3 \times 3$ convolutional operations and one rectified linear unit (ReLU) at each layer. In the encoding path, a max-pooling with a stride of 2 is applied between two layers for feature map down-sampling. The number of feature channels for the 5 encoding layers are 64, 128, 256, 512 and 512 respectively.

Correspondingly, a de-convolutional operation is performed in the decoder between two consecutive layers to up-sample the feature maps. The number of feature channels for the 5 decoding layers are 512, 512, 256, 128 and 64 respectively. A $1 \times 1 \times 1$ convolutional layer is applied at the end of the decoder to convert the dimensions of the output to the same as the input. The discriminator has the same structure as the generator's encoder, but a $1 \times 1 \times 1$ convolutional layer is appended to the end for reducing the dimensionality for image discrimination. The cost function of our image synthesis network consists of an adversarial loss L_{GAN} (Eq. (1)) that has been used in WGAN-GP [13], and a pixel-wise reconstruction loss L_{L1} (Eq. (2)).

$$L_{GAN}(G, D) = \mathbb{E}_x[D(G(x)] - \mathbb{E}_x[D(x)] \tag{1}$$

$$L_{L1}(G) = \mathbb{E}_x[|x - G(x)|] \tag{2}$$

where $G : x \rightarrow \hat{x}$ is the generator that maps an input image x to its target reconstruction image $\hat{x}$. D indicates the discriminator that classifies if an image is real or fake. The generator intends to fool the discriminator by producing realistic images, and the discriminator aims to identify the fake ones from the real images. The reconstruction loss L_{L1} minimises the pixel-wise intensity differences between the real and the synthesised images. A weight α is applied to balance the L_{L1} and L_{GAN} losses (Eq. (3)). The image synthesis network is therefore trained by optimising the following objective:

$$\min_{G} \max_{D} L_{GAN}(G, D) + \alpha L_{L1}(G) \tag{3}$$

2.2 Task-Guided Branch

For the age estimation application, we intend to synthesise MRI images of a particular age if not enough images are available for that age group. The GAN-based image synthesis network introduced in Sect. 2.1 can be used to generate new images by sampling the learned latent feature space (output of the encoder in the generator) and then inputting into the decoder in the generator. Ideally, if an age-specific manifold in the latent space can be identified, an MRI image of a particular age can be generated by linear interpolation in the manifold using the latent feature vectors of images in nearby age groups. However, the latent space is high dimensional, which controls variety of image characteristics (e.g. geometric, intensity, context, gender etc.). It makes it extremely difficult to identify the age-specific manifold.

Hence, we propose to integrate an age-specific branch to the end of the generator in the GAN model that guides the latent feature learning to be more age-specific. In our age estimation application, as a widely used method in image analysis [10], a 3D VGG-like [14] regression network is added to the GAN. It consists of 5 down-sampling layers and 4 fully connected (FC) layers. Each down-sampling layer includes two $3 \times 3 \times 3$ convolutional operations, each of them is followed by a ReLU activation function. As a commonly used improvement, a residual block [15] is used in each layer for faster convergence. The number of feature channels in the 5 down-sampling layers are 16, 32, 64, 128 and 256 respectively. The feature map of the last down-sampling layer is flattened to a vector and fed into the FC layers. Each FC layer is also followed by a ReLU function, and the number of units in the FC layers are 2048, 1024, 64, 1, respectively. The final FC layer produces the estimated age. The $L2$ loss is used to measure the error between the estimated age and the ground truth age, as expressed in Eq. (4).

$$L_{TASK}(G) = \mathbb{E}_{G(x)}[\left(y - R(G(x))\right)^2] \tag{4}$$

y is the ground truth age for an input image x. $R : G(x) \rightarrow \hat{y}$ is the estimated age using the regression model R. $G(x)$ is the synthesised image from the previously described GAN model. Both the GAN model and the regression branch are optimised simultaneously end-to-end. By combining both the GAN loss and the regression loss (with a weight β), the final objective function of our method is:

$$\min_{G} \max_{D} L_{GAN}(G, D) + \alpha L_{L1}(G) + \beta L_{TASK}(G) \tag{5}$$

2.3 Latent Space Interpolation for Image Synthesis

After model training, we synthesise new images by interpolating the latent feature space. Specifically, to generate an image of a particular age y^s, we randomly select two images (one younger and one older) in the training set with an age difference of less than 1 to y^s. The synthesised image S_{img} is generated using Eq. (6) and the associated age is calculated using Eq. (7).

$$S_{img} = G_{decoder}\left(\varepsilon G_{encoder}\left(x^i\right) + (1 - \varepsilon)G_{encoder}\left(x^j\right)\right) \tag{6}$$

$$y^s = \varepsilon y^i + (1 - \varepsilon)y^j \tag{7}$$

where x^i and x^j are the two randomly selected images with the age of y^i and y^j respectively. ε is a randomly determined value between 0 and 1. $G_{encoder}$ is the trained encoder in the generator that generates the latent feature for a given image, and $G_{decoder}$ is the decoder in the generator that generates a synthesised image from the interpolated latent feature vector. Many pairs of S_{img} and y^s can then be generated and used as the additional training data for age estimation model training.

3 Experiments and Results

3.1 Dataset

We evaluated the proposed method using a public brain MRI dataset from the Autism Brain Imaging Data Exchange (ABIDE) I [16] and II [17]. Standard minimal pre-processing was run on the data, following the steps described in [18], including brain extraction, bias field correction and nonlinear registration [19–21] to the MNI template space. The dataset used in this paper contains 1150 images of healthy subjects with age from 6 to 24. Note that, after the pre-processing steps, variations of the brain volume and geometric differences across subjects were removed. This prevents the model from learning the most distinguishable feature (i.e. brain volume) for age estimation. Features like brain volume can be easily obtained by image segmentation and integrated to an age estimation model. Here we train the model to learn local features that is much more difficult to be observed by human. After these pre-processing steps, the image size is 182 × 218 × 182 voxels, and we resized them to 96 × 96 × 96 voxels due to the limitation of GPU memory. The main aim of the experiments was to demonstrate the effectiveness of the proposed task-guided branch rather than to achieve the state-of-the art age estimation performance, hence the down-sampled images were used to reduce computational cost and memory consumption. We also applied the zero-mean and min-max methods as pre-processing steps to normalise the image intensity to the range of [0, 1]. To avoid model bias, we balanced the numbers of male and female for each age group (age interval of 1 year), resulting in 862 images (431 male and 431 female). Note that the number of images for each age group is not balanced (e.g. 6 samples for age 5 to 6, 86 samples for age 9 to 10). Subsequently, we divided them into training, validation and testing sets with the ratio of approximately 75%, 5% and 20% (i.e. 640, 62 and 160) respectively. For a fair comparison, all experiments performed in this paper used the same validation and test sets.

3.2 Experiments

For method comparison, we firstly trained a baseline regression model (named as REG) using the VGG-based architecture as described in Sect. 2.2 for age estimation based on the 640 training images.

As the second comparative method, we trained a GAN-based model (Sect. 2.1) without the age regression branch. As the number of images in different age groups are highly unbalanced, we generated synthesised images using the method described in Sect. 2.3 to ensure at least 50 images for each age group after combining the real and synthesised images, resulting in a total of 940 images for training. Subsequently, an age regression network (same as REG) was trained using the new dataset (940 images). We name this method REG-GAN.

For our proposed method, we trained the GAN-based model with a regression branch by optimising the objective function in Eq. (5). Synthesised images were then generated for each age group in the same way as the REG-GAN method, and followed by training an age estimation model using the new dataset. We name our proposed method REG-GAN-TGB.

For a fair comparison, both REG-GAN and REG-GAN-TGB were trained for 100 epochs with the batch size of 20. Adam optimizer [22] was used. The learning rate was 0.0001 and multiplied by 0.8 for every 10 epochs. The parameter tuning was performed using the validation set. As a starting point, we tried to make each part of the loss function (Eq. (5)) having similar values (around 0.5) after several epochs. Several experiments were performed by varying the weights around that initial values using the validation set. The final values of α and β in Eq. (5) were experimentally determined as 10 and 0.1 respectively.

All of the final age estimation models in REG, REG-GAN and REG-GAN-TGB were trained for 200 epochs with the batch size of 20. The learning rate was 0.0001 and decayed by multiplying 0.8 in every 10 epochs.

3.3 Results

We report quantitative results measured by Mean Squared Error (MSE) and Mean Absolute Error (MAE) of age estimation, and the corresponding image reconstruction loss (L_{L1}) of GAN, as listed in Table 1. The age estimation errors of 160 testing images using the three methods for each of the age groups are presented in Fig. 2. The error bars indicate the mean and standard deviation of MSE. The number of real images for training and testing are also presented as bins in Fig. 2.

Table 1 shows that the age estimation result (MSE and MAE) of REG-GAN method is even worse than the baseline REG method. This indicates that a synthesised image using REG-GAN method is not correlated to the associated age, which has a negative impact to the regression model training. In contrast, our proposed method REG-GAN-TGB achieved better performance than the other methods with statistical significance (measured by Wilcoxon signed rank test with $p < 0.01$). It demonstrates the effectiveness of the added task-guided branch, which helps the GAN model to synthesis more meaningful images for different age groups.

Table 1. Comparison of our proposed method (REG-GAN-TGB) to the baseline method (REG) and REG-GAN method. Mean $\pm$ standard deviation of L_{L1} loss (image reconstruction error), MSE and MAE of age estimation are reported.

Method	L_{L1} loss	MSE	MAE
REG	N/A	5.1922 ± 9.0424	1.7060 ± 1.5153
REG-GAN	0.0168 ± 0.0044	6.7092 ± 11.6758	1.9356 ± 1.7266
REG-GAN-TGB	0.0264 ± 0.0066	**4.0699** ± 6.2613	**1.5415** ± 1.3055

From the MSE of age estimation for each of the age groups presented in Fig. 2, it can be seen that the error increases when the number of real images decreases for all three methods. Overall, the proposed REG-GAN-TGB method consistently improved the baseline REG method for most age groups, while the REG-GAN method performed worse than the baseline method for most age groups. The larger errors for some age groups (e.g. 18–22) were due to the limited number of real images for training and image synthesis. It also led to the relatively larger standard deviation values of MSE in Table 1 and Fig. 2.

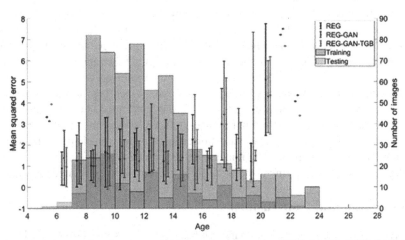

Fig. 2. MSE of age estimation for each of the age groups using REG (black), REG-GAN (blue) and REG-GAN-TGB (red). Error bars indicate the mean and standard deviation values. Bins show the number of training (grey) and testing (pink) images of real data.

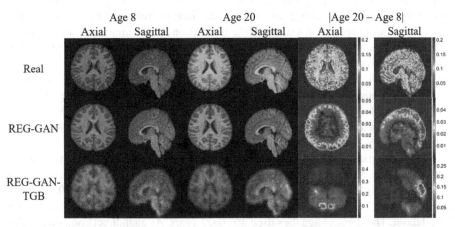

Fig. 3. Visual examples of real and synthesised images using REG-GAN and REG-GAN-TGB for age 8 and 20 respectively. |Age 20-Age 8| column shows the absolute intensity difference of the average image of 20 real/reconstructed images between age 8 and 20 for each method.

In Fig. 3 columns 1 to 4, we show some visual examples of real images and synthesised images using the REG-GAN and REG-GAN-TGB methods for age 8 and 20 in both axial view (slice 50) and sagittal view (slice 50). For each of the methods, we also show the image intensity difference of using the average image of 20 real/reconstructed images for age 20 subtracted by the average image of 20 real/reconstructed images for age 8 (column 5 and 6 in Fig. 3). It can be seen that the intensity difference image obtained from the real MRIs does not highlight specific brain regions that reflect the age difference. It is also observed that the visual image quality generated by the REG-GAN method is much better than our method. However, the intensity difference between age 8 and 20 generated by REG-GAN is quite small (maximum of 0.05 out of 1) and mainly highlighted the whole brain border regions. In contrast, although our proposed method generated more blurry images than REG-GAN, it produced a much larger intensity difference (maximum of 0.49 out of 1) in the occipital lobe region than other brain regions (red arrow in Fig. 3), indicating an active change from age 8 to 20. The occipital lobe contains the primary and associated visual cortex which is primarily responsible for visual processing. This finding needs to be further confirmed in brain science field. Overall, we have demonstrated that our image synthesis method not only helps to boost the performance of downstream age estimation task but also extremely powerful to identify age-related image regions. The image synthesis process using our method can be considered as a low pass filtering that effectively extracts the key age-related information.

4 Discussion and Conclusions

In this paper, we have proposed an image synthesis solution using a GAN-based model. Based on an age estimation task using 3D brain MRI, we have demonstrated that by integrating a regression branch with an additional loss term to the GAN model, the learned latent feature space and the synthesised images are more age-specific. With the augmented data from our proposed method, the performance of the age estimation model is significantly improved. However, like most machine learning generative models, the synthesised images are still in the same data distribution of the training set, which cannot be used to generate out-of-distribution samples. The main benefit of the proposed solution is to balance the dataset to minimised the bias in model learning.

It is also noteworthy to emphasise that we removed the brain volume variations for the age estimation modelling as we were keen to explore new local image features that are more difficult to be identified by human. Using our method, we have identified the occipital lobe brain region that is highly correlated to the age difference between 8 and 20.

For the comparison of our method to other state-of-the-art methods, we investigated other studies that use brain MRIs for age estimation [23]. Most of them were evaluated on an age range of 18 to 90 years, and the MAE were around 4 to 7. In 2020, He et al. [24] proposed a method to estimate the age for children on a subset of a public brain MRI dataset. The range of age they used was 0–22 years. As an indirect comparison, the MAE of their 3D method on that dataset in the age groups of 6–10, 11–15, 16–22 were 1.12, 1.19, 1.85, respectively. The MAE of our proposed REG-GAN-TGB method on our dataset in the same age range were 1.32, 1.37 and 2.11 respectively.

For future work, we plan to test our method on more datasets and comprehensively compare the performance with other state-of-the-art methods. We will further investigate the impact of the task-guided branch in the latent space and attempt to detect disease biomarkers by visualising different features of the generated images. For example, use this approach in a Parkinson dataset to explore whether the method could again highlight meaningful brain regions that correlate to the disease progression.

References

1. Shin, H.-C., et al.: Medical image synthesis for data augmentation and anonymization using generative adversarial networks. In: Gooya, A., Goksel, O., Oguz, I., Burgos, N. (eds.) SASHIMI 2018. LNCS, vol. 11037, pp. 1–11. Springer, Cham (2018). https://doi.org/10.1007/978-3-030-00536-8_1
2. Frid-Adar, M., Diamant, I., Klang, E., Amitai, M., Goldberger, J., Greenspan, H.: GAN-based synthetic medical image augmentation for increased CNN performance in liver lesion classification. Neurocomputing (2018). https://doi.org/10.1016/j.neucom.2018.09.013
3. Pesteie, M., Abolmaesumi, P., Rohling, R.N.: Adaptive augmentation of medical data using independently conditional variational auto-encoders. IEEE Trans. Med. Imaging. (2019). https://doi.org/10.1109/TMI.2019.2914656
4. Goodfellow, I., et al.: Generative adversarial nets. In: Advances in Neural Information Processing Systems, pp. 2672–2680 (2014)
5. Kingma, D.P., Welling, M.: Auto-encoding variational bayes. In: 2nd International Conference on Learning Representations, ICLR 2014 - Conference Track Proceedings (2014)
6. Isola, P., Zhu, J.Y., Zhou, T., Efros, A.A.: Image-to-image translation with conditional adversarial networks. In: Proceedings - 30th IEEE Conference on Computer Vision and Pattern Recognition, CVPR 2017 (2017). https://doi.org/10.1109/CVPR.2017.632.
7. Mirza, M., Osindero, S.: Conditional generative adversarial nets. arXiv Prepr. arXiv:1411.1784 (2014)
8. Antipov, G., Baccouche, M., Dugelay, J.-L.: Face aging with conditional generative adversarial networks. In: 2017 IEEE International Conference on Image Processing (ICIP), pp. 2089–2093 (2017)
9. Wang, Z., Tang, X., Luo, W., Gao, S.: Face aging with identity-preserved conditional generative adversarial networks. In: Proceedings of the IEEE Computer Society Conference on Computer Vision and Pattern Recognition (2018). https://doi.org/10.1109/CVPR.2018.00828
10. Xia, T., Chartsias, A., Tsaftaris, S.A., Initiative, A.D.N., et al.: Consistent brain ageing synthesis. In: International Conference on Medical Image Computing and Computer-Assisted Intervention, pp. 750–758 (2019)
11. Liu, X., et al.: Data augmentation via latent space interpolation for image classification. In: 2018 24th International Conference on Pattern Recognition (ICPR), pp. 728–733 (2018)
12. DeVries, T., Taylor, G.W.: Dataset augmentation in feature space. In: 5th International Conference on Learning Representations, ICLR 2017 - Workshop Track Proceedings (2019)
13. Gulrajani, I., Ahmed, F., Arjovsky, M., Dumoulin, V., Courville, A.: Improved training of wasserstein GANs. In: Advances in Neural Information Processing Systems (2017)
14. Simonyan, K., Zisserman, A.: Very deep convolutional networks for large-scale image recognition. arXiv Prepr. arXiv:1409.1556 (2014)
15. He, K., Zhang, X., Ren, S., Sun, J.: Deep residual learning for image recognition. In: Proceedings of the IEEE Conference on Computer Vision and Pattern Recognition, pp. 770–778 (2016)
16. Di Martino, A., et al.: The autism brain imaging data exchange: towards a large-scale evaluation of the intrinsic brain architecture in autism. Mol. Psychiatry. **19**, 659–667 (2014)
17. Di Martino, A., et al.: Enhancing studies of the connectome in autism using the autism brain imaging data exchange II. Sci. Data **4**, 1–15 (2017)
18. Alfaro-Almagro, F., et al.: Image processing and Quality Control for the first 10,000 brain imaging datasets from UK Biobank. Neuroimage **166**, 400–424 (2018)
19. Jenkinson, M., Bannister, P., Brady, M., Smith, S.: Improved optimization for the robust and accurate linear registration and motion correction of brain images. Neuroimage **17**, 825–841 (2002)

20. Andersson, J.L., Jenkinson, M., Smith, S.: Non-linear optimisation. FMRIB Analysis Group Technical Reports. TR07JA1 (2007)
21. Jenkinson, M., Beckmann, C.F., Behrens, T.E.J., Woolrich, M.W., Smith, S.M.: Fsl. Neuroimage. **62**, 782–790 (2012)
22. Kingma, D.P., Ba, J.: Adam: a method for stochastic optimization. arXiv Prepr. arXiv:1412.6980 (2014)
23. Sajedi, H., Pardakhti, N.: Age prediction based on brain MRI image: a survey. J. Med. Syst. **43**, 279 (2019)
24. He, S., et al.: Brain Age Estimation Using LSTM on Children's Brain MRI. In: 2020 IEEE 17th International Symposium on Biomedical Imaging (ISBI), pp. 1–4 (2020)

First Trimester Gaze Pattern Estimation Using Stochastic Augmentation Policy Search for Single Frame Saliency Prediction

Elizaveta Savochkina[1($\boxtimes$)], Lok Hin Lee[1], Lior Drukker[2],
Aris T. Papageorghiou[2], and J. Alison Noble[1]

[1] Department of Engineering Science, University of Oxford, Oxford, UK
elizaveta.savochkina@eng.ox.ac.uk
[2] Nuffield Department of Women's and Reproductive Health,
University of Oxford, Oxford, UK

Abstract. While performing an ultrasound (US) scan, sonographers direct their gaze at regions of interest to verify that the correct plane is acquired and to interpret the acquisition frame. Predicting sonographer gaze on US videos is useful for identification of spatio-temporal patterns that are important for US scanning. This paper investigates utilizing sonographer gaze, in the form of gaze-tracking data, in a multi-modal imaging deep learning framework to assist the analysis of the first trimester fetal ultrasound scan. Specifically, we propose an encoder-decoder convolutional neural network with skip connections to predict the visual gaze for each frame using 115 first trimester ultrasound videos; 29,250 video frames for training, 7,290 for validation and 9,126 for testing. We find that the dataset of our size benefits from automated data augmentation, which in turn, alleviates model overfitting and reduces structural variation imbalance of US anatomical views between the training and test datasets. Specifically, we employ a stochastic augmentation policy search method to improve segmentation performance. Using the learnt policies, our models outperform the baseline: KLD, SIM, NSS and CC (2.16, 0.27, 4.34 and 0.39 versus 3.17, 0.21, 2.92 and 0.28).

Keywords: Fetal ultrasound · First trimester · Gaze tracking · Single frame saliency prediction · U-Net · Data augmentation

1 Introduction

Visual attention prediction, commonly known as saliency detection, is the task of inferring the objects or regions that attract human attention in an image [21]. Saliency prediction is important to aid replication of human perception and has recently been explored in computer vision [7,14,17]. Cai et al. [6] proposed the SonoEyeNet model for automated detection of the abdominal circumference

© Springer Nature Switzerland AG 2021
B. W. Papież et al. (Eds.): MIUA 2021, LNCS 12722, pp. 361–374, 2021.
https://doi.org/10.1007/978-3-030-80432-9_28

plane (ACP) utilizing eye-tracking heat maps, hereafter referred to as *saliency maps*, extracted with a pre-trained SonoNet model [2]. Encoder-decoder networks are popular for modern semantic and instance segmentation models where each pixel is labelled with a class [13]. Such networks can be used to predict the likelihood that each pixel is fixated upon by modelling saliency.

Medical image data preparation is seen as one of the key components to affect the performance of a deep learning medical imaging model. In particular, data augmentation can address overfitting [1] and class imbalance [12]. It is a common approach to use augmentation transformation such as scaling, rotation and translation that make training images look natural and realistic. However, recent works [9,15] have found that different types of transformations that cause images to look "less realistic" lead to improved generalization.

The RandomAugment (RA) paper [9] introduces a search space with the regularization strength that can be tailored based on model and dataset size. Our dataset has a high structural (anatomical view) variability amongst different US frames due to ultrasound artifacts and characteristics that are a result of fetal position, variable maternal body wall thickness, and sonographer expertise. With a larger number of samples, there is a higher demand for data cleaning and a higher chance for distorted images to occur.

'Mixup' [20] is a type of regularization that alleviates sensitivity to distorted examples by including artificial data into model training. As our task is specific to medical imaging, the RA policy search can be extended from single-image transformation strategy such as RA and include mixed-example image strategy adopted from [20]. Lee et al. [15] combined RA transformations and included non-linear mixed examples for classification of standard planes and named the model Mix.RA which is later used in our research. We, therefore, include non-linear transformations to the RandomAugment policy search.

The specific clinical problem that motivates this work is first trimester ultrasound imaging. There is little work on this topic. See for example, Ryou et al. [16] who performs automated image analysis of multiple fetal anatomies from a single 3D ultrasound scan and references therein.

Motivated by the importance of the ultrasound screening in early stages of pregnancy as well as recent trends in cognitive science, we propose a simple U-Net network variant for single frame visual saliency prediction. We employ Random Augmentation with Mixup as part of the data augmentation strategy.

Contribution. The contributions of this study are two-fold. First, we consider single frame saliency prediction from a first trimester multi-modal ultrasound dataset. Our model predicts the gaze not only for the standardized planes used to measure the Nuchal Translucency (NT) and the Crown-Rump Length (CRL), but for all structures and planes that come into sonographer view. Second, we demonstrate that stochastic augmentation policy search can be used for segmentation purposes and is able to accommodate the structural variations of the US anatomical views.

2   Data Augmentation Strategy

Figure 1 presents an overview of the proposed method which consists of data generation, data augmentation, grid search and an encoder-decoder network for saliency prediction. Inspired by Cubuk et al. [9], we adopt a simple grid search with fixed magnitude schedule and a total of $K = 19$ transformations which include additional non-linear transformations by Lee et al. [15]. Each augmentation policy is defined by n, which is the number of transformations from K an image undergoes, and m, which is the magnitude distortion of each transformation. These transformations are then applied to the mixed-example images [20] with which we share the m hyperparameter. Henceforth, we refer to RandomAugment with non-linear transformations as Random Augmentation (RA) and Random Augmentation strategy with Mixup as Mix.RA.

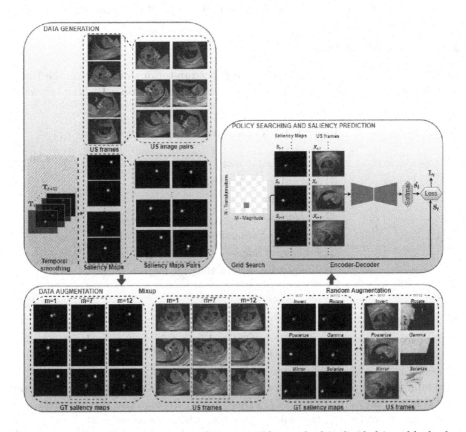

Fig. 1. Overview of our proposed architecture. The method is divided into blocks for better visualization starting from data generation, followed by data augmentation, policy searching and saliency prediction.

2.1 Data

We use a novel dataset of clinical fetal ultrasound exams with real time sonographer gaze tracking data. The exams were performed on a GE Voluson E8 scanner (GE, USA) while the video signal of the machine's monitor is recorded 30 Hz. Gaze is simultaneously recorded 90 Hz with a Tobii Eye Tracker 4C (Tobii, Sweden). Ethics approval was obtained for data recording and data are stored according to local data governance rules. For our experiments, we use 115 fetal first trimester videos. When sonographers perform a scan, they search for a standard plane, freeze the image, take measurements and save the best acquisition of the standard biometry plane. The data are split into training, validation and test datasets, with 80/20 training/test split by number of images. The total amount of frames after data cleaning is 45,630 US frames with corresponding saliency maps. Dataset division results in 70 videos (29,250 frames) for training, 17 videos (7,290 frames) for validation and 29 videos (9,126 frames) for testing.

2.2 Data Preparation

Table 1 summarizes the data preparation procedure prior to employing the Random Augmentation and Random Augmentation with Mixup.

Table 1. Data Preparation Procedure

Data preparation step
1) Discard irrelevant frames
2) Clean data
Gaze Manipulation:
3) Aggregate gaze points
4) Filter gaze maps
5) Temporally smooth the gaze maps

(1) Discarding irrelevant frames: All video frames that did not correspond to 2D B-mode live scanning (e.g. Doppler, 3D/4D or frozen frames) or had no gaze data were discarded. *(2) Data Cleaning:* We inspected 150 fetal US videos, manually deleted corrupt and double-image frames leaving us with 115 US videos and a total number of 45630 US frames. *(3) Aggregate gaze points:* Gaze points outside the US image fan were discarded. *(4) Gaze maps filtering:* The saliency maps are computed by smoothing the binary gaze maps with Gaussian kernel. Gaussian filter has the standard deviation of 13.5 and the window size of 81. The standard deviation of the Gaussians is equivalent to ca. 1° visual angle to account for the radius of visual acuity and the uncertainty of the eye tracker measurements [8]. *(5) Temporal Smoothing:* We sum gaze points for each frame to represent the gaze distribution that sonographer might have over the similar frames.

2.3   Encoder-Decoder Network

The encoder consists of four (Conv → MaxNorm → BatchNorm → Relu → Conv → Dropout) blocks, with channels (16, 32, 64 and 128). The decoder consists of 4 (ConvTranspose → MaxNorm → Dropout → Conv → Conv) blocks, with channels (128, 64, 32 and 16). Each block has a dropout of 0.5, resulting in a model with 218k trainable parameters. As the primary focus of our work is on data augmentation strategies, the skeleton of the encoder-decoder network will not be discussed in detail.

2.4   Mixed-Example Data Augmentation

Mixup smooths out network performance in linear interpolates of input feature vector. Our original dataset $D = \{(X_i, Y_i)\}$ consists of a series of i ultrasound frames X and their corresponding ground truth saliency maps Y, where a single US image is used as an input into the encoder-decoder network. In order to employ the Mixup, we generate two sets of US and GT saliency map images at random and pair them into a dataset $D_p = \{(x_1, x_2)_{\frac{i}{2}}, (y_1, y_2)_{\frac{i}{2}}\}$. The training distribution is extended using element-wise weighted averaging of two random examples [20].

Although linear intensity averaging does not produce realistic examples of US images, the artificial image generated has a smooth transition from two images into one. In contrast, other non-linear methods [18] slice image features into blocks making the edges of the mixed images to appear which is not a good representation of distorted examples in the test set and defeat the method benefits for our application.

As we pair US images, their corresponding GT set of saliency maps undergo the same transformation. After concatenation, the resulted US image and GT saliency map (SM) are denoted as $\tilde{x}$ and $\tilde{y}$ illustrated in Fig. 2.

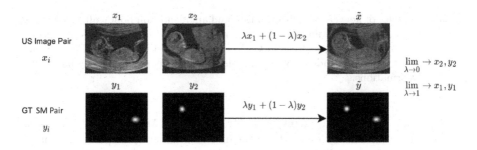

Fig. 2. The procedure for linear mixed-example data augmentation using US image pair and corresponding GT saliency map image pair and the final artificial mixed-example image pairs

Formally, given initial US images $x_{1,2}$ and their corresponding GT saliency maps $y_{1,2}$, the generated artificial mixed-example images $\tilde{x}$ and $\tilde{y}$ are:

$$\tilde{x} = \lambda x_1 + (1 - \lambda)x_2$$
$$\tilde{y} = \lambda y_1 + (1 - \lambda)y_2 \qquad (1)$$

where, $\lambda \sim Beta(\frac{m}{10}, \frac{m}{10})$ for each pair of examples and m is a learned hyperparameter varied from 0–12. As m approaches 10, λ are more uniformly distributed across 0–1 and the interpolation intensity of artificial images increases.

2.5   Random Augmentation

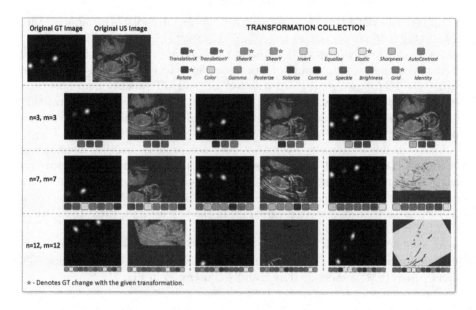

Fig. 3. Examples of how each transformation and augmentation policy affect US input images and their corresponding GT saliency maps. Each color represents a transformation where original images (US and GT) are transformed using a number of transformations (n) at a magnitude of (m). *Star*: denotes change of GT saliency maps with given transformations.

We employ a random augmentation strategy, RandAugment, to tackle overfitting which we believe arose not from the limited number of 1st trimester images but from the structural differences in the anatomical views. Unless the dataset is annotated and ultrasound (US) frames are each given a class label to distinguish between the types and the number of frames that belong to each class, we are unable to balance the structural variation of each frame in the training set with a test set. To avoid turning the problem into a classification task, we attempt to combat the structural variations of the US anatomical views solely with diverse random augmentation.

As mentioned in the overview of our proposed method, we adopt a grid search with fixed magnitude schedule and a total of $K = 19$ available transformations. Each augmentation policy is defined by n, which is the number of transformations from K that an image undergoes, and m, which is the magnitude distortion of each transformation. These transformations are the applied to the mixed-example images with which we share the m hyperparameter. Following [15], the inclusion of non-linear transformations such as grid distortion, speckle and elastic transformation are beneficial to use on US images. Figure 3 depicts 6 affine, 10 histogram and 3 non-linear transformations examples where each US frame and corresponding GT saliency map change with respect to the type of transformation.

2.6 Training

The models are implemented in Tensorflow 2.1 and image manipulations are performed with Pillow 7.1.2 and OpenCV 3.4.9 libraries. Each training run varied from 8 h to 5 d on a single Nvidia GTX 2060 Ti with a total of 40 models run. Networks were trained up to 120 epochs with a batch size of 50 via adaptive moment estimation (Adam) with an initial learning rate of 0.001.

For training we use frames that appear 3 s before the first freeze-frame. To keep the full 3 s before the freeze-frame and account for temporal smoothing, with the refresh monitor of 30 fps, we extract 90 US frames (3sec × 30 fps = 90 frames) with 12 additional frames to smooth frame 0. A frame persists in our eye for 0.1–0.4 s [19] which translates to 3–12 frames of 30 HZ ultrasound video. To store the maximum amount of information we use 12 US frames. Therefore, the batches of 102 US frames and 102 corresponding ground truth saliency map frames are each combined to form 1 US input frame and 1 GT saliency map frame. GT and US images are normalized and randomly shuffled with 1–1 correspondence.

All images were resized to 288-244 pixels for data augmentation. Random horizontal and vertical flipping were used in all RA policies as a baseline augmentation. We also compare our augmentation policies with the augmentation policy in previous works [2,6,11] as an example of a conventional augmentation policy, consisting of random rotation with an angle uniformly sampled from [−25, 25] degrees and random horizontal flipping. We denote this as the baseline augmentation policy. The performance of networks trained with augmentation was evaluated with values of n, m where n, $m = \{1, 3, 5, 7, 9, 12\}$ using a grid search to find optimal n, m values.

2.7 Saliency Map Prediction

Given an image and a gaze point set $(\mathbf{X}, G)\epsilon D$, we generate a visual saliency map $\mathbf{S}\epsilon[0, 1]^{H_D \times W_D}$, where $S_{i,j}$ is the probability that pixel $X_{i,j}$ is fixated upon [10]. The saliency map is then used as the target for the predicted probability map $\hat{\mathbf{S}}$. We generate $\mathbf{S}$ as a sum of Gaussians around the gaze points in G, normalized such that $\sum_{i,j} S_{i,j} = 1$. The saliency map yields the training target $\mathbf{S}^* \epsilon [0, 1]^{H_D \times W_D}$ which is used as part of the augmented mini-batch described in detail in Sect. 2.4. Finally, the training loss is computed via the Kullback-Leibler divergence (KLD) between the predicted and true distribution:

$$\mathbf{L}_s(\mathbf{S}^*, \hat{\mathbf{S}}) = D_{KL}(\mathbf{S}^* \parallel \hat{\mathbf{S}}) = \sum_{i,j} S_{i,j}^* \cdot (log(S_{i,j}^*) - (log(\hat{S}_{i,j})) \qquad (2)$$

The hyperparameters n and m that yield the best segmentation performance are used during final model evaluation.

2.8   Evaluation Metrics

We evaluate the models using the metrics of the MIT Saliency Benchmark [4]. Several measures have been proposed for evaluating saliency models which are distribution-based and location-based. Pearson's Correlation Coefficient (CC), Kullback-Leibler divergence (KL) and Similarity or histogram intersection (SIM) fall under the first category. Normalized Scanpath Saliency (NSS) falls under the second category. KLD and SIM are more sensitive to false negatives than to false positives, while NSS and CC treat them symmetrically [5].

3   Experiments and Results

3.1   Quantitative Results

Figure 4 shows the average test scores for the Random Augmentation, Mix.RA and the baseline with 2 augmentations. Random Augmentation includes RandomAugment and non-linear transformations and Mix.RA is the Random Augmentation strategy with Mixup. Both augmentation strategies, RA and. Mix.RA, outperform the baseline on all metrics. RA with values $n, m = \{7, 9\}$ performs best on KLD and CC metrics, whilst RA with values $n, m = \{3, 9\}$ scores the highest on SIM metric. RA with Mixup with values $n, m = \{3, 7\}$ receives the best score on NSS metric. Results are further discussed in Sect. 4.

For our segmentation task, Mixup complements the RA and generally produces very similar or better results for the saliency metrics up to the magnitude, $m = 9$. Higher magnitudes and number of augmentations show a decline in saliency performance which can be seen in Fig. 4.

3.2   Representative Examples

Figure 5 shows examples of the predictions of the Mix. RA strategy and the comparative models on test data. Specifically, the RA strategy is represented by the best tuples of values, $n, m = \{7, 9\}$, and Mix.RA is represented by $n, m = \{3,7\}$. Our strategies are compared to the ground truth gaze distribution. Since the training and validation data were divided scan-wise fulfilling the case for

RANDOM AUGMENTATION

Baseline

KLD	3.17
SIM	0.21
NSS	2.92
CC	0.28

m → (columns), n ↓ (rows)

KLD

	1	3	5	7	9	12
1	-	3.09	2.82	2.64	2.5	2.58
3	3.00	2.72	2.72	2.25	2.27	2.23
5	2.30	2.53	2.33	2.26	2.21	2.21
7	2.75	2.25	2.42	2.21	**2.16**	2.17
9	2.66	2.40	2.27	2.21	2.23	2.20
12	2.60	2.40	2.25	2.22	2.25	2.26

SIM

	1	3	5	7	9	12
1	-	0.24	0.24	0.24	0.25	0.23
3	0.25	0.24	0.24	0.24	**0.27**	0.25
5	0.23	0.24	0.24	0.25	0.25	0.25
7	0.24	0.26	0.24	0.25	0.25	0.24
9	0.25	0.23	0.25	0.24	0.22	0.22
12	0.24	0.25	0.23	0.23	0.23	0.22

NSS

	1	3	5	7	9	12
1	-	3.39	3.67	3.85	3.92	3.60
3	3.81	3.66	3.97	4.00	4.21	4.10
5	3.40	3.67	3.92	4.11	4.07	4.15
7	3.76	4.17	3.92	4.09	4.19	4.22
9	3.83	3.61	3.95	4.09	4.01	4.04
12	3.64	3.96	3.89	3.96	3.89	3.94

CC

	1	3	5	7	9	12
1	-	0.31	0.30	0.34	0.53	0.34
3	0.34	0.33	0.37	0.37	0.38	0.38
5	0.30	0.33	0.36	0.37	0.37	0.38
7	0.34	0.38	0.36	0.38	**0.39**	**0.39**
9	0.35	0.33	0.37	0.38	0.38	0.38
12	0.33	0.36	0.36	0.37	0.37	0.38

MIXED-EXAMPLE RANDOM AUGMENTATION

Baseline

KLD	3.17
SIM	0.21
NSS	2.92
CC	0.28

m → (columns), n ↓ (rows)

KLD

	1	3	5	7	9	12
1	-	2.64	2.37	2.35	2.44	2.63
3	2.62	2.49	2.30	2.28	2.33	2.29
5	2.50	2.26	2.33	2.25	2.28	2.21
7	2.51	2.28	2.29	2.26	2.33	2.34
9	2.42	2.31	2.32	2.34	2.31	2.29
12	2.63	2.33	2.37	2.32	2.33	2.34

SIM

	1	3	5	7	9	12
1	-	0.24	0.21	0.25	0.24	0.18
3	0.24	0.21	0.26	0.26	0.24	0.23
5	0.25	0.23	0.24	0.22	0.21	0.23
7	0.23	0.23	0.22	0.23	0.20	0.24
9	0.24	0.23	0.22	0.20	0.20	0.22
12	0.25	0.22	0.22	0.23	0.20	0.21

NSS

	1	3	5	7	9	12
1	-	3.47	3.57	4.10	3.72	2.91
3	3.78	3.36	4.26	**4.34**	3.77	3.79
5	3.88	4.00	3.47	3.77	3.70	4.01
7	3.50	3.94	3.67	3.77	3.62	4.01
9	3.81	3.86	3.80	3.55	3.51	3.84
12	3.80	3.51	3.63	3.79	3.44	3.74

CC

	1	3	5	7	9	12
1	-	0.32	0.33	0.37	0.34	0.28
3	0.34	0.32	0.38	0.38	0.35	0.36
5	0.35	0.37	0.35	0.36	0.36	0.37
7	0.32	0.36	0.34	0.35	0.34	0.36
9	0.35	0.35	0.35	0.34	0.34	0.37
12	0.34	0.33	0.34	0.35	0.34	0.36

Fig. 4. Results of visual saliency prediction with RA and Mix.RA compared to baseline with 2 augmentations. Next to the training loss (KLD), models are evaluated on the metrics normalized scanpath saliency (NSS), Pearson's correlation coefficient (CC) and histogram intersection (SIM) (for references see [3]). Best performing augmentation strategies are marked in bold.

90 consecutive and temporally smoothed frames, the frames are entirely unseen by the network. Moreover, the network is agnostic as to which sonographer is performing the scan.

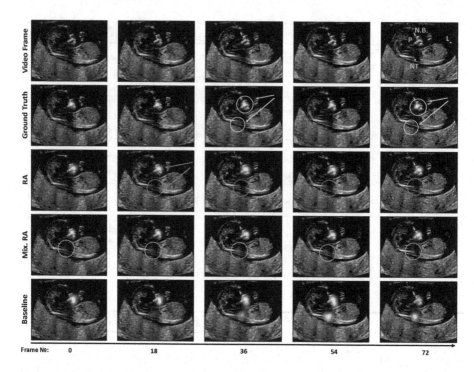

Fig. 5. Five frames from an exemplary search sequence. The rows show the input frames, the ground truth saliency annotations, saliency predictions of our encoder-decoder network with Mix. RA against RA and baseline models, respectively. The relevant anatomical structures denoted in the last input frame (top right) include palate (P), nasal bone (N.B.), limbs (L) and nuchal translucency (NT). The ground truth is circled in yellow, RA secondary predictions are circled in red and Mix.RA secondary predictions are circled in blue. (Color figure online)

From the circled ground truth frame (yellow), we see that the sonographer primarily focuses on the palate (P) and checks the nuchal translucency (NT) for guidance during scanning.

At frame zero, RA and Mix.RA predict sonographer focusing on the end of the palate where Mix.RA also predicts the gaze at the end of the nuchal translucency (circled in blue). The baseline model has more spread-out saliency prediction around the fetal profile - lips, chin and NT.

Over the next 3 exemplary frames, RA and Mix.RA strategies predict temporally smooth saliency maps maintaining similar positions of the palate. Mix.RA predicts fixations around the NT (circled in blue), whilst, RA assigns low saliency values to NT (circled in red). The ground truth fixations are mainly on the palate with low probability assigned to the NT. Starting from frame 36, the baseline augmentation prediction heavily over-estimates the gaze of sonographer looking at the palate and NT.

On the last frame, the sonographer fixates on the bottom end of the palate and focuses less on the NT. Both, RA and Mix.RA models show a better prediction of the sonographer fixating on the palate and less on the NT. In contrast, the baseline predicts the saliency maps with maxima around the NT and varying maxima between the nasal bone and the palate.

4   Discussion

The results presented show both augmentation strategies, RA and Mix.RA outperform the baseline on all saliency metrics. As Mixup only complements the RA producing similar or better results up to the magnitude, $m = 9$. Further augmentation with values $n, m = 12$ reduces the performance of the saliency prediction. In contrast, at low n, m values, both strategies display low diversity of augmented data making inferior predictions due to little input data variation. The best augmentation strategy is directly dependent on the size of the model and the dataset. Due to the fact that our dataset is larger than [15] we required a higher number of augmentations to determine the tuple that performs best for the saliency metrics.

The results show that the baseline augmentation method prediction overestimates the gaze of the sonographer looking at the nuchal translucency, whilst for RA the prediction of the sonographers gaze is more focused on the NT for the most frames. Both models, RA and Mix.RA, show better predictions of gaze location for sonographers.

Quantitatively, RA performs better without adding Mixup. With high augmentation values, n, m, the image becomes too distorted if used with mixed-example augmentation. Analysing the Random Augmentation strategy, the tuple combination $n, m = \{7, 9\}$ performs best using the KLD and CC metrics, and $n, m = \{3, 9\}$ using the SIM metric. The KLD metric is an integral part of our analysis as it is the only metric that measures the difference between two probability distributions, and is used for segmentation and saliency maps detection [5]. Considering the two of our ground truth fixations, if any ground truth fixation locations are missed, KLD is highly penalized which makes it the most relevant metric for our task.

The NSS metric is the best performing metric for Mix.RA, NSS is the least sensitive to false positives which is clearly seen in Fig. 5 for the secondary prediction. SIM and CC metrics are similar with the CC score being penalised due to false negatives and SIM penalizing the misalignment of density of predicted saliency maps. Unless the shape of the Gaussian blur of our predicted saliency map exactly corresponds to the GT shape, SIM will see a dramatic decline in its score. High values of similarity appear in RA strategy due to the metric being very sensitive to false positives which is reflected in Fig. 5. RA secondary prediction receives either no saliency prediction or a very low probability which is symmetric to the ground truth.

In future work, a natural extension would be to parameterize the mixed-example generation with an additional third hyperparameter z which is separate from the m, n used for paramteterizing the RA strategy.

$$\tilde{x} = \lambda x_1 + (1 - \lambda)x_2$$
$$\tilde{y} = \lambda y_1 + (1 - \lambda)y_2 \tag{3}$$

where $\lambda \sim Beta(\frac{z}{10}, \frac{z}{10})$ for each pair of examples and z is a learned hyperparameter varied from 0–12. This would allow for greater degrees of freedom in specifying the final augmentation at the expense of an increased search space during hyperparameter optimization.

Further, the current first trimester single frame saliency predictions might be improved by incorporating a temporal module. Due to a complete randomisation of frames, the current model is only able to predict future frames having seen the frames which are independent of each other. The network can receive two very similar ultrasound frames with the same gaze point location on each. As no temporal information about the sequence of past frames is provided, the model is not able to differentiate between the fast moving segments and slow moving segments. As a solution, in future work we will explore different forms of recurrent neural networks (RNNs). As part of data preparation, we will introduce a variable length ground truth (GT) compression in order to differentiate between the static and fast-moving US frames. This method will help improve the standard temporal smoothing technique by compressing GT saliency maps with rates dependant on the change in US frames.

5   Conclusion

We have presented a single frame saliency prediction for first trimester ultrasound images. We have shown that, for segmentation purposes, using a simple hyper-parameter grid search method we are able to find an augmentation strategy that outperforms standard augmentation methods. For saliency map segmentation, RA with values $n, m = \{7, 9\}$ showed a decrease in KLD-score of 1.01 over that of conventional ultrasound augmentation strategy and an increase of 0.11 on CC metric. An increase of 0.06 in the SIM metric is found in RA with $n, m = \{3, 9\}$. Finally, RA with Mixup showed an increase of 1.42 for the NSS metric.

Importantly, both models localize well the two structures, the palate and the nuchal translucency which are important for clinical measurement. Automation of clinical measurement will be the subject of future work.

Acknowledgements. This work is supported by the ERC (ERC-ADG-2015694581, project PULSE) and the EPSRC (EP/R013853/1 and EP/T028572/1). AP is funded by the NIHR Oxford Biomedical Research Centre.

References

1. Al, W.A., Yun, I.D.: Reinforcing medical image classifier to improve generalization on small datasets (2019)

2. Baumgartner, C.F., et al.: Sononet: real-time detection and localisation of fetal standard scan planes in freehand ultrasound. IEEE Trans. Med. Imaging **36**(11), 2204–2215 (2017)
3. Borji, A.: Saliency prediction in the deep learning era: An empirical investigation. arXiv preprint arXiv:1810.03716 10 (2018)
4. Bylinskii, Z., et al.: Mit saliency benchmark (2015)
5. Bylinskii, Z., Judd, T., Oliva, A., Torralba, A., Durand, F.: What do different evaluation metrics tell us about saliency models? IEEE Trans. Pattern Anal. Mach. Intell. **41**(3), 740–757 (2018)
6. Cai, Y., Sharma, H., Chatelain, P., Noble, J.A.: Sonoeyenet: standardized fetal ultrasound plane detection informed by eye tracking. In: 2018 IEEE 15th International Symposium on Biomedical Imaging (ISBI 2018), pp. 1475–1478. IEEE (2018)
7. Chang, M.M.L., Ong, S.K., Nee, A.Y.C.: Automatic information positioning scheme in AR-assisted maintenance based on visual saliency. In: De Paolis, L.T., Mongelli, A. (eds.) AVR 2016. LNCS, vol. 9768, pp. 453–462. Springer, Cham (2016). https://doi.org/10.1007/978-3-319-40621-3_33
8. Chatelain, P., Sharma, H., Drukker, L., Papageorghiou, A.T., Noble, J.A.: Evaluation of gaze tracking calibration for longitudinal biomedical imaging studies. IEEE Trans. Cybern. **50**(1), 153–163 (2018)
9. Cubuk, E.D., Zoph, B., Shlens, J., Le, Q.V.: Randaugment: practical automated data augmentation with a reduced search space. In: Proceedings of the IEEE/CVF Conference on Computer Vision and Pattern Recognition Workshops, pp. 702–703 (2020)
10. Droste, R., et al.: Ultrasound image representation learning by modeling sonographer visual attention. In: Chung, A.C.S., Gee, J.C., Yushkevich, P.A., Bao, S. (eds.) IPMI 2019. LNCS, vol. 11492, pp. 592–604. Springer, Cham (2019). https://doi.org/10.1007/978-3-030-20351-1_46
11. Droste, R., Cai, Y., Sharma, H., Chatelain, P., Papageorghiou, A.T., Noble, J.A.: Towards capturing sonographic experience: cognition-inspired ultrasound video saliency prediction. In: Zheng, Y., Williams, B.M., Chen, K. (eds.) MIUA 2019. CCIS, vol. 1065, pp. 174–186. Springer, Cham (2020). https://doi.org/10.1007/978-3-030-39343-4_15
12. Eaton-Rosen, Z., Bragman, F., Ourselin, S., Cardoso, M.J.: Improving data augmentation for medical image segmentation (2018)
13. Hu, J., et al.: S-unet: a bridge-style u-net framework with a saliency mechanism for retinal vessel segmentation. IEEE Access **7**, 174167–174177 (2019)
14. Itti, L.: Automatic foveation for video compression using a neurobiological model of visual attention. IEEE Trans. Image Process. **13**(10), 1304–1318 (2004)
15. Lee, L.H., Gao, Y., Noble, J.A.: Principled ultrasound data augmentation for classification of standard planes (2021)
16. Ryou, H., Yaqub, M., Cavallaro, A., Papageorghiou, A.T., Noble, J.A.: Automated 3d ultrasound image analysis for first trimester assessment of fetal health. Phys. Med. Biol. **64**(18), 185010 (2019)
17. Setlur, V., Takagi, S., Raskar, R., Gleicher, M., Gooch, B.: Automatic image retargeting. In: Proceedings of the 4th International Conference on Mobile and Ubiquitous Multimedia, pp. 59–68 (2005)
18. Summers, C., Dinneen, M.J.: Improved mixed-example data augmentation. In: 2019 IEEE Winter Conference on Applications of Computer Vision (WACV), pp. 1262–1270. IEEE (2019)

19. Varley, J.: Persistence of Vision. Penguin (1988)
20. Zhang, H., Cisse, M., Dauphin, Y.N., Lopez-Paz, D.: mixup: beyond empirical risk minimization. arXiv preprint arXiv:1710.09412 (2017)
21. Zhang, Z., Xu, Y., Yu, J., Gao, S.: Saliency detection in 360 videos. In: Proceedings of the European Conference on Computer Vision (ECCV), pp. 488–503 (2018)

Classification

Dopamine Transporter SPECT Image Classification for Neurodegenerative Parkinsonism via Diffusion Maps and Machine Learning Classifiers

Jun-En Ding[1]($\boxtimes$), Chi-Hsiang Chu[2], Mong-Na Lo Huang[3],
and Chien-Ching Hsu[4]

[1] Research Center for Information Technology Innovation,
Academia Sinica, Taipei, Taiwan
`ding1119@citi.sinica.edu.tw`
[2] Department of Statistics, National Cheng-Kung University, Tainan, Taiwan
[3] Department of Applied Mathematics, National Sun Yat-sen University,
Kaohsiung, Taiwan
[4] Department of Nuclear Medicine, Kaohsiung Chang Gung Memorial Hospital,
Chang Gung University College of Medicine, Taoyuan City, Taiwan

Abstract. Neurodegenerative parkinsonism can be assessed by dopamine transporter single photon emission computed tomography (DaT-SPECT). Although generating images is time consuming, these images can show interobserver variability and they have been visually interpreted by nuclear medicine physicians to date. Accordingly, this study aims to provide an automatic and robust method based on diffusion Maps and machine learning classifiers to classify the SPECT images into two types, namely Normal and Abnormal DaT-SPECT image groups. In comparison with deep learning methods, our contribution is to propose an explainable diagnosis process with high prediction accuracy. In the proposed method, the 3D images of N patients are mapped to an $N \times N$ pairwise distance matrix and are visualized in diffusion Maps coordinates. The images of the training set are embedded into a low-dimensional space by using diffusion maps. Moreover, we use Nyström's out-of-sample extension, which embeds new sample points as the testing set in the reduced space. Testing samples in the embedded space are then classified into two types through the ensemble classifier with Linear Discriminant Analysis (LDA) and voting procedure through twenty-five-fold cross-validation results. The feasibility of the method is demonstrated via Parkinsonism Progression Markers Initiative (PPMI) dataset of 1097 subjects and a clinical cohort from Kaohsiung Chang Gung Memorial Hospital (KCGMH-TW) of 630 patients. We compare performances using diffusion maps with those of three alternative manifold methods for dimension reduction, namely Locally Linear Embedding (LLE), Isomorphic Mapping Algorithm (Isomap), and Kernel Principal Component Analysis (Kernel PCA). We also compare results using 2D and 3D CNN methods. The diffusion maps method has an average accuracy of 98% for the PPMI and 90% for the KCGMH-TW dataset with twenty-five fold

© Springer Nature Switzerland AG 2021
B. W. Papież et al. (Eds.): MIUA 2021, LNCS 12722, pp. 377–393, 2021.
https://doi.org/10.1007/978-3-030-80432-9_29

cross-validation results. It outperforms the other three methods concerning the overall accuracy and the robustness in the training and testing samples.

Keywords: Diffusion distance · Diffusion maps · Linear Discriminant Analysis · Manifold learning · Nonlinear dimensionality reduction · Parkinson's disease

1   Introduction

Parkinson's disease (PD) is a neurodegenerative disorder, and its pathological feature is the loss of dopaminergic neurons in the substantia nigra [1]. Currently, the diagnosis of PD is mainly based on the clinical symptoms (tremor at rest, rigidity, bradykinesia, gait disturbance) and the response to medication (levodopa or dopamine aganist). Parkinsonian syndromes refer to some diseases with clinical symptoms which is similar to PD. The etiologies of Parkinsonian syndromes may also be related to the degeneration of dopamine neurons (such as multiple system atrophy, progressive supranuclear palsy, dementia with Lewy bodies, and corticobasal degeneration) or non-neurodegenerative (such as essential tremor, secondary parkinsonism related to hydrocephalus, stroke, drugs, toxins, trauma, brain tumor, or infection) [2,3].

Dopamine transporters (DaT) are located on the presynaptic dopaminergic nerve terminal and play an important role in regulating extracellular dopamine level via reuptake dopamine into the nerve terminal. The loss of dopaminergic neurons in neurodegenerative parkinsonism leads to the reduction of DaT. Many radiotracers for single photon emission computed tomography (SPECT) such as $[^{99m}Tc]$TRODAT-1 [4], ^{123}I-beta-CIT, ^{123}I FP-CIT, which can bind specifically with DaT, were developed to diagnose PD and neurogenerative parkinsonism.

There are studies for diagnosis of Parkinson's disease based on features extracted from SPECT images, such as shape and volume, etc., combining with statistical tests or machine learning classifiers [5,9]. However, features like the shape of the abnormal striatum may be quite different and in certain cases the features cannot be captured correctly, especially when most images require high-dimensional analysis, their proposed analysis can be time consuming and may have a large estimation error due to noisy images.

In recent years, due to rapid advancements in data storage and hardware technology, neural network methods have become a powerful technique for prediction and image feature space extraction. One of the most well-known neural networks is the convolutional neural network (CNN). CNNs have been widely applied for medical image analysis, including MRI and fMRI, well-known CNN models using transfer learning architecture such as AlexNet [17], VGG-16 [18], VGG-19 and Deep Convolution Network (DCNN). The series of VGGNet was first proposed VGG-16 in 2014-ILSVRC competition followed by VGG-19 as two successful architectures on ImageNet. VGG-16 and VGG-19 use different frameworks, and their models make an improvement on AlexNet by replacing large kernel-sized filters with multiple small kernel-sized filters resulting in 13 and 16 convolution layers for VGG-16 and VGG-19 respectively. In addition, more and more deep learning classifications of ^{123}I SPECT scans use transfer learning from deep neural networks pretrained on nonmedical images [20]. However, the deep learning results are often require a large training set to improve their classification accuracy. In general, biomedical images have a high dimension feature space, which further complicates the classification process. Consequently, it is desirable to perform dimension reduction to improve the efficiency of the training and testing process. Many well-known non-linear dimension reduction methods such that Kernel PCA, multidimensional scaling (MDS), Isometric feature mapping, and locally linear embedding were used for image classification to improve speed [6].

In comparison with previous studies, we use a manifold learning methodology, namely the Diffusion Maps (DM) [7] to perform dimension reduction of the SPECT image for classification purposes. Diffusion Maps do not require complicated parameter adjustment and therefore have a shorter training time and require fewer samples to capture more important information in low-dimensional space. In addition to classification or early diagnosis and visualization, a general standard specification for non-linear dimensionality reduction methods is spectral decomposition. Through mapping the images into a low-dimensional space with new coordinates corresponding to the eigenspace associated with the largest few eigenvalues, it is hoped that the points that have a relationship between samples can be as close as possible after dimensionality reduction, while still keep the initial data structure. The method which refers to Diffusion Maps hopes to find the geometric description of the corresponding low-dimensional data through the diffusion process. In summary, we make three major contributions in this paper:

- We propose a robust method for images embedding associated with their striatum similarity and combine with the ensemble classifier.
- We provide a diagnosis procedure for every subject with twenty-five voting to predict and construct an interpretable two-model confusion matrix.
- We conduct extensive experiments on two real-world datasets. One is obtained from the benchmark of Parkinson's Progression Markers Initiative (PPMI) database [8,10], and the other one is Taiwan clinical database. In our method, we can use low-cost computing time than deep learning method and get better and more robust results.

2 Datasets of PPMI and Clinical Cohort

2.1 PPMI Dataset

Data for this study were obtained from the PPMI database, a longitudinal, multicentre study to assess the progression of clinical features, imaging, and biologic markers on PD patients and healthy controls (HC). All the PD subjects were at an early stage of Hoehn and Yahr stage I or II at baseline. The diagnosis of PD was confirmed from the PPMI imaging core that the screening DaT-SPECT (^{123}I FP-CIT) is consistent with DaT deficit. Further details of the PPMI database can be found at (http://www.ppmi-info.org). The PPMI dataset in our study contained 876 PD and 414 HC subjects.

We divide our KCGMH-TW samples into two classes (normal and abnormal). We perform analyses with data based on a single ($1 \times 128 \times 128 \times 3$, i.e., the center image in Fig. 1) or three ($3 \times 128 \times 128 \times 3$, i.e., the three images of the middle row in Fig. 1) from KCGMH-TW in our investigation. The PPMI ($1 \times 109 \times 91 \times 3$) dataset has been divided into two classes (HC and PD).

2.2   Clinical Dataset of KCGMH-TW

To attest the effectiveness of the proposed method in KCGMH-TW clinical dataset, we enrolled 730 patients who underwent [^{99m}Tc] TRODAT-1 brain SPECT between January 2017 and June 2019 in the Kaohsiung Chang Gung Memorial Hospital, Taiwan (KCGMH-TW). The Chang Gung Medical Foundation Institutional Review Board approved this retrospective study and waived the requirement for obtaining informed consent from the patients. Each patient was intravenously injected with a 925-MBq dose of [^{99m}Tc] TRODAT-1 (Institute of Nuclear Energy Research, Taiwan). [^{99m}Tc] TRODAT SPECT images were acquired using a hybrid SPECT/CT system (Symbia T; Siemens Medical Solution). SPECT images were obtained with 30 s per step acquiring 120 projections over a circular 360° rotation using low-energy, high-resolution parallel-hole collimators. A 128×128 matrix and a $\times 1.45$ zoom were used. The CT images were acquired without contrast medium; they used the following parameters: 130 kV; 45 mAs (Image Quality Reference mAs, CARE Dose 4D; Siemens Medical Solutions); rotation time, 1.5 s; collimation, 2×2.5 mm. CT images were reconstructed to a 512×512 image matrix with a very smooth kernel, H08s (Siemens Medical Solutions) for SPECT attenuation correction. Raw SPECT data were reconstructed into transaxial slices using flash 3D (OSEM reconstruction method with 3D collimator beam modeling) with 8 subsets and 8 iterations and corrected with the H08s CT attenuation map. Images were smoothed using a 3D spatial Gaussian filter (fullwidth at half maximum, 6 mm). The reconstructed transaxial slice thickness was 3.3 mm.

We select nine consecutive SPECT transaxial images showing the whole striatal radioactivity. Each patient has a separate 9 slices SPECT images, where it is more visible as shown in Fig. 1. The pixel size of each cell is $128 \times 128 \times 3$ (RGB), so we take the middle best three slices of striatal images by nuclear medicine physicians [9] and combine into a single image from the nine slices SPECT as our research target.

2.3   Labeling Criterion

All DaT-SPECT images were visually interpreted by three experienced board-certified nuclear medicine physicians according to Society of Nuclear Medicine practice guideline [11]. The labeling criteria were established manually as follows.

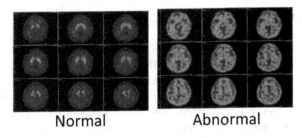

Normal　　　　　　　**Abnormal**

Fig. 1. Nine consecutive transaxial images of dopamine transporter single photon emission computed tomography (DaT-SPECT) showing the whole striatal radioactivity were displayed in a 3 × 3 slices for visual interpretation. (**Left**) The normal images of DaT-SPECT show symmetrical, well-delineated comma-shaped radioactivity in the bilateral striata. (**Right**) The abnormal images of DaT-SPECT show reduced radioactivity of the bilateral striata (nearly equal to the background radioactivity of the brain). The putamen is usually more severely affected than caudate nucleus resulting in a circular or oval shape.

(1) Normal DaT-SPECT images: the normal striata on transaxial images should look crescent- or comma-shaped and should have symmetric well-delineated borders. (2) Abnormal DaT-SPECT images: the abnormal striata have reduced intensity on one or both sides, often shrinking to a circular or oval shape. The putamen is usually more severely affected than the caudate nucleus.

Blinded to patients' clinical information except age and gender, three nuclear medicine physicians visually interpreted the DaT-SPECT images independently. The final consensus result of normal or abnormal image was assigned if at least 2 physicians achieved an agreement. The PPMI dataset images ($1 \times 109 \times 91 \times 3$) were divided into two classes (HC and PD). Similarly, the clinical dataset images of KCGMH-TW were divided into two classes (normal and abnormal). The training/validation/testing dataset were summarized in Table 1. We analyzed KCGMH-TW dataset images based on single image ($1 \times 128 \times 128 \times 3$, i.e., the center image in Fig. 1) or three ($3 \times 128 \times 128 \times 3$, i.e., the three images in the middle row in Fig. 1).

Table 1. Dataset statistics

Dataset (training/validation/testing)	Two classes	Mean age
PPMI (872/225/193)	HC (n = 414) PD (n = 876)	-
KCGMH-TW (504/126/100)	Normal (n = 353) Abnormal (n = 377)	68.4 68.2

3   Methodology

In this study, we firstly propose the use of diffusion maps for dimension reduction of the data, and then find the corresponding classification methods for disease diagonsis. Accuracy of the proposed methodology is evaluated through a cross-validation procedure for the training samples and later for the testing samples. For making a cross-validation study, we split a total of 630 original data samples to normal and abnormal group. Then we randomly divide each group into five folds and cross combine them into 25 folds of paired data. And then each fold of partitions contains one of the five folds according to individual normal and abnormal group respectively as validation sample. The divided data will be trained across the combinations and validated until each combination sample set has been trained. Our proposed method is applied to each of the 25 training sets and conducts Nystrom's out-of-sample expansion [12] on the new (validation) sample to project the new sample into the low-dimensional space obtained through the training set, and later the validation sample points are classified by the optimal classifier obtained in the training step accordingly. Finally, a new test sample set (n = 100) for prediction is added for comparision and illustration, where each subject in this test set will have a total of twenty-five votes.

3.1   Training Sample Reduction via Diffusion Maps

Our framework uses a graph model treating data points (samples) as nodes connected by edges with distances, defined by a weighting scheme with w_{ij} denoting distance of node i to node j, for all $i, j = 1, \cdots, n$. Given n data point set $\{X_i\}_{i=1}^{n}$, where $X_i \in \mathbb{R}^{N \times N}$, $i = 1, 2, ..., n$. The data point is embedded to a manifold surfaces $\mathcal{M}$ in high-dimensional space and later is mapped into a lower dimensional space through a diffusion maps as shown in Fig. 2. For illustration let an undirected graph $G(V, E, W)$, in which V and E are the set of vertices and edges, respectively, be defined that each node is connected by the weighted edges matrix $W = (w_{ij})$, where w_{ij} is the weight connecting the edge i and j.

In diffusion maps, the weights between node i and j is usually defined through a kernel function $\mathcal{K}(X_i, X_j)$ such as the Gaussian [13] kernel with $w_{ij} = \mathcal{K}(X_i, X_j)$, where $\mathcal{K}$ is a positive and symmetric kernel matrix using Euclidean distance $\|X_i - X_j\|^2$ for the X_i and X_j metric

$$\mathcal{K}_{ij} = \mathcal{K}(X_i, X_j) = \exp(-\frac{\|X_i - X_j\|^2}{\alpha}). \tag{1}$$

The α is a scale parameter which measures the X_i and X_j in manifold neighborhood distance, we can construct a matrix with normalized rows entry with unit length as

$$P = D^{-1}\mathcal{K} = (p(X_i, X_j)), \tag{2}$$

where the ith element of degree matrix D is computed by $d_{ii} = \sum_{j=1}^{n} \mathcal{K}_{ij}$ and denoted as $D = \mathbf{diag}(d_{11}, d_{22}, ..., d_{nn})$. The resulting matrix P is actually a Markov transition matrix with every entry to be nonnegative and has all row sums to be equal to 1. The position of the data point X_i and X_j is connected on intrinsic manifold $\mathcal{M}$ by the weights, $p(X_i, X_j) = \mathcal{K}_{ij}/d(x_i)$, which has been normalized by $d(x_i)$ for each row i.

If points X_i and X_j are similar, the distance between the two points is closer, and has a higher probability jumping from node i to nearby node j. The similarity of the two data points can be evaluated by the following distance measure

$$D^2(X_i, X_j) = \sum_{u \in X} \|p(X_i, u) - p(X_j, u)\|^2 = \sum_k \|P_{ik} - P_{kj}\|^2. \tag{3}$$

Now let Y_i be the $n \times 1$ distance vector with elements composed of distances from node i to all nodes $j = 1, \ldots, n$ (including itself), mapped from the original data point $X_i, i = 1, \ldots, n$, namely

$$Y_i := \begin{bmatrix} p(X_i, X_1) \\ p(X_i, X_2) \\ \vdots \\ p(X_i, X_n) \end{bmatrix} = P_{i\cdot}^T \tag{4}$$

Note that the Euclidean distance between Y_i and Y_j can be expressed as

$$\|Y_i - Y_j\|_E = \sum_k \|p(X_i, u) - p(X_j, u)\|^2 = \sum_k \|P_{ik} - P_{kj}\|^2 = D^2(X_i, X_j).$$

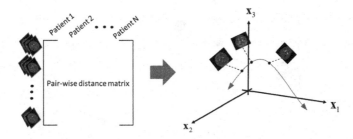

Fig. 2. Compute pairwise of SPECT features distance and then mapping to low-dimensional space for image embedding.

3.2   Nyström's Out-Of-Sample Extension

Manifold learning methods usually require recalculation of the kernel matrix with new added samples, and the above steps are repeated for the entire dataset. It becomes more difficult when new sample points are added sequentially. Nyström's out-of-sample extension [14] allows the original sample to be extended by adding new samples and embedding them into the existing low-dimensional space to form the manifold geometry of the new sample points.

The benefits of Nystrom's out-of-sample procedure extending the new samples are illustrated as follows: (1) The compliance of machine learning is divided into training set and validation set in standard verification steps; (2) When new samples are added, it is not necessary to repeatedly calculate diffusion distance of the whole Markov matrix which also solves the problem of computation time; (3) Maintain the training set geometric structure from the original samples. We can extend new sample point $X_{new} \in \mathbb{R}^{N \times N}$ from the validation set, and then recalculate the Euclidean distance and Kernel matrix, the new element of Kernel matrix has an augmented vector $\mathcal{K}_{new}^T = (\mathcal{K}_{new,1}, \ldots, \mathcal{K}_{new,n})$ with

$$\mathcal{K}_{new,j} = \mathcal{K}(X_{new}, X_j) = \exp(-\frac{\|X_{new} - X_j\|^2}{\alpha}). \qquad (5)$$

The augmented vector $P_{new}^T = D_{new}^{-1} \mathcal{K}_{new}^T$ is normalized vector of $\mathcal{K}_{new}^T$ by dividing its row sum $D_{new} = \sum_j^n \mathcal{K}_{new,j}$. So the augmented Markov transition matrix P_{n+1} can be rewritten as

$$P_{n+1} = \begin{pmatrix} P_n \\ P_{new}^T \end{pmatrix}, \quad \mathcal{K}_{n+1} = \begin{pmatrix} \mathcal{K}_n \\ \mathcal{K}_{new}^T \end{pmatrix}. \qquad (6)$$

Finally, project a new sample to the diffusion space of chosen k eigenvector above by

$$\psi_l(X_{new}) = \frac{1}{\lambda_l} \sum_{j=1}^n p(X_{new}, X_j)\psi_l(X_j), \, l = 1, ..., k. \qquad (7)$$

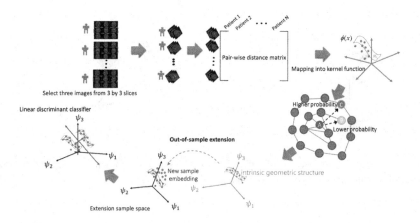

Fig. 3. The process of the high-dimensional reduction with out-of-sample projection for classifications.

Each point for each patient is projected onto the original trained low-dimensional diffusion space, and it goes on to use the selected classifier to carry on the classification as shown in Fig. 3. The algorithm for constructing the corresponding diffusion maps and out-of-sample extension is presented in Algorithm 1 below.

Algorithm 1. Diffusion Maps and out-of-sample extension algorithm.

Require:

  Data of images $\{X_i\}_{i=1}^n \in \mathbb{R}^{N \times N}$;

Ensure:

  Projected new coordinate vector $Y'_{new} \in \mathbb{R}^{N \times k}$;

 1: **Normalized data:** $\tilde{X}_i, i = 1, 2, .., n.$
   $\tilde{X}_i = (X_i - \bar{X})/S(X_i)$, which $\bar{X}_i = \frac{1}{N}\sum_{j,k} X_i(j,k)$
   $S(X_i) = \frac{1}{N^2-1}\sum_{j,k}(X_i(j,k) - \bar{X}_i(j,k))^2$;

 2: **Construct Kernel matrix:** $K = (K_{i,j}), K_{i,j} = K(\tilde{X}_i, \tilde{X}_j)$;

 3: **Build Markov matrix:** $P = D^{-1}K$,
   $D = \mathbf{diag}(d_{11}, d_{22}, ..., d_{nn})$, where $d_{ii} = \sum_{j=1}^n K_{ij}$;

 4: **Spectral decomposition** P **matrix with corresponding eigenpairs**
   $\{\lambda_j, \psi_j\}_{j=1}^n$ **as coordinate:**
   $Y_i'^T := (\lambda_1\psi_1(i), \lambda_2\psi_2(i)..., \lambda_k\psi_k(i))$

 5: **Out-of-sample extension**

 6: **Compute new extension vectors:**
   $P_{new}^T = d_{new}^{-1}K_{new}^T$, where $K_{new}^T = (K_{new,1}, ..., K_{new,n})$;

 7: **Repeat Step 1 to Step 4 and compute extented Markov matrix:**
   $P_{n+1}^T = (P_n, P_{new}^T)$, where $K_{n+1}^T = (K_n, K_{new}^T)$;

 8: **Project new samples on eigenvecotrs as coordinate:**
   $\psi_l(X_{new}) = \frac{1}{\lambda_l}\sum_{j=1}^k p(X_{new}, X_j)\psi_l(X_j), l = 1, ..., k$;

 9: **return** $\psi_l(X_{new}), l = 1, ..., k$;

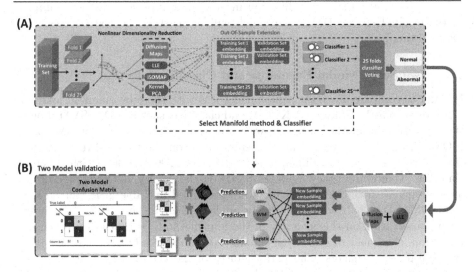

Fig. 4. The workflow of the two steps model diagnosis architecture with twenty-five folds cross-validation. (A) Procedure for nonlinear dimension reduction through manifold method and diagnosis with given classifier. (B) Procedure for new test samples embedding and diagnoses with DM and LLE as well as the corresponding two model confusion matrix.

4   Experiments

In this work, we implement the same data preprocessing procedure in our clinical data KCGMH-TW and PPMI dataset. Our approach mainly consists of two steps as demonstrated in Fig. 4:

- First step: several manifold learning methods are used for dimension reduction on the training sets for low-dimensional embedding, and then out-of-sample extension is applied to the new samples of the validation sets [16]. Next, classify the embedded samples and aggregate the ensemble classifiers with the twenty-five voting results to making prediction on the corresponding test set.
- Second step: according to the results of the first step, we use the DM and the classifier with the best performance to predict new test samples. Each patient is diagonalized through the twenty-five classification voting procedure and the prediction results with different manifold learning methods can be compared through the two model confusion matrices.

In the first step, three classifiers are compared, namely LDA, SVM and Logistic regression. The classifier with the best performance is chosen to enter the second step process. In this work, the DM parameter is set to be $\alpha = 8$ and time step $t = 1$ in Markov transition matrix $P^{(t)}$. In out-of-sample step, there is no need to recompute the entire diffusion matrix, simply project the new samples to the original chosen DM space which will be used as the feature space for diagonsis analysis.

4.1   Two Steps Model Ensemble and Classifer Selection

In order to compare diagnosis results using DM with those by manifold learning methods such as LLE, ISOMP, KPCA, and classifiers such as LDA, SVM, Logistic Regression in another scenario, we compare the performances of DM, LLE, ISOMAP, and KPCA diagonalized results based on the twenty-five folds classifications under four low-dimensionality (30, 100, 200 and 300) with different classifiers. Overall accuracy averages based on the twenty-five folds predictions after voting with the three classifiers are presented in Table 2. It can be seen that (DM, LDA) pair with 200 dimensionality seems to be the best. An interesting observation on the accuracy of the four methods, is that after dimension reduction, the linear discriminant classifier (LDA) works better than the other two non-linear classifiers. The performances of the DM and LLE are superior to the other two methods, therefore in the following we use DM and LLE with LDA for comparisons of the performances for testing the new 100 samples.

For each test patient, based on the two-step procedure above, there are twenty-five ensemble votes for prediction, then we can calculate the proportion of machine predicted abnormal probability: $p_k = \frac{V_k}{\sum_{k=1}^{} V_k}$, with V_k denoting the kth voting results, where $V_k = 0$ or 1 representing normal or abnormal respectively, $k = 1, \ldots, 25$. So we build a threshold to $(0,1)$ $e.g.,(\mathbb{1}_{\{p_k > 0.5\}})$ for testing the voting prediction.

Table 2. Performance comparisons of the four manifold learning methodology and three ensemble classifiers on KCGMH-TW

Method	LDA				SVM			Logistic		
	Dimension	Acc.	Sens.	Spec.	Acc.	Sens.	Spec.	Acc.	Sens.	Spec.
DM	30	0.88	0.93	0.82	0.86	0.93	0.79	0.85	0.90	0.80
	100	0.89	0.94	0.84	0.86	0.89	0.83	0.88	0.90	0.86
	200	**0.90**	**0.91**	**0.91**	0.85	0.86	0.84	0.87	0.88	0.86
	300	**0.91**	**0.91**	**0.91**	0.87	0.92	0.82	0.88	0.92	0.84
LLE	30	0.84	0.87	0.81	0.83	0.88	0.77	0.81	0.84	0.79
	100	0.89	0.96	0.81	0.83	0.88	0.77	0.81	0.84	0.79
	200	0.89	0.93	0.83	0.87	0.93	0.81	0.85	0.87	0.82
	300	0.89	0.93	0.83	0.87	0.81	0.93	0.85	0.87	0.82
ISOMAP	30	0.80	0.81	0.79	0.81	0.83	0.79	0.79	0.77	0.81
	100	0.86	0.89	0.84	0.85	0.86	0.84	0.85	0.84	0.85
	200	0.87	0.89	0.85	0.86	0.86	0.85	0.87	0.87	0.87
	300	0.87	0.89	0.85	0.86	0.86	0.85	0.87	0.87	0.87
KPCA	30	0.79	0.80	0.79	0.76	0.80	0.71	0.75	0.85	0.70
	100	0.83	0.83	0.82	0.77	0.82	0.71	0.77	0.83	0.70
	200	0.86	0.90	0.81	0.79	0.85	0.72	0.78	0.84	0.71
	300	0.87	0.93	0.81	0.80	0.89	0.73	0.78	0.84	0.71

4.2   Classification

We compare the performances of the two-step manifold approach with 2D-CNN, 3D-CNN on KCGMH-TW and PPMI datasets in twenty-five folds prediction. The best performing 200-dimensions for dimension reduction is chosen as a baseline, and then we compare the performances with different well-known CNN models such as AlexNet [17], VGG-16 [18], VGG-19 and Deep Convolution Network (DCNN). We implement the KCGMH-TW SPECT dataset with three middle images from the 9×9 slices in a high-order brain image tensor as input to the 3D-Convolutional Neural Network model [19]. The volumes of each SPECT sample therein have a size of $3 \times 128 \times 128 \times 3$ voxels. For the CNN approach, we set 193 images as the test set and 100 images in the manifold method test in Table 3.

We consider the early abnormal in our KCGMH-TW for training ($n = 630$) and testing ($n = 100$) sample set and implement our procedure in the PPMI dataset (testing, $n = 197$). The classification results are reported in Table 3. DM+LDA method generally achieves the best performance on two datasets with single and three types. Compare to our approach methods, 2D-CNN often have excellent prediction than 3D-DCNN such that VGG-16 in PPMI is 93% and DCNN in KCGMH-TW single image is 87% but also has higher variability of 0.13 and 0.22. The comparison of our method DM+LDA in PPMI single image and KCGMH-TW three images classification has average accuracy 98% and 90% performance with lower variability is 0.02 and 0.05.

Table 3. Binary classification in different datasets

Datasets	Single image	Model	Acc.	Prec.	Senc.	Spec.
PPMI	HC vs PD	DCNN	0.89 (±0.54)	0.92	0.85	0.90
		AlexNet	0.88 (±0.27)	0.89	0.78	0.92
		VGG-16	0.93 (±0.13)	0.94	0.95	0.92
		VGG-19	0.89 (±0.21)	0.91	0.93	0.86
		Diffusion Maps + LDA	**0.98** (±0.02)	0.98	0.96	0.97
KCGMH-TW	Normal vs Abnormal	DCNN	0.87 (±0.22)	0.88	0.89	0.84
		AlexNet	0.83 (±0.35)	0.86	0.84	0.80
		VGG 16	0.85 (±0.44)	0.84	0.89	0.76
		VGG 19	0.87 (±0.11)	0.88	0.92	0.81
		Diffusion Maps + LDA	0.86 (±0.04)	0.85	0.93	0.77
Datasets	**Three images**	**Model**	**Acc.**	**Prec.**	**Senc.**	**Spec.**
KCGMH-TW	Normal vs Abnormal	3D-DCNN	0.82 (±0.10)	0.83	0.84	0.79
		Diffusion Maps + LDA	**0.90** (±0.05)	0.88	0.95	0.84

From the overall results in Fig. 5, we draw prediction accuracies of the DM+LDA and CNN methods for each of the twenty-five folds respectively. In Fig. 5(a)(b), performances of using a single PPMI image for each subject to predict 193 testing cases as well as those of using a single KCGMH-TW images are displayed. It can be seen that the DM+LDA has robust performance in the 25 folds results than those of VGG-16, VGG-19, AlexNet, and DCNN. The PPMI dataset has high image quality compared to our clinical KCGMH-TW dataset. To reduce effects of the noise in our KCGMH-TW, we also consider use of three images for DM+LDA for comparisons. We find that DM+LDA procedure has higher accuracies as well as less variations with three images, see Fig. 5(c) and Fig. 5(d). It may be due to that three images have included the most obvious symmetrical strata in high order tensor space (Fig. 6).

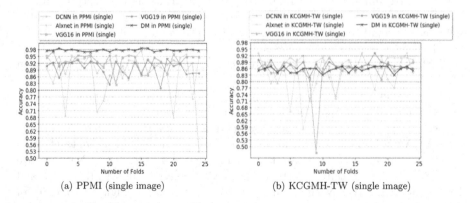

(a) PPMI (single image)   (b) KCGMH-TW (single image)

Fig. 5. The performace of DM and CNN comparison on PPMI and KCGMH-TW dataset.

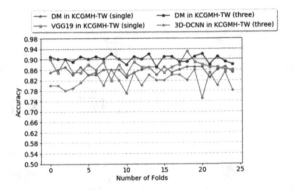

Fig. 6. Performance of DM in single image, three images and 3D-DCNN methods

5 Diagnosis and Discussion

In this section, we focus on machine prediction on the 100 test subjects and diagnosis of misclassification samples. Our diagnostic framework as exhibited in Fig. 4 (B), provides a general interpretable misclassification table for diagnosis in the final step with the confusion table. It is helpful to see whether an incorrect diagnosis is due to the subject's age or not.

5.1 Two Model Confusion Matrices

From Sect. 4.1, it is obvious that the procedure with DM and LDA classifier has higher accuracy in classification after the twenty-five folds voting. As LLE also has quite good performance with the twenty-five folds procedure, we examine the diagnosis results based on the two models and form a two-model confusion matrix. We firstly set the doctor label as true label (i.e., two classes) and the ensemble DM and LLE diagnoses in two classes as in Fig. 7(a), later check how many misclassifications are observed under the true label normal or abnormal. The diagonal elements in the two model confusion matrix table are the numbers of correct predictions (positive is with label **1**, negative is with **0**) for the DM and LLE ensembles respectively. The numbers of misclassified samples are shown in the non-diagonal sub-table. For example in the left table in Fig. 7(b) left column is true label of normal, DM and LLE predict ID:678 patient is early abnormal. In the right column Fig. 7(b), ID:677 is abnormal and has been predicted incorrectly as normal by LLE and correctly as abnormal by DM, while ID:678 is predicted incorrectly as normal by both models. In the left column of the table Fig. 7(c), three of the subjects with true label as normal are diagnosed correctly as normal (ID: 657, 674, 714) by DM and incorrectly as abnormal by LLE. While in the right column of the table Fig. 7(d), seven of the subjects (ID: 652, 669, 675, 693, 694, 709, 712) with true label as abnormal are diagnosed incorrectly by both model as normal.

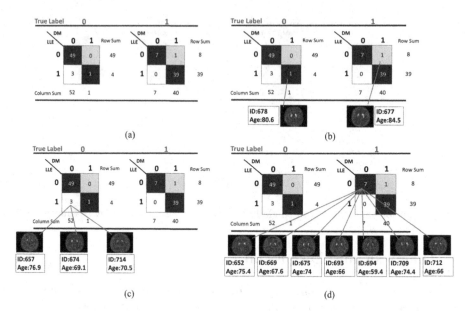

Fig. 7. Two-model confusion matrices with DM and LLE ensembles

Diagnosis Table: Sometimes we care more about false negative situation as the seven cases as indicated in Fig. 7(d). For these misclassifications, we provide more detailed information with the doctor's diagnoses and the probability of LDA matching as demonstrated in Fig. 8, in order to identify early abnormal in the true designation of three classes. We can also examine whether our model misclassifications is due to either aging or shrinking striatum abnormal. For example, in the case of the misclassification ID:709, whose age is 74.4, higher than the average age 68.3 of our dataset.

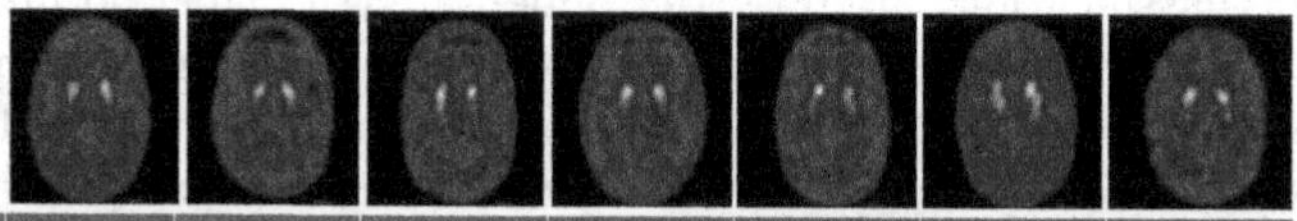

	ID:652	ID:669	ID:675	ID:693	ID:694	ID:709	ID:712
Age	75.4	67.6	74	66	59.4	74.4	66
Number of LDA predict Abnormal	2	0	0	0	0	11	0
Number of LDA predict Normal	23	25	25	25	25	14	25
Proportion of LDA predict Abnormal	0.08	0.00	0.00	0.00	0.00	0.44	0.00

Fig. 8. The diagnosis table for the seven false negative samples on their age and proportion of the twenty-fold voting results.

5.2 Visualization

Traditional analysis of the SPECT image is for the physician to visualize the symmetry of the left and right striatum. However, the early abnormal symptoms are not so obvious and hard to detect. To find mild changes of Parkinson's disease, we use DM for dimension reduction on 630 original SPECT images in low-dimensional space, and visualize the embedded test samples in the corresponding three-dimensional manifold using the first three eigenvectors ψ_1, ψ_2, ψ_3 with the largest three eigenvalues. They are shown to be like an U curve with abnormal samples appeared on the right side of the curve as shown in Fig. 9. The colors from dark purple to dark red indicate the distance between sample points of each patient. Moreover, we label the true diagnosis of each patient (two classes) to examine the spatial status. The data point (ID:342) with the largest distance from the origin on the right extreme is the most serious abnormal case. In contrast, the point (ID:250) on the left extreme is normal, and the early abnormal appears in the middle position, such as ID: 18.

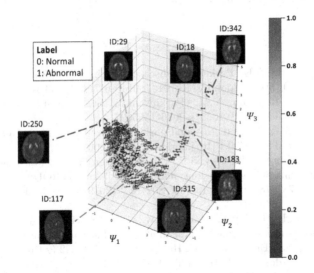

Fig. 9. Trajectory of 630 SPECT images embed in three-dimentional space and annotated the true label on each data points. The manifold is colored by SPECT similarity distance of DM and eigenvectors mapping.

6 Conclusion

In addition to maintaining excellent geometric structure in low dimensions, DM is computationally less expensive than deep learning methods and does not require too many parameter adjustments.

In Sect. 4.2, the average accuracy of DM +LDA classification for KCGMH-TW normal vs. abnormal (three images) is up to 90% and up to 98% for PPMI dataset (single). The overall performance on KCGMH-TW is quite accurate and robust with lower variation than other deep learning methods. For understanding precisions of diagnoses of the two manifold learning methods, in Sect. 5.1 we have constructed confusion matrix of the two most compatible models. We examined those samples which were misclassified by our methods in more details and found out that many were diagnosed as early abnormal by the doctors, and sometimes there were differences on the diagnoses among doctors. Through the confusion matrix we are able to provide some explainable reasons for the diagnosis discrepancies among the doctors and our methods. Finally, through DM, we can embed the sample images into lower dimensional space and visualize how the normal and abnormal sample images scatter around in the three-dimensional eigenspace corresponding to the largest three eigenvalues of the DM method. In future works, we will include other existing diagnosis variables with extracted features to cross-examine the diagnosis results and see if we may improve the diagnosis accuracy further. Moreover, we will investigate, if more than 3 slices of images for each subject are used, it would be helpful for detection of early abnormal cases with more information.

References

1. Kalia, L.V., Lang, A.E.: Parkinson's disease. The Lancet **9996**, 896–912, Elsevier Ltd, 386 (2015). https://doi.org/10.1016/S0140-6736(14)61393--3
2. Keener, A.M., Bordelon, Y.M.: Parkinsonism. Semin Neurol. **36**(4), 330–334 (2016). https://doi.org/10.1055/s-0036-1585097
3. Hayes, M.T.: Parkinson's disease and parkinsonism. Am J.. Med. **132**(7), 802–807 (2019). https://doi.org/10.1016/j.amjmed.2019.03.001
4. Kung, H.F., Kung, M.P., Wey, S.P., Lin, K.J., Yen, T.C.: Clinical acceptance of a molecular imaging agent: a long march with [99mTc]TRODAT. Nuclear Med. Biol. **34**(7), 787–789 (2007) 132(7), 802–807 (2019). https://doi.org/10.1016/j.nucmedbio.2007.03.010
5. Prashanth, R., Dutta Roy, S., Mandal, P.K., Ghosh, S.: High-accuracy classification of parkinson's disease through shape analysis and surface fitting in ^{123}I-Ioflupane SPECT imaging. IEEE J. Biomed. Health Inf. **21**, 794–802 (2016) 132(7), 802–807 (2019). https://doi.org/10.1109/jbhi.2016.2547901
6. Faaeq, A., Guruler, H., Peker, M.: Image classification using manifold learning based non-linear dimensionality reduction. In: 26th IEEE Signal Processing and Communications Applications Conference, pp. 1–4 (2018). 132(7), 802–807 (2019). https://doi.org/10.1109/SIU.2018.8404441
7. Coifman, R.R., Lafon, S., Lee, A.B., Maggioni, M., Warner, F., Zucker, S.: Geometric diffusions as a tool for harmonic analysis and structure definition of data: diffusion maps. Proc. Natl. Acad. Sci. **102**(21), 7426–7431 (2005). https://doi.org/10.1073/pnas.0500334102
8. Cummings, J.L., HencNormalliffe, C., Schaier, S., Simuni, T., Waxman, A., Kemp, P.: The role of dopaminergic imaging in patients with symptoms of dopaminergic system neurodegeneration. Brain **134**(11), 3146–66 (2011) https://doi.org/10.1093/brain/awr177

9. Faro, A., Giordano, D., Spampinato, C., Ullo, S., Di Stefano, A.: Basal ganglia activity measurement by automatic 3-D striatum segmentation in SPECT images. IEEE Trans. Instrumentation Measur. **60**(10), 3269–3280 (2011). https://doi.org/10.1109/TIM.2011.2159315

10. Quan, J., Xu, L., Xu, R., Tong, T., Su, J.: DaTscan SPECT image classification for Parkinson's disease. arXiv. 1–9 (2019) https://doi.org/arXiv:1909.04142v1

11. Djang, D,S.W. et al.: SNM practice guideline for dopamine transporter imaging with ^{123}I-ioflupane SPECT 1.0. J. Nuclear Med. **53**, 154–163 (2012) https://doi.org/10.2967/jnumed.111.100784

12. ScNormallar, A., Rokach, L., Amit, A.: Diffusion ensemble classifiers. In: IJCCI 2012 - Proceedings of the 4th International Joint Conference on Computational Intelligence, pp. 443–450 (2012). https://doi.org/10.5220/0004102804430450

13. De La Porte, J., Herbst, B.M., Hereman, W., Van Der Walt, S.J.: An introduction to diffusion maps. In: The 19th Symposium of the Pattern Recognition Association of South Africa (2008)

14. Bengio, Y., et al.: Out-of-sample extensions for LLE, Isomap, MDS, eigenmaps, and spectral clustering. In: Advances in Neural Information Processing Systems

15. Prashanth, R., Dutta Roy, S., Mandal, P.K., Ghosh, S.: Automatic classification and prediction models for early Parkinson's disease diagnosis from SPECT imaging. Expert Syst. Appl. **41**, 3333–3342 (2014). https://doi.org/10.1016/j.eswa.2013.11.031

16. Raeper, R., Lisowska, A., Rekik, I.: Joint correlational and discriminative ensemble classifier learning for dementia stratification using shallow brain multiplexes. In: Frangi, A.F., Schnabel, J.A., Davatzikos, C., Alberola-López, C., Fichtinger, G. (eds.) MICCAI 2018. LNCS, vol. 11070, pp. 599–607. Springer, Cham (2018). https://doi.org/10.1007/978-3-030-00928-1_68

17. Krizhevsky.: Imagenet classification with deep convolutional neural networks, Alex and Sutskever, Ilya and Hinton, Geoffrey E. Neural Inf. Process. Syst. **25** (2012). https://doi.org/10.1145/3065386

18. Simonyan, K., Zisserman, A.: Very deep convolutional networks for large-scale image recognition. In: ICLR, pp. 1–14 (2015). https://arxiv.org/abs/1409.1556

19. Esmaeilzadeh, S., Yang, Y., Adeli, E.: End-to-end parkinson disease diagnosis using brain MR-images by 3D-CNN, Arxiv (2018). https://arxiv.org/abs/1806.05233

20. Kim, D.H., Wit, H., Thurston, M.: Artificial intelligence in the diagnosis of Parkinson's disease from ioflupane-123 single-photon emission computed tomography dopamine transporter scans using transfer learning. Nucl. Med. Commun. **39**(10), 887–893 (2018). https://doi.org/10.1097/MNM.0000000000000890

BRAIN2DEPTH: Lightweight CNN Model for Classification of Cognitive States from EEG Recordings

Pankaj Pandey[1] and Krishna Prasad Miyapuram[2(✉)]

[1] Computer Science and Engineering, IIT Gandhinagar, Ahmedabad, India
pankaj.p@iitgn.ac.in
[2] Centre for Cognitive and Brain Sciences, IIT Gandhinagar, Ahmedabad, India
kprasad@iitgn.ac.in

Abstract. Several Convolutional Deep Learning models have been proposed to classify the cognitive states utilizing several neuro-imaging domains. These models have achieved significant results, but they are heavily designed with millions of parameters, which increases train and test time, making the model complex and less suitable for real-time analysis. This paper proposes a simple, lightweight CNN model to classify cognitive states from Electroencephalograph (EEG) recordings. We develop a novel pipeline to learn distinct cognitive representation consisting of two stages. The first stage is to generate the 2D spectral images from neural time series signals in a particular frequency band. Images are generated to preserve the relationship between the neighboring electrodes and the spectral property of the cognitive events. The second is to develop a time-efficient, computationally less loaded, and high-performing model. We design a network containing 4 blocks and major components include standard and depth-wise convolution for increasing the performance and followed by separable convolution to decrease the number of parameters which maintains the tradeoff between time and performance. We experiment on open access EEG meditation dataset comprising expert, nonexpert meditative, and control states. We compare performance with six commonly used machine learning classifiers and four state of the art deep learning models. We attain comparable performance utilizing less than 4% of the parameters of other models. This model can be employed in a real-time computation environment such as neurofeedback.

Keywords: EEG · CNN · Deep Learning · Meditation · NeuroFeedback · Neural signals

1 Introduction

Deep Learning (DL) has sparked a lot of interest in recent years among various research fields. The most developed algorithm, among several deep learning methods, is the Convolutional Neural network (CNN) [12,31]. CNNs have made a

© Springer Nature Switzerland AG 2021
B. W. Papież et al. (Eds.): MIUA 2021, LNCS 12722, pp. 394–407, 2021.
https://doi.org/10.1007/978-3-030-80432-9_30

revolutionary impact on computer vision, speech recognition, and medical imaging to solve challenging problems that were earlier difficult using traditional techniques. One of the complex problems was classification. In the short span of 4 years from 2012 to 2015, the ImageNet image-recognition challenge, which includes 1000 different classes in 1.2 million images, has consistently shown reduced error rates from 26% to below 4% using CNN as the major component [20]. Identification of brain activity using CNNs has established remarkable performance in several brain imaging datasets, including functional MRI, EEG, and MEG. For example, Payan and Montana classified Alzheimer's and healthy brains with 95% accuracy using 3D convolution layers on the ADNI public dataset containing 2265 MRI scans [24]. In the recent work, Dhananjay and colleagues implemented three layers of CNN architecture to predict the song from EEG brain responses with 84% accuracy, despite having several challenges such as having low SNR, complex naturalistic music, and human perceptual differences [29]. However, all these models are building complex and deep networks without considering the limitation of time and resources. Two main concerns observed with deep and wide architectures are having the millions of parameters that lead to high computational cost and memory requirement. Therefore, this opens up the opportunity to navigate the research to develop lightweight domain-specific architectures for resource and time constraint environments. One such scenario is a real-time analysis of EEG signals.

EEG brain recordings have a high temporal resolution as well as a wide variety of challenges such as low signal-to-noise ratio, noise can be of different shape, for example, artifacts from eye movement, head movement, and electrical line noise [19]. Hence, to extract significant features, it requires a sophisticated and efficient method that considers the spatial information in depth. In recent times, deep learning methods have been showing a significant improvement over traditional machine learning algorithms. And, DL models have been expanding in real-time computation also. Real-time analysis requires fast train as well as test time. The major property that should hold in lightweight architecture is to design the blocks, which reduces trainable parameters while maintaining state-of-the-art performance [30]. A recent paper [29] on the classification of EEG signals has introduced a significant performance by generating the time-frequency images from EEG signals, but the proposed model is made of several components, which loads the model with 5.8 million parameters. In this study, we have proposed a novel pipeline to generate 2D images from EEG signals and develop a model that produces comparable performance with state-of-the-art networks with minimum time for training and testing and suitable for EEG classification tasks.

2   Related Studies

2.1   Cognitive Relevance

Meditation is a mental practice that enhances several cognitive abilities such as enhancing attention, minimizing mind wandering, and developing sustained attention [7]. EEG is the most widely used technique in the neuroscientific study

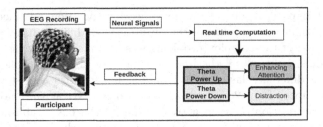

Fig. 1. NeuroFeedback protocol: EEG signals are passed to the neural computational toolbox, including preprocessing and analysis stages. Analysis pipeline comprises required computational algorithms, which generate feedback for the participant in real-time with minimum lag.

of meditation. EEG signals are decomposed into five frequency bands, including delta (1–4Hz), theta (5–8 Hz), alpha (8–12 Hz), beta (13–30 Hz), and gamma (30–70 Hz). Several meditation studies report the importance of theta and alpha waves for enhancing attention and are associated with several cognitive processes [5]. These findings led the researchers to design a neurofeedback protocol [6] to modulate oscillatory activities on the naive participants; a mechanism derives from the focused-attention meditation technique. Neurofeedback is a process to provide visual/audio feedback to a participant while recording his/her neural responses in real-time, which develops skills to self-regulate electrical activity of the brain, as shown in Fig. 1. Previous studies on neurofeedback have shown promising results to enhance performance, including athletes (archers improve their shooting performance), musicians, professional dancers, and non-artists to attain skills resembling visual artists [26]. NeuroFeedback requires a time-efficient and high-performing computational technique in terms to classify the different stages of brain responses.

2.2 Deep Learning Networks

There are four state-of-the-art architectures that have been the best way to understand the significance of performance and light-weight architecture.

Deep CNN Architectures: Deep CNNs (DCCN) have been introduced to generate local low-level and mid-level feature learning in the initial layers and high-level and global learning in the deeper layers. High-level features are the combination of low and mid-level features. VGG is one of those DCNNs that decreased the ImageNet error rate from 16% to 7% [28]. VGG expresses Visual Geometry Group. They employed small convolution filters of size 3 × 3 in place of the large filters comprising 11 × 11, 5 × 5 filters and observed the significant performance with varying depth of the network. The use of small filters also reduced the no of parameters, thus decreasing the number of computations. ResNet50 is the next powerful and successful network after VGG, which stands for residual networks having 50 layers [15]. ResNet was introduced to address two problems. The network has the ability to bypass a layer if the present layer could decrease performance and leads to overfitting, and this process is referred to as identity mapping. Another significance is to allow an alternate shortcut path for the gradient to flow that could avoid vanishing gradients problem [12]. ResNet decreased the error rate of 7%(VGG) to 3.6%.

Lightweight CNN Architectures: MobileNet V1 was an early attempt to introduce the lightweight model [16]. The remarkable performance of this model was achieved by substituting the standard convolution operation with depthwise separable convolution. This enhances the feature representation, which makes the learning more efficient. The two primary components of depthwise separable convolution are depthwise and pointwise convolution, which are introduced to adjust channels and reduce parameters [17]. The stacking of these two components generates novelty in the model. After this, another model proposed was Mobilenet V2 [25]. Mobilenet V2 is the extension of Mobilenet V1. Sandler and colleagues identified that non-linear mapping in lower-dimension increased information loss. To address this problem, a significant module was introduced with three consecutive operations. Initially, the dimension of feature maps is expanded using 1×1 convolution, followed by, a depthwise convolution of 3×3 to retain the abstract information. And in the last part, all the channels are condensed into a definite size using 1×1 pointwise convolution. These transformations are processed into a bottleneck residual block which is the core processing unit in place of standard convolution.

In our study, we use standard and depthwise convolution to enhance the performance and depthwise separable convolution to make the model time-efficient. The novelty of our work lies in the proposed pipeline to develop 2D plots from EEG signals having 3 RGB channels representing power spectral, which preserves the oscillatory information along with spatial position of electrodes. We classify three cognitive states comprising expert, non-expert meditative states, and control states (no prior experience of meditation). This paper discusses the following components a) Our proposed pipeline b) Experimentation on Dataset c) Comparative studies of ML and DL models d) Ablation Study.

3 Data and Methods

3.1 Data and Preprocessing

We used two open-access repositories consisting of Himalayan yoga meditators and controls [3,4]. In this research, we used EEG data of 24 meditators and 12 control subjects. Twenty-four meditators were further divided into two groups comprising twelve experts and twelve non-experts. Data were captured using 64 channels Biosemi EEG system at the Meditation Research Institute (MRI) in Rishikesh, India. Experimental design and complete description are mentioned in the paper [5]. The expert group had an experience of a minimum of 2 h of daily meditation for one year or longer, whereas non-experts were irregular in their practice. Control subjects had no prior meditation experience and were asked to pay attention to the breath's sensations, including inhalation and exhalation. A recent study [23] has shown the significant differences between expert and non-expert meditators and refers to as two distinct meditative states. Here, we refer to meditative and control states as cognitive states. As it includes cognitive components, such as involvement of attention during inhalation and exhalation of breathing, practitioners engaging in mantra meditation which enhances elements of sustained attention and language [2,5,21]. EEG data corresponding to breathing and meditation were extracted and preprocessed using Matlab, and EEGLAB software [11]. We classified three states emerging from expert, non-expert, and control groups, respectively.

EEG signals were downsampled 256 Hz. A high pass linear FIR filter 1 Hz was applied followed by removing the line noise artifacts at frequencies of 50,100,150,200, 250. Artifact correction and removal were performed using Artifact Subspace Reconstruction (ASR) method. Bad channels were removed and spherical interpolation was performed for reconstructing the removed signal, an essential step to retain the required signal. Data were re-referenced to average. Independent Components Analysis (ICA) was applied to classify the brain components and to remove the artifacts, including eye blink, muscle movement, signals generated from the heart, and other non-biological sources.

3.2   Methods

We divide the classification pipeline into two processing units. The first is to create power spectral density 2D plots from neural responses, known as "Neural Timeseries to 2D" and the second is to define the classification model term as "2D to Prediction" as shown in Fig. 2.

1. **Neural Timeseries to 2D:** We divide the process of creating images from EEG signals into the following three steps.
   (a) *Window Extraction:* EEG time-series signals are extracted into windows of 2, 4, and 6 s. For example, if a signal contains 24 s, we get 12, 6, and, 4 no of windows respectively. Varying window length identifies the information content and plays a significant role in discriminating the classes. In some applications, we have the luxury to extract varying windows sizes depending on the task.
   (b) *Power Spectral Analysis:* Power spectral density (PSD) is estimated for the extracted window using the Welch method [13]. Welch's method is also known as the periodogram method for computing power spectrum, the time signal is divided into successive blocks followed by creating a periodogram for each block, and estimating the density by averaging. Oscillatory cortical activity related to meditation primarily observes in two frequency bands, theta (5–8 Hz) and alpha (9–12 Hz) [5]. These bands are further subdivided into theta1 (5–6 Hz), theta2 (7–8 Hz), and alpha1 (9–10 Hz), alpha2 (11–12 Hz). For every channel, PSD is computed for the mentioned four bands.

(c) *Topographic (2D-3 Channel) Plot:* We use the topoplot function of the EEGLAB that transforms the 3D projection of electrodes in a 2-D circular view using interpolation on a fine cartesian grid [14]. Topographic plots were earlier implemented in Bashivan's work [1], and they combined plots from three bands to form one image of three channels, whereas we create one image of size 32 × 32 for each band having three RGB channels, and this might help to understand the significance of each band in the specific cognitive task. This plot preserves the relative distance between the neighboring electrodes and their underlying interaction, generating task-based latent representation using convolution.

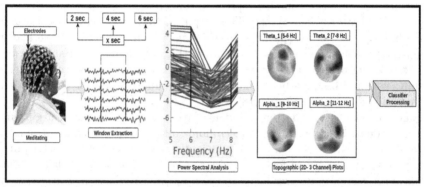

(a)Neural TimeSeries Signals to 2D: Images are formed using power spectral analysis.

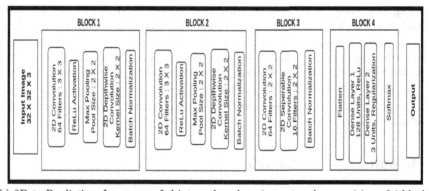

(b) 2D to Prediction: Images are fed into a deep learning network comprising of 4 blocks.

Fig. 2. Proposed pipeline for classification

2. **2D to Prediction:** Our model comprises 4 Blocks as shown in Fig. 2(b). The first two blocks are introduced to capture deep feature information and the third block for reducing the computation. The major components in these three blocks are regular Conv2D, depthwise spatial Conv2D, depthwise separable convolution, max pooling, ReLu(Rectified Linear Unit) activation function, and batch normalization.

Conv2D to learn richer features at every spatial scale. Depthwise convolution which acts on each input channel separately and extremely efficient to generate succinct representation [9]. Depthwise separable convolutions factorize a regular convolution into a depthwise convolution and a (1,1) convolution called a pointwise convolution [16]. This was initially introduced for generic object classification [27] and later used in Inception models [18] to reduce the computation in the first few layers. Kernel sizes for the initial two blocks are 3 × 3 and 2 × 2 for standard 2D convolution with 64 filters and 2 × 2 for Depthwise convolution. The third block comprises standard Conv2D and depthwise separable convolution of kernel size 2 × 2 with 64 and 12 filters respectively. ReLu activation is introduced in the initial two layers to prepare the network to learn complex representations and to generate faster convergence and better efficiency [22]. The output of ReLu is processed by the Max Pooling operations for spatial sub-sampling, which downsample the feature maps by generating features in patches of the feature map [9]. Batch normalization is performed at last in the three blocks to minimize internal covariate shift, which subsequently accelerates the training of deep neural nets and enables higher learning rates [18].

Fourth block contains flatten, two dense layers and a softmax activation function. A dense layer is employed to combine all the features learned from previous layers where every input weight is directly connected to the output of the next layer in a feed-forward fashion. Since we are doing multiclass classification, the output layer has a softmax activation function. Softmax as an activation function is used because the model requirement is to predict only the specific class, which results in high probability. To find out the loss of the model, categorical cross-entropy is used to predict the probability of a class. Deep learning models are trained using tensorflow(keras) [10] and machine learning algorithms employ scikit learn library [8]. GPU NVIDIA GTX 1050 (4 GB RAM) are used for this study and the batch size are set to 30 because of memory constraints and kept the maximum epoch to 30 to maintain the timing constraint as well as to avoid overfitting and for optimal training and validation loss.

4   Results and Discussion

This section discusses five measures a) performance of our model b) comparison of time and parameters c) training and validation loss d)visualization of layers e) ablation study

4.1   Performance

Baseline Methods: We compared the performance with commonly used classifiers and state of the art deep learning models. We trained six machine learning classifiers and tried several hyperparameters as shown in Table 1 and reported the maximum accuracy for all except NLSVM because it was around chance level. We experimented with four state of the art deep learning models VGG16 [28], ResNet50 [15], MobileNet [16], MobileNetv2 [25] and keep block4 of our model as the last block in all the models. These models are best suited for classification tasks. We reported the cross-subject average accuracy using leave one out validation, eleven subjects from each condition were used for training and one subject from each condition was used for testing and then reported the average accuracy of 12 iterations. For example, in the 2-s window, 9909 samples were used for training and 1101 for validation and 813 samples for testing, this was iterated for twelve times.

Table 1. Machine learning classifiers

Classifier	Parameters
Linear SVM (SVM)	Penalty parameter C values: 0.1,0.5,1.5,5,20,40,80,120,150
Non linear SVM (NLSVM)	Kernel: 'rbf', upper bound on the fraction of margin errors nu : 0.3,0.4,0.5, 0.6
AdaBoost (AB)	Algorithm = SAMME.R, no of estimators : 50, 100,200,250, 300,400
Logistic regression (LR)	Regularisation : l1, solver: saga
Linear discriminant analysis(LDA)	Threshold rank estimation = 0.0001
Random forest (RF)	No of estimators: 50,100,150,200,250,300,400, min samples leaf:5, criterion: entropy

We obtained maximum accuracy 64.3% in theta1, 60.8% in theta2, 56.5% in alpha1, and 56.1% in alpha2 for a 2-s window using Random Forest Classifier as shown in Fig. 3. Deep learning models completely outperformed traditional ML techniques, we obtained 97.27% in VGG16, 91.01% in ResNet50, 90.75% in MobileNet, 88.73% in MobilenetV2 and 94.57% in Brain2Depth(Ours) for theta1 in 2 s window as shown in Fig. 4. The minimum accuracy in traditional classifiers might be due to utilizing the images directly for training by flattening the images into a features matrix. There might be an improvement if we could have tried with power spectral features but didn't anticipate performance comparable with DL models. This cannot preserve the spatial position as the topo plot does. However, when compared our model with DL models showed comparable performance in all windows.

4.2   Parameters and Time

Parameters define the complexity of a model, we tried to keep our model simple and explainable so that each block can be understood easily. We compared the parameters and time for training and testing. Brain2Depth demonstrated comparable performance while using parameters which were only 0.52% of VGG16,

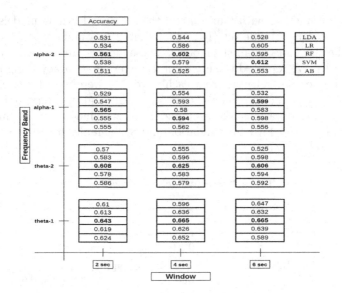

Fig. 3. Performance of ML Models: Classification of three cognitive states; expert, non-expert meditative states, and control. Each box indicates five accuracy values representing classifiers mentioned in the right top corner. Bold text represents the maximum value in the box.

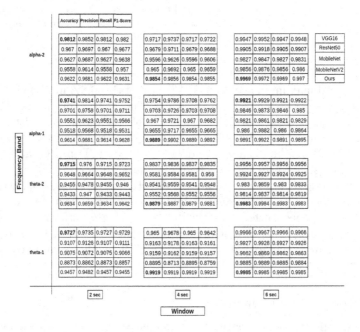

Fig. 4. Performance of CNNs: Each column of a box represents accuracy, precision, recall, and f1-score, and row indicates the CNN network mentioned on the right top corner.

0.32% of ResNet50, 2.28% of MobileNet, and 3.16% of MobileNetV2. The training and testing time of our model for 9909 samples and 813 samples were 84.882 and 0.152 s whereas the mobilenet took the minimum time in all four models, which were 214.989 and 0.44 s. Fast training may help to develop a model fast with millions of images and make a quick deployment. Minimum test time is most important in real-time prediction.

Table 2. The number of trainable parameters in each network with train and test time, respectively.

Model	Parameters	Training time(s)	Testing time(s)	Testing per sample(ms)
VGG16	14780739	652	0.473	0.582
ResNet50	23850371	847.483	1.119	1.376
MobileNet	3360451	214.989	0.44	0.541
MobileNetV2	2422339	258.503	0.702	0.863
Ours	**76627**	**84.882**	**0.152**	**0.187**

Table 3. Test performance of each iteration on theta1 band of 2-s window: Training performed on 33 subjects and testing on 3 subjects, including all conditions. Bold values represents two best performance.

Model	1	2	3	4	5	6	7	8	9	10	11	12
VGG16	**0.839**	**0.915**	**0.957**	**0.985**	**0.996**	1	**0.998**	**0.988**	**0.998**	**0.999**	**0.997**	1
RESNET50	0.701	0.617	0.834	0.934	0.975	0.989	0.975	0.939	.989	.998	.988	0.99
Mobilenet	0.619	0.689	0.786	0.935	0.974	0.996	0.981	**0.946**	0.986	0.993	**0.998**	0.988
MobilenetV2	0.626	0.674	0.7	0.864	0.935	0.975	0.985	0.931	0.985	0.996	0.994	0.982
Ours	**0.797**	**0.852**	**0.865**	**0.946**	**0.987**	**0.997**	**0.998**	0.915	1	**0.999**	0.996	**0.998**

4.3 Training vs Performance Tradeoff

ResNet, MobileNet, and MobileNetV2 performed a little low when compared with VGG and Brain2Depth for a 2-s window in theta1 . We further explored the subjectwise performance to understand the differences. We found that accuracy dropped significantly for one and two iterations as shown in the Table 3. In the next step, we investigated training and validation loss to verify whether this happened because of overfitting or underfitting. Figure 5 shows that training loss is high for three comparatively with others too. This shows that it may require more training rounds for this specific case however our model demonstrates that with moderate training it may perform well. And ResNet is a heavy architecture it may also require a large number of training samples. Hence with optimal time, our network can be trained efficiently even though the training sample can be noisy or varying samples.

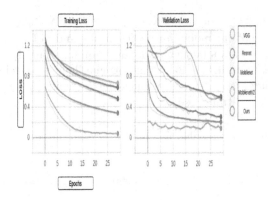

Fig. 5. Training and validation loss of the first iteration in theta1 band of 2-s window with respect to all networks.

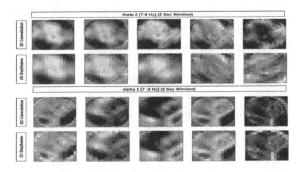

Fig. 6. Visualization of Block1 Layers: Intermediate representation of standard and depthwise convolution layers in theta2 (7–8 Hz) and alpha1 (9–10 Hz) bands for expert meditators.

4.4 Visualization

We have visualized the intermediate CNN outputs in block1 for two frequency bands i.e. theta2 and alpha1. Figure 6 shows the outputs of five filters for 2D convolution and depthwise convolution in the expert condition. Layers have learned the different features, more specifically in theta2, the frontal region has shown the heightened contribution as compared with alpha1. This has also been studied in meditation research on the role of the frontal midline region in the experienced meditators [5].

Table 4. Ablation Study: Performance evaluation on 2-s window with (A) several changes in the number of filters and kernel sizes from block1 to block3. (B) change the order of Relu, Batch Normalization, and Max pooling (C) modify layer with another layer.

A

BLOCK 1		Performance (2 sec window)				
2D Conv Filter | Kernel	2D Depthwise Kernel Size	Parameters	theta1	theta2	alpha1	alpha2
64 | (2,2)	(2,2)	≈ 75 K	0.946	0.956	0.958	0.959
64 | (2,2)	(3,3)	≈ 65 K	0.943	0.955	0.959	0.953
64 | (3,3)	(3,3)	≈ 77 K	0.937	0.951	0.954	0.956
32 | (3,3)	(2,2)	≈ 53 K	0.919	0.94	0.943	0.946
32 | (3,3)	(3,3)	≈ 57 K	0.919	0.945	0.932	0.946
BLOCK 2						
64 | (3,3)	(3,3)	≈ 77 K	0.938	0.955	0.958	0.954
64 | (2,2)	(2,2)	≈ 70 K	0.907	0.935	0.948	0.945
32 | (3,3)	(2,2)	≈ 49 K	0.898	0.921	0.921	0.933
BLOCK 3						
2D Conv Filter | Kernel	2D Seperable Filter | Kernel					
64 | (3,3)	16 | (2,2)	≈ 86 K	0.956	0.964	0.963	0.969
64 | (3,3)	32 | (2,2)	≈ 96 K	0.95	0.962	0.96	0.959
64 | (3,3)	16 | (3,3)	≈ 81 K	0.949	0.964	0.962	0.967

B

Order	Parameters	theta1	theta2	alpha1	alpha2
_,ReLu, BN, _,MP	≈ 76 K	0.938	0.957	0.954	0.961
,BN, ReLu, MP,	≈ 76 K	0.933	0.959	0.958	0.952

C

Block	Layer Modify	Parameters (Millions)	theta1	theta2	alpha1	alpha2
1	Replace Depthwise with Conv2D[64, (2,2)]	≈ 0.103	0.952	0.964	0.966	0.964
2	Replace Depthwise with Conv2D[64, (2,2)]	≈ 0.103	0.96	0.97	0.97	0.971
3	Replace Seperable with Depthwise(2,2)	≈ 0.111	0.954	0.965	0.966	0.973
3	Remove Seperable	≈ 0.151	0.955	0.972	0.967	0.971

4.5 Ablation and Performance Studies

We provide detailed ablation and performance studies. We exploited our model in the following three ways and reported the significant observations.

1. *Variation in Filters:* Different level of granularity in the learning representation depends on the size and number of filters. Small-size filters generate fine-grained information, whereas large-size filters represent coarse-grained information. We varied the number of filters and their sizes into three blocks as shown in Table 4A. We kept the two blocks intact and changed the left block and observed the performance. No. of parameters ranged from 49K to 91K, and observed performance changes with varying parameters. Even with 53K parameters in block1, we found a significant performance above 90% in all the bands. This shows that our presented model can be further customized according to the need.

2. *Swapping of components:* We swapped the positions of batch normalization, Relu, and max-pooling. We didn't observe any significant differences as shown in Table 4B.

3. *Change in layers:* In each block, we replaced one layer type with another layer as shown in Table 4C. We replaced depthwise convolution with standard convolution. We observed a slight improvement by $(1-2)\%$ in all the frequency bands however parameters got double and increased the training timing by 7 s. Hence, this suggests that the model can be efficiently fine-tuned according to the availability of the resources and other constraints such as time.

5   Conclusion

This study exhibits a pipeline for the EEG classification task, incorporating steps to create topo images from EEG signals and a lightweight CNN model. In several medical domains, heavy deep architectures are not required, and a simple model is needed to produce similar results with less time. Our proposed study shows state-of-the-art performance while using only 0.52% and 3.16% parameters of the VGG and MobileNetV2 network, leading to a significant reduction in train and test time. Our model can be efficiently deployed in several real time protocols and effectively suited for resource constraint environment.

Acknowledgement. We thank SERB and PlayPower Labs for supporting PMRF Fellowship. We thank FICCI to facilitate this PMRF Fellowship. We thank Pragati Gupta and Nashra Ahmad for their valuable feedback.

References

1. Bashivan, P., Rish, I., Yeasin, M., Codella, N.: Learning representations from eeg with deep recurrent-convolutional neural networks. arXiv preprint arXiv:1511.06448 (2015)
2. Basso, J.C., McHale, A., Ende, V., Oberlin, D.J., Suzuki, W.A.: Brief, daily meditation enhances attention, memory, mood, and emotional regulation in non-experienced meditators. Behav. Brain Res. **356**, 208–220 (2019)
3. Braboszcz, C., Cahn, B.R., Levy, J., Fernandez, M., Delorme, A.: Increased gamma brainwave amplitude compared to control in three different meditation traditions. PLoS ONE **12**(1), e0170647 (2017)
4. Brandmeyer: Bids-standard: bids-standard/bids-examples. https://github.com/bids-standard/bids-examples/tree/master/eeg_rishikesh
5. Brandmeyer, T., Delorme, A.: Reduced mind wandering in experienced meditators and associated eeg correlates. Exp. Brain Res. **236**(9), 2519–2528 (2018). https://doi.org/10.1007/s00221-016-4811-5
6. Brandmeyer, T., Delorme, A.: Closed-loop frontal midline θ neurofeedback: a novel approach for training focused-attention meditation. Front. Hum. Neurosci. **14**, 246 (2020)
7. Brandmeyer, T., Delorme, A., Wahbeh, H.: The neuroscience of meditation: classification, phenomenology, correlates, and mechanisms. In: Progress in Brain Research, vol. 244, pp. 1–29. Elsevier (2019)
8. Buitinck, L., et al.: API design for machine learning software: experiences from the scikit-learn project. In: ECML PKDD Workshop: Languages for Data Mining and Machine Learning, pp. 108–122 (2013)
9. Chollet, F.: Xception: Deep learning with depthwise separable convolutions. In: Proceedings of the IEEE Conference on Computer Vision and Pattern Recognition, pp. 1251–1258 (2017)
10. Chollet, F., et al.: Keras (2015). https://github.com/fchollet/keras
11. Delorme, A., Makeig, S.: Eeglab: an open source toolbox for analysis of single-trial EEG dynamics including independent component analysis. J. Neurosci. Methods **134**(1), 9–21 (2004)

12. Dhillon, A., Verma, G.K.: Convolutional neural network: a review of models, methodologies and applications to object detection. Prog. Artif. Intell. **9**(2), 85–112 (2020)
13. EEGLAB: Sccn: sccn/eeglab. https://github.com/sccn/eeglab/blob/develop/functions/sigprocfunc/spectopo.m
14. EEGLAB: Sccn: sccn/eeglab. https://sccn.ucsd.edu/~arno/eeglab/auto/topoplot.html
15. He, K., Zhang, X., Ren, S., Sun, J.: Deep residual learning for image recognition. In: Proceedings of the IEEE Conference on Computer Vision and Pattern Recognition, pp. 770–778 (2016)
16. Howard, A.G., et al.: Mobilenets: Efficient convolutional neural networks for mobile vision applications. arXiv preprint arXiv:1704.04861 (2017)
17. Hua, B.S., Tran, M.K., Yeung, S.K.: Pointwise convolutional neural networks. In: Proceedings of the IEEE Conference on Computer Vision and Pattern Recognition, pp. 984–993 (2018)
18. Ioffe, S., Szegedy, C.: Batch normalization: Accelerating deep network training by reducing internal covariate shift. arXiv preprint arXiv:1502.03167 (2015)
19. Jiang, X., Bian, G.B., Tian, Z.: Removal of artifacts from EEG signals: a review. Sensors **19**(5), 987 (2019)
20. Khan, A., Sohail, A., Zahoora, U., Qureshi, A.S.: A survey of the recent architectures of deep convolutional neural networks. Artif. Intell. Rev. **53**(8), 5455–5516 (2020)
21. Lee, D.J., Kulubya, E., Goldin, P., Goodarzi, A., Girgis, F.: Review of the neural oscillations underlying meditation. Front. Neurosci. **12**, 178 (2018)
22. Nair, V., Hinton, G.E.: Rectified linear units improve restricted boltzmann machines. In: ICML (2010)
23. Pandey, P., Miyapuram, K.P.: Classifying oscillatory signatures of expert vs non-expert meditators. In: 2020 International Joint Conference on Neural Networks (IJCNN), pp. 1–7. IEEE (2020)
24. Payan, A., Montana, G.: Predicting alzheimer's disease: a neuroimaging study with 3d convolutional neural networks. arXiv preprint arXiv:1502.02506 (2015)
25. Sandler, M., Howard, A., Zhu, M., Zhmoginov, A., Chen, L.C.: Mobilenetv 2: Inverted residuals and linear bottlenecks. In: Proceedings of the IEEE Conference on Computer Vision and Pattern Recognition, pp. 4510–4520 (2018)
26. Sho'ouri, N., Firoozabadi, M., Badie, K.: The effect of beta/alpha neurofeedback training on imitating brain activity patterns in visual artists. Biomed. Sig. Process. Control **56**, 101661 (2020)
27. Sifre, L., Mallat, S.: Rigid-motion scattering for image classification. Ph. D. thesis (2014)
28. Simonyan, K., Zisserman, A.: Very deep convolutional networks for large-scale image recognition. arXiv preprint arXiv:1409.1556 (2014)
29. Sonawane, D., Miyapuram, K.P., Rs, B., Lomas, D.J.: Guessthemusic: song identification from electroencephalography response. In: 8th ACM IKDD CODS and 26th COMAD, pp. 154–162 (2021)
30. Wu, J., Tang, T., Chen, M., Wang, Y., Wang, K.: A study on adaptation lightweight architecture based deep learning models for bearing fault diagnosis under varying working conditions. Expert Syst. Appl. **160**, 113710 (2020)
31. Zhang, Q., Zhang, M., Chen, T., Sun, Z., Ma, Y., Yu, B.: Recent advances in convolutional neural network acceleration. Neurocomputing **323**, 37–51 (2019)

D'OraCa: Deep Learning-Based Classification of Oral Lesions with Mouth Landmark Guidance for Early Detection of Oral Cancer

Jian Han Lim[1(✉)], Chun Shui Tan[1], Chee Seng Chan[1], Roshan Alex Welikala[3], Paolo Remagnino[3], Senthilmani Rajendran[2], Thomas George Kallarakkal[4], Rosnah Binti Zain[4,5], Ruwan Duminda Jayasinghe[6], Jyotsna Rimal[7], Alexander Ross Kerr[8], Rahmi Amtha[9], Karthikeya Patil[10], Wanninayake Mudiyanselage Tilakaratne[4,6], John Gibson[11], Sok Ching Cheong[2], and Sarah Ann Barman[3]

[1] Centre of Image and Signal Processing, Faculty of Computer Science and Information Technology, Universiti Malaya, 50603 Kuala Lumpur, Malaysia
[2] Head and Neck Cancer Research Team, Cancer Research Malaysia, 47500, Subang Jaya, Malaysia
[3] Digital Information Research Centre, Faculty of Science, Engineering and Computing, Kingston University, Surrey KT1 2EE, UK
[4] Department of Oral and Maxillofacial Clinical Sciences, Faculty of Dentistry, Universiti Malaya, 50603 Kuala Lumpur, Malaysia
[5] Faculty of Dentistry, MAHSA University, 42610 Bandar Saujana Putra, Jenjarom, Malaysia
[6] Centre for Research in Oral Cancer, Department of Oral Medicine and Periodontology, Faculty of Dental Sciences, University of Peradeniya, Peradeniya 20400, Sri Lanka
[7] Department of Oral Medicine and Radiology, BP Koirala Institute of Health Sciences, Dharan 56700, Nepal
[8] Oral and Maxillofacial Pathology, Radiology and Medicine, New York University, New York, NY 10010, USA
[9] Faculty of Dentistry, Trisakti University, Kota Jakarta Barat, Jakarta 11440, Indonesia
[10] Oral Medicine and Radiology, Jagadguru Sri Shivarathreeshwara University, Mysuru 570015, Karnataka, India
[11] Institute of Dentistry, University of Aberdeen, Aberdeen AB25 2ZR, UK

Abstract. Oral cancer is a major health issue among low- and middle-income countries due to the late diagnosis. Automated algorithms and tools have the potential to identify oral lesions for early detection of oral cancer. In this paper, we aim to develop a novel deep learning framework named D'OraCa to classify oral lesions using photographic images. We are the first to develop a mouth landmark detection model for the oral images and incorporate it into the oral lesion classification model as a guidance to improve the classification accuracy. We evaluate the performance of five different deep convolutional neural networks and MobileNetV2 was chosen as the feature extractor for our proposed mouth

© Springer Nature Switzerland AG 2021
B. W. Papież et al. (Eds.): MIUA 2021, LNCS 12722, pp. 408–422, 2021.
https://doi.org/10.1007/978-3-030-80432-9_31

landmark detection model. Quantitative and qualitative results demonstrate the effectiveness of the mouth landmark detection model in guiding the classification model to classify the oral lesions into four different referral decision classes. We train our proposed mouth landmark model on a combination of five datasets, containing 221,565 images. Then, we train and evaluate our proposed classification model with mouth landmark guidance using 2,455 oral images. The results are consistent with clinicians and the F_1 score of the classification model is improved to 61.68%.

Keywords: Deep learning · Classification · Oral lesions · Mouth landmark

1 Introduction

Oral cancer is one of the most common malignant tumor with high risk in low- and middle-income countries (LMICs). There were an estimated 354,864 new cases of cancers of the oral cavity, and 177,384 deaths in 2018 [7]. Smoking, alcohol use and chewing of betel quid are the major risk factors for oral cancer [1,26,29,33]. Many people are unaware that cancer could arise in the oral cavity because of poor awareness of cancer-related symptoms. The early detection of oral cancer is essential for better survival. Oral cancer is often preceded by lesions termed as oral potentially malignant disorders (OPMDs), which are easily visible for early detection without the need of special instruments. Based on the appearances of oral lesions, specialists can make decisions on next course of action according to their clinical experience [40]. However, due to the limited effort towards screening and early detection, most patients affected by oral cancer are diagnosed at advanced-stages [29].

Artificial Intelligence (AI) has been adopted in various industries to improve the efficiency as well as to reduce the cost. Recent advances in deep learning techniques have improved the performance of AI models in various domains that can achieve or even outperform human level performance in cognition related tasks [28]. Deep learning has also gained popularity and made remarkable progress in the medical field to perform clinical diagnosis such as classifying skin lesions [12,34], detecting pneumonia from chest X-rays [4,31] and enhancing visualization of pathologies [15,23,25]. The development of deep learning techniques has yielded impressive results in the medical field, but it is not meant to replace humans, rather to assist humans and improve the efficiency.

For the past few years, early detection of oral cancer using deep learning techniques has been a significant research area. Specifically, deep learning algorithms are trained to capture fine-grained features of oral lesions and identify the specific visual patterns of oral cancer. The previous works are mainly based on different types of images such as multidimensional hyperspectral images [19], computed tomography (CT) images [44], microscopic images [3,13,22], autofluorescence images [37,39] and photographic images [14,41,42]. In this work, we propose a novel deep learning framework to classify the oral lesions from photographic images into four different referral decision classes. Our proposed framework consists of a mouth landmark detection module and oral lesion classification module (Fig. 1). We design a new mouth landmark model to detect the location of the mouth and use it as an explicit feature to guide the classification model.

The contributions are twofold: i) To the best of our knowledge, we are the first to develop mouth landmark detection model that can detect the location of the mouth from the oral images. Existing facial landmark detection models do not work on oral images which do not consist of the entire human face. ii) We propose a novel oral lesion classification framework, namely D'OraCa with mouth landmark guidance for early detection of oral cancer. Experiments show that the performance of the classification model improves significantly with the mouth landmark guidance (Tables 4 and 5).

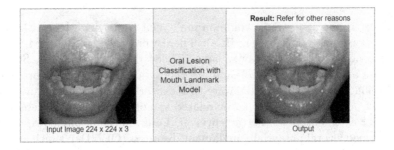

Fig. 1. Proposed oral lesion classification model with mouth landmark guidance that takes an oral images as input, detects the location of the mouth, and outputs the referral decision. On this example, proposed model correctly detects the mouth landmark and classifies the oral lesion as 'Refer for other reasons'.

2 Related Work

This section reviews the most relevant works related to the current research on oral lesion classification models and mouth landmark detection models.

2.1 Mouth Landmark Detection

There are no existing works on mouth landmark detection for oral images. However, there are a few studies on mouth features detection for front views of closed or slightly open mouth images. In [5], the authors focused on finding out the mouth candidates by segmenting the image based on skin-color. The input image must be a human face taken from the front view. It is not applicable for mouth images. Pantic et al. [30] proposed a mouth detection method with rule-based reasoning to extract the four mouth feature points based on template

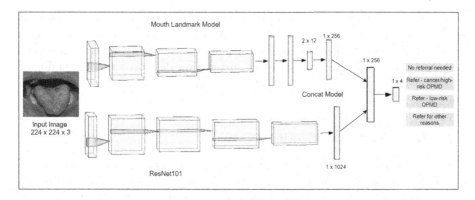

Fig. 2. Overview of the proposed architecture. Our network consists of two components. (Top) The first is the mouth landmark detection module, which detects the location of the mouth through a deep CNN and generates the mouth landmark feature. (Bottom) The second is the classification module. This module feeds oral images into ResNet-101 to obtain the fixed-size feature vector and fuses the two features to generate the final classification results.

matching. The proposed method depends on illumination conditions, assumes that the mouth-color pixel is red and segments based on it. The author mentioned that the proposed method can deal only with limited out-of-plane head rotations and expressionless mouth appearance. However, the oral images are usually taken at a different angle and the mouth is opened slightly larger to capture the oral cavity as shown in Fig. 4.

We also refer to facial landmark detection model as our related work due to the similar research area. The goal of facial landmark detection is to detect key points in human faces such as the eye corner, eyebrows, nose, chin and mouth. It is quite similar to our mouth landmark detection where the input image is only mouth area instead of the entire human face. Before the advent of deep learning, conventional facial landmark detection were mainly based on the template fitting method [2,45,48] and the cascaded regression-based method [8,20,38,43]. The template fitting method builds the face shape templates to fit the input images and estimates the landmark locations. While, the cascaded regression-based method estimates the landmark locations using image features with an initial guess and refines them using a cascade of machine learning models.

With the fast development of deep learning techniques in computer vision, deep learning based methods [9,11,21,27,46,47] have significantly boosted and outperformed both the template fitting method and cascaded regression-based method, creating a new state-of-the-art in facial landmark detection task. Most of them leverage deep convolutional neural networks (CNN) to learn facial features and predict the facial landmark in an end-to-end fashion. For instance, Yu et al. [46] proposed a deep deformation network and Lv et al. [27] presented a deep regression architecture with two-stage re-initialization for facial landmark detection. In [11], a style-aggregated network has been proposed to deal with the large intrinsic variance of image styles for facial landmark detection. Chandran et al. [9] proposed an attention-driven architecture for facial landmark detection on very high resolution facial images without downsampling. Motivated by the development of facial landmark detection task, we develop a mouth landmark detection model based on deep CNN to detect the mouth key points in oral images.

2.2   Oral Lesion Classification

The previous works in oral lesion classification can be categorized based on the types of input images. We found that previous research was mainly limited to highly standardized images such as multidimensional hyperspectral images, CT images, microscopic images and autofluorescence images. Jeyaraj et al. [19] proposed a partitioned CNN algorithm to classify multidimensional hyperspectral images of the oral cavity into normal, benign or cancerous region. Xu et al. [44] developed a three-dimensional CNN algorithm for the early diagnosis of oral cancer. The proposed algorithm performed binary classification on CT images of oral cavity to profile oral tumors as benign or malignant. In [22], the authors showed that the fuzzy classifier were able to classify normal and oral cancer stages using the combination of texture based features from the histopathological images. Similar work has been done in [13] by using CNN to identify seven tissue classes from the histopathological images. Aubreville et al. [3] proposed a novel CNN-based approach for oral squamous cell carcinoma (OSCC) diagnosis on confocal laser endomicroscopy (CLE) images. Song et al. [37] and Uthoff et al. [39] presented a CNN binary classification method for oral cancer based on autofluorescence and white light images.

There are a few existing works involving the use of photographic images which is the most relevant to our work. The oral images can be captured directly using a smartphone and did not require specialized instruments. Fu et al. [14] developed a cascaded CNN model to perform binary classification on early detection of OSCC from photographic images. While, Welikala et al. [41] focused on detection and classification of oral lesions from photographic images using the Faster R-CNN [32] and ResNet-101 [16] network. Three separate models were built to explore different binary and multi-class image classification tasks. The authors further extended the work in [42] to compare the performance of five common CNN architectures on the binary classification of 'referral' vs. 'non-referral'. Transfer learning was applied on the CNN architectures pretrained on the ImageNet dataset [10] and fine-tuning to the smaller oral image dataset.

3   Methodology

In this section, we present our novel architecture for classification of oral lesions with mouth landmark guidance as shown in Fig. 2. In our proposed architecture, we integrate the mouth landmark detection model with the deep learning-based image classification model to classify oral lesions for the early detection of oral cancer. Firstly, the mouth landmark detection model is employed to detect the location of mouth in an image. This explicit information is to tell where should the classification model focus on to look for the cancerous signs. The mouth landmark detection model will be further discussed in Sect. 3.1, then followed by the deep learning-based image classification model to predict the referral decision classes in Sect. 3.2. We compare the performance of the classification model with or without mouth landmark guidance. The objective is to prove and experiment on the hypothesis of the mouth landmark features might help the image classification model to focus on the mouth area in the image to increase the accuracy of the model.

3.1 Mouth Landmark Detection Module

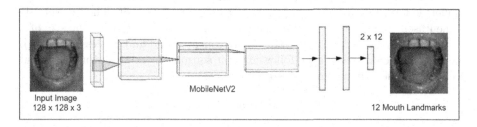

Fig. 3. Architecture of the proposed Mouth Landmark Detection Model: The oral image is fed into MobileNetV2, followed by two fully-connected layers to output 12 landmarks (green dots), indicating the location of mouth. Best viewed in color. (Color figure online)

The proposed mouth landmark detection module leverages the benefit of deep CNN to extract the image features and predict the mouth landmarks. This module is illustrated in Fig. 3. Technically, the input oral image I is fed into the deep CNN to extract the features. The features are then encoded by two fully-connected layers and a softmax layer to output the N number of mouth landmarks. The formula can be represented as:

$$p\,(M\,|\,I) = softmax(F_1(f_{CNN}(I)))\tag{1}$$

where f_{CNN} represents the deep CNN encoder, F_1 denotes the two fully-connected layers, I is the input oral image and $M = \{m_i\}_{i=1}^{Z}$ with $m_i \in \mathbb{R}$: $0 \leq m_i \leq 1$ is the output mouth landmark key points. The mouth landmark detection model is trained to minimize the mean squared error (MSE) loss L_{mse} as:

$$L_{mse}\,(Y_m, M) = \frac{1}{Z}\sum_{i}^{Z}(y_{m,i} - m_i)^2\tag{2}$$

where Y_m represents the ground-truth landmark and $Z = 2 \times N$ is the number of mouth landmark key points, each landmark consists of two points to represent x-coordinate and y-coordinate.

3.2   Oral Lesion Classification Module

The classification module feeds the oral images into ResNet-101 to obtain the fixed-size feature vector. This feature vector is fused with the mouth landmark key points and classifies the image into the four different referral decision classes. The referral decision classes are "No referral needed", "Refer - cancer/high-risk OPMD", "Refer - low-risk OPMD" and "Refer for other reasons". The architecture of our network is summarized in Fig. 2. For the classification module, we have chosen ResNet-101 as the feature extraction network due to its superiority in the image classification tasks. ResNet-101 is a CNN with a much deeper layers, which consists of 101 layers with residual blocks that having shortcut connections to solve the vanishing gradient problem in training.

In order to guide the classification model with the proposed mouth landmark detection module, we encode the mouth landmark key points M using a fully-connected layer into a feature vector f_m with the size of 1×256. The ResNet-101 is then used to encode the oral image into a feature vector f_o with the size of 1×1024. Both feature vectors f_m and f_o are concatenated and processed through the last fully-connected layer followed by a softmax layer to output the final prediction. The formula can be represented as:

$$p(R \mid I) = softmax(F_2(f_m \oplus f_o)) \tag{3}$$

where $\oplus$ represents concatenation, F_2 denotes the last fully-connected layer and R is the predicted referral decision. The classification model is trained to minimize the cross-entropy loss L_{ce} as:

$$L_{ce}(Y_r, R) = -\sum_{i}^{C} y_{r,i} log(r_i) \tag{4}$$

where Y_r represents the ground-truth referral decision and C is the number of referral decision classes.

4   Experiments

4.1   Dataset and Metrics

There is no publicly available mouth landmark dataset for us to train our proposed mouth landmark detection model. Therefore, we make use of the existing facial landmark datasets, augmented the data to form our mouth landmark dataset for training and evaluation. We combine the face images from HELEN [24], 300W [35], AFW [48], IBUG [36], LFPW [6] and 300-VW [36] datasets to form a total of 221,565 face images. Each face images consists of 68 landmarks to indicate the location of eye corner, eyebrows, nose, chin and mouth. We preprocess the face images to extract only the mouth region with 20 mouth landmarks. We separate 180,000 images for training set, 20,000 images for testing set and 21,565 images for validation set.

To train and evaluate our proposed oral lesion classification model, we built a well-annotated oral image dataset which consists of 2,455 images collected from clinical experts from across the world. Each image was annotated by 1 to 7 expert clinicians to produce the referral decision, type of lesions, bounding box of the lesions, site, outline, etc. Each image was also linked to its metadata such as gender, age, smoking, alcohol use and chewing of betel quid. The annotations from multiple expert clinicians were processed with a novel strategy proposed by [41] to form a single set of annotations for the classification task. In this work, we only used the referral decision label as our classification objective. The dataset was split into 1,963 images for training set, 248 images for testing set and 244 images for validation set. The number of images for each referral decision class was shown in Table 1.

Table 1. Number of images according to the referral decision class

Referral decision	Training	Validation	Testing	Total
No referral needed	394	49	50	493
Refer - cancer/high-risk OPMD	509	63	64	636
Refer - low-risk OPMD	548	68	69	685
Refer for other reasons	512	64	65	641
Total	1963	244	248	2455

4.2 Mouth Landmark Detection Result

Table 2. Comparison between different deep CNNs on the mouth landmark testing set. The bold numbers represent the best result.

Methods	Mean square error (MSE)	
	(20 landmarks)	(12 landmarks)
Custom network	0.04567	0.04716
MobileNetV2	0.04454	**0.04239**
MobileNetV3	0.04658	0.05294
ResNet-50	0.04415	0.04948
ResNet-101	0.04543	0.04948

We evaluate the performance between different deep CNNs as the feature extractor for our proposed mouth landmark detection model. We compare the performance of MobileNetV2 [18], MobileNetV3 [17], ResNet-50, ResNet-101 [16] and a custom network. The custom network is built using 5 convolutional layers, 2 fully-connected layers and we apply max pooling after convolutional layers. These models were evaluated using MSE in 12 and 20 landmarks to measure the average squared difference between the estimated landmark values and the actual landmark value.

As shown in Table 2, ResNet-50 and MobileNetV2 achieved the lowest MSE in 20 landmarks and 12 landmarks detection task respectively. MobileNetV2 (12 landmarks) was chosen as the deep CNN for our proposed mouth landmark model and integrated into the classification model. This was due to the lowest MSE achieved by MobileNetV2 and its lightweight model compared to the other methods. As the proposed mouth landmark detection model will be built into a mobile app in the future, a smaller size and faster inference time are required. As shown in Table 3, MobileNetV2 has the lowest number of parameters, smallest model size and fastest inference time.

Table 3. Comparison between different deep CNNs on the model size, number of parameters and inference time.

	Custom network	MobileNetV2	ResNet-50
No of parameters	7 million	2 million	23 million
Model size	30 MB	9 MB	90 MB
Inference time/Image	0.009 s	0.007 s	0.1 s

The qualitative results are shown in Fig. 4. Our proposed mouth landmark detection model can generate the correct mouth landmark in different angles of the mouth for oral images. For example, the top right image in Fig. 4 is showing the mouth captured from the right angle and the proposed model still can detect the correct mouth landmark. However, there are also some failure cases produced by our proposed model as shown in Fig. 5.

4.3   Oral Lesion Classification Result

Due to the lower number of oral images for the classification task, we implemented data augmentation on the dataset to generate more training samples through image pre-processing such as horizontal flip, horizontal shift and zoom. Note that data augmentation was not carried out on the validation and testing set. We used ResNet-101 pretrained on the ImageNet dataset as our deep learning model and performed transfer learning to our dataset, which as a result significantly reduced the training time and avoided overfitting the model.

To show the efficacy of our proposed mouth landmark guidance in the oral lesion classification model, we presented the quantitative result of the classification model with/without mouth landmark guidance on the test set in Table 4 and 5. Table 4 shows the result of the oral lesion classification model without mouth

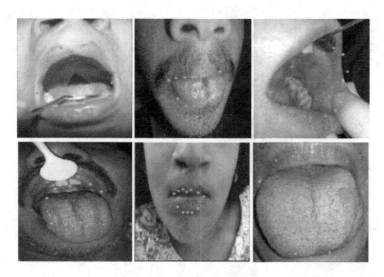

Fig. 4. Qualitative results of the proposed mouth landmark detection model on a few images. It is noticed that the model is able to generate the correct mouth landmarks (green dots). Best view in color. (Color figure online)

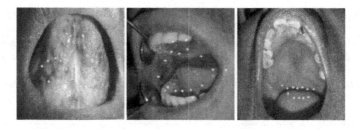

Fig. 5. Incorrect mouth landmarks (green dots) generated by the proposed mouth landmark detection model. Best view in color (Color figure online)

Table 4. Oral lesion classification result without mouth landmark guidance, where TP, FP, TN and FN are true positive, false positive, true negative and false negative, respectively.

Class	TP	FP	TN	FN	Precision	Recall	F_1 score
No referral needed	20	13	185	30	60.61%	40.00%	48.19%
Refer - cancer/high-risk OPMD	49	37	147	15	56.98%	76.56%	65.33%
Refer - low-risk OPMD	34	27	152	35	55.74%	49.28%	52.31%
Refer for other reasons	40	28	155	25	58.82%	61.54%	60.15%
Macro-average					58.04%	56.84%	56.50%

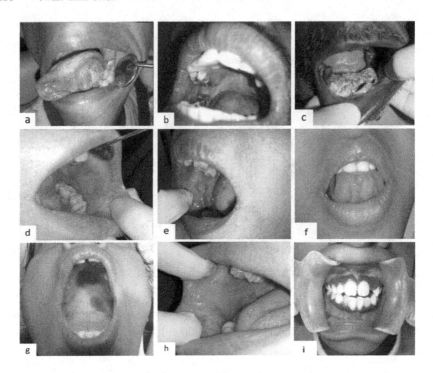

Fig. 6. Result of oral lesion classification model with mouth landmark guidance. (a), (b) and (c) are correctly classified as 'Refer - cancer/high-risk OPMD'. (d) and (e) are correctly classified as 'Refer - low-risk OPMD'. (f) and (g) are correctly classified as 'Refer for other reasons'. (h) and (i) are correctly classified as 'No referral needed'.

Table 5. Oral lesion classification result with mouth landmark guidance.

Class	TP	FP	TN	FN	Precision	Recall	F_1 score
No referral needed	24	18	180	26	57.14%	48.00%	52.17%
Refer - cancer/high-risk OPMD	46	25	159	18	64.78%	71.88%	68.14%
Refer - low-risk OPMD	43	25	154	26	63.24%	62.32%	62.77%
Refer for other reasons	42	25	157	23	62.69%	64.62%	63.64%
Macro-average					61.96%	61.70%	61.68%

landmark guidance. The model can achieve a precision of 58.04%, a recall of 56.84% and a F_1 score of 56.50%. Table 5 shows the result of the oral lesion classification model with mouth landmark guidance. The model can achieve a precision of 61.96%, a recall of 61.70% and a F_1 score of 61.68%. With mouth landmark guidance, the F_1 score of each referral decision classes were improved significantly, especially the F_1 score of "Refer - low-risk OPMD" class increased from 52.31 to 62.77 with a 20% improvement. The qualitative results from the oral lesion classification model with mouth landmark guidance are provided in Fig. 6. The results are consistent with clinicians.

5   Conclusion

We presented a novel deep learning framework to classify the oral lesions from photographic images into four different referral decision classes. We also developed a mouth landmark detection model that can detect the location of the mouth from the oral images. We showed that the oral classification accuracy improved significantly with the guidance of the mouth landmark detection model. The model was trained and validated on a well-annotated oral image dataset containing 2,455 images. In conclusion, the initial results show the effectiveness of deep learning in early detection of oral cancer and we believe our proposed method can greatly contribute to the medical field. In future, we plan to improve the model by building a larger dataset with well-annotated labels and make use of the risk factors information to train the model.

Acknowledgements. This work was supported by the Medical Research Council under grant MR/S013865/1.

References

1. Amarasinghe, H., Johnson, N., Lalloo, R., Kumaraarachchi, M., Warnakulasuriya, S.: Derivation and validation of a risk-factor model for detection of oral potentially malignant disorders in populations with high prevalence. Br. J. Cancer **103**(3), 303–309 (2010)
2. Asthana, A., Zafeiriou, S., Cheng, S., Pantic, M.: Robust discriminative response map fitting with constrained local models. In: Proceedings of the IEEE Conference on Computer Vision and Pattern Recognition, pp. 3444–3451 (2013)
3. Aubreville, M., et al.: Automatic classification of cancerous tissue in laserendomicroscopy images of the oral cavity using deep learning. Sci. Rep. **7**(1), 1–10 (2017)
4. Ayan, E., Ünver, H.M.: Diagnosis of pneumonia from chest x-ray images using deep learning. In: 2019 Scientific Meeting on Electrical-Electronics & Biomedical Engineering and Computer Science (EBBT), pp. 1–5. IEEE (2019)
5. Bao, P.T., Nguyen, H., Nhan, D.: A new approach to mouth detection using neural network. In: 2009 IITA International Conference on Control, Automation and Systems Engineering (case 2009), pp. 616–619. IEEE (2009)
6. Belhumeur, P.N., Jacobs, D.W., Kriegman, D.J., Kumar, N.: Localizing parts of faces using a consensus of exemplars. IEEE Trans. Pattern Anal. Mach. Intell. **35**(12), 2930–2940 (2013)
7. Bray, F., Ferlay, J., Soerjomataram, I., Siegel, R.L., Torre, L.A., Jemal, A.: Global cancer statistics 2018: globocan estimates of incidence and mortality worldwide for 36 cancers in 185 countries. CA: Cancer J. Clin. **68**(6), 394–424 (2018)
8. Cao, X., Wei, Y., Wen, F., Sun, J.: Face alignment by explicit shape regression. Int. J. Comput. Vis. **107**(2), 177–190 (2014)
9. Chandran, P., Bradley, D., Gross, M., Beeler, T.: Attention-driven cropping for very high resolution facial landmark detection. In: Proceedings of the IEEE/CVF Conference on Computer Vision and Pattern Recognition, pp. 5861–5870 (2020)

10. Deng, J., Dong, W., Socher, R., Li, L.J., Li, K., Fei-Fei, L.: Imagenet: a large-scale hierarchical image database. In: 2009 IEEE Conference on Computer Vision and Pattern Recognition, pp. 248–255. IEEE (2009)
11. Dong, X., Yan, Y., Ouyang, W., Yang, Y.: Style aggregated network for facial landmark detection. In: Proceedings of the IEEE Conference on Computer Vision and Pattern Recognition, pp. 379–388 (2018)
12. Esteva, A., et al.: Dermatologist-level classification of skin cancer with deep neural networks. Nature **542**(7639), 115–118 (2017)
13. Folmsbee, J., Liu, X., Brandwein-Weber, M., Doyle, S.: Active deep learning: Improved training efficiency of convolutional neural networks for tissue classification in oral cavity cancer. In: 2018 IEEE 15th International Symposium on Biomedical Imaging (ISBI 2018), pp. 770–773. IEEE (2018)
14. Fu, Q., et al.: A deep learning algorithm for detection of oral cavity squamous cell carcinoma from photographic images: a retrospective study. EClinicalMedicine **27**, 100558 (2020)
15. Gulshan, V., et al.: Development and validation of a deep learning algorithm for detection of diabetic retinopathy in retinal fundus photographs. Jama **316**(22), 2402–2410 (2016)
16. He, K., Zhang, X., Ren, S., Sun, J.: Deep residual learning for image recognition. In: Proceedings of the IEEE Conference on Computer Vision and Pattern Recognition, pp. 770–778 (2016)
17. Howard, A., et al.: Searching for mobilenetv3. In: Proceedings of the IEEE/CVF International Conference on Computer Vision, pp. 1314–1324 (2019)
18. Howard, A.G., et al.: Mobilenets: Efficient convolutional neural networks for mobile vision applications. arXiv preprint arXiv:1704.04861 (2017)
19. Jeyaraj, P.R., Nadar, E.R.S.: Computer-assisted medical image classification for early diagnosis of oral cancer employing deep learning algorithm. J. Cancer Res. Clin. Oncol. **145**(4), 829–837 (2019)
20. Kazemi, V., Sullivan, J.: One millisecond face alignment with an ensemble of regression trees. In: Proceedings of the IEEE Conference on Computer Vision and Pattern Recognition, pp. 1867–1874 (2014)
21. Kowalski, M., Naruniec, J., Trzcinski, T.: Deep alignment network: a convolutional neural network for robust face alignment. In: Proceedings of the IEEE Conference on Computer Vision and Pattern Recognition Workshops, pp. 88–97 (2017)
22. Krishna, M.M.R., et al.: Automated oral cancer identification using histopathological images: a hybrid feature extraction paradigm. Micron **43**(2–3), 352–364 (2012)
23. Laukamp, K.R., et al.: Fully automated detection and segmentation of meningiomas using deep learning on routine multiparametric MRI. Eur. Radiol. **29**(1), 124–132 (2019)
24. Le, V., Brandt, J., Lin, Z., Bourdev, L., Huang, T.S.: Interactive facial feature localization. In: Fitzgibbon, A., Lazebnik, S., Perona, P., Sato, Y., Schmid, C. (eds.) ECCV 2012. LNCS, vol. 7574, pp. 679–692. Springer, Heidelberg (2012). https://doi.org/10.1007/978-3-642-33712-3_49
25. Li, R., et al.: Deep learning based imaging data completion for improved brain disease diagnosis. In: Golland, P., Hata, N., Barillot, C., Hornegger, J., Howe, R. (eds.) MICCAI 2014. LNCS, vol. 8675, pp. 305–312. Springer, Cham (2014). https://doi.org/10.1007/978-3-319-10443-0_39
26. Llewellyn, C.D., Linklater, K., Bell, J., Johnson, N.W., Warnakulasuriya, S.: An analysis of risk factors for oral cancer in young people: a case-control study. Oral Oncol. **40**(3), 304–313 (2004)

27. Lv, J., Shao, X., Xing, J., Cheng, C., Zhou, X.: A deep regression architecture with two-stage re-initialization for high performance facial landmark detection. In: Proceedings of the IEEE Conference on Computer Vision and Pattern Recognition, pp. 3317–3326 (2017)
28. Mintz, Y., Brodie, R.: Introduction to artificial intelligence in medicine. Minim. Invasive Ther. Allied Technol. **28**(2), 73–81 (2019)
29. Nagao, T., Warnakulasuriya, S.: Screening for oral cancer: future prospects, research and policy development for Asia. Oral Oncol. **105**, 104632 (2020)
30. Pantic, M., Tomc, M., Rothkrantz, L.J.: A hybrid approach to mouth features detection. In: 2001 IEEE International Conference on Systems, Man and Cybernetics. e-Systems and e-Man for Cybernetics in Cyberspace (Cat. No. 01CH37236), vol. 2, pp. 1188–1193. IEEE (2001)
31. Rajpurkar, P., et al.: Chexnet: Radiologist-level pneumonia detection on chest x-rays with deep learning. arXiv preprint arXiv:1711.05225 (2017)
32. Ren, S., He, K., Girshick, R., Sun, J.: Faster r-CNN: towards real-time object detection with region proposal networks. IEEE Trans. Pattern Anal. Mach. Intell. **39**(6), 1137–1149 (2016)
33. Rimal, J., Shrestha, A., Maharjan, I.K., Shrestha, S., Shah, P.: Risk assessment of smokeless tobacco among oral precancer and cancer patients in eastern developmental region of Nepal. Asian Pac. J. Cancer Prev.: APJCP **20**(2), 411 (2019)
34. Saba, T., Khan, M.A., Rehman, A., Marie-Sainte, S.L.: Region extraction and classification of skin cancer: A heterogeneous framework of deep CNN features fusion and reduction. J. Med. Syst. **43**(9), 1–19 (2019)
35. Sagonas, C., Antonakos, E., Tzimiropoulos, G., Zafeiriou, S., Pantic, M.: 300 faces in-the-wild challenge: database and results. Image Vis. Comput. **47**, 3–18 (2016)
36. Sagonas, C., Tzimiropoulos, G., Zafeiriou, S., Pantic, M.: 300 faces in-the-wild challenge: The first facial landmark localization challenge. In: Proceedings of the IEEE International Conference on Computer Vision Workshops, pp. 397–403 (2013)
37. Song, B., et al.: Automatic classification of dual-modalilty, smartphone-based oral dysplasia and malignancy images using deep learning. Biomed. Opt. Express **9**(11), 5318–5329 (2018)
38. Tzimiropoulos, G.: Project-out cascaded regression with an application to face alignment. In: Proceedings of the IEEE Conference on Computer Vision and Pattern Recognition, pp. 3659–3667 (2015)
39. Uthoff, R.D., et al.: Point-of-care, smartphone-based, dual-modality, dual-view, oral cancer screening device with neural network classification for low-resource communities. PloS ONE **13**(12), e0207493 (2018)
40. Van der Waal, I., de Bree, R., Brakenhoff, R., Coebegh, J.: Early diagnosis in primary oral cancer: is it possible? Medicina oral, patologia oral y cirugia bucal **16**(3), e300–e305 (2011)
41. Welikala, R.A., et al.: Automated detection and classification of oral lesions using deep learning for early detection of oral cancer. IEEE Access **8**, 132677–132693 (2020)
42. Welikala, R.A., et al.: Fine-tuning deep learning architectures for early detection of oral cancer. In: Bebis, G., Alekseyev, M., Cho, H., Gevertz, J., Rodriguez Martinez, M. (eds.) ISMCO 2020. LNCS, vol. 12508, pp. 25–31. Springer, Cham (2020). https://doi.org/10.1007/978-3-030-64511-3_3
43. Xiong, X., De la Torre, F.: Supervised descent method and its applications to face alignment. In: Proceedings of the IEEE Conference on Computer Vision and Pattern Recognition, pp. 532–539 (2013)

44. Xu, S., et al.: An early diagnosis of oral cancer based on three-dimensional convolutional neural networks. IEEE Access **7**, 158603–158611 (2019)

45. Yu, X., Huang, J., Zhang, S., Yan, W., Metaxas, D.N.: Pose-free facial landmark fitting via optimized part mixtures and cascaded deformable shape model. In: Proceedings of the IEEE International Conference on Computer Vision, pp. 1944–1951 (2013)

46. Yu, X., Zhou, F., Chandraker, M.: Deep deformation network for object landmark localization. In: Leibe, B., Matas, J., Sebe, N., Welling, M. (eds.) ECCV 2016. LNCS, vol. 9909, pp. 52–70. Springer, Cham (2016). https://doi.org/10.1007/978-3-319-46454-1_4

47. Zhang, Z., Luo, P., Loy, C.C., Tang, X.: Facial landmark detection by deep multi-task learning. In: Fleet, D., Pajdla, T., Schiele, B., Tuytelaars, T. (eds.) ECCV 2014. LNCS, vol. 8694, pp. 94–108. Springer, Cham (2014). https://doi.org/10.1007/978-3-319-10599-4_7

48. Zhu, X., Ramanan, D.: Face detection, pose estimation, and landmark localization in the wild. In: 2012 IEEE Conference on Computer Vision and Pattern Recognition, pp. 2879–2886. IEEE (2012)

Towards Linking CNN Decisions with Cancer Signs for Breast Lesion Classification from Ultrasound Images

Ali Eskandari$^{(\boxtimes)}$ (iD), Hongbo Du, and Alaa AlZoubi

School of Computing, University of Buckingham, Buckingham, UK
{2007269,hongbo.du,alaa.alzoubi}@buckingham.ac.uk

Abstract. Convolutional neural networks have shown outstanding object recognition performance, especially for visual recognition tasks such as tumor classification in 2D ultrasound (US) images. In Computer-Aided Diagnosis (CAD) systems, interpreting CNN's decision is crucial for accepting the system in the clinical use. This paper is concerned with 'visual explanations' for decisions from CNN models trained on ultrasound images. In particular, we investigate the link between the CNN decision and the calcification cancer sign in breast lesion classification task. To this end, we study the output visualization of two different breast lesion recognition CNN models in two folds: Firstly, we explore two existing visualization approaches, Grad-CAM and CRM, to gain insight into the function of feature layers. Secondly, we introduce an adaptive Grad-CAM, called EGrad-CAM, which uses information entropy to freeze feature maps with no or minimal information. Extensive analysis and experiments using 1624 US images and two breast classification models show that calcification feature contributes to the CNN classification decision for both malignant and benign lesions. Furthermore, we show many feature maps in the final convolution layer are not contributing to the CNN decision, and our EGrad-CAM produces similar visualization output to Grad-CAM using 24%–87% of the feature maps. Our study demonstrates that the CNN decision visualization is a promising direction for bridging the gap between CNN classification decision of US images of breast lesions and cancer signs.

Keywords: Deep learning visualization · Breast cancer · Ultrasonography · Calcification cancer signs · Cancer recognition

1 Introduction

Breast cancer is the most common cancer in women and is affecting millions of women every year in the world [1]. Early detection of breast lesion is essential for a successful diagnosis. Comparing to MRI and CT modalities, ultrasonography is non-radiative, less harmful (intrusive) and low-cost technique for cancer screening and diagnostics. However, analyzing US images is challenging and radiologists of different experiences might describe signs of malignancy differently. The American College of Radiology

© Springer Nature Switzerland AG 2021
B. W. Papież et al. (Eds.): MIUA 2021, LNCS 12722, pp. 423–437, 2021.
https://doi.org/10.1007/978-3-030-80432-9_32

(ACR) developed a guideline called Breast Imaging-Reporting and Data System (BI-RADS) [2] that describes different cancer signs such as calcification. Radiologists use BI-RADS to report the malignancy stage of the lesion in a standardized manner.

Automatic diagnostics of breast cancer is critical in early detection and decrease the mortality rate, especially for females. Several CAD systems have been developed for classifying US images of breast lesions [3–6]. Recently, convolutional neural networks have shown outstanding performance in object recognition especially for the largescale visual recognition tasks and their ability to capture and learn discriminative features such as color, textures and shape [7]. Several deep learning solutions have been developed and used successfully for breast lesion classification in US images [3, 6]. However, these models have the limitation of not being able to explain the classification decision or identify the image parts that contribute to the output decision. Few previous works have been conducted on natural images to provide visual explanations for CNN decisions [8–10]. However, there is a stronger need to explain the CNN decision in medical applications due to ethical requirements and society's concerns towards "black box" style decision-making by automated computerized intelligent systems.

Calcifications and their appearance in the lesion can help the doctor judging whether the case is benign or possibly cancer. The main aim of this work is to investigate the link between the CNN decision of breast lesion classification and the calcification feature. The key contributions of this paper include: (1) investigate the importance of the calcification cancer feature in the CNN decision of classifying breast lesions; (2) use both Grad-CAM [8] and CRM [10] visualization methods to identify the importance of calcification regions in classifying benign and malignant cancer types; (3) develop a new adaptive Grad-CAM, EGrad-CAM, method which uses entropy of the feature maps as an indicator of information and visualize feature maps with high entropy values; (4) analyzing the calcification sign importance using two different CNN models. Our analysis is the first which directly investigates the link between the CNN model decision and calcification malignancy sign for breast lesion classification.

2 Background and Related Work

2.1 Breast Cancer Signs

BI-RADS guideline [2] is a risk assessment tool that is widely used to ensure quality in the diagnosis of cancerous breast tumors. Shape, orientation, margin, boundary, echogenicity pattern, posterior acoustic pattern, and calcification are the most prominent sonographic characteristics of a mass in US images and used to aid radiologists in grading clinical findings. One of the main applications of the BI-RADS is predicting the likelihood of malignancy based on the descriptive terms of the categories. Calcification is an important manifestation of malignant lesion that may or may not exist in the tumor and appears as an extensive hyperechoic region. This sign is of great significance in improving the early detection of the breast cancer. Usually, lesions with suspicious calcifications might have further clinical assessment (e.g., biopsy). Therefore, finding the link between this feature and the CNN decisions can thus significantly boost the understanding of the features learned and used by the network.

2.2   CNN Classification Models

Several CNN models for the object classification in natural images have been developed such as VGG19 [11] and GoogleNet [12]. A review of deep neural network architectures and their applications for object recognition is provided in [13]. Both VGG19 and GoogleNet models have shown outstanding performance in object classification in natural images. Where VGG19 architecture [11] trained on the ImageNet dataset, consists of 47 layers, 16 convolutional, 3 fully connected layers, and approximately 144 million weight parameters. GoogleNet architecture [12] has 9 inception modules, and 22 layers deep (27 including the pooling layers), a GAP layer at the end of the last inception module, and has approximately 7 million parameters. However, training such deep CNN architectures require large datasets. Several research studies using CNN-based deep learning architectures for different medical applications have been proposed [14]. To overcome the limitation of the medical image datasets availability, transfer learning techniques were used with pretrained CNN models (e.g., [11, 12]). Previous studies have been conducted on US images to classify breast lesions using transfer learning and achieved a high performance [3–5]. Generic CNN models for automatic thyroid and breast lesions classification were developed in [3]. VGG19 model was tuned using US images producing BNet model for classifying breast lesions. A data set of 672 breast images were used to train BNet achieving an accuracy of 89%. We thus adopt BNet and GoogleNet architectures to build breast lesion classification models and evaluate the importance of calcification regions.

2.3   CNN Decision Understanding and Visualization

CNN models are complex, and their architectural design is getting deeper and deeper increasing the number of trainable parameters. Therefore, it is hard to trace and understand the features learned by such models. Hence few attempts have been made to find the connection between the image regions (or pixels) and the CNN predictions. In recent years different visualization techniques have been introduced to gain a better insight into CNN models [8–10]. Zhou et al. [9] presented a Class Activation Mapping (CAM) which produce a heatmap showing the importance of each pixel in the CNN classification decision. CAM method requires a GAP layer to visualize the weighted of the resulting feature maps at the pre-softmax layer. A gradient-weighted class activation mapping (Grad-CAM) [8] method was developed as an efficient generalization of CAM. Comparing to CAM, Grad-CAM method can be applied with CNN model without the GAP layer. Grad-CAM++ [15] enhances Grad-CAM visualization in a pixel-wise way to improve multiple class occurrences and refine the process of generating feature maps. Another approach that focuses on gradient-free way to visualize CNN introduced in Score-CAM [16] and Ablation CAM [15]. There are other approaches for interpreting CNNs which only uses the response of a feed forward propagation with attention mechanism that focuses on a specific region in an image [17], or like Soft Proposal Networks [15] which is an effective model in localization tasks, but these two former models require retraining and modification of the network. In this research, we use both Grad-CAM method and CRM (developed for medical images) to investigate the visualization decision of CNN models for breast lesion classification. In Grad-CAM, based on

Eq. (1) the gradient of the score for class c, S_c (pre-softmax layer), will be computed for the feature activations $f_k(x, y)$, where k is the index of feature map, at a spatial location (x, y), i.e., α_k^c.

$$\alpha_k^c = \Sigma_{x,y} \frac{\partial S_c}{\partial f_k(x, y)} \tag{1}$$

Then, the final heatmap of Grad-CAM, $M_c^{Grad-CAM}$, will be defined as a weighted sum of all these feature maps, which follows by a ReLU function to remove the potential impact of negative weights in the desired class.

$$M_c^{Grad} = ReLU\left(\Sigma_k \alpha_k^c f_k(x, y)\right) \tag{2}$$

The Class-selective Relevance Mapping (CRM) [10] method was introduced to measure the importance of the activation at a spatial location (x, y) in the feature maps produced from the last convolution layer. CRM calculates a prediction score S_c in Eq. (3) and $S_c(l, m)$, after removing spatial element (l, m) from the last convolutional layer, in Eq. (3).

$$S_C = \Sigma_{x,y} \Sigma_k \omega_k^C f_k(x, y), \quad S_c(l, m) = \Sigma_{x,y \neq l,m} \Sigma_k \omega_k^C f_k(x, y) \tag{3}$$

Here, ω_k^c is the weight of the k-th feature map of the corresponding output node from the class c. Then, CRM is defined based on a linear sum of incremental mean squared error (MSE) between S_c and $S_c(l, m)$, computed using Eq. (4), M^{CRM}, representing the heatmap output.

$$M^{CRM} = \sum_{c=1}^{N} \{(S_c - S_c(l, m))\}^2 \tag{4}$$

In many automated medical decision-making systems, visualizing and interpreting CNNs output is a must to accept the system in clinical use. Few previous works attempted to use CNN visualizing techniques for medical images recognition [5, 18, 19]. A review of interpretability techniques for radiology AI is presented at [20]. Heechan et al. [21] used attention guided CNN for the classification of breast cancer histopathology images and used an attention mechanism for localization. The authors in [22] developed an attention gate model with an unsupervised manner that can learn to focus on target structures in the fetal ultrasound scan plane. A neural network that performs simultaneous anatomical region detection (classification) and localization in 2D ultrasound images by using Soft Proposal Networks was presented in [23]. Baumgartner et al. [24] developed a feature attribution technique based on Wasserstein Generative Adversarial Networks that can detect class specific regions at a high resolution. Tanaka et al. [5] classified malignant and benign breast masses in US images and attempted to visualize important regions in the image using the Deconvnet method. Zhou, L.-Q. et al. [18] attempt to examine the feasibility of using CNN to predict negative axillary lymph node metastasis in US images using CAM method with In-ceptionV3, Inception-ResNetV2 and ResNet-101 architectures. Xie, B. et al. [19] presented a method to detect five common abnormal fetal brains in US images using Grad-CAM. Although these methods have successfully used different CNN visualization methods for US and other medical modalities applications, none of these techniques has investigated the link between the domain know cancer signs and the CNN decision. In addition, none of these techniques considered the amount of information encoded in each feature maps prior to the visualization.

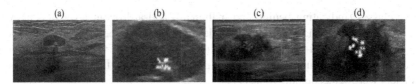

Fig. 1. (a) Benign and (c) malignant original US image with RoI lesion indicated in green bounding box; (b, d) extracted RoI lesion image with calcification points (red points). (Color figure online)

3   Materials and Methods

3.1   Data Collection

Two datasets of US images of breast lesions were collected from two hospitals in China. The first dataset, Dataset A, consists of 798 benign and 326 malignant was used to build CNN models for breast lesion classification. The second dataset, Dataset B, consists of 300 benign and 200 malignant cases that were used to understand the CNN decision of lesion classification and calcification region importance analysis.

Each image in both datasets consists of one breast lesion and the type of the lesion was confirmed by the Fine Needle Aspiration. The region of interest (RoI) of each lesion was manually cropped by an experienced radiologist. Then, an RoI image bounding box of the lesion was generated. The lesion RoI images were rescaled to 224 × 224 and used to build CNNs and understand their decision. Additionally, we have asked a radiologist with more than 5 years of experience to label all 500 images according to the cancer sign descriptors defined in the BI-RADS guideline. 76 images have calcification sign, including 7 benign cases and 69 malignant cases. 1,111 calcification regions were localized manually by an expert, 86 regions on benign cases and 1,025 regions on malignant cases. Figures 1a and b show examples of benign and malignant cases with bounding boxes localizing the lesions, and Figs. 1c and d show benign and malignant cases with calcification points illustrated in red. We were only able to collect the ground truth labels of calcification sign due to the difficulty of labelling other cancer signs at the region level.

3.2   Breast Lesion Classification Model

We developed two different CNN models for breast lesion classification. Both models trained to classify benign and malignant cancer types. As described in Sect. 3.3, the models were used to generate heatmaps for the importance of all pixels in the RoI image lesion in the CNN classification decision.

BNet Model. Gaining inspiration from the DCNN architecture [3] we aim to take advantage of its structure; which is generic in classifying different cancers types with images captured from different US machines. We used the architecture of the CNN model (BNet) [3], with the same architectural parameter settings to train a model for breast lesion classification using dataset A. The layers trained using the CNN [11] and the ImageNet dataset [7] were adapted for breast cancer recognition. A new fully connected layer for the two classes (indicating benign and malignant) and a dropout layer of 25% were used. We set the network parameters as follow: iteration number = 9080, initial learn rate = 0.0001, and mini-batch size = 8. These configurations were set to ensure that the parameters are fine-tuned for the breast lesion classification task. The other network parameters were set as default values [11].

GNet Model. The parameters of the CNN model [12] were pre-trained on the ImageNet dataset [7] for the task of object recognition from natural images. The layers trained using the CNN [12] and the ImageNet dataset [7] were adapted for breast cancer recognition. The architecture of the CNN model [12] was adapted by replacing and fine-tuning the last fully connected layer, softmax layer and output layer. Since the images of breast cancer are labelled by either of two classes, a new fully-connected layer (FC) was added for the two classes benign and malignant. A softmax layer (S) and a classification output layer (O) were also added, where the output of the last fully connected layer was fed to a 2-way softmax layer. The adapted architecture called GNet. Finally, we set the network parameters as follow: 150 epochs representing 2250 iterations, initial learn rate = 0.0001, and mini-batch size = 64 to ensure that the parameters are fine-tuned for the breast lesion classification task. The other network parameters are set as default values [12].

3.3   Calcification Cancer Sign Analysis

A lesion's calcification is a small fleck of calcium usually seen as small bright regions on ultrasonography. It presents a strong feature in breast lesions as it has different shapes and brightness level. However, there is no evidence that breast classification CNN models use this feature to recognize the lesion type: benign or malignant. Our method uses both models developed in Sect. 3.2 to analyze the importance of each pixel in the RoI lesion image including the calcification regions and their contribution to the CNN decision. To generate the importance score of each pixel, we used the Grad-CAM, CRM and EGrad-CAM visualization methods. For all calcification points in dataset B, we applied a 3×3 window with the centroid of calcification points and use the mean of their importance score to analyze the calcification feature.

Grad-CAM. We used Grad-CAM [8] to visualize the regions in the RoI lesion image that contribute to the class prediction. First, BNet and GNet models were used to recognize US RoI image. For each image, a class label (benign or malignant) was assigned and Grad-CAM was used as described in Eq. 1 and 2. Using the gradient and the highest probability classification score, parts of the image lesion were identified and their scores indicating their contribution to the decision. In particular, for each RoI image, we visualized *'relu5_4'* layer of BNet with the activation size of $14 \times 14 \times 512$, and *'inception_5b-output'* of GNet with the activation size of $7 \times 7 \times 1024$. A heatmap representing the importance score of each pixel in the RoI image was generated. The heatmap scores were normalized between 0–1; where 0 indicates the pixel (or region) has no contribution to the CNN decision. Finally, the scores of calcification regions were extracted and analyzed for both benign and malignant cases.

CRM. Similarly, we used CRM [10] to visualize the regions in the RoI lesion image that contribute to the class prediction. However, CRM method requires a GAP layer, therefore, it was only implemented to visualize *'inception_5b-output'* of GNet. Using CRM heatmap (Eq. 4), the scores of calcification regions were extracted and analyzed for both benign and malignant cases.

Entropy Grad-CAM (EGrad-CAM). In this section, we present our approach for adapting Grad-CAM by visualizing the feature maps that encode relatively high information. The output of the last convolutional feature maps consists of local features which have links with different lesion parts. However, some of these feature maps may not have information that contributes to the CNN decision. We aim to exploring an adaptive way of selecting and visualizing these feature maps instead of simply aggregating all of them. The result of applying the filters to an input image is captured by the feature maps of a CNN, hence, examining the feature maps show which features of the network learned. Network activations (or feature maps) are computed by forward propagating the input image through the network up to the specified layer. Therefore, prior to the gradient calculation in the Grad-CAM approach, we employed Shannon entropy [25] as a metric to measure the amount of information in each feature maps in the desired layer. In particular, for each RoI image, we calculate the entropy of each feature map (feature map elements were normalized between 0 to 255) in *'relu5_4'* layer of BNet and *'inception_5b-output'* of GNet. Equation (5) represents the entropy calculation, where p_n indicates the probability of the grayscale values n appearing in the feature map.

$$ H = -\sum_{n=0}^{255} p_n \log_2 p_n \tag{5} $$

Feature maps that have entropy higher than 0 or predefined threshold were aggregated and used to produce the visualization heatmap scores. The visualization output using EGrad-CAM, $M_c^{EGrad-CAM}$, presented in Eq. 6.

$$ M^{EGrad-CAM} = \begin{cases} ReLU\left(\Sigma_k \alpha_k^c f_k(x,y)\right) & H \geq threshold \\ Freeze\ FM & otherwise \end{cases} \tag{6} $$

We will show through experiments that after freezing all feature maps of 0 entropy, EGrad-CAM produces similar visualization output to Grad-CAM with a smaller number of feature maps in both *'relu5_4'* and *'inception_5b-output'*. Moreover, as presented in Sect. 4, to find the percentage of feature maps that contribute to the final heatmap, the number of feature maps with entropy greater than the specified threshold is divided by the total number of feature maps, ie. 512 in *'relu5_4'* and 1024 in *'inception_5b-output'*.

4   Experimental Results

This section explores the importance of the calcification feature to the classification of breast lesions by building two CNN models on Dataset A and visualizing them to find the importance of the calcification in their classification decision. All experiments were run on an Intel Xeon(R) W-2102 CPU @ 2.90 GHz with 16.0 GB RAM.

4.1 Breast Lesion Classification

We used the RoI US images in Dataset A (326 malignant and 798 benign) to train both BNet and GNet architectures. To determine the classification accuracy, we split the US images into training and testing sets at a ratio of 90% to 10% for each class. The BNet model achieved an accuracy of 91.1%, a true positive rate (TPR) of 96.9% and a true negative rate (TNR) of 88.8% in classifying breast lesions. The GNet model achieved an accuracy of 87.5%, a TPR of 84.6% and a TNR of 88.4% in classifying breast lesions. Further, we tested both models using all 500 images in Dataset B. BNet achieved an accuracy of 86.8% with 91.5% and 84% TPR and TNR rates, respectively, and GNet achieved an accuracy of 84% with 92% and 79% TPR and TNR rates, respectively.

As part of our analysis, we evaluated the "inter-class" differences and "intra-class" similarity between benign and malignant cases by visualizing the breast tumor regions contributing to the classification output of US images in Dataset B. First, we used CRM, Grad-CAM and EGrad-CAM (threshold = 0) methods to generated visualization heatmaps from 'inception_5b-output' layer in GNet. For each visualization method, we generated a heatmap for each correctly classified US RoIs image in dataset B. Then, we averaged their heatmaps to produce a single heatmap representing each of the correctly classified benign and malignant cases. Figure 2 shows the visualization output of CRM, Grad-CAM and EGrad-CAM applied to 'inception_5b-output' layer.

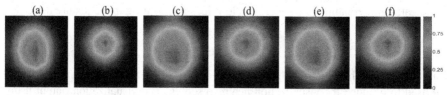

Fig. 2. Average heat map scores of correctly classified benign and malignant cases by GNet, (a) CRM benign cases, (b) CRM malignant cases; (c) Grad-CAM benign cases, (d) Grad-CAM malignant cases; (e) EGrad-CAM benign cases, (f) EGrad-CAM malignant cases. (Color figure online)

As we can see from Fig. 2, in all three visualization techniques, the distribution of the highest important areas within the lesion differs between benign and malignant cases. Note that an area with higher importance (red to yellow) on the heatmap represents the most contributed areas in CNN's final decision to the corresponding predicted label. Figure 2 shows GNet uses different parts of the lesion image to make a correct prediction on both benign and malignant cases. In addition, when comparing Grad-CAM and EGrad-CAM (threshold = 0), both methods produce the same visualization output while EGrad-CAM uses an average of 66% of feature maps in benign cases and 71% of feature maps in malignant cases, the mean of contributed feature maps over the correctly classified images in dataset B were divided by all feature maps within the layer. This shows that a large number of the feature maps in 'inception_5b-output' layer have no or minimal impact on the GNet decision.

Similarly, we used Grad-CAM and EGrad-CAM methods to generated visualization heatmaps from 'relu5_4' layer in BNet. Figure 3 shows the visualization out-put of

Grad-CAM and EGrad-CAM applied on BNet. As shown in Fig. 3, the distribution of the important areas differs between benign and malignant lesions. The lesion boundary textures on benign cases, especially the bottom boundaries, are more likely to contribute to BNet decisions, while in malignant cases the middle textures appear to play an important role in the classification decision. Comparison of Grad-CAM and EGrad-CAM (threshold = 0) shows the same visualization output in both classes, however, an average of only 60% of feature maps in the benign cases and 47% of feature maps in the malignant cases were used to produce the classification decision. This shows that a large number of the feature maps in *'relu5_4'* layer of BNet have no or minimal impact on the BNet decision. It also confirms that both breast lesion classification models (GNet and BNet) have feature maps in their final layer which have minimal impact on the CNN decision.

Despite both GNet and BNet were trained and tested using the same dataset, Figs. 2 and 3 show both models learned and used different image regions to classify both benign and malignant cases. This may refer to the differences in their structure and the way of automatically extracting and learning discriminative features.

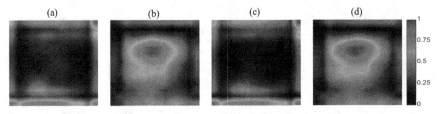

Fig. 3. Average heat map scores of correctly classified cases by BNet; (a) Grad-CAM benign, (b) Grad-CAM malignant; (c) EGrad-CAM benign, (d) EGrad-CAM malignant.

4.2 Breast Lesion Calcification Analysis

The importance of each pixel in the US RoI image was discussed and analyzed on the lesion level in Sect. 4.1. In this section, we aim to study the importance of the calcification regions in both benign and malignant lesions using BNet and GNet classification models. The 76 calcification cases (7 benign and 69 malignant) from Dataset B were used for this analysis. We used our method in Sect. 3.3 to estimate the importance of all pixels representing the calcification regions. The importance scores of 1111 calcification regions, 86 on benign cases and 1025 on malignant cases were estimated using Grad-CAM, CRM and EGrad-CAM.

Grad-CAM Calcification Analysis. In this section, we aim to analyse the importance of calcification points using both GNet and BNet models and compare their scores produced by Grad-CAM on correctly and misclassified breast lesion cases.

GNet Calcification Analysis. By calculating the heatmap from the final layer in GNet (*'inception_5b-output'*), we estimated the importance scores of the calcification regions (or pixels). Figure 6 shows Grad-CAM-GNet heatmap output for benign and malignant

Table 1. Correctly classified Benign (B) and Malignant (M) lesions with calcification by BNet and GNet with Grad-CAM, and GNet with CRM

Score range	GNet Grad-CAM scores ratio			BNet Grad-CAM scores ratio			GNet CRM scores ratio		
	All	B	M	All	B	M	All	B	M
0	0.01	0.00	0.01	0.07	0.61	0.03	0.002	0.00	0.00
(0–0.25)	0.11	0.26	0.10	0.38	0.33	0.39	0.31	0.59	0.29
[0.25–0.5)	0.25	0.38	0.24	0.35	0.06	0.38	0.28	0.28	0.28
[0.5–0.75)	0.29	0.29	0.29	0.13	0.00	0.14	0.25	0.13	0.26
[0.75–1)	0.34	0.07	0.36	0.06	0.00	0.05	0.16	0.00	0.17
1	0.00	0.00	0.00	0.00	0.00	0.00	0.00	0.00	0.00
≥ 0.5	0.63	0.36	0.65	0.19	0.00	0.19	0.41	0.13	0.43

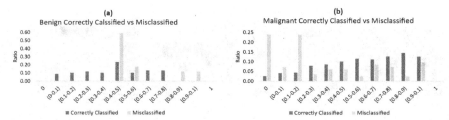

Fig. 4. Grad-CAM-GNet analysis for calcification points in (a) benign and (b) malignant cases.

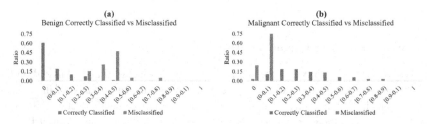

Fig. 5. Grad-CAM-BNet analysis for calcification points in (a) benign and (b) malignant cases.

images lesions. Table 1 shows the ratio of calcification points in defined scores intervals. The table shows over 99% of calcification points have contributed to the breast lesion classification. Where is a 63% of correctly classified lesions with calcification points have Grad-CAM score greater than or equal to 0.5. Table 1 also shows calcification points have a higher importance in classifying malignant cases comparing to benign. In addition, we compared Grad-CAM scores of calcification points for the misclassified benign and malignant cases. Figure 4a shows that the misclassified benign lesions have calcification points with high scores comparing to the correctly classified lesion. In fact, all misclassified benign lesions have calcification points with scores greater than

0.4, while on correctly classified ones we see the distribution of scores in the lower ranges. Such high scores of calcification points might contribute to incorrect prediction of the GNet model of benign lesions. On the other hand, Fig. 4b shows that around 70% of calcification points coming from the misclassified malignant lesions have a score lower than 0.5. These results indicate that calcification points play a decisive role in lesion classification, which also shows the link between the domain know cancer sign (calcification) and the CNN decision.

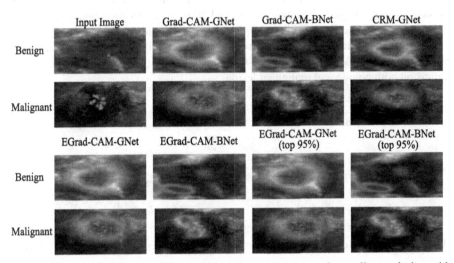

Fig. 6. Output visualization of different methods for a benign and a malignant lesion with calcification points.

BNet Calcification Analysis. Similar to GNet, we generated heatmaps from *'relu5_4'* layer of BNet. Then, we estimated the importance scores of the calcification regions. Figure 6 shows Grad-CAM-BNet heatmap output for benign and malignant images lesions. Table 1 shows the ratio and number of calcification points in defined scores intervals. The table shows over 93% of calcification points have contributed to the lesion classification of the GNet model. Where 19% of correctly classified lesions with calcification points have Grad-CAM score greater than or equal to 0.5. The result shows none of the points in the correctly classified benign lesions has a score higher than 0.5. Figure 5 shows calcification points have higher scores for misclassified benign lesions and lower scores for misclassified malignant lesions. Figure 5a shows the misclassified benign lesions have calcification points with high scores comparing to the correctly classified lesion. On the other hand, Fig. 5b shows that around 70% of calcification points coming from the misclassified lesions have a score lower than 0.5. These results confirm the importance of calcification regions in the CNN decision.

CRM Calcification Analysis. Similar to Grad-CAM, we analyzed the calcification points using CRM on the GNet model. Figure 6 shows CRM-GNet heatmap output for benign and malignant images lesions. By calculating the heatmap scores of the 'inception_5b-output' layer, the ratio and number of calcification points in defined scores intervals were calculated as illustrated in Table 1. The table shows over 99.8% of calcification points have contributed to the lesion classification of the GNet model. Where 41% of correctly classified lesions with calcification points have CRM scores greater than or equal to 0.5.

Table 1 also shows calcification points have a higher importance in classifying malignant cases comparing to benign. Figure 7a shows that the misclassified benign lesions have calcification points with relatively high CRM scores comparing to the correctly classified lesions. Figure 7b shows that around 50% of calcification points have CRM scores below 0.2. The distributions of CRM scores of calcification regions coming from both benign and malignant lesions (correctly and misclassified lesions) show a similar pattern to Grad-CAM. However, Grad-CAM assigned higher scores for calcification regions on both benign and malignant lesions.

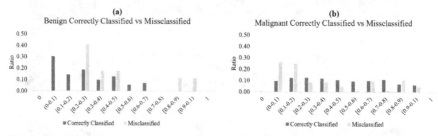

Fig. 7. (a) CRM-GNet analysis for calcification points in (a) benign and (b) malignant cases.

EGrad-CAM Calcification Analysis. We are primarily interested in exploring an adaptive way of selecting and visualizing feature maps that encode information affected by a sub-region in the input US image of the lesion, and our final experimental evaluation applies our EGrad-CAM method, to the lesions with calcification feature. We analyzed the importance of calcification points using both GNet and BNet models and compare their scores produced by EGrad-CAM on correctly and misclassified lesions. We used EGrad-CAM to select and visualize the feature maps from 'relu5_4' and 'inception_5b-output'. First, all feature maps with zero entropy were excluded from calculating the importance scores of each pixel in the US lesion image using Eq. 5 and 6. With this setting, over the 69 correctly classified breast lesion images, an average of 28% of the feature maps from 'inception_5b-output' of GNet have no contributions to the calcification scores. Similarly, 52% of the feature maps from 'relu5_4' layer of BNet have no contributions to the calcification scores for the 72 correctly classified lesions. As a result, EGrad-CAM with 0-entropy produced the same heatmaps scores as Grad-CAM but with a large reduction in the number of analyzed feature maps.

We have repeated our experimental evaluation of EGrad-CAM, using feature maps that have top 95% entropy values to study the impact of freezing feature maps with entropy close to zero on the overall visualization output. With this setting, an average of 33% of the feature maps from 'inception_5b-output' and 73% of the feature maps from 'relu5_4' have a minimum contribution to the calcification scores. In conclusion, the distributions of EGrad-CAM (top 95%) scores of calcification regions coming from both benign and malignant lesions show a similar pattern to Grad-CAM and CRM. However, EGrad-CAM uses a significantly lower number of feature maps. Figure 6 shows EGrad-CAM-GNet and EGrad-CAM-BNet heatmaps output for benign and malignant image lesions. Figure 6 also shows BNet model visualized by EGrad-CAM and Grad-CAM ignored the calcification regions in making the correct prediction on the benign lesion, while used the calcification areas to help in the prediction of malignant one. On the GNet, although there may be some calcification points have been captured on the benign lesion, they have low scores comparing to the calcification points on the malignant lesion.

5 Conclusion and Discussion

Despite the recent development and success of using convolutional neural networks for breast lesion classification in US images, understanding the CNN decision is still an open question. So far, work has been limited to build CNN models using transfer learning or from scratch to classify breast lesions into benign or malignant. However, no link between the CNN decision and the domain known cancer signs (such as calcification) have been investigated. In this paper, we have explored different CNN visualization methods on CNN models trained using breast lesion in US images. The first part of our work is to build two CNN models (BNet and GNet) for breast lesion classifications. Both models have shown very good performance in classifying benign and malignant lesions using two datasets. Grad-CAM, CRM and EGrad-CAM were used to visualize the important regions in the breast lesion that contribute to the CNN decision. All three methods have shown different visualization output of benign and malignant tumors. The second part of the work is concerned with finding the link between trained CNN features and calcification regions. Our investigation identified a direct link between the calcification feature and the CNN decision on classifying lesion types. We have shown, with extensive experiments, calcification sign is a strong feature learned by CNN to classify malignant lesions and relatively low contribution to benign lesion classification. This observation of CNN decision behavior aligns with the medical understanding as the calcification feature tends to be a significant sign of malignancy. We also proposed an adaptive Grad-CAM method, EGrad-CAM, based on information entropy of feature maps which can be readily integrated into various CNN visualization methods. EGrad-CAM produced the same results as Grad-CAM with a smaller number of feature maps from the final convolution layers. The possible explanation for this reduction is that some of the feature maps in the investigated layers have minimal information, and minimal impact on the CNN decision, therefore increasing attention should be made when designing CNN architecture for breast lesion classification.

In addition to the calcification sign, CNN may also learn other features that contribute to the lesion classification decision. Therefore, we plan to extend our work in several ways. Firstly, we wish to further support ongoing research into CNN decision by analyzing other cancer signs such as lesion echogenicity and margin. We also wish to further evaluate our EGrad-CAM on other types of lesions from ultrasound images.

Acknowledgments. This research is sponsored by TenD Innovations.

References

1. Bray, F., Ferlay, J., Soerjomataram, I., Siegel, R.L., Torre, L.A., Jemal, A.: Global cancer statistics 2018: GLOBOCAN estimates of incidence and mortality worldwide for 36 cancers in 185 countries. CA: Cancer J. Clin. **68**(6), 394–424 (2018). https://doi.org/10.3322/caac.21492
2. Mercado, C.L.: Bi-rads update. Radiol. Clin. **52**(3), 481–487 (2014)
3. Zhu, Y.-C., et al.: A generic deep learning framework to classify thyroid and breast lesions in ultrasound images. Ultrasonics **110**, 106300 (2021)
4. Wang, Y., Choi, E.J., Choi, Y., Zhang, H., Jin, G.Y., Ko, S.-B.: Breast cancer classification in automated breast ultrasound using multiview convolutional neural network with transfer learning. Ultrasound Med. Biol. **46**(5), 1119–1132 (2020)
5. Tanaka, H., Chiu, S.-W., Watanabe, T., Kaoku, S., Yamaguchi, T.: Computer-aided diagnosis system for breast ultrasound images using deep learning. Phys. Med. Biol. **64**(23), 235013 (2019)
6. Moon, W.K., Lee, Y.-W., Ke, H.-H., Lee, S.H., Huang, C.-S., Chang, R.-F.: Computer-aided diagnosis of breast ultrasound images using ensemble learning from convolutional neural networks. Comput. Meth. Program. Biomed. **190**, 105361 (2020)
7. Deng, J., Dong, W., Socher, R., Li, L.-J., Li, K., Fei-Fei, L.: ImageNet: a large-scale hierarchical image database. In: 2009 IEEE Conference on Computer Vision and Pattern Recognition, pp. 248–255. IEEE (2009)
8. Selvaraju, R.R., Cogswell, M., Das, A., Vedantam, R., Parikh, D., Batra, D.: Grad-cam: visual explanations from deep networks via gradient-based localization. In: Proceedings of the IEEE International Conference on Computer Vision, pp. 618–626 (2017)
9. Zhou, B., Khosla, A., Lapedriza, A., Oliva, A., Torralba, A.: Learning deep features for discriminative localization. In: Proceedings of the IEEE Conference on Computer Vision and Pattern Recognition, pp. 2921–2929 (2016)
10. Kim, I., Rajaraman, S., Antani, S.: Visual interpretation of convolutional neural network predictions in classifying medical image modalities. Diagnostics **9**(2), 38 (2019)
11. Simonyan, K., Zisserman, A.: Very deep convolutional networks for large-scale image recognition. arXiv preprint arXiv:1409.1556 (2014)
12. Szegedy, C., et al.: Going deeper with convolutions. In: Proceedings of the IEEE Conference on Computer Vision and Pattern Recognition, pp. 1–9 (2015)
13. Liu, W., Wang, Z., Liu, X., Zeng, N., Liu, Y., Alsaadi, F.E.: A survey of deep neural network architectures and their applications. Neurocomputing **234**, 11–26 (2017)
14. Byra, M., et al.: Breast mass classification in sonography with transfer learning using a deep convolutional neural network and color conversion. Med. Phys. **46**(2), 746–755 (2019)
15. Rony, J., Belharbi, S., Dolz, J., Ayed, I.B., McCaffrey, L., Granger, E.: Deep weakly-supervised learning methods for classification and localization in histology images: a survey. arXiv preprint arXiv:1909.03354 (2019)
16. Wang, H., et al.: Score-CAM: score-weighted visual explanations for convolutional neural networks. In: Proceedings of the IEEE/CVF Conference on Computer Vision and Pattern Recognition Workshops, pp. 24–25 (2020)
17. Fukui, H., Hirakawa, T., Yamashita, T., Fujiyoshi, H.: Attention branch network: learning of attention mechanism for visual explanation. In: Proceedings of the IEEE/CVF Conference on Computer Vision and Pattern Recognition, pp. 10705–10714 (2019)
18. Zhou, L.-Q., et al.: Lymph node metastasis prediction from primary breast cancer US images using deep learning. Radiology **294**(1), 19–28 (2020)
19. Xie, B., et al.: Computer-aided diagnosis for fetal brain ultrasound images using deep convolutional neural networks. Int. J. Comput. Assist. Radiol. Surg. **15**(8), 1303–1312 (2020). https://doi.org/10.1007/s11548-020-02182-3

20. Reyes, M., et al.: On the interpretability of artificial intelligence in radiology: challenges and opportunities. Radiol. Artif. Intell. 2(3), e190043 (2020). https://doi.org/10.1148/ryai.202019 0043
21. Yang, H., Kim, J.-Y., Kim, H., Adhikari, S.P.: Guided soft attention network for classification of breast cancer histopathology images. IEEE Trans. Med. Imaging 39(5), 1306–1315 (2019)
22. Schlemper, J., et al.: Attention gated networks: learning to leverage salient regions in medical images. Med. Image Anal. 53, 197–207 (2019)
23. Toussaint, N., et al.: Weakly supervised localisation for fetal ultrasound images. In: Stoyanov, D., et al. (eds.) DLMIA/ML-CDS -2018. LNCS, vol. 11045, pp. 192–200. Springer, Cham (2018). https://doi.org/10.1007/978-3-030-00889-5_22
24. Baumgartner, C.F., Koch, L.M., Tezcan, K.C., Ang, J.X., Konukoglu, E.: Visual feature attribution using Wasserstein GANs. In: Proceedings of the IEEE Conference on Computer Vision and Pattern Recognition, pp. 8309–8319 (2018)
25. Shannon, C.E.: A mathematical theory of communication. Bell Syst. Tech. J. 27(3), 379–423 (1948)

Improving Generalization of ENAS-Based CNN Models for Breast Lesion Classification from Ultrasound Images

Mohammed Ahmed[✉], Alaa AlZoubi, and Hongbo Du

The University of Buckingham, Buckingham MK18 1EG, UK
{1200526,Alaa.alzoubi,Hongbo.du}@buckingham.ac.uk

Abstract. Neural Architecture Search (NAS) is one of the most recent developments in automating the design process for deep convolutional neural network (DCNN) architectures. NAS and later Efficient NAS (ENAS) based models have been adopted successfully for various applications including ultrasound image classification for breast lesions. Such a data driven approach leads to creation of DCNN models that are more applicable to the data set at hand but with a risk for model overfitting. In this paper, we first investigate the extent of the ENAS model generalization error problem by using different test data sets of ultrasound images of breast lesions. We have observed a significant reduction of overall average accuracy by nearly 10% and even more severe reduction of specificity rate by more than 20%, indicating that model generalization error is a serious issue with ENAS models for breast lesion classification in ultrasound images. To overcome the generalization error, we examined the effectiveness of a range of techniques including reducing model complexity, use of data augmentation, and use of unbalanced training sets. Experimental results show that different methods for the tuned ENAS models achieved different levels of accuracy when they are tested on internal and two external test data sets. The paper demonstrates that ENAS models trained on an unbalanced dataset with more benign cases tend to generalize well on unseen images achieving average accuracies of 85.8%, 82.7%, and 88.1% respectively for the internal and the two external test data sets not only on specificity alone, but also sensitivity. In particular, the generalization of the refined models across internal and external test data is maintained.

Keywords: ENAS · Ultrasound image · Breast lesion classification · Deep learning · Reduce generalization error · Imbalanced dataset

1 Introduction

Breast cancer is the most common type of cancers among women with the highest fatality rate in the world [1]. 2.3 million new cases and 685 thousand deaths were reported in 2020 alone [1]. Early detection of breast cancer is vital for saving patient lives through appropriate and effective treatments. Digital Mammography (DM) and Ultrasound (US) are two commonly used imagery systems for the detection. Between the two image

© Springer Nature Switzerland AG 2021
B. W. Papież et al. (Eds.): MIUA 2021, LNCS 12722, pp. 438–453, 2021.
https://doi.org/10.1007/978-3-030-80432-9_33

modalities, US imaging is considered as being safer, more versatile, and more sensitive to cancerous lesions located in intensive areas [2]. However, analysing US images is challenging, and often results in diagnostic differences among radiologists of different backgrounds and experiences. Recently, Computer-Aided Diagnosis (CAD) systems were applied to medical image analysis applications [3]. A CAD system for ultrasound image classification provides a wide decision support platform for the diagnosis of various types of cancer including breast lesions [4].

In recent years, deep learning neural networks, particularly convolutional neural networks (CNN), have showed outstanding performance in a variety of computer vision applications such as object detection, segmentation, and recognition [5]. For image classification, instead of relying on specifically designed algorithms to extract certain types of features, CNNs learn distinguishing features directly from input images and classify the images, accordingly, providing an end-to-end solution to the problem. Developing neural network classifiers involves designing a network architecture (also known as network topology) followed by using a learning algorithm to tune the weights attached to the links between neurons on different layers. Two main approaches are followed for CNN architectures [6]: manually designed (or handcraft) architectures and automatically searched optimal architectures. Several well-known architectures, such as LeNet for classifying images of handwriting digits [7] and AlexNet for classifying daily life objects from photographic images, have been manually designed [8]. Recently, many attempts have been made on cancer recognition from ultrasound images using deep learning CNN (DCNN) models. Majority of the existing work are concerned with adapting the state-of-the-art CNN architectures with or without transfer learning from pre-trained models on the ImageNet dataset [9]. In [10], three milestone architectures for classifying medical images, i.e. LeNet, AlexNet, and GoogLeNet, were tested and their performances compared with a dataset of 37,698 images of 5 classes of different modalities (CT scan, MRI, X-ray, PET, and US). Based on the comparison, AlexNet architecture was further modified by removing the last convolution layer and ignoring dropouts in both fully connected layers.

For breast lesion classification from US images, a modified GoogLeNet without two auxiliary classifiers was used as the backbone architecture for training classification models [11]. In [12], a simple CNN architecture (CNN3) was purposely designed for differentiating breast messes also from ultrasound images. The network consists of three convolutional layers and two fully connected layers. Each convolutional layer is followed by Relu activation function and max pooling with stride 2 for dimension reduction. All convolutional layers used 3×3 filters. The global average pooling was applied before the first fully connected layer. The CNN3 model was compared with three pre-trained models (inception V3, ResNet50, and xception) and conventional handcrafted features. Better performance of deep learning models against the handcrafted features was reported. Masud et al. [13] proposed a specifically designed new CNN architecture that consists of one convolutional layer followed by batch normalization, max pooling, and one fully connected layer, and showed that the architecture outperformed several known pre-trained CNN models (AlexNet, DarkNet19, GoogelNet, MobileNet-v2, TesNet18, ResNet50, VGG16 and Xception). In [14], an attempt was made to investigate the power of transfer learning by comparing a VGG16 model trained from scratch with the fine-tuned pre-trained VGG16 model. In [15], several pre-trained state-of the-art CNN architectures, i.e. InceptionV3, VGG16, VGG19 and ResNet50, were broadly compared for breast lesion classification from ultrasound images. The study found that InceptionV3 models outperformed the other models.

It has been acknowledged that designing successful CNN architectures manually is a challenging task and involves careful setting of many hyperparameters like the depth of architecture, the number of filters, the sizes of the filters, connectivity between layers, etc. Therefore, in-depth knowledge and experience in CNN designs are required. Neural Architecture Search (NAS) [16] offers a new alternative approach that automatically searches for an optimal CNN architecture through reinforcement learning (RL) from the dataset of interest. Further research works under this approach include the scheme proposed in [17], ENAS [18] and PNAS [19]. Efficient Neural Architecture Search (ENAS) [18] aims at reducing the burden on resource requirements of NAS [17] and PNAS [19] by simplifying the architecture and using specific searching strategies. The micro search space of ENAS [18] directs the controller to search for optimal cells of operations while fixing a CNN architecture as a stack of blocks of normal and reduction cells with a predefined configuration. The macro search space of ENAS aims at generating an optimal network architecture of basic operations and connections. With either search strategy, after the optimal CNN architecture is found, the CNN architecture-based classifier is trained from scratch using the same training data set at hand. ENAS is one of the most efficient NAS methods, and is gaining recognition for CNN architecture search ([20, 21]) for medical image analysis. A more recent adaptation of ENAS micro search for building classification models to recognize breast lesion from ultrasound images was reported in [22], showing that ENAS models outperformed CNN3 and AlexNet trained from scratch.

Training a DCNN model with limited training records and yet achieving high accuracy on an external dataset is desirable. Many existing approaches quantify model generalization based on measuring model's accuracy on an internal test set held out from the training examples, but it is likely that the accuracy of a model that performs well on the internal test set will drop on an external test set [23]. Overfitting is one of the main reasons for dropping performance of CNN models on external datasets. Several methods such as data augmentation, regularization, batch normalization, reducing model complexity and early stopping, have been proposed to reduce overfitting of CNN models [24]. In [25], the authors studied the generalization error of transfer learning models for breast cancer classification. They selected two known CNN architectures (AlexNet and GoogleNet) for their study with mammogram images. Zeimarani et al. [26] applied data augmentation, L2 regularization and dropout to reduce model overfitting of a specially designed CNN architecture for breast cancer classification from ultrasound images. Published work on ENAS model overfitting issue is extremely limited and studying ENAS model overfitting for breast lesion recognition from ultrasound images is non-existent. The researchers in [27] proposed a method to reduce overfitting of ENAS models for natural images by establishing skip connections, but their work was concerned with the macro search space for operations rather than the micro approach for optimal cells. This investigation, as far as we are aware, is the first attempt to study ENAS model overfitting issue in this context.

2 Materials and Methods

2.1 Datasets

Three different sets of ultrasound images of breast lesions (Modelling, TenD_test and BUSI_test) from two data sources have been used in this study. Data sets Modelling and TenD_test are provided by our collaborators (TenD Innovations), and BUSI_test set was obtained from the public domain [28]. Modelling data set contains 1,102 US images (726 of benign lesions and 376 of malignant lesions). This set is used for architecture search and modelling. TenD_test Data set consists of 500 US images (300 of benign and 200 of malignant lesions). This data set is used as a separate external test set. Both sets contain images captured from US machines of different makes and models (i.e. Siemens, Toshiba, GE, and Philips). The class labels for the images in both data sets are provided by experienced radiologists after referring to the pathology reports on the lesions. The initially collected data set BUSI_test contains 780 ultrasound images (133 of normal breasts without lesions, 437 of benign lesions and 210 of malignant lesions). To be consistent with the other two data sets, the images of normal breasts without lesions and the images with artefacts such as markers, lines and drawings are excluded from BUSI_test. A collection of 565 images (355 of benign and 210 of malignant lesions) was eventually selected. This updated BUSI_test set is then used as the second external test set for evaluating CNN models. All regions of interest (RoIs), i.e. lesion regions, were manually cropped by experienced radiologists. Figure 1 shows some examples of cropped RoI images of benign and malignant lesions from the Modelling and BUSI-test datasets.

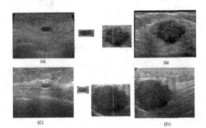

Fig. 1. Cropped example RoI images (in red boxes): Modelling dataset ((a) Benign and (b) Malignant), BUSI_test ((c) Benign and (d) Malignant). (Color figure online)

2.2 Optimizing CNN Architecture with ENAS Approach

This investigation adopts the micro search of ENAS [18] to find the optimal CNN architecture for the problem at hand. A designated subset of Modelling data set was used for the search (see Sect. 3 for more details). The micro search approach consists of two stages. The first stage searches for optimal normal (N) and optimal reduction (R) cells within a pre-configured architecture based on validation accuracy. Five operations, i.e. depth-wise separable with filter sizes 3 × 3 and 5 × 5, average pooling and max pooling with window size 3 × 3, and identity, are taken as the basic operations for the ENAS RNN controller to use in generating candidate cells. The backbone architecture consists of one stem convolution and 7 layers (cells) where the stem convolution is a normal convolutional layer with filter size 3 × 3. Each cell (either normal or reduction) consists of 5 nodes and each node is an element wise addition of two different operations as showed in Fig. 2(a). For the searching stage, the number of filters started from 20 and doubled after each reduction layer, a batch size 8 and 150 epochs and all other hyperparameters are set as the original ENAS [18]. At each epoch, the RNN controller generates 10 architectures, and each generated architecture is then evaluated on a mini-batch of validation set. The validation accuracy is used as a reward to improve controller for generating better architectures at the next round. At the second modelling stage, optimal normal cells and optimal reduction cells are chosen according to validation accuracy, the CNN architecture is created by stacking cells as layers according to a configuration formula [18]. In this study, 17 layers (cells) are stacked in the following fashion: 5 Normal cells + 1 Reduction cell + 5 Normal cells + 1 Reduction cell + 5 Normal cells. This architecture is known as ENAS17. Then we used the same data set for the search to train an ENAS17 model from scratch through 100 epochs with the same setting for all other hyperparameters as the original ENAS (see Sect. 3 for further details). Figure 2(a) shows the internal structure of the optimal normal cell and that of the optimal reduction cell, respectively. Figure 2(b) illustrates the ENAS17 architecture composed by the optimal cells.

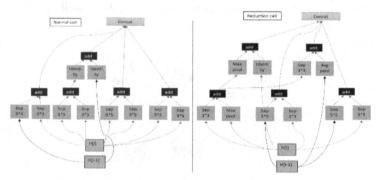

(a) Optimal Norman cells (left) and Optimal Reduction cell (right) generated by ENAS for Breast Lesion classification

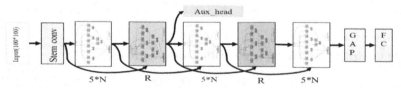

(b) Optimal CNN architecture found by ENAS where 5*N: a stack of 5 normal cells, R: reduction cell, GAP: global average pooling, and FC: fully connected layer.

Fig. 2. Optimal CNN architecture searched by ENAS with 17 layers.

2.3 Reducing ENAS Model Generalization Error

The main goal of machine learning is to design models that have high accuracy rates and can generalize well on data sets that have not been seen by the models before. Models trained on a limited amount of training data tend to overfit the training data by remembering their specifics. Such models may not perform well on unseen data and hence difficult to generalize. Since DCNN models have many parameters to tune during the learning process, they tend to overfit to the training examples unless more variations of training data are introduced. It is for this reason that training DCNN models often requires many training examples. In this study, we investigate three overfitting reduction techniques including: 1) reducing model complexity; 2) data augmentation; and 3) modelling with unbalanced data. We also investigate the effects of each technique on model's generalization errors on independently sampled data sets.

Reducing Architecture Complexity. Since over parameterized models are likely to overfit and not perform well on unseen data, reducing model complexity is one of the common approaches used for reducing the effect of overfitting. To simplify the initially trained ENAS17 model, rather than simplifying the basic operations considered, we purposely maintain the optimal cells and reduce the number of layers of the searched CNN architecture from 17 down to 7. In other words, we designed an ENAS7 architecture that consists of 7 layers (cells). The number of layers was reduced to 7 to maintain a good level of architecture depth to learn simple and complex features from the lesion image. Figure 3 presents the simplified ENAS7 CNN architecture which has nearly half of the number of parameters of the ENAS17 architecture, and the tested accuracy (internal test) is only reduced by a fraction of one percent, as reported in [20]. It is interesting to observe whether the model overfitting can be reduced by the model complexity reduction.

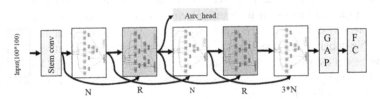

Fig. 3. Simplified ENAS7 architecture.

Impact of Data Augmentation. One method of reducing chances for model overfitting is to train CNN models with a large amount of training data. When the number of training examples is limited, data augmentation is a frequently adopted method to enlarge the training set. However, because augmentation is based on existing examples, it is possible that specifics of the limited number of original training examples can be amplified and over learnt by the model. Therefore, the effect of data augmentation on ENAS model overfitting still needs to be tested. This investigation adopts two types of data augmentation methods as presented in [4]. The geometric methods include rotation and mirroring which alter the original RoI image orientation while preserving the shape of the lesions within the images. We used 3 rotation angles, i.e. 90, 180, and 270°

anticlockwise around the center of the RoI image. The mirroring operation produces a reflective duplication of a RoI image by flipping the image across its vertical axis. We also used the singular value decomposition (SVD) compression scheme to generate images that maintain the geometry of the original RoI image while losing some image details. The method generates three images with 45%, 35%, and 25% ratios of the selected top singular values from each RoI image, approximately preserving the important property of the original image. Due to time constraint, we have not attempted other recent techniques such as those mentioned in [29].

To investigate the effects of the different types of data augmentation on ENAS model overfitting reduction, both ENAS17 and ENAS7 models are trained under the following scenarios of data augmentation. In the first scenario, no augmentation is applied. In the second scenario, only geometric data augmentation methods, i.e. rotations and mirroring, are used for expanding the training set. In the third scenario, the training set is enlarged by using the SVD methods with mirroring. In the last scenario, the training set is enlarged by using all data augmentation methods as described earlier.

Exploiting Unbalanced Datasets. A data set with unbalanced classes means that the distribution of different classes within the set is unequal. For instance, a training set may contain more cases of benign than malignant ones. In fact, this is more realistic; the vast majority of lesions are benign and only a small minority of cases are malignant. For some application domains, using the same number of samples from each class is more desirable for building an unbiased classification model. Therefore, a commonly adopted approach in machine learning is to use down-sampling or up-sampling techniques to bring the cases of different classes to a balanced level [30]. However, we have observed from several published works in the literature (e.g. [4, 30]) and our own experiences of developing DCNN models from ultrasound images that using datasets of balanced benign and malignant classes often leads to models with higher sensitivity and lower specificity [30]. It is interesting to investigate upon which class the ENAS models tend to overfit more in both balanced and unbalanced class situations, and whether maintaining a reasonable unbalance of benign and malignant cases will bring about positive effect in reducing model overfitting. For this study, we trained ENAS17 and ENAS7 models on balanced and unbalanced class data sets where the number of benign cases near doubles the number of malignant cases.

3   Experiments and Results

Experiments have been conducted to evaluate the generalization rate of ENAS CNN models for breast cancer classification from ultrasound images. In this section, we will first present the generalization of ENAS models on external test datasets (TenD_test and BUSI_test). We then examine the effects of reducing generalization errors of ENAS based models through the methods mentioned in the previous section. The experimental study starts first by using the ENAS Micro approach to search for optimal CNN architectures. For the searching stage, we used the same setting as the default in ENAS [18] as described in Sect. 2.2. For the searching stage, we used a randomly sampled subset of the Modelling data set called BModeling with 262 images of benign and 262 images of malignant lesions. Using stratified random sampling, we split the BModeling dataset into 20% for testing and 80% for training. The remaining 80% were further splitted into 90% for searching for the optimal cell and 10% for validation. After generating a set of cells by ENAS controller, for designing the final CNN architecture we selected optimal cells (Normal and Reduction) that have highest validation test accuracies.

3.1 Evaluating ENAS17 Model Generalization

The first experiment is to evaluate the generalization of the classification models based on the optimized ENAS17 architecture. The models were trained from scratch by using the BModeling data set. To gain an understanding of how well the models perform on the data from the same data set, a 5-fold cross-validation protocol was followed. The test partitions are known here as internal set. To ensure there are sufficient training examples, at each iteration of the cross-validation, the training partitions were enlarged by using all the data augmentation methods mentioned in the previous section while the test partition is not enlarged to avoid any distortion of test results. At each fold, we also use the trained model to classify the two external data sets (TenD_test and BUSI_test). We then took the average performance of the models over the test partitions as the result of the internal test. We take the averages of true positive rates, true negative rates and overall accuracies by all models during the cross validation on TenD_test and BUSI_test respectively. The degree of generalization error can be indicated by the difference between the average accuracy of the internal tests and the average accuracy of external tests.

Table 1 (i.e. ENAS17 with use of all the augmentation techniques) shows that a reasonably high level of average accuracy of 89.3% is achieved on the internal tests. The average true positive rate at 92% is even higher that the overall accuracy on the internal tests. However, the table also clearly shows that overall accuracies of the ENAS17 models on the two external test sets dropped by more than 10%. Despite the relatively good generalization on the true positive rate (in fact the performance is better on the external test sets than the internal set), the true negative rates on both data sets were significantly reduced by 21% on the BUSI_test and 26% on the TenD_test. Generalization is clearly an issue for the ENAS17 models. The worse effect on the true negative rate indicates that using balanced data set may have disadvantaged benign lesion classification results, consistent to the observations on the issue made by the existing research works.

3.2 Reducing Generalization Error for ENAS Models

Reducing Architectural Complexity of ENAS Models. To reduce generalization error for ENAS17 models, we simplified the architecture by removing 10 normal cells as described in the previous section (the simplified ENAS7 architecture is depicted in Fig. 3). The architecture has a reduced number of weight parameters about 50% of the number for ENAS17. We then trained the ENAS7 models on the same BModeling data set with the same hyperparameter settings as for ENAS 17 models so that we can focus on the effect of model complexity. We tested the ENAS7 models through the same evaluation protocol. Table 1 (i.e. ENAS7 with use of all the augmentation techniques) shows the performances of the ENAS7 models on the internal and the external test sets. The overall accuracies on the three data sets are similar to those by ENAS17 models with a similar drop of performance on the external data sets. The results are not as expected. The simpler models have better performances on the true negative rates but at the expense of reduced true positive rates across different sets. In other words, the simplification has not significantly reduced generalization errors despite some marginal improvements in TNR on internal and external test sets.

Table 1. Effect of data augmentation on generalization of ENAS 17 and ENAS7.

Models	Training scenario	Test sets	TNR	TPR	Acc.	C.I. at (95%)
ENAS 17	Without any use of augmentation techniques	Internal	87.8	84.0	85.9	(79.3, 92.5)
		BUSI_test	81.0	70.1	75.6	(72.0, 79.1)
		TenD_test	64.8	95.4	80.1	(76.6, 83.6)
	With use of all the augmentation techniques	Internal	86.7	92.0	89.3	(83.5, 95.2)
		BUSI_test	65.1	93.3	79.2	(75.9, 82.5)
		TenD_test	60.7	97.5	79.1	(75.6, 82.7)
	With use of only SVD and mirroring techniques	Internal	83.9	87.1	85.5	(78.8, 92.2)
		BUSI_test	68.6	90.9	79.7	(76.4, 83.1)
		TenD_test	63.3	96.5	79.9	(76.4, 83.4)
	With use of rotation techniques	Internal	85.9	89.7	87.8	(81.6, 94)
		BUSI_test	64.7	93.7	79.2	(75.8, 82.5)
		TenD_test	61.7	97.9	79.8	(76.3, 83.3)
ENAS 7	Without any use of augmentation techniques	Internal	86.2	86.3	86.3	(79.7, 92.8)
		BUSI_test	76.3	83.9	80.1	(76.8, 83.4)
		TenD_test	63.9	93.6	78.8	(75.2, 82.4)
	With use of all the augmentation techniques	Internal	90.9	86.7	88.8	(82.8, 94.8)
		BUSI_test	69.7	88.0	78.9	(75.5, 82.2)
		TenD_test	65.0	96.7	80.9	(77.4, 84.3)
	With use of only SVD and mirroring techniques	Internal	86.7	85.1	85.9	(79.3, 92.5)
		BUSI_test	63.3	89.0	76.2	(72.7, 79.7)
		TenD_test	58.1	97.1	77.6	(73.9, 81.2)
	With use of rotation techniques	Internal	87.4	85.5	86.5	(80, 93)
		BUSI_test	63.4	93.0	78.2	(74.8, 81.6)
		TenD_test	63.0	97.4	80.2	(76.7, 83.7)

Effects of Data Augmentation. To explore the effect of the data augmentation methods on overfitting in ENAS models, we examined different training set settings prepared for the different scenarios as mentioned in Sect. 2.3. Table 1 summarizes the performance metrics in different scenarios of augmenting training examples for both ENAS17 and

ENAS7 models. There are no clear-cut results. The table shows that the models trained on the balanced training set without data augmentation are performing better on BUSI_test in terms of true negative rate with a lesser performance drop of 6.8% for ENAS17 models and 10% lesser drop for ENAS7 models.

Table 2. Reducing generalization error of ENAS 17 and ENAS 7 with unbalanced dataset.

Models	Training set scenario	Test sets	TNR	TPR	Acc.	C.I. at (95%)
ENAS 17	Without any use of augmentation techniques	Internal	90.3	65.3	77.8	(72.3, 83.3)
		BUSI_test	89.2	66.0	77.6	(74.2, 81)
		TenD_test	79.8	89.7	84.8	(81.6, 87.9)
	With use of all the augmentation techniques	Internal	89.5	72.6	**81.1**	(75.9, 86.3)
		BUSI_test	80.5	84.1	**82.3**	(79.2, 85.4)
		TenD_test	79.8	94.8	**87.3**	(84.4, 90.2)
	With use of only SVD and mirroring techniques	Internal	89.3	75.6	**82.5**	(77.5, 87.5)
		BUSI_test	82.4	84.8	**83.6**	(80.6, 86.7)
		TenD_test	77.5	95.3	**86.4**	(83.4, 89.4)
	With use of rotation techniques	Internal	89.7	65.0	77.3	(71.8, 82.8)
		BUSI_test	77.7	79.1	78.4	(75, 81.8)
		TenD_test	81.8	92.3	87.1	(84.1, 90)
ENAS 7	Without any use of augmentation	Internal	84.1	81.2	82.6	(77.6, 87.6)
		BUSI_test	84.6	68.0	76.3	(72.8, 79.8)
		TenD_test	68.5	95.1	81.8	(78.4, 85.2)
	With use of all the augmentation techniques	Internal	92.0	69.2	80.6	(75.4, 85.8)
		BUSI_test	83.8	75.4	79.6	(76.3, 83)
		TenD_test	80.8	94.6	87.7	(84.8, 90.6)
	With use of only SVD and mirroring techniques	Internal	92.3	64.6	78.5	(73, 83.9)
		BUSI_test	87.3	66.7	77.0	(73.5, 80.5)
		TenD_test	81.2	86.3	83.8	(80.5, 87)
	With use of rotation techniques	Internal	91.0	69.8	**80.4**	(75.1, 85.6)
		BUSI_test	81.0	80.8	**80.9**	(77.6, 84.1)
		TenD_test	79.1	95.2	**87.2**	(84.2, 90.1)

Reducing Generalization Error by Using Training Set of Unbalanced Classes. As explained in Sect. 2.1, we intend to investigate the effect of unbalanced training set to model overfitting. For this experiment, the unbalanced data set Modeling has been used as the training set for the second stage of the ENAS approach. Separate ENAS17 and ENAS7 architectures are optimally searched, and ENAS17 and ENAS7 models are trained and tested through a 5-fold cross-validation. We want to examine the combined effects of an unbalanced data set when different augmentation methods are used. The results are shown in Table 2. The results indicate that using unbalanced training set for ENAS architectures does improve the performance of ENAS models on external data sets by significantly reducing the generalization error rate of ENAS models. For the ENAS17 models, using all augmentation methods and SVD with mirroring balanced the performance the best across the internal and external test sets. For the trained ENAS7 models, using the rotation methods delivers more stable performances across the internal and external test sets.

3.3 Comparison with Existing Methods

Finally, we aim to compare the extents of model generalization for ENAS architectures with the extents of model generalization for other CNN architectures. For the comparison, we selected VGG16 [31], Resnet50 [32] and Inception V3 [33], well-established CNN architectures pre-trained on the ImageNet data set for object classification, and CNN3 [12], a purposely designed CNN architecture for breast lesion classification from US images. We trained the CNN3 models from scratch with all default hyperparameters settings by using our data sets under different scenarios that were used for training the ENAS models. As for VGG16, ResNet50 and Inception V3 we adapted the pre-trained them by modifying only the last fully connected layer to cope with binary classification.

We first trained CNN3, VGG16, ResNet50 and Inception V3 on the balanced set BModeling and then the unbalanced set Modeling with training sets enlarged by using all data augmentation methods presented in Sect. 2.3, and then evaluated both models on external datasets. Table 3 summarizes the performances of all the models under the same conditions. The results show that all models show signs of overfitting when the internal test results are compared against the external test results. In particular, other CNN models trained with balanced classes show much worse generalization errors on true negative rates than ENAS models although the overfitting effects on the overall accuracies are slightly greater by ENAS models than other CNN models. Other CNN models trained from unbalanced data sets also show significantly reduced generalization errors like ENAS models. It is also clear that ENAS models have consistently outperformed other CNN models in terms of overall accuracies across internal and external tests indicating the potentials of ENAS models in this area of application. The comparative study shows that all CNN models overfit to the limited training data despite the use of data augmentation. It is true that ENAS models trained from balanced class data sets intend to overfit to the data more and hence performed not as good for the unseen external test sets, but the overall higher levels of accuracy by the ENAS models counterbalance the loss of generalization rate.

4 Discussion

Three methods have been attempted to improve generalization of ENAS models on external datasets. Each of these methods has a different degree of impact on the ENAS model generalization. By reducing model complexity by nearly 50%, ENAS7 models deliver a slightly reduced accuracy, but the simplified models do not show significant reduction of generalization errors. On the effect of data augmentation, test results have showed that the models trained on data obtained by the rotation data augmentation technique performed similar to those models trained on data obtained by other data augmentation techniques. In some specific cases, some models such as ENAS7 models trained on unbalanced data set performed even better for external tests than the models based on other techniques of data augmentation.

Table 3. Comparison ENAS generated CNN models with existing CNN models.

Dataset used	Models	Test sets	TNR	TPR	Acc	C.I. at (95%)
BModeling (Balanced)	CNN3	Internal	80.5	75.6	78.1	(70.2, 85.9)
		BUSI test	53.6	97.7	75.7	(72.1, 79.2)
		TenD test	27.5	98.3	62.9	(58.7, 67.2)
	VGG16	Internal	81.7	81.7	81.7	(74.4, 89.1)
		BUSI test	54.8	92.3	73.5	(69.9, 77.2)
		TenD test	48.9	97.4	73.1	(69.2, 77)
	ResNet50	Internal	60.4	84.3	72.4	(63.8, 80.9)
		BUSI test	60.7	85.0	72.9	(69.2, 76.5)
		TenD test	42.8	91.4	67.1	(63, 71.2)
	InceptionV3	Internal	79.8	53.0	66.4	(57.4, 75.4)
		BUSI test	70.1	52.0	61.1	(57, 65.1)
		TenD test	66.7	63.0	64.8	(60.6, 69)
	ENAS17	Internal	86.7	92.0	89.3	(83.5, 95.2)
		BUSI test	65.1	93.3	79.2	(75.9, 82.5)
		TenD test	60.7	97.5	79.1	(75.6, 82.7)
	ENAS7	Internal	90.9	86.7	88.8	(82.8, 94.8)
		BUSI test	69.7	88.0	78.9	(75.5, 82.2)
		TenD test	65.0	96.7	80.9	(77.4, 84.3)
Modeling (Unbalanced)	CNN3	Internal	86.8	71.8	79.3	(74, 84.7)
		BUSI test	71.7	80.8	76.2	(72.7, 79.7)
		TenD test	66.4	91.7	79.1	(75.5, 82.6)
	VGG16	Internal	88.0	54.4	71.2	(62.6, 79.8)
		BUSI test	85.4	63.3	74.4	(70.8, 78)
		TenD test	85.1	55.6	70.4	(66.4, 74.4)

(continued)

Table 3. (*continued*)

Dataset used	Models	Test sets	TNR	TPR	Acc	C.I. at (95%)
	ResNet50	Internal	87.1	34.9	61.0	(51.7, 70.3)
		BUSI test	93.8	24.4	59.1	(55, 63.1)
		TenD test	76.7	63.2	69.9	(65.9, 74)
	InceptionV3	Internal	79.6	37.7	58.6	(49.2, 68)
		BUSI test	81.6	28.1	54.9	(50.8, 59)
		TenD test	76.4	40.4	58.4	(54.1, 62.7)
	ENAS 17	Internal	89.5	72.6	81.1	(75.9, 86.3)
		BUSI test	80.5	84.1	82.3	(79.2, 85.4)
		TenD test	79.8	94.8	87.3	(84.4, 90.2)
	ENAS 7	Internal	92.0	69.2	80.6	(75.4, 85.8)
		BUSI test	83.8	75.4	79.6	(76.3, 83)
		TenD test	80.8	94.6	87.7	(84.8, 90.6)

The most effective method for reducing generalization error of ENAS models is to use the unbalanced training data set at the modelling stage of ENAS process. When unbalanced data set is used for training classifiers, the models typically over-learn from the majority class due to increased priority and significance. We exploited this negative effect to balance out the loss of accuracy by overfitting in the ENAS models by increasing the number of benign cases for our dataset. To further investigate the effect of different class balance ratios on reducing generalization error, we conducted a few controlled tests on various benign vs malignant ratios. Training the ENAS models with training examples at the ratio of 1.20:1, the generalization error did not improve. Then at the ratio to 1.40:1, we noted that the generalization error of TNR reduced around 8% on TenD_test and 3% on BUSI_test. Therefore, we were encouraged in using a higher ratio of 1.93:1 in data set A, and a good generalization on the external data sets was shown in Table 2. The optimal ratio is still a research question for the time being.

This study attempts only a few methods for reducing model overfitting. Several other methods may also help. For instance, dropout rate is one of the common techniques used. Dropout means that each neuron of the fully connected layer is activated with a pre-defined ratio during training. In this experiment, within a small scale, we evaluate the effects of the dropout rates on ENAS17 and ENAS7 models, and then determine the optimal dropout rate for the models. We trained both models with different dropout rates such as (0%, 10%, 20%, 35%, and 50%). It must be recognized that the dropout rate of 20% on the fully connected layer was defined as the default in the original ENAS modelling. In this experiment, unbalanced data sets with all augmentation methods were used for training ENAS17 and ENAS7 models. We used a single split 80% for training and 20% for testing for the internal test whereas TenD_test and BUSI_test sets were still used as independent data sets. Our test results show that different dropout rates do have only marginal impacts on the effect of overfitting. The results also showed that the true positive rate and overall accuracy decreased a lot more for the ENAS17 models trained with dropout rates of 10% and 50% than those of the ENAS17 models trained with the default dropout rate of 20%. We therefore decided to maintain 20% as an optimal dropout rate ENAS17 models. The results also show that the ENAS7 models with dropout rate 35% perform better than ENAS7 with 20%. More investigation is still needed on this aspect of reducing model overfitting.

5 Conclusion

This paper has explored different approaches for reducing generalization error of ENAS models and improving their performance for breast cancer classification from ultrasound images. The first part of our work evaluates the ENAS models' performance not only on internal test set but also independently sampled external test sets collected from different medical centers. Our study has shown that generalization is an issue with ENAS models. The second part of our work is concerned with mitigating the overfitting of ENAS models through various means. The most effective technique for reducing generalization error for ENAS models is the unbalanced dataset technique which has improved the generalization of ENAS models in overall testing accuracy and TNR for both ENAS17 and ENAS7 models. The ENAS17 models trained on unbalanced data achieved more stable performance on internal as well as external test sets. The possible explanation for these results is that some of the important features of benign lesions may be lost during training, and therefore increasing benign cases for the training set would counterbalance such a loss.

In the future, we plan to modify the ENAS architecture in terms of operations used in search space and backbone structure to improve the network architecture's capability to generate more accurate and more robust CNN models for breast lesion classification from ultrasound images. We also plan to investigate more advanced data augmentation techniques for reducing generalization errors.

Acknowledgments. This research is sponsored by TenD Innovations.

References

1. The International Agency for Research on Cancer (IARC) report. World Cancer Day 2021: Spotlight on IARC research related to breast cancer. International Agency for Research on Cancer. https://www.iarc.who.int/featured-news/world-cancer-day-2021/. Accessed 12 May 2021
2. Stavros, T.A., Thickman, D., Rapp, L., Dennis, M.A., Parker, S.H., Sisney, G.: Solid breast nodules : use of sonography to distinguish lesions. Radiology **196**, 123–134 (1995)
3. Xie, X., Niu, J., Liu, X., Chen, Z., Tang, S., Yu, S.: A Survey on incorporating domain knowledge into deep learning for medical image analysis. Med. Image Anal. **69**, 101985 (2021)
4. Zhu, Y.C., et al.: A generic deep learning framework to classify thyroid and breast lesions in ultrasound images. Ultrasonics **110**, 106300 (2021)
5. Goodfellow, Y., Bengio, Y., Courville, A.: Deep Learning. MIT Press, USA (2016)

6. Wistuba, M., Rawat, A., Pedapati, T.: A survey on neural architecture search, vol. 20, pp. 1–21 (2019). [Online]: http://arxiv.org/abs/1905.01392. Accessed 4 Mar 2021

7. LeCun, Y., Bottou, L., Bengio, Y., Haffner, P.: Gradient-based learning applied to document recognition. Proc. IEEE **86**(11), 2278–2324 (1998)

8. Krizhevsky, A., Sutskever, I., Hinton, G.E.: ImageNet classification with deep convolutional neural networks. Adv. Neural. Inf. Process. Syst. **25**, 1097–1105 (2012)

9. Litjens, G., et al.: A survey on deep learning in medical image analysis. Med. Image Anal. **42**, 60–88 (2017)

10. Tajbakhsh, N., et al.: Convolutional neural networks for medical image analysis: full training or fine tuning? IEEE Trans. Med. Imaging **35**(5), 1299–1312 (2016)

11. Han, S., et al.: A deep learning framework for supporting the classification of breast lesions in ultrasound images. Phys. Med. Biol. **62**(19), 7714–7728 (2017)

12. Xiao, T., Liu, L., Li, K., Qin, W., Yu, S., Li, Z.: Comparison of transferred deep neural networks in ultrasonic breast masses discrimination. Biomed. Res. Int. **2018**, 1–9 (2018)

13. Masud, M., Eldin Rashed, A.E., Hossain, M.S.: Convolutional neural network-based models for diagnosis of breast cancer. Neural Comput. Appl. **5**, 1–12 (2020). https://doi.org/10.1007/s00521-020-05394-5

14. Hijab, A., Rushdi, M.A., Gomaa, M.M.: Breast cancer classification in ultrasound images using transfer learning. In: Proceedings of Fifth International Conference on Advances in Biomedical Engineering (2019)

15. Zhang, H., Han, L., Chen, K., Peng, Y., Lin, J.: Diagnostic efficiency of the breast ultrasound computer-aided prediction model based on convolutional neural network in breast cancer. J. Digit. Imaging **33**(5), 1218–1223 (2020). https://doi.org/10.1007/s10278-020-00357-7

16. Zoph, B., Le, Q.V.: Neural architecture search with reinforcement learning, pp.1–16 (2017). [Online]: http://arxiv.org/abs/1611.01578. Accessed 12 May 2021

17. Zoph, B., Le, Q.: Learning transferable architectures for scalable image recognition. In: Proceedings of the IEEE Conference on Computer Vision and Pattern Recognition (CVPR), pp. 8697–8710 (2018)

18. Pham, H., Guan, M.Y., Zoph, B., Le, Q.V., Dean, J.: Efficient neural architecture search via parameters sharing. In: Proceedings of International Conference on M.L, pp. 4095–4104 (2018)

19. Liu, C., et al.: Progressive neural architecture search. In: Ferrari, V., Hebert, M., Sminchisescu, C., Weiss, Y. (eds.) ECCV 2018. LNCS, vol. 11205, pp. 19–35. Springer, Cham (2018). https://doi.org/10.1007/978-3-030-01246-5_2

20. Gessert, N., Schlaefer, A.: Efficient neural architecture search on low-dimensional data for OCT image segmentation. arXiv preprint arXiv:1905.02590 (2019)

21. Dong, N., Xu, M., Liang, X., Jiang, Y., Dai, W., Xing, E.: Neural architecture search for adversarial medical image segmentation. In: Shen, D., et al. (eds.) MICCAI 2019. LNCS, vol. 11769, pp. 828–836. Springer, Cham (2019). https://doi.org/10.1007/978-3-030-32226-7_92

22. Mohammed, A., Du, H., AlZoubi, A.: An ENAS based approach for constructing deep learning models for breast cancer recognition from ultrasound images. In: Proceedings of MIDL Conference. arXiv preprint arXiv:2005.13695 (2020)

23. Recht, B., Roelofs, R., Schmidt, L., Shankar, V.: Do ImageNet classifiers generalize to ImageNet? In: Proceedings of International Conference on Machine Learning, pp. 5389–5400 (2019)

24. Rice, L., Wong, E., Kolter, Z.: Overfitting an adversarially robust deep learning. In: Proceedings of International Conference on Machine Learning, pp. 8093–8104. PMLR (2020)

25. Samala, R.K., Chan, H.P., Hadjiiski, L.M., Helvie, M.A., Richter, C.D.: Generalization error analysis for deep convolutional neural network with transfer learning in breast cancer diagnosis. Phys. Med. Biol. **65**(10), 105002 (2020)

26. Zeimarani, B., Costa, M.G.F., Nurani, N.Z., Bianco, S.R., De Albuquerque Pereira, W.C., Filho, C.F.F.C.: Breast lesion classification in ultrasound images using deep convolutional neural network. IEEE Access. **8**, 133349–133359 (2020)

27. Jiang, Y., Zhao, C., Dou, Z., Pang, L.: Neural architecture refinement: a practical way for avoiding overfitting in NAS. arXiv preprint arXiv:1905.02341 (2019)

28. Al-Dhabyani, W., Gomaa, M., Khaled, H., Fahmy, A.: Dataset of breast ultrasound images. Data Brief **28**, 104863 (2020)

29. Johnson, J.M., Khoshgoftaar, T.M.: Survey on deep learning with class imbalance. J. Big Data **6**(1), 1–54 (2019). https://doi.org/10.1186/s40537-019-0192-5

30. Buda, M., Maki, A., Mazurowski, M.A.: A systematic study of the class imbalance problem in convolutional neural networks. Neural Netw. **106**, 249–259 (2018)

31. Simonyan, K., Zisserman, A.: Very deep convolutional networks for large-scale image recognition. arXiv preprint arXiv:1409.1556 (2014)

32. He, K., Zhang, X., Ren, S., Sun, J.: Deep residual learning for image recognition. In: Proceedings of the IEEE Conference on Computer Vision and Pattern Recognition, vol. 2016, pp. 770–778 (2015)

33. Szegedy, C., Vanhoucke, V., Ioffe, S., Shlens, J., Wojna, Z.: Rethinking the inception architecture for computer vision. In: Proceedings of the IEEE Conference on Computer Vision and Pattern Recognition, vol. 2016, pp. 2818–2826 (2016)

Image Enhancement, Quality Assessment, and Data Privacy

Comparison of Privacy-Preserving Distributed Deep Learning Methods in Healthcare

Manish Gawali[1]([✉]), C. S. Arvind[2], Shriya Suryavanshi[1], Harshit Madaan[1], Ashrika Gaikwad[1], K. N. Bhanu Prakash[2], Viraj Kulkarni[1], and Aniruddha Pant[1]

[1] DeepTek Inc, Pune, India
manish.gawali@deeptek.ai
[2] Singapore Bioimaging Consortium - A*Star, Singapore, Singapore

Abstract. Data privacy regulations pose an obstacle to healthcare centres and hospitals to share medical data with other organizations, which in turn impedes the process of building deep learning models in the healthcare domain. Distributed deep learning methods enable deep learning models to be trained without the need for sharing data from these centres while still preserving the privacy of the data at these centres. In this paper, we compare three privacy-preserving distributed learning techniques: federated learning, split learning, and SplitFed. We use these techniques to develop binary classification models for detecting tuberculosis from chest X-rays and compare them in terms of classification performance, communication and computational costs, and training time. We propose a novel distributed learning architecture called SplitFedv3, which performs better than split learning and SplitFedv2 in our experiments. We also propose alternate mini-batch training, a new training technique for split learning, that performs better than alternate client training, where clients take turns to train a model.

Keywords: Privacy-preserving · Distributed deep learning · Federated learning · Split learning · SplitFed · Medical imaging

1 Introduction

There is a shortage of labeled data available in the healthcare domain, and even if it is available, healthcare data is commonly distributed and needs to be aggregated at a centralized storage site so that deep learning models can be trained. However, most of the healthcare centers and laws at the country level such as the General Data Protection Regulation (GDPR) [19] and the Health Insurance Portability and Accountability Act (HIPAA) [2] are rightfully protective of the data and do not allow free sharing of data across computer networks and national boundaries. Distributed learning methodologies solve this problem by enabling models to train using data from various healthcare centers without compromising the privacy of the data at these centers.

© Springer Nature Switzerland AG 2021
B. W. Papież et al. (Eds.): MIUA 2021, LNCS 12722, pp. 457–471, 2021.
https://doi.org/10.1007/978-3-030-80432-9_34

1.1 Federated Learning

Federated learning (FL) [7,10,11] is a distributed learning method that enables training of neural network models across multiple devices or servers without the need for movement of data. This is in contrast to centralized training where all the data samples from various data sources have to be collected at a centralized processing site. In FL, multiple federated rounds are performed to obtain a robust model. The workflow for one federated round (Fig. 1) consists of the following steps: (i) pushing the global model from the main server to the clients (healthcare centers), (ii) training models on all healthcare center servers, and sending the local updates to the main server. (iii) The main server aggregates the updates received from the centers and upgrades the global model using federated averaging algorithm. This new global model is robust as it has learned from a large and diverse set of data [14]. Also, each healthcare center benefits as it can use a model which has also learned from some other healthcare center's data.

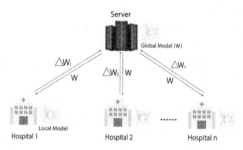

Fig. 1. Federated Learning - n hospitals collaborate to train a global model. The global model W, which is present at the server is transmitted to all hospitals, where each hospital trains a model with their own data. The server collects the updates ΔW_i's from all hospitals and aggregates them to produce a single global update. This global update is used to tweak the global model W.

1.2 Split Learning

Split learning (SL) [3] consists of training a machine learning model across multiple hosts by splitting the model into multiple segments. In the simplest split learning configuration called label-sharing configuration in which labels for data are present on the server, each client (healthcare center) performs one step of forward propagation step till a particular layer called the cut layer [3] as shown in Fig. 2. The outputs at the cut layer are sent to the server, where the forward propagation is carried out on the rest of the network to generate predictions. The training loss is calculated at the server, using labels and predictions. A back-propagation step is performed on the network that is present at the server, i.e. up to the cut layer. The gradients are sent back to the client so that a backpropagation step can be carried out on the first segment of the model. This process is repeated multiple times to obtain a final model. The server cannot access the raw local client data in the training process, thus preserving the privacy of the client.

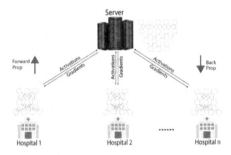

Fig. 2. Split Learning - The neural network architecture is segmented into two parts. The first segment resides at the hospital (client) and the second segment (common across all hospitals) resides on the server. n hospitals collaborate to train a global server-side model and n client-side models.

1.3   SplitFed

SplitFed learning (SFL) is a new decentralized machine learning methodology proposed by Thapa et al. [18], which combines the strengths of FL and SL. In the simplest configuration called the label sharing configuration, the entire neural network architecture is 'split' into two parts. Instead of training the client networks sequentially, Thapa et al. proposed training the client networks parallelly, which is a property drawn from FL. There are two variants of splitfed: SplitFedv1 (SFLv1) and SplitFedv2 (SFLv2). In SFLv1, clients perform a forward propagation step in parallel on their respective data and send the activations obtained at the cut layer to the main server. The main server performs forward propagation on the server-side network for all client activations in parallel. Subsequently, the server performs a backpropagation step and sends back the gradients to respective clients. At this time, the main server updates the server-side network using a weighted average of gradients obtained from backpropagation step. The clients perform a backpropagation step using the gradients obtained from the server and send the updates to fed server as shown in Fig. 3. Fed server averages the updates received from all clients and sends out a single update to all clients. The clients use this aggregate update to tweak their models. Therefore, the client and server-side networks are synchronized. In SFLv2, the training of the server-side network is sequential; i.e., clients perform forward propagation and backpropagation one by one sequentially. The client networks are synchronized at the end of each epoch by averaging all client updates at the fed server.

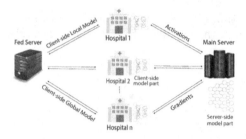

Fig. 3. SplitFed - The neural network architecture is segmented into two parts: client-side model and server-side model. n hospitals collaborate to train a global server-side or n server-side models and global client-side or n client-side models depending upon the variant. Fed server averages updates from client-side models and main server averages updates for server-side models.

2   Related Work

Sheller et al. [17] implemented federated learning in the medical domain for the first time. They demonstrated that U-Net models trained on the BraTS dataset using federated learning and models trained by traditional centralized method had similar dice scores. The concept of differential privacy was applied by Li et al. [8] for federated learning. Li et al. used a segmentation model for the BraTS dataset to show that incorporation of differential privacy slows down the convergence of the FL model.

Gupta et al. [12] introduced split learning and applied the U-shaped split configuration in the medical domain. They compared SL with two techniques, centrally hosted and non-collaborative configuration, for two sets of problems: binary classification (fundus images) and multi-class classification (chest X-rays). With an increase in the number of clients, the performance of split learning remained stable, whereas the performance for the non-collaborative technique declined continuously.

Liu et al. [9] used the federated learning framework for different deep learning architectures to detect COVID-19 using chest X-rays. Roth et al. [14] demonstrated that models trained on mammography data from multiple data sources using federated learning perform better than standalone models trained on data from a particular data source.

Prior works have compared distributed learning methods with centralized training but not with other distributed learning methods for application in the medical domain. In this comparative study, we evaluate the cost (in terms of classification performance, training time, communication, and computational costs) of using distributed learning in practice. Further, we contribute to this field by introducing a novel distributed learning architecture called SplitFedv3 (SFLv3) and a new training method called alternate mini-batch training. We implement these innovations and compare them with existing distributed learning techniques and training methods.

3   Data and Methods

This section describes the datasets and experimental setup for distributed learning methods.

3.1   Data

We obtained chest X-ray scans from five different sources. Three of these were private datasets, which we refer to as DT_1, DT_2, and DT_3. The remaining two were publicly available research sets MIMIC [5] referred to as DT_4 and Padchest [1] referred to as DT_5. A team of board-certified radiologists manually annotated these X-ray images using a custom built annotation tool. X-rays which showed indications of infiltrates, nodular shadows, cavitation, breakdown, lymph nodes, pleural effusion, bronchiectasis, fibrosis, scar, granuloma, nodule, pleural thickening, calcification, calcified lymph nodes, calcified pleural plaques were labelled as TB-suspect. Images which did not show these indications were labelled as TB-negative. In addition to these labels, the radiologist also drew polygon masks around the region of interest in which these manifestations were observed. Table 1 describes the dataset distribution and number of training, validation and test data taken from various sources. For each data source, the percentage of images belonging to the class TB-suspect (prevalence) in the training set is 50%. The prevalence in the validation and test sets is 10%. For experimentation, two different image resolution data was considered (i) 224×224 for densenet architecture (ii) 768×768 for U-Net architecture.

Table 1. Distribution of TB CXR images

Data	DT_1	DT_2	DT_3	DT_4	DT_5	Total
Train	3772	1150	1816	880	1090	8708
Validation	500	500	500	500	500	2500
Test	500	500	500	500	500	2500

3.2   Topology and Neural Network Architectures

The experimental network topology consists of one server and five clients, where each client has data from a single data source. The clients are virtual workers i.e. they reside on the same machine as the server. We chose this topology as it is close to the practical setting where hospitals (clients) are likely to have non-I.I.D data.

All of our experiments were done using PySyft [15]. We performed two sets of experiments for classification by varying the model architecture. For the first set, we used DenseNet-121 architecture [4]. For the second set, we used the U-Net architecture [13] with Xception as the backbone. The U-Net architecture is traditionally used for segmentation problems, but we used it for a classification task by deriving probabilistic output from segmentation output. For both sets of experiments, we used binary cross-entropy as the loss function and the Adam optimizer [6] with standard parameters ($\beta_1 = 0.9$ and $\beta_2 = 0.999$) and learning rate of 10^{-4}. The batch size was 64 in DenseNet experiments and 4 for U-Net experiments. DenseNet models were trained for 10 epochs, whereas U-Net models were trained for 5 epochs. These models were trained for the stated number of epochs as they converge within those number of epochs. We saved the model with the least validation loss on the validation set and evaluated it on the test set. The specifications for the machine used for the experiments were 8 GB RAM, Ubuntu 18.04 OS, Tesla T4 16 GB GPU.

3.3   Federated Learning Settings

For federated learning models, we used federated averaging algorithm [10] to update the global neural network model at the end of each federated round (epoch). We do not address the concept of differential privacy for the experiments.

3.4   Split Learning Settings

We experimented with two split learning configurations: the vanilla split learning/label sharing (LS) configuration and the U-shaped split-learning/non-label sharing (NLS) configuration as shown in the Fig. 4. In the LS configuration, the input images remain with the clients and the labels go to the server, whereas in the NLS configuration, the input images and the labels are both present with the clients.

We trained the split learning model using the alternate client (AC) training and the alternate mini-batch (AM) training techniques. In alternate client training, the clients train their networks on their entire data sequentially, and the server network, which is common for all clients, updates sequentially as well. In alternate mini-batch training, a client updates its network on one mini-batch, after which the client next in order takes over. As the number of data samples can vary for each client, if some client finishes up with its mini-batches, then it has to wait until the next epoch starts, during which other clients can continue training on mini-batches sequentially. So, sequential updates on mini-batches distinguish the server-side training in alternate mini-batch training from the server-side training in alternate client training. The advantage of alternate mini-batch training over alternate client training is that it avoids sequential training over client data for the large trainable server-side model, rendering the model training in a more randomized manner like the centralized training setting.

In the DenseNet experiments, the network was split such that first 4 layers are at the client end and the rest of the network is at the server for the label-sharing configuration. For the non-label sharing configuration, the last fully connected layer is present at the client-side in addition to first 4 layers. In the U-Net experiments, the network was split such that first 6 layers are at the client end

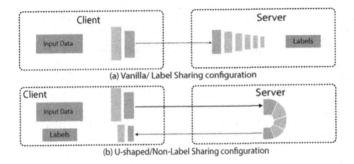

Fig. 4. Split Learning Configurations. In vanilla/label sharing configuration client-side has raw input data and server-side has labels while in U-shaped/non-label sharing configuration both the raw input data and labels are present at client-side

and the rest of the network is at the server for the label-sharing configuration. For the non-label sharing configuration, the segmentation head (consisting of the last 3 layers) is at the client-side in addition to the first 6 layers.

We do not use any form of weight synchronization; all client network segment weights are unique after training. We pass an image from a particular data source from train, validation, and test sets through the corresponding client network. For example, an image from the DT_5 data source, whether it be from train, validation or test set, would be passed for forward propagation through the client network residing on the client having the DT_5 data.

3.5 SplitFed Learning Settings

We have excluded SFLv1 from our experiments due to the unavailability of a supercomputer. We propose a novel architecture called SplitFedv3 which has the potential to outperform SL and SFLv2. As a large trainable part of the network is at the server in SL and SFLv2 and a sequential order is followed for training the model at the clients, "catastrophic forgetting" [16] can happen, where the trained server-side model (for SL and SFLv2) and trained client-side model (for SL) favors the client data it recently used for training and is prone to forget the 'learnings' from the previous client's data. The major advantage of SFLv3 (as shown in Algorithm 1) is that the client-side networks are unique for each client and the server-side network is an averaged version, the same as in SplitFedv1. The problem of catastrophic forgetting is avoided due to averaging of the server-side network. In SFLv2 and SFLv3, the split happens at the same position in the networks, as described in the split learning settings for the DenseNet and U-Net experiments. For SplitFed, we used only the alternate client training technique, and we experimented with both, the LS and the NLS configurations.

3.6 Evaluation Metrics

The distributed learning techniques are evaluated on the following metrics: performance, training time, data communication, and computation. To set a

Algorithm 1. SplitFedv3 algorithm for label sharing configuration. SplitFed network(W) is divided into two parts W^C and W^S. The learning rate η is same for client-side and server-side model. Training client-side and server-side models at round t.

1: **procedure** MAINSERVER_TRAIN   ▷ C_t : set of n_t clients participating at round t
2:    **for** each client $i \in C_t$ in parallel **do**   ▷ $\mathbf{W}^C_{i,t}$: Client-side model of client i at round t
3:        $(\mathbf{A}_{i,t}, \mathbf{Y}_i) \leftarrow$ ClientForwardProp$(\mathbf{W}^C_{i,t})$ ▷ $\mathbf{A}_{i,t}$: Activations from client-side model of client i at round t
4:        Pass $\mathbf{A}_{i,t}$ through $\mathbf{W}^S_t$ (Forward Prop)
5:        Compute $\hat{\mathbf{Y}}_i$
6:        Loss calculation with $\mathbf{Y}_i$ and $\hat{\mathbf{Y}}_i$ ▷ $\mathbf{Y}_i$: true labels, $\hat{\mathbf{Y}}_i$: predicted labels
7:        Calculate $\nabla \ell_i(\mathbf{W}^S_t, \mathbf{A}^S_t)$ (Back Prop)   ▷ $\nabla \ell_i(\mathbf{W}^S_t, \mathbf{A}^S_t)$: Gradient of $\mathbf{A}_{i,t}$
8:        Send $d\mathbf{A}_{k,t} := \nabla \ell_k(\mathbf{A}^s_t; \mathbf{W}^S_t)$ to client i
9:        ClientBackprop$(d\mathbf{A}_{i,t})$
    endfor
10:    Server-side model update: $\mathbf{W}^S_{t+1} \leftarrow \mathbf{W}^S_t - \eta \frac{n_t}{n} \sum_{j=1}^n \nabla \ell_i(\mathbf{W}^S_t; \mathbf{A}^S_t)$
11:
12: **procedure** CLIENTFORWARDPROP$(\mathbf{W}^C_{i,t})$
13:    Set $\mathbf{A}_{i,t} = \phi$
14:    **for** each local epoch from 1 to E **do**   ▷ E : total number of local epochs at client end
15:        **for** batch $b \in \mathcal{B}$ **do**   ▷ $\mathcal{B}$: set of local data batches
16:            Forward propagation on $\mathbf{W}^C_{i,b,t}$
17:            Concatenate activations from final layer of $\mathbf{W}^C_{i,b,t}$ to $\mathbf{A}_{i,t}$
18:            Concatenate respective true labels to Y_i
        end
19:    Send $\mathbf{A}_{i,t}$ and $\mathbf{Y}_i$ to the main server
20:
21: **procedure** CLIENTBACKPROP$(\mathbf{W}^C_{i,t})$
22:    **for** batch $b \in \mathcal{B}$ **do**
23:        Calculate gradients $\nabla \ell_i(\mathbf{W}^C_{,b,t})$ (Back Prop)
24:        $\mathbf{W}^C_{i,t} \leftarrow \mathbf{W}^C_{i,t} - \eta \nabla \ell_i(\mathbf{W}^C_{i,b,t})$
    end

---

benchmark for performance, we trained a model using the traditional centralized method for both sets of experiments. For evaluating performance, we use threshold diagnostic metrics: AUROC, AUPRC, and threshold-dependent techniques such as F1-score and kappa. Elapsed training time, data communication, and computation are valuable metrics for distributed learning methodologies as they provide information on the feasibility of using a method in practice. We calculate all these three metrics for one epoch of model training.

4   Results

In this section, the performance and feasibility of distributed learning methods across various facets is evaluated and discussed.

4.1   Classification Performance

No distributed learning method achieves the benchmark performance as the centralized model for the DenseNet and U-Net experiments (refer Figs. 5, 6, 7, 8, 9, 10 and Table 2). For DenseNet experiments (label sharing, non-label sharing, and alternate client training) and U-Net experiments (label sharing, alternate client training), SFLv3 performs better than split learning and SFLv2 because the large trainable server-side model is an averaged version of all server-side models which have learned using a particular client's data. Similarly, using alternate mini-batch training improves the performance of DenseNet (label sharing, non-label sharing) and U-Net (label sharing) split learning models as it adds randomization in the learning process of the server-side model as opposed to sequential alternate client training in split learning. Further, U-Net distributed learning models tend to perform better than their DenseNet counterparts. There is a noticeable drop in AUPRC, F1-score and, kappa values in DenseNet experiments. The U-Net federated learning model has the best overall performance, considering all four performance metrics.

Table 2. Performance of distributed learning methods

Methods	Performance							
	DenseNet				U-Net			
	AUROC	AUPRC	F1 score	Kappa	AUROC	AUPRC	F1 score	Kappa
Centralized	0.9568	0.7629	0.72	0.69	0.9569	0.8088	0.75	0.71
FL	0.9114	0.652	0.58	0.52	0.9422	0.7456	0.74	0.71
SL_LS_AC	0.8931	0.5291	0.46	0.37	0.9282	0.7208	0.7	0.65
SL_LS_AM	0.9016	0.6105	0.5	0.42	0.9382	0.7322	0.68	0.64
SL_NLS_AC	0.872	0.5227	0.42	0.31	0.8779	0.6478	0.68	0.63
SL_NLS_AM	0.9347	0.7104	0.62	0.57	0.9036	0.5918	0.62	0.56
SFLv2_LS_AC	0.8634	0.5005	0.42	0.31	0.9146	0.7069	0.66	0.61
SFLv2_NLS_AC	0.8996	0.5998	0.53	0.46	0.8999	0.6912	0.65	0.6
SFLv3_LS_AC	0.918	0.6158	0.56	0.49	0.9319	0.7253	0.73	0.69
SFLv3_NLS_AC	0.9046	0.5906	0.55	0.47	0.9272	0.6314	0.7	0.66

4.2   Elapsed Training Time

Elapsed training time is the wall clock time for training a model for 1 epoch. The time taken to train the centralized and different distributed learning models is shown in Table 3. SL, SFLv2, and SFLv3 models take almost the same time to train depending upon the configuration (label sharing or no label sharing). FL models take significantly less time to train than split learning, SFLv2, and SFLv3, for both sets of experiments as individual local deep learning models train in parallel at clients.

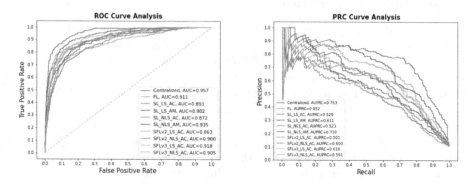

Fig. 5. DenseNet AUROC and AUPRC curves

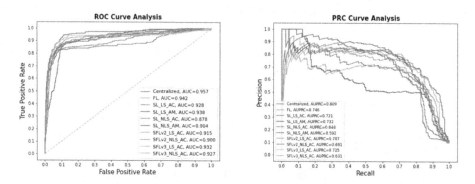

Fig. 6. U-Net AUROC and AUPRC curves

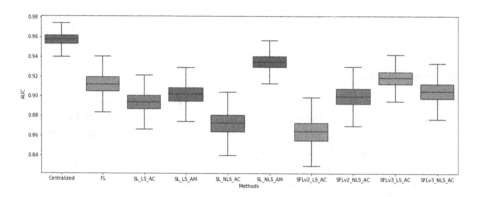

Fig. 7. DenseNet AUC ROC confidence intervals

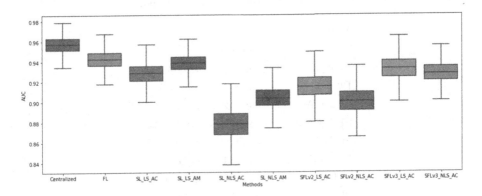

Fig. 8. U-Net AUC ROC confidence intervals

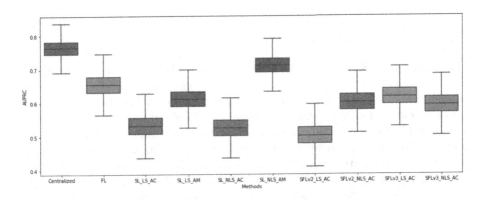

Fig. 9. DenseNet AUPRC confidence intervals

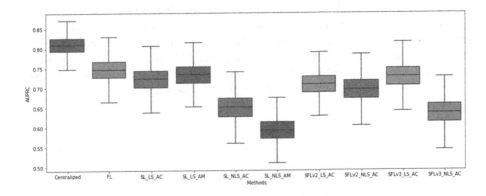

Fig. 10. U-Net AUPRC curves

Table 3. Time Taken for training: 1 epoch

Methods	Time	
	DenseNet	U-Net
Centralized	1 min 40 s	44 min 51 s
FL	2 min 13 s	57 min 51 s
SL_LS_AC	5 min 23 s	170 min 27 s
SL_LS_AM	5 min 22 s	170 min 23 s
SL_NLS_AC	5 min 29 s	279 min 17 s
SL_NLS_AM	5 min 42 s	279 min 4 s
SFLv2_LS_AC	5 min 24 s	170 min 40 s
SFLv2_NLS_AC	5 min 43 s	280 min 45 s
SFLv3_LS_AC	5 min 23 s	172 min 21 s
SFLv3_NLS_AC	5 min 44 s	279 min 24 s

4.3 Data Communication

The amount of back-and-forth data communication that takes place between the server and all clients is shown in Table 4. One epoch consists of training a model on train data and validating it on validation data for saving the weights. The data communication in federated learning consists of sending a model back and forth between the server and clients, whereas data communication for SL models consists of transfer of activations and gradients in training mode and transfer of activations in evaluation mode (validation). More data transfer occurs in non-label sharing configuration than the label-sharing configuration of SL as there is an additional overhead of activations to be sent from the server to the client. SFLv2 has an additional expense of sending the client network models back-and-forth before and after averaging. Here, the client model segments are small in size (in the range of bytes) and have no significant effect on data communication for both DenseNets and U-Nets. In SFLv3, the server model segment needs to be averaged, but as it resides on the server, there is no need for transfer of the server model segment. The amount of data transfer in SL, SFLv2, and SFLv3 is enormous. Unless a strong network with high bandwidth is used, these methods seem infeasible to be used in practice. The data transfer in Federated Learning is low, which makes it suitable for use in practical settings.

Table 4. Data communication (in GB) for 1 epoch

Methods	Data communication	
	DenseNet	U-Net
Centralized	-	-
FL	0.13	0.54
SL_LS_AC	14.89	774.05
SL_LS_AM	14.89	774.05
SL_NLS_AC	18.61	1474.2
SL_NLS_AM	18.61	1474.2
SFLv2_LS_AC	14.89	774.05
SFLv2_NLS_AC	18.61	1474.2
SFLv3_LS_AC	14.89	774.05
SFLv3_NLS_AC	18.61	1474.2

Table 5. Computation (Flops) for DenseNet and U-Net experiments for 1 epoch. The unit for server and average client computations is TFlops and for averaging models is MFlops.

Methods	DenseNet			U-Net		
	Server	Avg Client	Averaging	Server	Avg Client	Averaging
Centralized	64.21	-	-	2129.17	-	-
FL	-	12.84	41.73	-	425.83	172.61
SL_LS_AC	61.53	0.53	-	2064.76	12.83	-
SL_LS_AM	61.53	0.53	-	2064.76	12.83	-
SL_NLS_AC	61.53	0.53	-	2062.84	13.26	-
SL_NLS_AM	61.53	0.53	-	2062.84	13.26	-
SFLv2_LS_AC	61.53	0.53	0.057	2064.76	12.83	0.116
SFLv2_NLS_AC	61.53	0.53	0.069	2062.84	13.26	0.117
SFLv3_LS_AC	61.53	0.53	41.66	2064.76	12.83	172.49
SFLv3_NLS_AC	61.53	0.53	41.68	2062.84	13.26	172.48

4.4 Computation

The computations that occur at the server (Server Flops) and clients (Client Flops) are in the range of TeraFlops. As each client has a different number of data samples, each client would have a different number of computations. We take an average of the computations for all clients and call this measure average client flops, which is in the range of TeraFlops. In federated learning, SFLv2 and SFLv3, the server needs to average out the models. Therefore, we have included averaging model flops as an additional parameter for comparison. Averaging model flops is in the range of MegaFlops. Since an additional part of the network resides on the client in the non-label sharing configuration, there are fewer computations than the label sharing configuration at the server.

The number of computations (Table 5) that take place at the client is significantly greater in FL than SL. A similar number of computations happen at the clients in SL, SFLv2, and SFLv3. These distributed learning techniques leverage the splitting property to keep a large trainable part of the network at the server, drastically reducing the computations at the client end.

5   Conclusion

Our comparative study demonstrated the cost and feasibility of using distributed learning methods in practice. The proposed distributed learning architecture, SplitFedv3, performs better in terms of the four performance metrics (AUC, AUPRC, F1 Score, and kappa) than SL and SplitFedv2. Moreover, the new alternate mini-batch training technique improves the performance of SL models. Apart from classification performance, metrics like training time, data communication, and computational costs play a vital role in deciding the feasibility of a particular distributed deep learning method in practical settings. The SL, SplitFedv2, and SplitFedv3 models take more time to train compared to the FL model and require more data communication. SL, SplitFedv2, and SplitFedv3 would need a high-speed network with large bandwidth to train in practical setting. However, the FL model has higher computational costs. To train an FL model, clients would require a good number of computational resources to carry out heavy computations. Unless clients have access to GPUs, the FL method would take a lot of time to carry out computations. In contrast, the clients in SL, SplitFedv2, and SplitFedv3 models would be able to carry out the small number of computations even without access to GPUs. For our comparative study, if we take all metrics such as performance, elapsed training time, data communication and computation into account, we found out that FL is the best distributed learning method, provided clients have adequate computing power.

References

1. Bustos, A., Pertusa, A., Salinas, J.M., de la Iglesia-Vayá, M.: Padchest: a large chest x-ray image dataset with multi-label annotated reports. Med. Image Anal. **66**, 101797 (2020)
2. Centers for Medicare & Medicaid Services: The Health Insurance Portability and Accountability Act of 1996 (HIPAA) (1996). http://www.cms.hhs.gov/hipaa/
3. Gupta, O., Raskar, R.: Distributed learning of deep neural network over multiple agents. J. Netw. Comput. Appl. **116**, 1–8 (2018)
4. Huang, G., Liu, Z., Van Der Maaten, L., Weinberger, K.Q.: Densely connected convolutional networks. In: Proceedings of the IEEE Conference on Computer Vision and Pattern Recognition, pp. 4700–4708 (2017)
5. Johnson, A.E., et al.: Mimic-cxr-jpg, a large publicly available database of labeled chest radiographs. arXiv preprint arXiv:1901.07042 (2019)

6. Kingma, D.P., Ba, J.: Adam: a method for stochastic optimization. arXiv preprint arXiv:1412.6980 (2014)
7. Konečný, J., McMahan, H.B., Ramage, D., Richtárik, P.: Federated optimization: Distributed machine learning for on-device intelligence. arXiv preprint arXiv:1610.02527 (2016)
8. Li, W., et al.: Privacy-preserving federated brain tumour segmentation. In: Suk, H.-I., Liu, M., Yan, P., Lian, C. (eds.) MLMI 2019. LNCS, vol. 11861, pp. 133–141. Springer, Cham (2019). https://doi.org/10.1007/978-3-030-32692-0_16
9. Liu, B., Yan, B., Zhou, Y., Yang, Y., Zhang, Y.: Experiments of federated learning for covid-19 chest x-ray images. arXiv preprint arXiv:2007.05592 (2020)
10. McMahan, B., Moore, E., Ramage, D., Hampson, S., Arcas, B.A.: Communication-efficient learning of deep networks from decentralized data. In: Artificial Intelligence and Statistics, pp. 1273–1282. PMLR (2017)
11. McMahan, B., Rampage, D.: Federated learning: collaborative machine learning without centralized training data. https://ai.googleblog.com/2017/04/federated-learning-collaborative.html
12. Poirot, M.G., Vepakomma, P., Chang, K., Kalpathy-Cramer, J., Gupta, R., Raskar, R.: Split learning for collaborative deep learning in healthcare. arXiv preprint arXiv:1912.12115 (2019)
13. Ronneberger, O., Fischer, P., Brox, T.: U-Net: convolutional networks for biomedical image segmentation. In: Navab, N., Hornegger, J., Wells, W.M., Frangi, A.F. (eds.) MICCAI 2015. LNCS, vol. 9351, pp. 234–241. Springer, Cham (2015). https://doi.org/10.1007/978-3-319-24574-4_28
14. Roth, H.R., et al.: Federated learning for breast density classification: a real-world implementation. In: Albarqouni, S., et al. (eds.) DART/DCL -2020. LNCS, vol. 12444, pp. 181–191. Springer, Cham (2020). https://doi.org/10.1007/978-3-030-60548-3_18
15. Ryffel, T., et al.: A generic framework for privacy preserving deep learning. arXiv preprint arXiv:1811.04017 (2018)
16. Sheller, M.J., et al.: Federated learning in medicine: facilitating multi-institutional collaborations without sharing patient data. Sci. Rep. **10**(1), 1–12 (2020)
17. Sheller, M.J., Reina, G.A., Edwards, B., Martin, J., Bakas, S.: Multi-institutional deep learning modeling without sharing patient data: a feasibility study on brain tumor segmentation. In: Crimi, A., Bakas, S., Kuijf, H., Keyvan, F., Reyes, M., van Walsum, T. (eds.) BrainLes 2018. LNCS, vol. 11383, pp. 92–104. Springer, Cham (2019). https://doi.org/10.1007/978-3-030-11723-8_9
18. Thapa, C., Chamikara, M.A.P., Camtepe, S.: Splitfed: when federated learning meets split learning. arXiv preprint arXiv:2004.12088 (2020)
19. Voigt, P., Von dem Bussche, A.: The eu general data protection regulation (gdpr). A Practical Guide, 1st edn. Springer International Publishing, Cham 10, 3152676 (2017)

MAFIA-CT: MAchine Learning Tool for Image Quality Assessment in Computed Tomography

Thiago V. M. Lima[1,2,3]([⊠]) [iD], Silvan Melchior[3], Ismail Özden[3],
Egbert Nitzsche[4], Jörg Binder[3], and Gerd Lutters[3]

[1] Department of Radiology and Nuclear Medicine, Luzerner Kantonsspital,
Lucerne, Switzerland
thiago.lima@luks.ch
[2] Institute of Radiation Physics, Lausanne University Hospital and University
of Lausanne, Lausanne, Switzerland
[3] Radiation Protection Group, Kantonsspital Aarau AG, Aarau, Switzerland
[4] Nuclear Medicine and PET Centre, Kantonsspital Aarau AG, Aarau, Switzerland

Abstract. Different metrics are available for evaluating image quality
(IQ) in computed tomography (CT). One of those is human observer
studies, unfortunately they are time consuming and susceptible to variability. With these in mind, we developed a platform, based on deep
learning, to optimise the work-flow and score IQ based human observations of low contrast lesions.

1476 images (from 43 CT devices) were used. The platform was evaluated for its accuracy, reliability and performance in both held-out tests,
synthetic data and designed measurements. Synthetic data to evaluate
the model capabilities and performance regarding varying structures and
background. Designed measurements to evaluate the model performance
in characterising CT protocols and devices regarding protocol dose and
reconstruction.

We obtained 99.7% success rate on inlays detection and over 96%
accuracy for given observer. From the synthetic data experiments, we
observed a correlation between the minimum visible contrast and the
lesion size; lesion's contrast and visibility degradation due to noise levels;
and no influence from external lesions to the central lesions detectability
by the model. From the measurements in relation to dose, only between
20 and 25 mGy protocols differences were not statistically significant
(p-values 0.076 and 0.408, respectively for 5 and 8 mm lesions). Additionally, our model showed improvements in IQ by using iterative reconstruction and the effect of reconstruction kernel.

Our platform enables the evaluation of large data-sets without the
variability and time-cost associated with human scoring and subsequently providing a reliable and relatable metric for dose harmonisation
and imaging optimisation in CT.

Electronic supplementary material The online version of this chapter (https://
doi.org/10.1007/978-3-030-80432-9_35) contains supplementary material, which is
available to authorized users.

B. W. Papież et al. (Eds.): MIUA 2021, LNCS 12722, pp. 472–487, 2021.
https://doi.org/10.1007/978-3-030-80432-9_35

Keywords: Computed tomography · Deep learning · Image quality

1   Introduction

A Swiss survey has shown that the annual effective dose arising from medical imaging is 1.4 mSv with computed tomography (CT) accounting for around 70% (Coultre et al. 2016). One of the key challenges in optimising imaging protocols is the choice of metrics to use. This is especially true since previously used metrics have shown to not be appropriate to currently used iterative reconstruction (IR) algorithms when evaluating image quality (Rotzinger et al. 2018).

A possible approach is to rely on metrics that account for lesion detectability based on observers, sometimes direct response from human observers (Singh et al. 2009, Guimarães et al. 2009) and other times modelling (Verdun et al. 2015). The problem with modelling solutions (which is the case for Channelized Hotelling Observers and other statistical discriminant approaches) is that they always over-perform their human counter parts and require some methods in order to match the human performance. This is normally obtained by either the introduction of noise or even benefiting from machine learning (Kopp et al. 2018). Human observer studies are valuable as they are able to directly measure clinical image quality. Unfortunately, these methods are time consuming, expensive, and the inter- and intra-observer variability is often large (Verdun et al. 2015).

Machine learning techniques, more specifically deep learning using convolutional neural networks (CNNs), take advantage of the most recent development in artificial intelligence and has shown promise in many problems in the world of medical image (Maier et al. 2019). Different tasks associated to medical images are largely benefiting from these techniques for example image detection and recognition, image registration, image reconstruction and computer aided diagnosis (Esteva et al. 2017, Kyono et al. 2018, Wang et al. 2018, Liu and Qi 2019, Liu et al. 2019).

We developed a software platform, benefiting from deep learning approaches, to optimise the work-flow on image quality scoring based human observations of low contrast lesions detectability. The purpose of this work is to describe the whole work-flow and the steps towards validation of this tool for CT image quality assessment.

2   Materials and Methods

From a previously acquired data-set of CT images (Sect. 2.1) different observers (ranging from 1 to 3) scored the visible lesions based on the difference detail curve (DDC), which quantifies the observer's detectability result of known lesion of different sizes and contrast (Sommer et al. 2017, Lima et al. 2018). These were either a medical physics expert or students (nonradiologists) from our department, one of whom had minimal prior experience with medical images. A dedicated platform (Sect. 2.2) was set to benefit from these labelled images in order to predict human response to the lesion detectability task (Sect. 2.3). Once the platform and model were developed, they were evaluated in different scenarios (Sect. 2.4).

2.1   Data Set and Phantom

Our project consisted of 1476 images (hereafter defined as volumes of interest) from 886 CT acquisitions obtained from 43 devices (31 from Siemens [Definition AS, Definition Edge, Definition Flash, Emotion, Force, go.Up and Sensation], 6 Philips [Brilliance and iCT], 5 GE [Optima and Revolution] and 1 Canon-Toshiba [Aquilion CXL]) scanned across Switzerland with defined protocols. The protocols used were set to cover 6 dose levels (reported $CTDI_{vol}$), no modulations (nor current, nor voltage) and at different voltage levels.

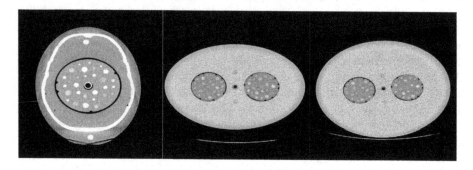

Fig. 1. Axial view of used phantoms. Head, small body and large body, respectively.

For the purpose of this work we used three physical (PMMA) phantoms developed in-house (Fig. 1). Two elliptical body phantoms with different effective diameters and an anthropomorphic head phantom. Two low contrast cylindrical inlays (made of translucent D150 epoxy casting resin), one where the lesions are made with contrast media (CM) material and the other where the lesions are made with the same resin used for the cylinder but with the concentration adjusted to match higher HU values (hereafter defined as native lesions). These inlays were shared between the two body phantoms. Each inlay containing 30 cylindrical rods of different diameters and contrasts (6 lesion sizes [3, 4, 5, 6, 8 and 9 mm diameter] for each of the 5 contrast levels [16, 32, 48, 64 and 80 HU above background]) (Lima et al. 2018). The head phantom consisted of one inlay similar to the one present in the body phantom but this time only with the native lesions. The combination of CT acquisition and the chosen inlay generated said volumes of interest, where for either body phantoms from each CT acquisitions two volumes were obtained. For the head phantom, since only one inlay was present, one volume was obtained from each CT acquisition. Therefore the number of obtained volumes were 1476 from the 886 CT acquisitions. The combination of phantoms and range of protocol parameter can be seen in Table 1.

Table 1. Phantoms used and scanned protocol parameters' range.

Phantom	Volumes of Interest	$CTDI_{vol}$ range (mGy)	Voltage range (kV)
Large Body (32 cm*)	587	7–20	100–120
Small Body (27 cm*)	587	3–13	100–120
Head (19 cm*)	302	30–60	120–140

* Effective diameter from ellipse

2.2 Platform

Figure 2 summarises the work-flow of the platform. We developed a detection algorithm based on RANdom SAmple Consensus (RANSAC) to automatically extract the inlay from the input scans, described in the appendix. The algorithm includes a read-out of a bar-code milled into the bottom of each inlay, identifying its type (i.e. CM or native).

Given the detected position and orientation of the inlay, the inlay description gives the position, size and contrast of each lesion. We cut them out within a defined region of interest (ROI, $d_x \times d_y \times d_z$), using mirror padding if needed. Each of the extracted lesions are then converted to a grayscale voxel-grid using an abdomen window (since the observers used this window for labelling the lesions) and then fed through a model, predicting the probability of being visible to an observer with previously specified parameters (Sect. 2.3).

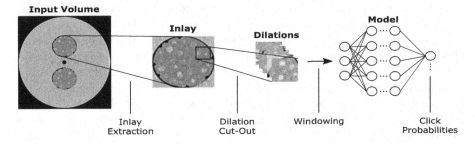

Fig. 2. Overview of our work-flow. We extract a specified inlay from the input image, cut out the lesions and feed them to our model. All steps are in three dimensions, the two-dimensional images are for visualisation purposes only.

2.3 Model Used

Architecture. For a given lesion and specified parameters for observer i, our model predicts the probability of being visible p_i (i.e. the probability of clicking on the lesion if the observer would label the current scan). The model consists of two parts: A convolutional neural network predicting a score s and an observer transformation mapping the score s to p_i.

For our neural network architecture, we use an adaption (see Fig. 3 and appendix) of SqueezeNet (Iandola et al. 2016), which reached high accuracy on a variety of tasks (Deng et al. 2009, Dibra et al. 2018, Kumar and Chellappa 2018) while being very light-weight. The network outputs a single number, the score s. Given this score, we model the probability of being visible to observer i as:

$$p_i = \sigma(\alpha_i s + \beta_i), \tag{1}$$

where σ is the sigmoid function

$$\sigma(x) = \frac{1}{1 + e^{-x}}. \tag{2}$$

So we have two degrees of freedom for each observer. β_i can be interpreted as the detection threshold for observer i, α_i as his consistency. The simplicity of this model allows for interpretability and post-training fit to new observers based on little data only. Since the last convolutional layer in SqueezeNet already allows for a linear transformation of the score (it has already a bias-term - additive offset), we fix $(\alpha_1, \beta_1) = (1, 0)$ to foster reproducibility of the obtained scores.

Training. During training of our multi-label classification task, we minimise the mean over the cross entropy losses for all observers which have labelled the given input lesion. Observer 1 has scored all 886 CT acquisitions, observer 2 has scored 569 and observer 3 has scored 353 CT acquisitions. Within each CT acquisition, and for each inlay (2 for each of the body phantoms and 1 for the head phantom), these labelling consisted in a binary choice (yes and no visibility) for each lesion.

We chose a rather large ROI in x and y directions, resulting in multiple lesions in the input. This large ROI allows the model to better analyse the surrounding of the lesion, but increases the task difficulty, because it needs to learn to focus on the central lesion only. In our experiments, this sometimes caused issues during training, resulting in a bad local minima. To overcome this problem, we use a simple curriculum learning scheme, where at the beginning of the training we mask out parts of the background such that only the lesion of interest (centre of the ROI) is visible to the model and then slowly remove the mask while training progresses. At test-time, no mask is employed anymore.

We use data-set augmentation (flipping and rotating) to improve the generizability of our model. We furthermore use bagging to decrease the prediction variance: we train k (in our experiments, $k = 5$) independent models on different folds of the train-data and average the predictions of all models at test-time.

2.4   Evaluation

The evaluation of our platform was divided into the work-flow, by using a held-out validation set, and performance. This was done firstly on synthetic data and secondly on real-world measurement on a clinical CT device.

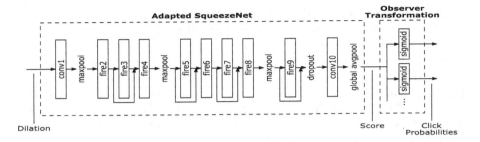

Fig. 3. A visualisation of our model, which is a combination of an adapted SqueezeNet and a custom observer transformation. The fire module architecture is described in Iandola et al. (2016)

Platform Validation. The first part of this evaluation consisted of assessing the accuracy of the inlay detection component of the platform. From the total number of images (1476), including the different inlays and phantom combinations (as described in Sect. 2.1), we calculated the number of correct matching inlays and errors. Individual inlays have physical markers to enable localisation (position and rotation) in addition to bar codes to enable identification (differentiation between CM and native material) (see Sect. 2.2). Correct assessment was defined when all markers and bar codes were correctly identified. The judgement was done manually, visually, and on the whole data-set (886 scans).

The second part consisted in assessing the click accuracy of the model. Since we trained k different models, each on a different fold of the data-set, we could use cross-validation for this step: each individual model is evaluated on its validation set. We furthermore extracted the learnt parameters α, β for the individual observers. All validations steps were performed on unseen labelled data (i.e. validation data did not include training data) and the model performance was evaluated versus its human counterpart.

Synthetic Data. Synthetic data was created in order to evaluate the model capabilities and performance by reproducing different conditions in terms of varying structures and background characteristics. Firstly, we evaluated the contribution of the evaluated lesion's contrast by varying the contrast (ΔHU) of the central lesion from 0 to 64 HU for different lesion sizes. Secondly, in terms of background noise level for different contrast ranges, and, thirdly from the influence of surrounding lesions to the detectability of the central lesion. For this we changed the surrounding lesions' contrast from 0 to 64 HU. For all steps, we compared the predicted visibility (probability of detection) by the evaluated parameter.

Measurements. A set of measurements was designed to evaluate the performance of the model. For this, the large body phantom (see Sect. 2.1) was scanned at a CT clinical device (SOMATON AS, Siemens Healthineers®, Erlangen, Germany) with different acquisition and reconstruction parameters.

We used the local abdomen protocol modified (no modulation) to a chosen dose ($CTDI_{vol}$) and reconstructed the images with both iterative reconstruction (IR) and filtered back projection (FBP) and different reconstruction kernels (ranging from soft tissue to bone). In total we obtained five dose acquisitions (5, 10, 15, 20 and 25 mGy) reconstructed with soft tissue kernel and IR (I30) and FBP (B30). For both the 5 and 10 mGy we also compared the effect of the reconstruction kernels (I30, I40, I50, I70).

We extended our platform to report results as DDC. Initially, our model reported a set of probabilities, one for each lesion size and contrast. Based on the results form the synthetic data (Fig. 6) a sigmoid function was fitted into this set of probabilities. We then define the DDC value at size s as the contrast at which the fitted curve is above 0.5, i.e. as the contrast at which the interpolated model rates the probability of clicking as more likely than not clicking. To visualise the full DDC, we do a piece-wise linear interpolation of these obtained contrast values.

In total for each acquisition/reconstruction combination, we performed five repetitions in order to estimate the obtained variance (from the device and model). A paired student t-test was used to calculate the statistical significance between the estimated contrast thresholds for two specific sizes - 5 and 8 mm lesions (detailed tables can be found in the appendix). These sizes were the reported by different radiologist across the sites where the CT images were taken. These corresponds to which lesion sizes radiologists were confident in report a clinical finding.

3   Results

3.1   Platform Validation

From the extraction tools we have obtained a 99.7% success rate in terms of correctly detecting the inlays (1471 cases where the inlay was correctly placed and 5 wrongly). Figure 4 shows an example of correct and incorrect inlay placement, respectively, (a) and (b).

In terms of the model evaluation (see Sect. 2.4), the model performance was assessed based on the ability to accurately reproduce the choices made by different human observers. Table 2 shows the obtained accuracy to given human observer and the model parameters. Observer 1, which the model was able to predict at highest accuracy, has also shown to be more consistent when scoring the images and he was the only observer that has scored the complete data-set. The larger variation was obtained for the less consistent observer (3).

Figure 5 exemplifies the obtained accuracy in term of a false negative and a false positive obtained by the model. In this example, for visual purposes, the ROI was averaged along the z-axis in order to reduce the noise and enhance the lesion being evaluated.

(a) Correct   (b) Incorrect

Fig. 4. Inlay extraction examples. The figure in the left, (a), shows the correctly place-ment of the inlay where physical edges and markup are correctly aligned. In Figure (b) the markup arrowheads are placed outside the physical indentations in the phantom, which indicates incorrect placement.

Table 2. Observer Evaluation Metrics. Numbers are mean (std) over k learned models in the bagging ensemble, measured on a held-out validation set.

Observer	Accuracy (%)	α	β
Observer 1	96.4 (0.22)	1.000 (0.000)	0.000 (0.000)
Observer 2	91.9 (0.23)	0.388 (0.019)	−0.889 (0.026)
Observer 3	87.8 (0.44)	0.158 (0.021)	1.814 (0.035)

3.2 Synthetic Data

In respect to lesion contrast, we observed a link between the minimum visible contrast and the lesion size, where smaller lesions require higher contrasts to become visible. Figure 6 shows the obtained probability of detection for given lesion size (a) and a visual example for the lesion of radius of 3 voxels in terms of variable contrast (b).

Secondly, we evaluated our mode in respect to the background noise. Figure 7 shows contribution of noise in degrading the visibility. It can also be seen that the smaller the lesion's contrast, this degradation starts at lower noise levels.

Additionally, the result shows the influence of the noise in the detectabil-ity due to the lesion size. In Fig. 7 b, it can be seen that for a larger lesion (5 voxels radius in this figure in comparison to 2 voxels) the degradation is smaller at similar noise levels and higher noise levels are needed between the different contrast levels. For higher contrast levels unrealistically high noise ($\sigma > 50$) is needed in order for the lesion stop being visible.

(a) False negative   (b) False positive

Fig. 5. Wrong Click Examples. Shown is the mean over the z-axis.

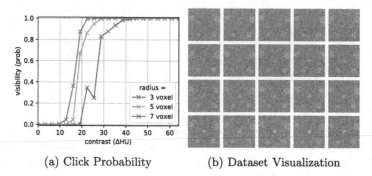

(a) Click Probability   (b) Dataset Visualization

Fig. 6. Synthetic experiments varying the contrast of the central lesion. The model is able to detect larger lesions at lower contrast. The noise was fixed to $\sigma = 20$. Left figure (a) shows the obtained probabilities, and, right (b) the visual description of the contrast variation from top left where contrast is 0 to bottom right where contrast is 64.

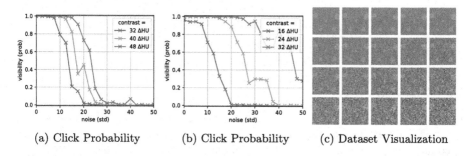

(a) Click Probability   (b) Click Probability   (c) Dataset Visualization

Fig. 7. Synthetic experiments for evaluating the effect of varying background noise. First it shows the visibility of given lesion by noise level. (a) Lesion with 2 voxel radius, (b) lesion with 5 radius. And (c) visual representation. Where top left noise is 0 to bottom right where noise is 50.

Thirdly, we evaluated the contribution from the surrounding lesions to the detectability of the central lesion. For this we changed the surrounding lesion' contrast from 0 to 64 HU (corresponding to the ΔHU ranges present in our phantom). Our results show that our model suffers no interference from these external lesions irrespective of their contrast. In Fig. 8 we plotted the predicted visibility for two contrast levels (10 and 50 ΔHU) for the 5 voxels radius lesion (as seen in Fig. 6) across the evaluated external contrast range. The chosen contrast levels correspond to no and full visibility, respectively (for 5 voxels radius lesion as seen in Fig. 6).

3.3 Measurements

From the designed measurements, firstly we evaluated the effect of the protocol dose in the predicted DDC. Figure 9 shows the dose dependency in detectability

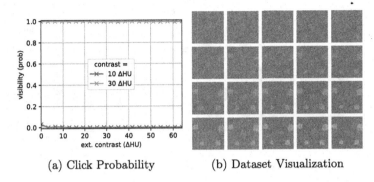

(a) Click Probability   (b) Dataset Visualization

Fig. 8. Synthetic experiments varying the contrast of the outer lesion with 5 voxel radius inner lesion at noise 20 (as in Fig. 6). Left (a) shows the visibility of the 5 voxels radius lesion by external contrast level. Right figure shows the visual representation of the increased external contrast. The chosen contrasts are at the beginning and end of the measured sigmoid in Fig. 6.

for both CM and native inlays of the phantom, where higher the protocol the dose more visible the lower contrast lesions are.

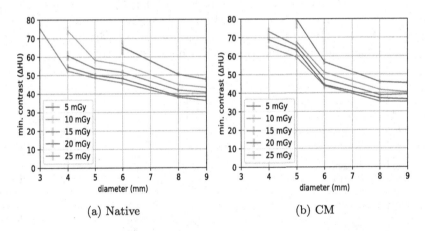

(a) Native   (b) CM

Fig. 9. Predicted DDC curves in respect to protocol's dose for both native and CM inlays.

Only between the 20 and 25 mGy protocols, with the native inlay, no statistical difference is obtained in terms of predicted image quality for the 5 and 8 mm size lesions (student t-test p-values 0.076 and 0.408 for 5 and 8 mm lesions, respectively).

In regards to the reconstruction method, Fig. 10 shows the obtained minimum contrast for two lesion sizes (5 and 8 mm) for different reconstruction methods (IR and FBP) and protocol dose (5, 10, 15, 20 and 25 mGy).

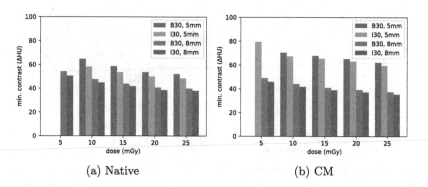

Fig. 10. DDC obtained by different protocol dose and reconstruction method

From these results it can be seen that the minimum observed contrast by our model, for both 5 and 8 mm lesion, is always lower for IR in comparison to the ones with FBP at the same protocol dose level.

Lastly we used the model to score the different acquisitions in respect to the reconstruction kernel. Figure 11 shows the obtained DDC for both 5 and 10 mGy. No significant difference was observed between the lower kernels (I30 and I40) at 5 mGy with p-values of 0.932, 0.982 and 0.952 for the 8 mm lesion in the native inlay, 5 and 8 mm in the CM inlay, respectively. Similar results were found for the 10 mGy protocol, with p-values of 0.820, 0.904, 0.650 and 0.972 for 5 and 8 mm lesion at the native and CM inlays, respectively.

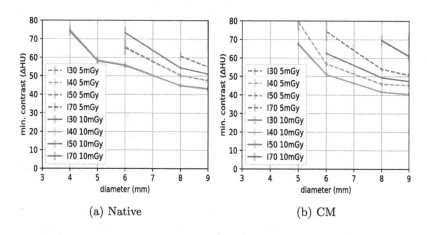

Fig. 11. Model scoring of DDC for different kernel and doses.

A different feature observed was the visibility of the lesions with the protocol reconstructed with the highest kernel (I70) at higher dose (10 mGy). Overall, the results showed that improved DDC were obtained at lower kernels.

4   Discussion

Objective image quality assessment in CT becomes feasible when human observation can be predicted, which will allow for efficient optimisation of scanning protocols and CT imaging systems (Yu et al. 2013). This would subsequently aid in harmonising patient exposure and its correlation to imaging optimisation (Samei et al. 2018).

The main issue when comparing image quality based on human scoring is that the intra- and inter-observer variability are not able to detect small variations in image quality which translates in becoming mostly useful when comparing large variations of dose or at very low image qualities (Goenka et al. 2016). Additionally to being time-consuming and expensive (Verdun et al. 2015, Yu et al. 2013).

The main issues with statistical discriminant approaches is that in order to match human performance a separate step is required (via generalised human performance tests) and that its performance is reduced in realistic setups where the testing dataset is different from the training set, e.g., reconstructed with different algorithms (Brankov et al. 2006, Kopp et al. 2018).

In this work we proposed a platform for evaluating the image quality in CT, based on deep learning techniques, to predict human scoring of low contrast lesions visibility. The main differential is that by using previously scored images using the DDC method, the model is trained on lesion detectability based on human responses of same image and lesion types as the evaluated images and not a surrogate. We validated this platform at different stages from the data preparation to the intrinsic characteristics of the model (Sect. 4.1). Then we evaluated the model performance in respect to synthetic data (Sect. 4.2) and designed measurements (Sect. 4.3).

4.1   Platform Validation

Our proposed platform was able to correctly detect the inlay location in 1471 cases, which represented 99.7% accuracy. Hansis and Lorenz (2015) reported a lower accuracy (96%) when employing the RANSAC approach based on multiple landmarks on detection and localisation of anatomical landmarks. Although a similar approach, the lower accuracy was due to higher complexities of their task and target. For the 5 cases, where the inlay was misplaced, manual placement was performed. Due to the small number of failure cases, we found that the obtained accuracy was sufficient for handling this large data-set.

Secondly, we validated the model in respect to the output given to three human observers used for training the model. As previously discussed, the model was trained with the response from different user low contrast detectability scores. In terms of the accuracy in Table 2, higher accuracy was obtained for the observer 1, which performed the highest number imaging scoring (100% of the data-set). Additionally, the model is also influenced by his consistency. The humans' scoring is based on a binary choice (yes and no visibility) in respect to a subjective threshold. This explains why in previous publications, when comparing to analytic methods, human scoring tend to be less consistent (Yu et al. 2015, Samei et al. 2018). The main benefit of using machine learning techniques for imaging scoring is that temporal variability of human scoring can be reduced since the contribution from external factors are excluded (monitor calibration, light conditions, tiredness, etc.).

4.2   Model Performance – Synthetic Data

Visibility is affected by lesion size, contrast and surroundings and in this work we evaluated the model performance against these parameters with synthetic data in terms of: the contrast of the lesion (Fig. 6), the noise level (Fig. 7) and by the presence of surrounding lesions (Fig. 8).

In Fig. 6 it can be seen, as expected, that the visibility starts at lower contrast values for larger lesions, and subsequently, the smaller lesions requires larger contrast in order to be detected. In Fig. 7 we evaluated the model's prediction in terms of noise and lesion size. Irrespective of the lesion size, the lower the lesion's contrast, the more they are affected by increasing noise levels. In respect to the lesion size, the visibility's degradation between the different lesion contrast start at higher noise levels for larger lesion sizes. The synthetic experiments showed that varying the noise with the larger legion of 5 voxel radius it is nearly impossible to make higher contrast regions vanish needing unrealistically high noise levels.

Lastly, we evaluated the presence of external lesions within the evaluated ROI. The model was designed to evaluate only the lesion at its center so the expectations were that no other variations of lesions within the ROI would alter its predictions. Figure 8 showed that no variation in the predicted visibility is observed at two distinct contrast levels for the 5 voxels radius lesion which indicates that the model is not affected by the presence of surrounding lesions at different contrast levels. These contrast levels were chosen because they represents the two extremes of the variability curve before and after the detection.

4.3   Model Performance – Measurements

Finally we evaluated if the model was sensitive enough for characterising CT protocols and devices in a designed set of measurements in terms of protocol dose, reconstruction kernel and method.

Our model was able to show significant differences on all evaluated tests. In relation to dose, we were able to obtain significant differences between variations of 5 mGy ($CTDI_{vol}$) for most of the lesion sizes evaluated. Only between 20 and 25 mGy protocols for the native inlay that differences were not statistical significant (p-values 0.076 and 0.408, respectively for 5 and 8 mm lesions). For the reconstruction method tests, our model showed the improvements in image quality of IR over FBP. Always at similar dose level protocols, IR significantly over-performed FBP in terms of visibility. For the reconstruction kernel tests, it showed the effect of kernel reconstruction in the detectability. Lower kernel improved the visibility of 5 mm lesions in both evaluated dose levels as expected.

4.4 Limitations

Our study had several limitations. Firstly, we have a limited number of human-predicted data scores in which the model was trained on, the estimation on uncertainty was defined by measurements and measurements were designed without benefiting from modulation, which is present for most clinical cases. Additionally, our phantom design does not take into account potential performance limiting factors such as z-axis resolution.

The use of non-modulated protocols was set for reducing software variances between vendors (Geyer et al. 2015) and the use of cylinders was to reduce variations across z-axis. The estimation of uncertainties based on multiple measurements was done to extract a meaningful value which correlates to the variable noise and resolution properties in CT in addition to the model fluctuations. A more in-depth description of different sources, and their contribution, of uncertainty (device, observer, fitting, prediction, ...) is to be evaluated at later stages.

5   Conclusion

In conclusion, the use of machine learning techniques provides a reliable method for human prediction of low contrast lesion detectability in scoring CT image quality. In our work we were able to develop a platform which handles all required imaging processing and score the image quality based on human scoring of low contrast visibility. This indicator enables the evaluation of large data-sets without the variability and time-cost associated with human scoring. This can provide a reliable and relatable metric for dose harmonisation and imaging optimisation in CT. Further studies are required to confirm the reported findings and to evaluate the performance at different clinical scenarios.

Acknowledgments. We thank Prof. S. Scheidegger, Mr. C. Sommer, Mr. M. Weyland and Ms. C. Durán from the Zurich University of Applied Sciences, ZHAW (Winterthur, Switzerland) for the enlightening discussions, comprehensive support and for the phantom development. Additionally, we thank Mr. Michael Barnard for revising our work to improve the grammar and readability.

References

Brankov, J.G., Wei, L., Yang, Y., Wernick, M.N.: Generalization evaluation of numerical observers for image quality assessment. In: 2006 IEEE Nuclear Science Symposium Conference Record, vol. 3, pp. 1696–1698 (2006). https://doi.org/10.1109/NSSMIC.2006.354225

Le Coultre, R., et al.: Exposure of the swiss population by radiodiagnostics: 2013 review. Radiat. Prot. Dosimetry **169**(1–4), 221–224 (2016)

Deng, J., Dong, W., Socher, R., Li, L.J., Li, K., Fei-Fei, L.: Imagenet: a large-scale hierarchical image database. In: 2009 IEEE Conference on Computer Vision and Pattern Recognition, pp. 248–255. IEEE (2009)

Dibra, E., Melchior, S., Balkis, A., Wolf, T., Oztireli, C., Gross, M.: Monocular rgb hand pose inference from unsupervised refinable nets. In: Proceedings of the IEEE Conference on Computer Vision and Pattern Recognition Workshops, pp. 1075–1085 (2018)

Esteva, A., et al.: Dermatologist-level classification of skin cancer with deep neural networks. Nature **542**, 115–119 (2017). https://doi.org/10.1038/nature21056

Fischler, M.A., Bolles, R.C.: Random sample consensus: a paradigm for model fitting with applications to image analysis and automated cartography. Commun. ACM **24**(6), 381–395 (1981)

Geyer, L.L., et al.: State of the art: iterative CT reconstruction techniques. Radiology **276**(2), 339–357 (2015). https://doi.org/10.1148/radiol.2015132766

Goenka, A.H., et al.: Image noise, cnr, and detectability of low-contrast, low-attenuation liver lesions in a phantom: effects of radiation exposure, phantom size, integrated circuit detector, and iterative reconstruction. Radiology **280**(2), 475–482 (2016)

Guimarães, L.S., et al.: Appropriate patient selection at abdominal dual-energy ct using 80 kv: relationship between patient size, image noise, and image quality. Radiology **257**, 732–742 (2009)

Hansis, E., Lorenz, C.: Landmark constellation models for medical image content identification and localization. Int. J. Comput. Assist. Radiol. Surg, 1–11 (2015). https://doi.org/10.1007/s11548-015-1328-5

Iandola, F.N, et al.: Alexnet-level accuracy with 50x fewer parameters and <0.5 mb model size (2016). arXiv preprint arXiv:1602.07360

Kopp, F.K., et al.: Cnn as model observer in a liver lesion detection task for x-ray computed tomography: a phantom study. Med. Phys. **45**(10), 4439–4447 (2018)

Kumar, A., Chellappa, R.: Disentangling 3D pose in a dendritic CNN for unconstrained 2D face alignment. In: Proceedings of the IEEE Conference on Computer Vision and Pattern Recognition, pp. 430–439 (2018)

Kyono, T., Gilbert, F.J., van der Schaar, M.: Mammo: a deep learning solution for facilitating radiologist-machine collaboration in breast cancer diagnosis (2018). ArXiv, abs/1811.02661

Lima, T.V.L., Schindera, S., Scheidegger, S., Lutters, G.: Connecting the missing piece: a retrospective evaluation of image quality and dose in respect to the parameters variability for a clinical CT protocol. In: ECR 2018 Eurosafe Imaging (2018). https://doi.org/10.1594/esi2018/ESI-0071

Liu, C.C., Qi, J.: Higher SNR PET image prediction using a deep learning model and MRI image. Phys. Med. Biol. **64**(11), 115004 (2019)

Liu, K., et al.: A gentle introduction to deep learning in medical image processing. Radiol. Artif. Intell. **1**(3), 1–8 (2019)

Maier, A., Syben, C., Lasser, T., Riess, C.: A gentle introduction to deep learning in medical image processing. Z Med. Phys. **29**, 86–101 (2019)

Rotzinger, D.C., et al.: Task-based model observer assessment of a partial model-based iterative reconstruction algorithm in thoracic oncologic multidetector CT. Sci. Rep. **8**(17734) (2018)

Samei, E., et al.: Medical imaging dose optimisation from ground up: expert opinion of an international summit. J. Radiol. Prot. **38**, 967–989 (2018)

Singh, S., et al.: Dose reduction and compliance with pediatric CT protocols adapted to patient size, clinical indication, and number of prior studies. Radiology **252**, 200–208 (2009)

Sommer, C., Icken, N., Özden, I., Lutters, G., Scheidegger, S.: Evaluation of low contrast resolution and radiation dose in abdominal CT protocols by a difference detail curve (DDC) method. Curr. Direct. Biomed. Eng. **3**(2), 517–519 (2017)

Verdun, F.R., et al.: Image quality in CT: from physical measurements to model observers. Physica Medica **31**(8), 823–843 (2015)

Wang, Y., et al.: Iterative quality enhancement via residual-artifact learning networks for low-dose CT. Phys. Med. Biol. **63**(21), 215004 (2018)

Yu, L., et al.: Prediction of human observer performance in a 2-alternative forced choice low-contrast detection task using channelized hotelling observer: impact of radiation dose and reconstruction algorithms. Medical Physics **40**(4), 475–482 (2013)

Echocardiographic Image Quality Assessment Using Deep Neural Networks

Robert B. Labs[1]([⊠]), Massoud Zolgharni[1,2], and Jonathan P. Loo[1]

[1] School of Computing and Engineering, University of West London, London, UK
robbie.labs@uwl.ac.uk
[2] National Heart and Lung Institute, Imperial College, London, UK

Abstract. Echocardiography image quality assessment is not a trivial issue in transthoracic examination. As the in vivo examination of heart structures gained prominence in cardiac diagnosis, it has been affirmed that accurate diagnosis of the left ventricle functions is hugely dependent on the quality of echo images. Up till now, visual assessment of echo images is highly subjective and requires specific definition under clinical pathologies. While poor-quality images impair quantifications and diagnosis, the inherent variations in echocardiographic image quality standards indicates the complexity faced among different observers and provides apparent evidence for incoherent assessment under clinical trials, especially with less experienced cardiologists. In this research, our aim was to analyse and define specific quality attributes mostly discussed by experts and present a fully trained convolutional neural network model for assessing such quality features objectively. A total of 1,650 anonymized B-Mode images with dissimilar frame lengths were stratified from most popular ultrasound vendors equipment and clinical quality scores were provided for each echo cine by Cardiologists at England's Hammersmith Hospital which fed our multi-stream architecture model. The regression model assesses the quality features for depth-gain, chamber clarity, interventricular (on-Axis) orientation and foreshortening of the left ventricle. Four independent scores are thus displayed on each frame which compares against cardiologists' manually assigned scores to validate the degree of objective accuracy or its absolute errors. Absolute errors were found to be ±0.02 and ±0.12 for model and inter observer variability, respectively. We achieved a computation speed of 0.0095 ms per frame on GeForce 970, with feasibility for 2D/3D real-time deployment. The research outcome establishes the modality for the objective standardization of 2D echocardiographic image quality and provides a consistent objective scoring mechanism for echo image reliability and diagnosis.

Keywords: Medical imaging · Echocardiography · Quality assessment

1 Introduction

1.1 Overview

During transthoracic (TTE) examinations, a quality image acquisition is a priced clinical enterprise mostly required for expert's assessment and quantification of cardiac functions. Due to its ubiquitousness and non-ionizing advantages, echocardiograms have

© Springer Nature Switzerland AG 2021
B. W. Papież et al. (Eds.): MIUA 2021, LNCS 12722, pp. 488–502, 2021.
https://doi.org/10.1007/978-3-030-80432-9_36

found its significance in antenatal, obstetric and general diagnosis of cardiac infarction. Although, echo image does provide rich information about myocardium, it does not present crisp edges of a well-defined resolution when compared to photographic images. This inherent limitation in echo image resolutions poses a challenge to clinical measurement and interpretation of image features, the reason it is solely consider suitable for experts. In addition, echo image acquisition requires significant skill, absence of which further exacerbates reliability and image quality. To get around these problems, quality assessment is carried out as a control to prevent suboptimal image quality from being quantified. Nevertheless, the method of image quality assessment is a subjective process, where an echocardiography specialist visually inspects the images and decides on what anatomical features present in the image to be pathologically relevant. This process is laced with a spread spectrum of opinion and decision variability [1] even when an image is reassessed by the same operator. These variabilities and uncertainties [2] are found to impair quantification accuracy of cardiac functions, diagnosis and the overall quality of patience care.

A two-dimension (2D) echocardiographic images therefore, requires an objective assessment with a view of achieving optimum image quality, reproducibility, accurate quantifications and to provide platforms for automated diagnosis. Much more of a necessity is the requirement to define a set of mandatory attributes which constitutes an objective standard in quality assessment procedure. Unfortunately, varying consensus still abound among clinicians and equipment manufacturers on what element of 2D echocardiographic images constitutes relevance to standard quality [3] while Cardiologists' are saddled with task for quantifying cardiac functions and provide accurate interpretations within the confine of varying clinical scenarios.

To this end, this research investigated and propose novel method and approach to quality assessment using legacy and domain attributes. These are only paramount to the quantification of the left ventricle (LV) functions, which serves the ultimate purpose of cardiac diagnosis or myocardial assessment. This research employs the use of multi-stream deep convolutional neural network to extract image's quality attributes and computes objective quality score which can be used to guide operators in obtaining optimum image quality in real-time.

1.2 Related Works

Prior to the use of deep convolutional models, several researchers proposed a series of methods for echo image quality assessment which were enhanced by the advent of deep convolutional neural networks. Some of the prominent research on real-time quality assessment using deep convolutional neural networks include four papers and our earlier research namely; [4–7] and [8]. The authors in [6, 7] presented an algorithm based on cardiac view detection. The approach successfully models the detection of chambers in apical A4C echocardiography and admitted that the approach did not guarantee good performance when images consist of significant noise or low contrast pathologies are assessed. In the same vein, the author in [7] presented an algorithm based on convolutional neural networks to tackle cardiac view classification using a multi-class detection approach. The result yields a contradicting detection results on images with low contrast-gain and high contrast-gain which reinforce the conclusion on a model that combines spatial and temporal extraction to guarantee better classification accuracy.

To the best of our knowledge, the most recent work on automated quality assessment, is the work by Luong, C. et al. [8], Labs, R. et al. [7], Dong, J. et al. [9] and Abdi et al. [10]. Abdi's work was based on a regression model to elicit the automatic scoring for five apical standard views. Quality scores were estimated based on the sequence of echo cine loops which include the end systole, end diastole to produce a single quality score per frame per view. The results were impressive with a prediction accuracy of 86% and a computation time of less than one second on a desktop computer. Unfortunately, Abdi's work used a weighted average of quality measure hence, the scores do not provide precise guidance to the aspect of image quality that needs to be optimized. On the other hand, Luong's [8] which investigated the mechanically ventilated TTE on hospitalised patients, yielded certain improvement in performance but with much larger dataset to represent wider population distribution. Luong's method of quality assessment was similar to Abdi's in the sense that the unified model produces a single score per image view across the nine apical standards considered. Luong admitted there exist no reference standard for the evaluation of echocardiographic image quality [8] but a scale of criterial used in many publications does not represent expert visual assessment and consensus on 2D echocardiographic image quality. Similarly, Dong [9] proposed a generic quality control framework on A4C. It considered application of image quality to fetal ultrasound to alleviate the challenges in antenatal investigation. The proposed method detailed the assessment of image quality using two features namely Gain and Zoom. It was considered as first comprehensive quality control system but significantly lacks adequacy for generalisation of quality attributes required for wider use case. Hence, suitability for quality assessment is inherently impaired.

In our study, we defined and modelled four quality attributes as this separately provide the most relevant quality assessment information for operators' feedback during the image acquisition and for generalised standard benchmarking purpose. This work is based on multi-stream regression model with selective qualitative attributes which are progressively distinguished from [10]. The advantage of this novel method of quality assessment provides the specific component of quality need to be optimised and guarantees clinical real-time feedback for optimum image quality in the lab. This means that during the acquisition phase, the operators can assess specific quality element independently, as would be indicated on each four attributes rather than obtaining a weighted average of quality components which is the existing and current assessment method obtainable in the most recent research papers.

To the author's knowledge, there are no published methods on attributes of quality and its assessment method in echocardiography modelling. Our novel approach and quality formulations can be used to assess, optimise and quantify echo images surgically, in real-time.

1.3 Main Contributions

Interpreting the results of the proposed architectures in the literature is not straightforward. This is because a direct comparison of the models' performance would require access to the same patient dataset. At present, no echocardiography dataset and the corresponding annotations for the image quality assessment is publicly available. We, therefore, aimed at evaluating the performance of deep learning models for the automated image quality assessment using an independent (PACS) echocardiography dataset which would be made available at IntSav repository.

In the view of the above, the main contributions of this research can be summarized as follows:

- Novel formulation of quality attributes for 2D echocardiographic images and novel method of its objective assessment
- Annotation of an independent echocardiography patient dataset showing four attributes of image quality namely: fore-shortening, chamber clarity, depth gain and axial orientation for A2C, A4C apical standard views.
- Public release of complete annotated patient dataset to allow future studies and external validation of the new approach or methods.
- Demonstrate the feasibility and applicability of four quality attributes framework which can be adapted for benchmarking, reference standard of evaluation and objective quality scoring of 2D echocardiographic cine loop.

2 Materials and Methods

2.1 Definition of Legacy Attributes of 2D Image Quality

Most of the subjective criterial used for the assessment of image quality [4, 9–11], in echocardiography and implemented under point of care workflow, can be classified under legacy attributes and domain attributes. Hence, we classified chamber clarity, image depth gain, artefacts and probe resolution density under legacy attributes. In the study, an objective consideration for chamber detection/chamber clarity and depth-gain were explored and we proposed as follows:

(i) **Chamber Clarity** is a quality attribute defined by several distinguishable pixel's formation in the echo image to reveal the left ventricle (LV), right ventricle (RV), right atrium (RA) and left atrium (LA). This attribute is summed up using the root mean square RMS contrast, Eq. (1), the visual perception of the subjective element of quality in echocardiography. Contrast provides the perceptual ability to distinguish between luminance levels and very common in medical imaging, to have echo data with low-contrast or very high-contrast yet bearing anatomical structures required for quantifications and diagnosis. Typical examples of varying contrast levels are given in Fig. 1. Low image contrast possesses a significant challenge where limited expertise is found. For cardiac images, the root mean squared (RMS) contrast; Eq. (1), $C_{i,j}$ which does not depend on angular frequency content or spatial distribution is best suited for 2D cardiac images and given by standard deviation of normalised pixel intensity $I_{i,j}$ for a given pathology, the area of interest; where (i, j) represents the *i-th* and *j-th* element of 2D image size M, N; in this case 227×227 used in our modelling. Contrast scores spans ranges from 0–9. 4.5 for the most obvious poor contrast, 6.0 for very high contrast and 9.0 for optimum contrast where relevant anatomical details are clearly visible. Chamber clarity is represented in our model as 'LC' attributes representing LV chamber clarity.

$$C_{i,j} = \sqrt{\frac{1}{MN} \sum_{i=0}^{N-1} \sum_{j=0}^{M-1} (I_{i,j} - \bar{I})^2} \tag{1}$$

(ii) **Depth-Gain** constitute a prominent attribute of 2D echocardiography [6], classified under legacy attribute. Image gain for anatomical features within the first few
centimetres in the image sector may become excessively high or the lower sector of
images becomes excessively low and marred with artefact which potentially affects
visibility or obscure relevant anatomical details as illustrated in Fig. 2. This is pecu-
liar to 2D echocardiography because of the way the ultrasound image is formed
through acoustic frequency and propagated in trabeculated tissues. Improper depth-
gain can induce significant lack of uniformity in the pixel intensities across the
image especially in the lower part of the image sector. Since echocardiographic
images are formed by reflected beams, they are susceptible to depth change, sector
sizes, and pathological differences. Equation (2) describes the intensity of reflected
beam, which is associated with depth gain; where $d^2\phi$ represent the luminous flux
of the infinitesimal area of source $d\sum$, dividing by the product of d_Σ, infinitesimal
solid angle $d\Omega_\Sigma$ and Θ_Σ angle between the normal Ω_Σ to the source $d\sum$. While
luminance is the photometric measure of the pixel luminous intensity per unit area
of light at a given area of interest, brightness therefore is the subjective impression
of the object of luminance $I_{i,j}$ and is measured in candela per square meters cd/m^2.
Since we are considering discrete signal samples in spatial domain, it unlikely that
2D echocardiographic image with varying luminous intensity can be referenced
using Minkowski metric [12]. Therefore, photometric intensity, Eq. (3) as σ_y^2, the
variance of y given a real valued sequence of $y = \{y_1 ... y_n\}$ with $\hat{y}$ as the mean
of y. For optimum values of depth-gain, a time gain compensation (TGC) controls
are often used to either decrease or increase in the near or far fields as appropriate
to the area of interest in apical standard view. Our model learns from ground truth
clinical scores and provides specific scores feedback to the operators depending
on pathology under consideration. Depth-gain score is represented as 'DG' in our
multi stream architecture.

$$I_{i,j} = \frac{d^2\Phi}{d\Sigma d\Omega_\Sigma cos\Theta_\Sigma} \tag{2}$$

$$\sigma_y^2 = \frac{1}{n-1} \sum_{i=1}^{n} (y_i - \hat{y})^2 \tag{3}$$

2.2 Definition of Domain Attributes of 2D Image Quality

As with legacy attributes discussed, domain attributes of quality specify the imaging
criteria which are essentially relevant and exist within cardiology domain [4]. We identified On-axis and Apical foreshortening as crucial attributes that contribute to quality

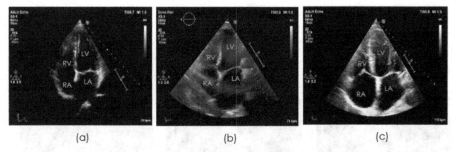

Fig. 1. Samples showing different assessment of chamber clarity of A4C. (a) chamber clarity in very low contrast image, (b) chamber clarity in average contrast image, (c) chamber clarity in optimum contrast images are all significant to diagnosis in TTE.

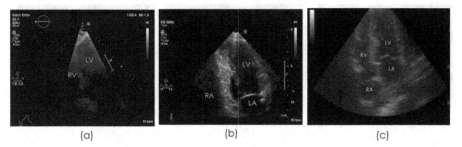

Fig. 2. Samples showing A4C's different assessment in image depth-gain; (a) poor depth-gain obscuring RV, RA & LA of the image sector. (b) inappropriate depth gain obscuring the RV & RA of the image sector, (c) showing four chambers in optimum depth gain. Its common in echocardiology to have a low depth gain images bearing significant anatomical details thereby making it relevant to cardiac diagnosis.

assessment and functional quantifications. We explore these in our study and propose as follows:

(i) **On-Axis Attributes:** In apical four chamber (A4C) standard projection, identifying four heart chambers is crucial to functional assessment and accurate diagnosis. However, there is a complex interaction of the probe's beam cutting through the heart at specific angle, represent significant skill challenge because heart's anatomical features do not present crisp edges as a reason of trabeculated endocardium [8]. The acceptable projection ensures that the LV is positioned at the heart's apex and its visible during the systolic cycles while the interventricular septum orients vertically and divides the sector into two, running through the anterior and posterior plane to yield a four-chamber apical view as illustrated in Fig. 3. On-Axis attributes is a function of probe's beam slicing through the centre of the apical chamber for optimum projection unless when the apex segment is cut off on purpose, which can be for a region of interest (ROI) analysis. We trained our model to recognise the degree of deviation and quantify the magnitude when ultrasound probe is medially or laterally translated which could cause image to be either off-axis significantly, mildly off-Axis or perfectly On-Axis.

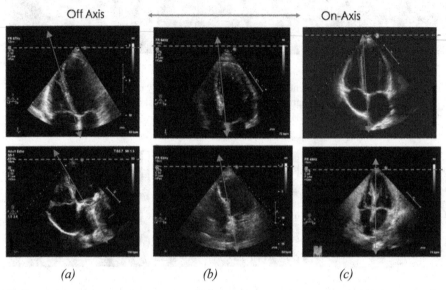

Fig. 3. Showing three samples, (a, b, c) of apical orientation: Significant Off-Axis image quality, Mildly off-axis image quality, On-axis image quality, respectively. An optimum on axis image quality shown the interventricular septum runs vertically down the middle of the screen indicated by blue arrows. (Color figure online)

(ii) **Apical foreshortedness** is common in echocardiography workflow [13] and represent a significant attribute in image quality assessment. We explored the presence and the magnitude of foreshortedness as a domain attribute by which A4C image quality is assessed. Foreshortening describes the non-linear perspective transformation, a kind of structural deformation where changes in size of the areas of interest (AOI) and distances becomes geometrically incongruent [14]. Figure 4, illustrates apical foreshortening during systolic cycle and foreshortening examples in single frame study. Hence, the view projection is defined by many-to-one mapping where distances to the image is inversely proportional to image size resulting in scaling deformation, vanishing point and angular distortions. This deformation introduces a third plane to a 2D image, representing a significant distortion of planar figure, chamber volume becomes inaccurate [15] and extremely essential anatomical features that relates to quantification and diagnosis are obscured. Thus, foreshortening prevents the detection of crucial pathologies in the apical region, like supra-apical infarction, thrombus and reliable clinical measurements [8]. We refer to this transformation in terms of the product of homogenous properties of the N+1 image in Eq. (4) (Fig. 5).

$$\begin{bmatrix} 1 & 0 & 0 & 0 \\ 0 & 1 & 0 & 0 \\ 0 & 0 & 1 & 0 \\ 0 & 0 & -\frac{1}{d} & 1 \end{bmatrix} \begin{bmatrix} x \\ y \\ z \\ 1 \end{bmatrix} = \begin{bmatrix} x \\ y \\ z \\ -\frac{1}{d} \end{bmatrix} => \left(-d\frac{x}{z}, d\frac{y}{z}, \right) \tag{4}$$

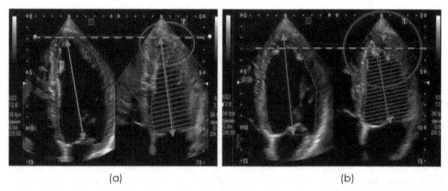

(a)   (b)

Fig. 4. Frame samples showing the effect of foreshortening in LV quantification. apex should remain the same during diastolic and systolic cycle (a) No foreshortening - the apex remains virtually the same during diastolic and systolic cycle, (b) Apex position alters during systolic cycle indicating apical foreshortening.

LV Foreshortening ⟵——————————————⟶ No LV Foreshortening

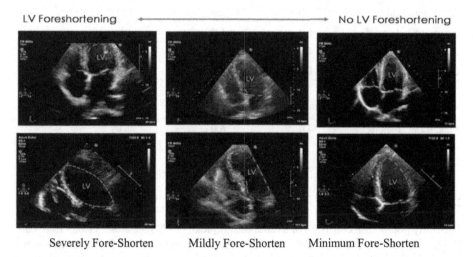

Severely Fore-Shorten   Mildly Fore-Shorten   Minimum Fore-Shorten

Fig. 5. Our model is trained to recognize 3 levels of foreshortening: severe foreshortening, mild foreshortening and zero foreshortening attributes.

2.3 Ground Truth Definition (Expert's Manual Score Criteria)

To model objective quality score for cardiac images, experts visually inspect each scan images using a developed interface that closely represent how the images are displayed during TTE in the laboratory. Two experts independently provided ground truth annotations using the criteria summarised in Table 1. On each quality attributes annotation, an image with maximum score range of 4.0, 6.0, 9.0 are considered as low quality, average quality and optimum quality, respectively. These manual scores are represented as ground truth (Q_{GTi}) in our model.

Table 1. Manual scores criteria for on-axis, clarity, depthgain & foreshorten assessment ground truth

Assessed element	Poor quality	Average quality	Optimum quality
Correct cardiac apex	2.0	4.0	6.0
Septum visible	1.0	1.5	2.0
Interatrial septum visible	1.0	1.0	1.0
Max. score (On-Axis)	**4.0**	**6.5**	**09**
2/4 chambers clarity	1.5	2.5	4.0
Mitral valve clarity	1.5	2.0	3.0
Tricuspid valve clarity	1.0	1.5	2.0
Max. score (LV clarity)	**4.0**	**6.0**	**09**
Apex signal gain	2.0	3.0	5.0
Basal signal gain	1.0	2.0	3.0
No excess gain artefacts	1.0	1.0	1.0
Max. score (depth-gain)	**4.0**	**6.0**	**09**
LV apex visibility	**1.0**	**2.0**	**3.0**
Normal-shaped diastole	1.5	2.0	3.0
Normal-shaped systole	1.5	2.0	3.0
Max. score (f-shorten)	**4.0**	**6.0**	**09**

2.4 Data Sources

The study population consisted of a random sample of (PACS2-Dataset) 1,039 Echocardiographic studies from patients with age ranges from 17 and 85 years, who were recruited from patients who had undergone echocardiography with Imperial College Healthcare NHS Trust. The acquisition of the images had been completed by experienced Echocardiographers using ultrasound equipment from GE healthcare (Vivid.1) and Philips Healthcare (iE33 xMatrix) manufacturers according to the standard protocols.

Ethical approval was obtained from the Health Regulatory Agency otherwise known as Integrated Research Application System (IRAS) identifier-243023. Patient automated anonymisation was performed to remove the patient-identifiable information. DICOM-formatted videos were then split into constituent frames, and 20 sequence frames were extracted from each echo cine loop while each frame bearing the same clinical score as its respective original echo cine score from each video to represent arbitrary stages of the heart cycle, resulting in 20,780 frames. The dataset randomly split into training (16,624 frames), validation (2,078 frames), and testing (2,078 frames) sub-datasets in a 60:20:20 ratio.

Consequently, the entire echo cine loops were independently studied between two cardiologists to create ground truth labelling annotation for ED/ES cycle and assigned clinical scores. In order to create an exclusive list for the 3 emerging classes, we implemented a visual aid software scorer in MATLAB, and allows each echo cine loop to playback the length of frames and using the score slider, a single score is assigned to the set of frames. We refer to the labelling as GT1 and GT2 from respective experts.

The disparity between the opinions was determined and found at 0.12 ± 0.08. Thus, each clip belongs to one of the four different attributes in our considerations and falls under one of the three quality ranges. A maximum clinical score of 4.5, 6.9 and 9.9 is taken as poor quality, standard quality and optimum quality respectively while scores less than 4.4 is ignored or render unsuitable for clinical measurement. All scores are normalised to range between (0 and 1) in our model.

2.5 Network Architecture

In the previous research [8], the multivariate network architecture was inspired by a couple of published works including Yang et al. [15] and Abdi et al.[7]. Although using a single network to elicit feature extraction on each quality attributes was investigated, the performance recorded on each attribute vary significantly as network model show high variance to attributes data on chamber clarity and foreshortening. Echo images also presents varying complexities in the data structure hence was considered inadequate to extract features across multiple domain attributes efficiently. Consequently, a multi-stream architecture based on model subclassing was considered with implementation to accept full image size via transformation algorithm. Our model outperforms state-of-the-art models on all chosen attributes.

The multi-stream regression architecture consists of four parallel DCNN, each stream architecture incorporates slightly different component layers built specifically to generalise the different complexities of each quality attributes encountered by adapting a single network architecture. The four models were simultaneously trained on four attributes datasets. The combined streams architecture was then optimised to extract specific quality elements per frame independently. The architecture is depicted in Fig. 6, with respective details for each network stream. We defined a cost function using mean absolute error (MAE), the l_1 loss function via adaptive moment estimation (ADM) algorithm and simultaneously computes the model accuracy of each stream. The annotations, ground truth (Q_{GT}) of the videos were used as the quality score for all constituent frames of that video for model development and predicted score (Q_P) quality were evaluated to determine range of errors with respect to inter observer variability disparity and experts' score.

Each network in the multi-stream architecture accepts image input size of 227 × 227 × 3 as a 20 frames sequence and computes the weighted matrix of each frame pixel in order to perform discrete convolution which yields 2D output feature map (F) as stated in Eq. (7). The combine model yields four objective quality scores on each image/frame which represents the quality attributes as enumerated in 2.1 and 2.2. Of the four streams model architecture, three streams consist of four convolutional layers, interleaved with MaxPool (MP) and Batch Normalization (BN) layers, as depicted in Fig. 6, were optimised for On-Axis, DepthGain and Foreshorten attributes while the fourth, consisting of three convolutional layers was specific to chamber clarity attributes. Each stream shares initial weights, defined in tensor flow API's loss weight's function and rectifier linear unit (ReLU) Eq. (5) was employed for inter-layer activation. While each stream yields a 2D features output, were flattened and are fed sequentially into a two layered, Long Short-Term Memory (LSTM) [16] to extract long term dependencies of images' temporal features. The final, linear layer provides the scores for each quality element through Sigmoid activation function defined in Eq. (6).

$$f(x)_{relu} = \max(0, x) \tag{5}$$

$$f(x)_{sigmoid} = 1/(1 + e^{-x}{}_i) \tag{6}$$

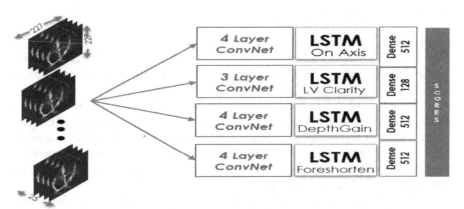

Layers	On-Axis	LV Clarity	DepthGain	Foreshorten
1st Layer	32:3x3	16:3x3	32:3x3	16:3x3
2nd Layer	32:3x3	BN+MP	32:3x3	16:3x3
	MP	32:3x3	BN+MP	BN+MP
3rd Layer	64:3x3	BN+MP	64:3x3	32:3x3
4th Layer	64:3x3	64:3x3	64:3x3	32:3x3
	BN+MP	Dense128	BN+MP	BN+MP
	Dense512		Dense512	Dense512

Fig. 6. Multi-Stream network architecture used in the research showing numbers of kernels, corresponding sizes for each layer. Each stream is optimized to extract specific quality attributes and provides separate predicted scores on each quality attribute.

$$F_{(i,j)k}^{l} = \sum_{i=0}^{n} \sum_{j=0}^{m} w_{i,mn}^{l} F_{(j+m)(k+n)}^{l-1} \tag{7}$$

2.6 Training, Batch Selection, Data Augmentation

Training Hyper-Parameters: The model consists of four regression models arranged in parallel and were simultaneously trained using 5-fold cross validation technique to ensure adequate learning on the dataset and performance was recorded for each model. The hyper parameters learning rate was set at 0.0002 with high momentum 0.95 and decay rate of 0.1 every 15 steps and were reproducibly initialized to minimise possible deviation in score performance. Training was initialised and completed as learning curves converged around 50 epochs.

Batch Selection: The hardware computational cost during training phase ran high as batch selection of 8 and 12 were experimented, memory utilization becomes significantly apparent at batch selection of 12 at a fixed length sequence of 20 than running a batch size of 8 at the same fixed length sequence. Hardware performance difference of 0.13% in terms of computational speed was a negligible trade-off, did not affect the model's ability to properly generalize new test samples.

Data Augmentation: Data augmentation was applied to allow optimum learning sequences for the models; a maximum translation of $[-0.05, +0.05]$ pixels and maximum rotation of $10°$ were applied randomly for horizontal, vertical and rotational angles, respectively. To prevent overfitting in the training phase, we applied batch normalization selectively, at specific convolution layer, early stopping and dropout (rate 0.32) for the training samples. Batch normalisation also helps stabilizes and speeds up convergence during the training phase.

Hardware and Software Resources: Model was implemented using TensorFlow backend. The experiment was carried out on a Z600 Intel i7 Quad Core mini server with 32 GB memory and additional GPU GeForce GTX 970 chipset's Maxwell architecture and featuring 4 GB RAM coupled to 1,664 CUDA cores.

2.7 Evaluation Metrics

Since the model uses multiplex variables for each score attributes, performance was evaluated via MAE and measured against absolute difference between cardiologist's score (Q_{GTi}) and model's predicted automatic scores (Q_{Pi}) for each respective attributes and models, we computed model's class error in Eq. (8) as:

$$Class_{err} = \frac{1}{n} \sum_{i=0}^{n} |Q_{GTi} - Q_{pi}| \tag{8}$$

Minimal error, therefore, indicates best fit and better model performance. The average accuracy was computed in Eq. (9) as:

$$Model_{acc} = 1 - (\frac{1}{n} \sum_{i=0}^{n} |Q_{GTi} - Q_{pi}|) * 100 \tag{9}$$

3   Results and Analysis

Given the complexity of varying pathological features in echo frames, our model could generalise on new echo frame with measured accuracy of 97.68% as shown in Table 2. The error distribution per quality attributes is depicted in Fig. 7, for On-Axis, Clarity, DepthGain and Foreshortening attributes, respectively. The model prediction speed was found to be 0.012 ms per frame for input pixel size of $227 \times 227 \times 3$, which is the assurance for real-time deployment and opportunity for enhancing clinical workflows.

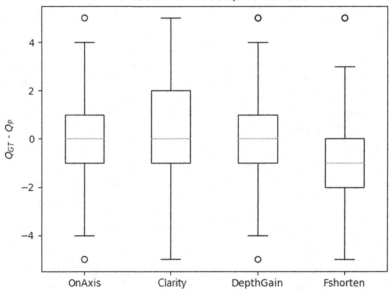

Fig. 7. Box Plot of the error distribution of test samples for each model. The error is computed as the difference between the prediction of the model and ground-truth. The x-axis shows four specific quality attributes of each model.

Table 2. Computed accuracy for the test samples

Models	On-axis	LV clarity	Depth gain	Fore-shortening	Average accuracy
Model accuracy	97.69%	97.51%	97.81%	97.70%	97.68%

3.1   Study Limitation and Future Work

Unlike our previous study [8] where we consider only three quality attributes of A4C, this study is based on four quality attributes suitable for clinical image quality assessment and satisfying wider clinical requirements.

This proposed approach yielded a superior performance in terms of deployability and use case compared to any existing approach in quality assessment which indicates the specific element of image quality that must be optimised in real-time and provides objectivity on quality assessment score for operators' feedback and guidance.

Since this is a novel approach in quality assessment methods, a comparison with any existing approach would make a fair judgement. Unfortunately, existing assessment was only based on weighted average method of quality scoring did not provide any parallel approach to measure by. Hence, its near impossibility to determine equivalence while different dataset of corresponding annotation was utilised. Therefore, we make our dataset with expert annotations on four quality attributes public at IntSav repository to allow external validation by other researchers and equipment manufactures.

Similarly, we have considered A4C and A2C frames as the primary apical view standards to demonstrate the feasibility of clinical application for quality assessment, A2C quantifications may not be a focus under clinical practice suitable for unified quantification, therefore, future study would include other relevant apical view standards like PLAX, PSAX, A5C. again, several global characteristics can be used to distinguish between the different levels of quality and assignment of image quality index. Here, we only considered 4 attributes of image quality for clinical studies and research. A future possible study would include selective criteria that would be suitable for point of care deployment and encompass major laboratory assessments criteria.

Finally, we used two annotations or ground truth labels provided by two expert cardiologist and one accredited annotator. Intra-observer variability can be examined by obtaining additional annotations from human experts and compared with the error in the predicted scores.

4   Conclusion

We have presented the clinical significance and feasibility of developing an automated quality assessment in 2D echocardiographic images that contribute to automated diagnosis and quantification in echocardiology. An automated image quality assessment technique would be significant as part of a system that could accelerate the learning curve for those training in echocardiography and automated quality control process which is required for both clinical and research purposes. This would provide a real-time guidance to less experienced operators, increase their chances of acquiring optimum quality images and enhance diagnostic accuracy of cardiac functions.

References

1. Nagata, Y., et al.: Impact of image quality on reliability of the measurements of left ventricular systolic function and global longitudinal strain in 2D echocardiography. Echo Res. Pract. **5**, 27–39 (2018). https://doi.org/10.1530/ERP-17-0047
2. Liao, Z., et al.: On modelling label uncertainty in deep neural networks: automatic estimation of intra- observer variability in 2d echocardiography quality assessment. IEEE Trans. Med. Imaging. **39**, 1868–1883 (2020). https://doi.org/10.1109/TMI.2019.2959209

3. Sassaroli, E., Crake, C., Scorza, A., Kim, D., Park, M.: Image quality evaluation of ultrasound imaging systems: advanced B-modes. J. Appl. Clin. Med. Phys. **20**, 115–124 (2019). https://doi.org/10.1002/acm2.12544

4. Sprawls, P.: Optimizing medical image contrast, detail and noise in the digital era. Med. Phys. Int. J. **2**(1), 1–8 (2014)

5. Abdi, A.H., et al.: Quality assessment of echocardiographic cine using recurrent neural networks: feasibility on five standard view planes. In: Descoteaux, M., Maier-Hein, L., Franz, A., Jannin, P., Collins, D.L., Duchesne, S. (eds.) MICCAI 2017. LNCS, vol. 10435, pp. 302–310. Springer, Cham (2017). https://doi.org/10.1007/978-3-319-66179-7_35

6. Abdi, A.H., et al.: Automatic quality assessment of apical four-chamber echocardiograms using deep convolutional neural networks. In: Styner, M.A., Angelini, E.D. (eds.) Orlando, Florida, United States, p. 101330S (2017). https://doi.org/10.1117/12.2254585.

7. Labs, R.B., et al.: Automated assessment of image quality in 2D echocardiography using deep learning. In: Conference proceedings ICRMIRO International Conference on Radiology, Medical Imaging and Radiation Oncology, Paris, France, 25–26 June 2020, Part XVII (2020)

8. Luong, C., et al.: Automated estimation of echocardiogram image quality in hospitalized patients. Int. J. Cardiovasc. Imaging **37**, 1–11 (2020). https://doi.org/10.1007/s10554-020-01981-8

9. Dong, J., et al.: A generic quality control framework for fetal ultrasound cardiac four-chamber planes. IEEE J. Biomed. Health Inform. **24**, 931–942 (2020). https://doi.org/10.1109/JBHI.2019.2948316

10. Abdi, A.H., et al.: Automatic quality assessment of echocardiograms using convolutional neural networks: feasibility on the apical four-chamber view. IEEE Trans. Med. Imaging. **36**, 1221–1230 (2017). https://doi.org/10.1109/TMI.2017.2690836

11. Zhang, J., et al.: Fully automated echocardiogram interpretation in clinical practice: feasibility and diagnostic accuracy. Circulation **138**, 1623–1635 (2018). https://doi.org/10.1161/CIRCULATIONAHA.118.034338

12. Nafchi, H.Z., Cheriet, M.: Efficient no-reference quality assessment and classification model for contrast distorted images (2018). ArXiv180402554 Cs. https://doi.org/10.1109/TBC.2018.2818402.

13. Ünlü, S., et al.: EACVI-ASE industry standardization task force. In: Badano, L.P., et al. (eds.) Impact of apical foreshortening on deformation measurements: a report from the EACVI-ASE Strain Standardization Task Force, Eur. Heart J. - Cardiovasc. Imaging. (2019). https://doi.org/10.1093/ehjci/jez189

14. Smistad, E., et al.: Real-time automatic ejection fraction and foreshortening detection using deep learning. IEEE Trans. Ultrason. Ferroelectr. Freq. Control., 1 (2020). https://doi.org/10.1109/TUFFC.2020.2981037.

15. Yang, J., Zhu, Y., Ma, C., Lu, W., Meng, Q.: Stereoscopic video quality assessment based on 3D convolutional neural networks. Neurocomputing **309**, 83–93 (2018). https://doi.org/10.1016/j.neucom.2018.04.072

16. Donahue, J., et al.: Long-term Recurrent Convolutional Networks for Visual Recognition and Description (2016). ArXiv14114389 Cs. http://arxiv.org/abs/1411.4389, Accessed 9 May 2021

Robust Automatic Montaging of Adaptive Optics Flood Illumination Retinal Images

Eva Valterova[1,2]([✉]) [ID], Franziska G. Rauscher[1] [ID], and Radim Kolar[2] [ID]

[1] Institute for Medical Informatics, Statistics, and Epidemiology, Leipzig University, Haertelstrasse 16-18, 041 07 Leipzig, Germany
valterova@vutbr.cz

[2] Faculty of Electrical Engineering and Communications, Department of Biomedical Engineering, Brno University of Technology, Technicka 12, 616 00 Brno, Czech Republic

Abstract. Adaptive optics (AO) flood illumination camera acquires retinal images with a limited field of view, which can be extended by image alignment into one wide field of view montage image. The image alignment into a montage requires efficient and accurate image registration. Since manual registration is demanding and disadvantageous, automatic registration is a beneficial improvement. We propose the first fully automated AO retinal image montage procedure. Here, we present three novel fully automated registration methods, which are based on two established image processing approaches. The first method utilizes scale invariant feature transform (SIFT) in combination with specific image preprocessing. The second method uses the phase correlation (PC) approach and the last method is a connection of PC and SIFT (PC-SIFT) algorithm. In total, 200 images acquired from the left and right eyes of 10 subjects were used for creating the wide field-of-view montage images and compared with manual montaging. The automated image montage was successfully achieved. Alignment accuracy evaluated by normalized mutual information metric showed that the PC-SIFT approach established the most accurate results, these are higher than manual montaging. Therefore, the AO montaging registration methods are able to achieve promising results in accuracy and time demand in comparison with manual montaging. Hence, the latter can be replaced by those fully automated procedures.

Keywords: Retina · Adaptive optics · Flood illumination · SIFT · Phase correlation · Image montaging

1 Introduction

Adaptive optics (AO) technology enables in vivo retinal imaging at cellular resolution. The imaging of individual retinal cell types, such as ganglion cells [14], photoreceptors [10], retinal pigment epithelial cells [18], or also vessel walls [13], offers observation of structural retinal changes and thus disease diagnosis and

© Springer Nature Switzerland AG 2021
B. W. Papież et al. (Eds.): MIUA 2021, LNCS 12722, pp. 503–513, 2021.
https://doi.org/10.1007/978-3-030-80432-9_37

monitoring [6,7,13]. Comprehensive retinal examination requires investigation of a large retinal area, which is limited by the field of view of an AO camera. The observed retinal area can be enlarged via image registration of single images acquired at different positions across the retina and montage into one wide field of view image.

The montage requires the determination of the overlap between images of adjacent retinal regions and accurate registration. One of the possible approaches is a manual montage with various graphics interfaces, e.g., Photoshop (Adobe Systems, Mountain View, California) [2] or ImageJ (National Institutes of Health, Bethesda, Maryland) [5]. Such an approach is highly labor intensive, time-consuming, and further complicated by variable image blur, noise, and contrast. On the other hand, an automatic approach offers fast and even more accurate results.

Limited literature exists on automated AO image registration. To our knowledge, there is only one paper focusing on automated AO flood illumination image registration, but not providing a quantitative analysis of the registration results [19]. Several approaches aim only at the registration of adaptive optics scanning laser ophthalmoscopy (AOSLO) images [3,4,9], but flood illumination AO images were never considered. However, flood illumination AO images differ from AOSLO images in several aspects. Such AO images have a larger field of view and thus lower resolution. Typically in the fovea, the single cones are not recognizable by flood illumination AO technology and thus the images are more blurred in this region. Furthermore, an AO image is captured by a flood illumination camera at once, but the AOSLO image is raster scanned across the retina, therefore varying more in contrast. Thus, montaging of the images acquired by the flood illumination AO camera is more challenging and methods previously used in retinal imaging can not be directly applied.

Most of the published AOSLO image montage methods rely on image transformation estimation from detected specific feature points [19], determined by principal component analysis scale invariant features (PCA-SIFT) [9], SIFT [4] or center locations of cones [3]. The outlier features are eliminated by random sample consensus (RANSAC) algorithm [3,4,9]. Furthermore, registration accuracy is typically evaluated by two metrics - normalized cross-correlation (NCC) and normalized mutual information (NMI) and by comparison with manual montaging [3,4].

Phase correlation (PC) method [8] has been widely used for image registration and image montaging. It provides an efficient and robust method for capturing the mutual shift, rotation, and scale. This approach is not dependent on keypoint detection and correspondence matching. This might be an advantage for flood illumination AO image montaging because the cone distribution has a hexagonal spatial distribution pattern throughout the retina [11]. This creates similar cone packing arrangements on different retinal locations, which can lead to false keypoint correspondence. Furthermore, flood illuminated AO images do not suffer from flexible distortion, which affects the raster scanned

AOSLO images. This makes the PC approach ideal for montaging using the rigid geometric transformation.

Although image registration in retinal imaging is a well-known problem and various approaches have been proposed (e.g., in the area of fundus camera images [1]), AO flood illumination image montaging has never been successfully achieved. Additionally, no accurate method is available for AO flood illumination image registration, especially for AO images capturing foveal avascular areas. In this paper, we propose an automatic approach of SIFT and PC method for flood illuminated AO image montaging. The SIFT approach was in detail described and tested in our previous work [16]. These approaches are compared together with their connection and manual alignment. Alignment accuracy is hereafter evaluated by NMI.

2   Materials and Methods

2.1   AO Image Acquisition and Problem Framework

The study adhered to the tenets of the Declaration of Helsinki and approval for the study was obtained from the Ethics Committee of the Medical Faculty of Leipzig University (209 /18-ek). Retinal images from the macular region were acquired from 10 healthy subjects by commercially available AO retinal camera with flood illumination (rtx1e, Imagine Eyes, Orsay, France). This camera captures images of pixel size 1500 × 1500 with a field of view size of 4° by 4°. Ten images were acquired from the left and the right eye, respectively. The adjacent images were captured at specified locations with sufficient overlap - one image in the center of the fovea, three images in the temporal direction from the fovea in 2° steps, and two further images in each meridian (nasal, inferior, and superior) with the same step size (see Fig. 3 - a). This large dataset consists of 200 images in total (10 single images from the left and right eye of 10 subjects) with retinal positions defined by image acquisition. Due to minor eye movements and unstable eye fixation, the preliminary defined position of each single image is not sufficiently accurate for image montage, but can be utilized as a priori information about the image initial position.

The set of 10 images captured within an eye forms the basis for a montage, where the origin of a coordinate system is determined by the image captured in the foveal location. The remaining 9 images are one after one, according to their distance from the fovea, aligned into the montage. The alignment transformation is estimated from overlapping parts with the central image or other adjacent images already aligned in the previous iteration. The overlapping parts of a reference image x_{ref} and a moving image x_{mov} are shown in Fig. 1 - a,b.

2.2   Phase Correlation

Phase correlation is a well-known method applied for image registration using Fourier transform (FT) [8]. The shift between two images can be estimated if

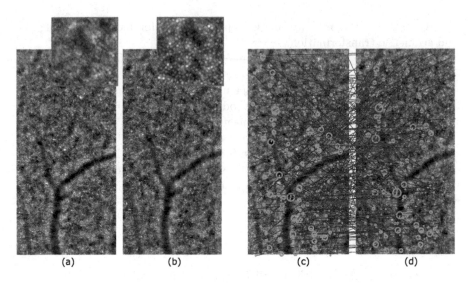

(a)   (b)   (c)   (d)

Fig. 1. Cropped overlapping parts (a) x_{ref} and (b) x_{mov} of two adjacent images. Red frames in (a) and (b) denotes zoom-in parts for highlighting the blur variability. The second pair of images(c–d) corresponds to (a–b) with marked results of SIFT. The green circles denote the detected features localization, scale, and rotation. Lines connect two matched keypoints. Blue lines correspond to incorrect matches and red lines to the correct ones. (Color figure online)

two image functions $f_1(x, y)$ and $f_2(x, y)$ differ only by displacement Δx and Δy

$$f_2(x, y) = f_1(x - \Delta x, y - \Delta y), \tag{1}$$

then the corresponding FTs are related as

$$F_2(u, v) = F_1(u, v)e^{-j2\pi(u\Delta x + v\Delta y)}. \tag{2}$$

The normalized cross-spectrum is thereafter expressed with the complex conjugate of one of the spectra (denoted with $*$)

$$\delta(x + \Delta x, y + \Delta y) = FT^{-1}\left\{\frac{F_2(u, v) \cdot F_1^*(u, v)}{|F_2(u, v) \cdot F_1^*(u, v)|}\right\} \tag{3}$$

The inverse Fourier transform of the normalized cross-spectrum defines the Dirac impulse δ at integer coordinates defining the spatial shift.

The rotation movement can be deduced in a similar manner. The image rotation is identical as the spectrum rotation. When the spectra $F_1(u, v)$ and $F_2(u, v)$ are converted into polar coordinates $M_1(\theta, \rho)$ and $M_2(\theta, \rho)$, the image rotation corresponds to the shift estimated by equation (3) from $M_1(\theta, \rho)$ and $M_2(\theta, \rho)$, respectively.

Before this spectral-based shift estimation, a windowing function must be used to eliminate the influence of a finite boundary effect. We have applied Hanning window to both x_{ref} and x_{mov} images. The images were also filtered by the median filter of window size 3 for noise reduction.

2.3   Scale Invariant Feature Transform (SIFT)

The SIFT registration approach [12] transforms the image into a large collection of keypoint feature vectors, invariant to image translation, rotation, scaling, and partially invariant to illumination changes. The feature descriptors are derived from Gaussian filtering differences and consist of feature localization, rotation, and scaling.

The AO images have to be preprocessed before the SIFT application to increase the similarity and unambiguity of keypoint detection in both images. The AO images are typically distorted by spatially variant image blur, which also varies across images (see Fig. 1 - a,b), thus the localization of a specific feature location is challenging. The spatially dependent blur effect can be suppressed by average filtering with window size 15. This filtering also blurs the single image cones. Further, the AO images also contain noticeably darker and lighter areas, which correspond to vessels and groups of higher reflective or absorptive photoreceptors, respectively. Those areas are barely affected by blur and mutually correspond in overlapping image parts. Therefore, these structures are a rich source of keypoints. The contrast of these areas was highlighted by histogram equalization, which also eliminates the non-uniform illumination and contrast of particular images. These modified images are used for keypoint detection. The identified keypoints in x_{ref} and x_{mov} are matched, as well as its detection, according [12] by *VLFeat* collection of algorithms [17].

An example of detected keypoints in reference and moving images are depicted in Fig. 1 - c,d by green circles depicting feature location, scale, and rotation. The matched keypoints are connected by lines. As shown, a lot of keypoints are matched incorrectly (blue lines). The correct match selection was based on Euclidean distance E_m between keypoint location in each pair m according Eq. 4. These Euclidean values were rounded to the order of tens, and then the value with the highest occurrence was set as a threshold p. The matches with Euclidean distance in the defined range around the threshold were assigned to a set of correct matches C (red lines - in Fig. 1).

$$E_m \in \langle p - 0.2p, p + 0.2p \rangle \Rightarrow m \in C \tag{4}$$

The transformation matrix of x_{mov} is computed from the correct matches and considers translation, rotation, and scaling. The translation and rotation corrects the image distortion caused by the eye and head movements, which occur between particular acquisitions. The scaling corrects possible small changes of image resolution due to changes of the eye refractive power, which might also change between individual acquisitions. The nonlinear transformation, correcting the projection of the sphere area on the planar detector, is not needed in the opposite of fundus image registration [1], because of the minimal surface curvature, ensured by the small FOV of the AO camera. The surface minimal surface curvature is given by the low ratio of spherical sector area (FOV) to the spherical radius of the eye. The example result is shown in Fig. 2 - d.

2.4 Phase Correlation Followed by Scale Invariant Feature Transform (PC-SIFT)

The phase correlation strengths rely on its computational simplicity, what implies low computation time. In addition, the estimated transformation is affected by each pixel (opposite the key location registration), which makes this approach robust to large shifts and variable illumination. That together make a phase correlation approach a powerful tool for AO image registration. However, its efficiency is limited to integer precision. On the other hand, SIFT can estimate the transformation of subpixel alignment and is less affected by variable blur with respect to PC approach. These findings directly encourage a combination of these two methods. It could be implemented by two-stage registration. The coarse stage uses the PC approach, thus estimating large shifts. The fine stage registration is implemented by SIFT approach.

2.5 Alignment Accuracy

The final montage accuracy is evaluated by normalized mutual information (NMI) between each pair of x_{ref} and x_{mov}. The NMI is given by

$$NMI = \frac{H(X) + H(Y) - H(X,Y)}{\sqrt{H(X)H(Y)}}, \tag{5}$$

where $H(X)$, $H(Y)$ and $H(X,Y)$ represent the marginal and joint probability density functions, respectively. The NMI ranges from 0 to 1, where 1 represents a perfect alignment and is widely used as a measure of image similarity in image alignment and registration [15].

3 Results and Discussion

The montage images were created from 200 images of 10 patients by four different approaches - PC, SIFT, PC-SIFT and manually. An example of the result for each method is shown in Fig. 2. From this it is visible that PC and SIFT methods proved accurate registration for single photoreceptor recognition, however, the individual photoreceptors are more blurry than photoreceptors in the aligned image by PC-SIFT or manual registration.

The original single adjacent images were aligned by a transformation matrix estimated by each method. An example of the wide-field montage created by PC-SIFT is shown in Fig. 3 - a,c together with a zoomed area, shown in Fig. 2. The overlapping parts of adjacent aligned images were averaged at each pixel - typically two or three images are averaged.

The NMI values were computed for each overlapping pair of x_{ref} and x_{mov}. The relative comparison with this metric shows that the PC-SIFT approach outperforms the manual alignment. The data showed higher alignment accuracy for AO images further from the fovea, for each registration method. The trend is shown in Fig. 3 - b, where the NMI of overlaps within the distance of 2° from the

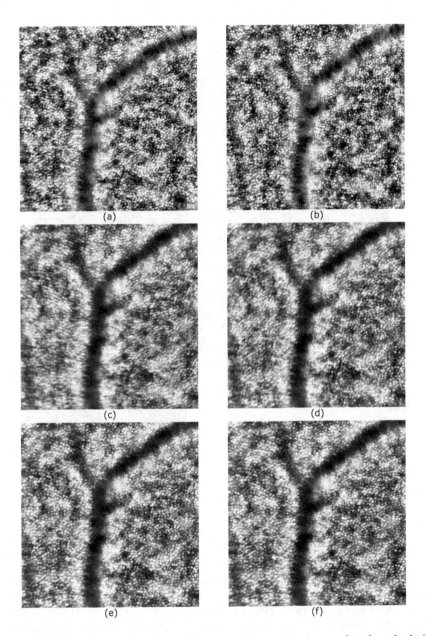

Fig. 2. (a) Cropped part of $x_{r}ef$ and (b) $x_{m}ov$. The registration results of methods (c) PC, (d) SIFT, (e) PC-SIFT and (f) manual registration on cropped overlapping parts of x_{ref} and x_{mov}. The example images shown here have 600×600 pixels resolution $\sim 0.2 \ mm^2$. The histogram equalization was applied on images for highlighting the images differences. (Color figure online)

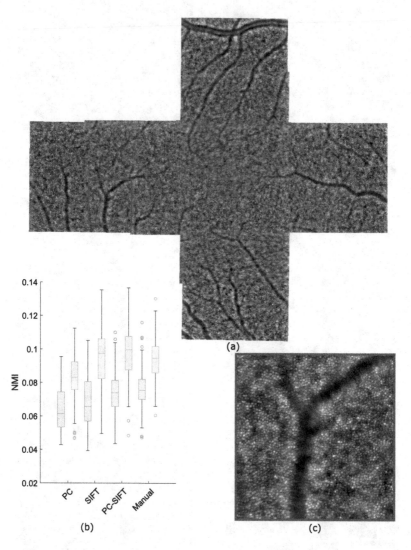

Fig. 3. (a) Averaged resulting montage after registration by PC-SIFT of 10 images from right eye with zoom in area (c) denoted by red frame. (b) Graph presenting the counted NMI for each method. The blue boxes depict the NMI counted from overlaps closer to the fovea then 2° and yellow boxes depict the NMI counted from overlaps further from the fovea. The central mark indicates the median, and the bottom and top edges of the box indicate the 25th and 75th percentiles, respectively. The whiskers extend to the most extreme data points, which do not consider outliers, and the outliers are plotted individually. (Color figure online)

Table 1. Averaged NMI of all overlapping regions.

Method	NMI	NMI standard deviation
PC	0.074	0.017
SIFT	0.084	0.021
PC-SIFT	0.087	0.02
Manual	0.086	0.015

fovea is depicted by blue boxes, and NMI of overlaps further from fovea than $2°$ is depicted by yellow boxes. The higher accuracy in alignment of images further from the fovea is presumably caused by the higher level of image blur in the foveal region and lack of blood vessels. This influences the determination of the keypoint correspondence in SIFT method and the precision of estimation of the Dirac impulse position in the PC approach.

Averaged values and standard deviations of NMI for the whole dataset are shown in Table 1. The overall evaluation showed the best results for the PC-SIFT method and comparable results with manual registration. Our results for AO flood illumination images are also in agreement with the results achieved by *Chen et al.*[4] for AOSLO images, where NMI values were in the range from 0.062 to 0.112.

It should be noted that the montaging results depend on the overlap of adjacent images. The image acquisition protocol in our application ensures that this overlap is typically 50%. Going to lower overlap would decrease the alignment precision as shown in *Chen et al.*, where the amount of overlapping for successful registration also depends on pathology and was between 11% and 35%. This suggests a possible image acquisition protocol modification by decreasing image overlap, which provides a larger FOV montage.

This new approach (PC-SIFT), consisting of connection and modification of known methods, shows that it can efficiently solve the problem of AO image montaging. Numerous studies, focused on AO flood illumination image analysis, cling to AO registration by manual montaging, which is labor intensive, time consuming, and complicated by image variability. Therefore, automated image montage is the crucial and necessary step for further advanced AO data analysis, which is now clearly and applicably available, based on the results of this study. Furthermore, the following novel possibilities of data analysis also arise. The results are comparable to published works focused on AOSLO images, even though registration of images from AO flood illumination images is more challenging, particularly in the foveal area. The AO image montages can be further utilized for photoreceptor detection followed by feature extraction, e.g., spatial distribution or density estimation leading to retina condition evaluation. In addition, the chosen registration approach will most likely make it possible to register even retinas with pathological changes, manifesting typically as discontinuous retinal structures, which will be also an aim of our further work.

4   Conclusion

This study establishes novel automated image registration for AO flood illumination images. We suggest three fully automated methods for AO image montage. Results of the three applied methods were compared with each other and with manual image montaging. The image alignment accuracy was evaluated by NMI metric for the dataset of 200 images. All tested approaches showed lower precision in the foveal area. The most accurate and robust alignment was achieved by the PC-SIFT approach, nevertheless, each method proved promising results for our large set of images. These highly promising results suggest that the algorithms can be used as a faster and fully automated alternative to manual montaging and will be used as a necessary step in our future work on AO flood illumination patient data analysis.

Acknowledgment. The authors express their sincere gratitude to Imagine Eyes, Orsay, France, for continuous support and loan of the rtx1e instrument for the measurements of this study.

References

1. Can, A., Stewart, C., Roysam, B., Tanenbaum, H.: A feature-based, robust, hierarchical algorithm for registering pairs of images of the curved human retina. IEEE Trans. Pattern Anal. Mach. Intell. **24**(3), 347–364 (2002). https://doi.org/10.1109/34.990136
2. Carroll, J., Neitz, M., Hofer, H., Neitz, J., Williams, D.R.: Functional photoreceptor loss revealed with adaptive optics. Proc. Natl. Acad. Sci. **101**(22), 8461–8466 (2004). https://doi.org/10.1073/pnas.0401440101
3. Chen, M., Cooper, R.F., Gee, J.C., Brainard, D.H., Morgan, J.I.W.: Automatic longitudinal montaging of adaptive optics retinal images using constellation matching. Biomed. Opt. Express **10**(12), 6476–6496 (2019). https://doi.org/10.1364/BOE.10.006476
4. Chen, M., Cooper, R.F., Han, G.K., Gee, J., Brainard, D.H., Morgan, J.I.W.: Multi-modal automatic montaging of adaptive optics retinal images. Biomed. Opt. Express **7**(12), 4899–4918 (2016). https://doi.org/10.1364/BOE.7.004899
5. Chew, A.L., Sampson, D.M., Kashani, I., Chen, F.K.: Agreement in cone density derived from gaze-directed single images versus wide-field montage using adaptive optics flood illumination ophthalmoscopy. Transl. Vision Sci. Technol **6**(6), 1–13 (2017). https://doi.org/10.1167/tvst.6.6.9
6. Georgiou, M., Kalitzeos, A., Patterson, E.J., Dubra, A., Carroll, J., Michaelides, M.: Adaptive optics imaging of inherited retinal diseases. Brit. J. Ophthalmol. **102**(8), 1028–1035 (2018). https://doi.org/10.1136/bjophthalmol-2017-311328
7. Gill, J.S., Moosajee, M., Dubis, A.M.: Cellular imaging of inherited retinal diseases using adaptive optics. Eye **33**(11), 1683–1698 (2019). https://doi.org/10.1038/s41433-019-0474-3
8. Kuglin, C.D.: Performance of the phase correlator in image guidance applications. Technical report, Control Data Corp Minneapolis MN Image Systems DIV (1976)
9. Li, H., Lu, J., Shi, G., Zhang, Y.: Automatic montage of retinal images in adaptive optics confocal scanning laser ophthalmoscope. Opt. Eng. **51**(5), 1–6 (2012). https://doi.org/10.1117/1.OE.51.5.057008

10. Liang, J., Williams, D.R., Miller, D.T.: Supernormal vision and high-resolution retinal imaging through adaptive optics. J. Opt. Soc. Am. A **14**(11), 2884–2892 (1997). https://doi.org/10.1364/JOSAA.14.002884

11. Lombardo, M., Serrao, S., Ducoli, P., Lombardo, G.: Eccentricity dependent changes of density, spacing and packing arrangement of parafoveal cones. Ophthal. Physiol. Opt. **33**(4), 516–526 (2013). https://doi.org/10.1111/opo.12053

12. Lowe, D.G.: Object recognition from local scale-invariant features. In: Proceedings of the Seventh IEEE International Conference on Computer Vision, vol. 2, pp. 1150–1157 (1999). https://doi.org/10.1109/ICCV.1999.790410

13. Paques, M., et al.: Adaptive optics ophthalmoscopy. Prog. Retinal EyeRes. **66**(1), 1–16 (2018). https://doi.org/10.1016/j.preteyeres.2018.07.001, https://linkinghub.elsevier.com/retrieve/pii/S1350946217300782

14. Prasse, M., Rauscher, F.G., Wiedemann, P., Reichenbach, A., Francke, M.: Optical properties of retinal tissue and the potential of adaptive optics to visualize retinal ganglion cells in vivo. Cell Tissue Res. **353**(2), 269–278 (2013). https://doi.org/10.1007/s00441-013-1602-1

15. Strehl, A., Ghosh, J.: Cluster ensembles-a knowledge reuse framework for combining multiple partitions. J. Mach. Learn. Res. **3**(Dec), 583–617 (2002)

16. Valterova, E.: Automatic adaptive optics retinal images montaging. In: Proceedings of the 27th Conference STUDENT EEICT 2021. Vysoké učení technické v Brně, Fakulta elektrotechniky a komunikačních technologií, Brno (2021)

17. Vedaldi, A., Fulkerson, B.: VLFeat: an open and portable library of computer vision algorithms (2008). http://www.vlfeat.org/

18. Williams, D.R.: Imaging single cells in the living retina. Vision Res. **51**(13), 1379–1396 (2011). https://doi.org/10.1016/j.visres.2011.05.002, https://linkinghub.elsevier.com/retrieve/pii/S0042698911001763

19. Xue, B., Choi, S.S., Doble, N., Werner, J.S.: Photoreceptor counting and montaging of en-face retinal images from an adaptive optics fundus camera. J. Opt. Soc. Am. A **24**(5), 1364–1372 (2007). https://doi.org/10.1364/JOSAA.24.001364

Radiomics, Predictive Models, and Quantitative Imaging

End-to-End Deep Learning Vector Autoregressive Prognostic Models to Predict Disease Progression with Uneven Time Intervals

Joshua Bridge[(✉)], Simon Harding, and Yalin Zheng

Department of Eye and Vision Science, Institute of Life Course and Medical Sciences,
University of Liverpool, Liverpool, UK
jbridge@liverpool.ac.uk

Abstract. We propose an end-to-end deep learning method combining implicit feature extraction and an autoregressive model to predict the future course of a disease or condition. By merging the feature extraction and autoregression into one deep learning model, we can simultaneously train both models together. Our novel approach begins by fine-tuning a pretrained convolutional neural network to extract features from previously obtained images of patients. A trainable autoregression mechanism then predicts the features of the future image and a fully connected layer gives a prognosis based on the predicted features. We utilize a novel time interval scaling, allowing the model to account for uneven time intervals and allowing us to choose the final time point that we wish to predict. Experiments on the Age-Related Eye Disease Study give a testing area under the receiver operating characteristic curve, sensitivity, and specificity of 0.966 (95% CI: 0.947, 0.984), 0.878 (0.810, 0.945), and 0.930 (0.914, 0.947), respectively. This shows that the model can predict progression with good performance.

Keywords: Prognosis · Autoregressive · Deep learning

1 Introduction

Prognostic models have been developed in various fields to predict the future outcome of disease [29]. These models are a vital component of personalized medicine, allowing clinicians to prepare treatments and allocate resources accordingly. Most developed prognostic models use traditional statistics, which often rely on features or variables being known and obtainable; this is not always the case in imaging data. More recently, deep learning has allowed automated feature extraction, avoiding the necessity to extract those features manually by a trained observer [3,27]. Models that utilize deep learning in prognostic models based on images fall into two main categories; explicit feature extraction and implicit feature extraction. The explicit feature extraction methods utilize deep learning to extract known features automatically, saving considerable

© Springer Nature Switzerland AG 2021
B. W. Papież et al. (Eds.): MIUA 2021, LNCS 12722, pp. 517–531, 2021.
https://doi.org/10.1007/978-3-030-80432-9_38

time. The extracted features can then be used in a traditional statistical model such as logistic regression or linear mixed-effects model in the case of longitudinal images. However, training automatic feature extraction algorithms require expertly annotated images and assume that we know which features to extract. Previously, these explicit feature extraction and classification algorithms have been trained in two steps, meaning that the classification error cannot be back-propagated to the feature extraction. Implicit feature extraction methods feed a single image, usually the last available image, into a convolutional neural network (CNN) and return a probability of progression. These methods allow the network to extract features implicitly. Previously, this approach has not considered multiple longitudinal images and possibly neglects important temporal patterns.

Here, we propose a method combining both implicit feature extraction and an autoregressive model into a single deep learning framework, allowing both parts to be trained simultaneously. In traditional statistics, the time series parameters must be estimated using a method such as ordinary least squares; however, in our method, we use deep learning to learn the parameters. A window function is employed to allow for more uneven time intervals. The proposed method requires no prior annotations and can be used on multiple longitudinal images.

2   Related Work

There are two main general approaches to developing deep learning prognostic models using images, explicit feature extraction combined with a statistical model and implicit feature extraction in an end-to-end model.

2.1   Autoregression

Autoregression has previously been used in deep learning in a different form for density estimation [32] or as a skip mechanism between hidden layers in generative autoencoders [15]. These models are autoregressive in the sense that each output from the previous layer is used as additional inputs in the next layer. In our model, autoregression acts as a layer on its own at the end of the network to predict the future feature vector in a sequence of images.

2.2   Explicit Feature Extraction

One approach to developing a prognostic model is first to extract the features, either manually or with an automatic algorithm, and then to use these features as variables in a traditional statistical model.

The field of ophthalmology, in particular, has several models taking this approach, applied to a range of imaging platforms. One of these is optical coherence tomography (OCT) [18], an imaging technique similar to ultrasound, but using light instead of sound. The technique allows the easy identification of individual retinal layers, which can then be accurately segmented and features, such as volumes, calculated. One of the first such models proposed by de Sisternes et al. [27] used a deep learning algorithm to extract area, volume, height, and reflectivity of subretinal structures from OCT images before a statistical model predicts the future onset of age-related macular degeneration (AMD) based on those features. Banerjee et al. [6] followed the same feature extraction approach, replacing the statistical model with a Recurrent Neural Network (RNN); they reported that this method provided improved performance over a random forest classifier.

There are also several prognostic models developed with this approach in neurology and cardiology. Hilario et al. [17] developed a Cox proportional hazards model [11] to predict the progression of diffuse gliomas based on features automatically extracted by commercial software using OCT images. Arenja et al. [4] also used commercial software to extract features to assess the prognostic value of magnetic resonance imaging in predicting the outcome of patients with heart disease.

While these models often achieve high predictive performance, they all rely on a set of feature extraction algorithms. These algorithms require a large amount of training data, labeled by experienced experts, which is expensive to produce and not always possible if features are unknown or difficult to quantify. By using two separate models for feature extraction and prediction, the feature extraction network is disconnected from the final classification, meaning the error cannot be backpropagated to improve feature extraction.

2.3   Implicit Feature Extraction

An alternative to the feature extraction and statistical model method is to use an end-to-end deep learning model. These models combine implicit feature extraction, with a convolutional neural network and classification, with a fully connected layer, into a single model, meaning we no longer need to label the features during training. Usually, the last available image is used as input into these models to predict the future outcome.

As with the previous approach, there are many developed models using this approach in the field of ophthalmology. Yan et al. [36] used InceptionV3 [30], pretrained on Imagenet [26], and combined the feature vector with genotype and phenotype data to be classified with a fully connected layer. Grassman et al. [14] combined six deep learning networks in a random forest ensemble method; they achieved improved results over any of the networks alone. This model classified patients into 13 separate groups, nine groups of risk of advanced AMD, three advanced AMD groups, and one ungradable group. Arcadu et al. [3] utilized multiple color fundus fields to predict diabetic retinopathy progression. A separate Inception V3 network [30] was fit to each field of the retina, and a prediction of progression was produced. The final overall prediction was made using a random forest classifier. Many of these fields are not routinely photographed, making this suitable for some applications only.

This approach has also been applied to neurology. Choi et al. [9] used a 3D CNN to predict conversion to Alzheimer's disease from MRI scans. The CNN method provided increased performance over a voxel-wise feature extraction and support vector machine method, suggesting this approach is preferable to using feature extraction.

More recently, methods able to deal with longitudinal data have been proposed [7]; these enable the network to better model the progression of the disease. Here, we propose a longitudinal method using vector autoregression (VAR) to predict what the future features may be.

3   Method

The proposed model is unique in utilizing a time series model to predict what the future extracted features may be using the previous features; the problem then becomes a simple classification. The features used are implicit and extracted using a CNN.

Given previous images $I_1, I_2, \ldots, I_n$ at time points $t_1, t_2, \ldots, t_n$, we aim to predict the diagnosis y_{n+1} of some future image I_n at time t_{n+1}. We also aim to account for uneven time intervals, where $t_{i+1}\text{-}t_i = t_{j+1} - t_j$ does not necessarily hold.

3.1   Overall Framework

While previous models aim to predict the future outcome, our model instead aims to predict the future features, which are then classified as progressing or non-progressing. We first fine-tune a pretrained CNN on each image at each time point; this results in a feature vector at each time point. The feature vectors are then multiplied by an interval scaling, which weights images closer to the outcome time as being more informative. We predict the future feature vector using a vector autoregressive model with trainable parameters. Finally, the predicted feature vector is classified with a fully connected layer. Each of these stages is trained in a single deep learning model, such that the classification error is backpropagated to improve the implicit feature extraction. This framework is displayed in Fig. 1.

3.2   CNN

We begin reducing each image at each time point image to a single vector representation. For this, we utilize a convolutional neural network (CNN). In our work, we chose InceptionV3 [30], pretrained on Imagenet [26] followed by global average pooling, as this network has been used by previous single time point methods and obtained excellent results [3,5,36]; however, our method may use any CNN of choice. InceptionV3 improves upon previous inception networks with factorized convolutions and heavy regularization to produce an accurate yet computationally efficient network. Previous experiments show excellent classification performance with a low receptive field resolution, suggesting the network

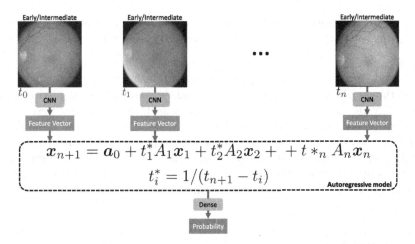

Fig. 1. Framework of the proposed method. The method takes color retinal images from multiple time points and extracts a feature vector for each image using a CNN. A future feature vector is then predicted using a vector autoregressive (VAR) model. An interval scaling is applied to account for uneven time points. Finally, a dense, fully connected layer, with sigmoid activation, returns the predicted probability of progression. In this example, images of age-related macular degeneration are shown; however, any disease and imaging modality may be used.

is suitable for images with small objects such as those encountered in medical imaging.

The CNN will reduce each image to a feature vector x_i of length F; these vectors can be treated as extracted features that the algorithm finds useful in making the final classification.

3.3   Time Series

To predict the future feature vector x_{n+1} at time t_{n+1}, we use time series analysis. Time series theory assumes that each feature is influenced by some random mechanism and is correlated with features at other time points. With these assumptions, we can use the previous features to predict the features of some future image. Here, we outline the time series method utilized in this paper, autoregression (AR), as well as the multivariate generalization, vector autoregression (VAR). These finite parameter models are derived as special cases of the general linear model [23].

Autoregressive models perform regression on previous values to predict the next value in the sequence. Autoregression is often used in finance and meteorology to predict stock prices and daily temperatures. The pth order autoregressive model denoted AR(p), is given by:

$$x_{n+1} = a_{n+1} + a_n x_n + \cdots + a_{n-p} x_{n-p} + e_t,$$

where x_n is the value we wish to predict using the previous values $\{x_i\}$, $\{a_i\}$ are parameters to be determined, and e_t is some stationary random noise.

VAR models extend the autoregressive model to the multivariate case by replacing the time series, $\{x_i\}$, with vectors $\{\boldsymbol{x}_i\}$ and the parameters, $\{a_i\}$, with matrices $\{A_i\}$. VAR allows us to consider the relationships between features.

Traditionally the parameters $\{A_i\}$ would be estimated using a method such as ordinary least squares; however, in our method, we create a trainable tensor and use deep learning to learn the parameters instead. Each feature vector, $\boldsymbol{x}_i$ of length F, obtained from the previous stage, is multiplied by a trainable tensor, A_i, of shape $F \times F$. Finally, the new feature vectors are summed along with an intercept vector $\boldsymbol{a}_0$.

3.4   Interval Scaling

Often patients may miss appointments or may have their screening intervals relaxed; this can create uneven intervals between visits. To account for this, interval scaling has previously been proposed [7]. These uneven time intervals are accounted for by further multiplying the feature vector by a scalar equal to:

$$t_i^* = 1/(t_{n+1} - t_i).$$

This also allows us to choose at which time point we wish to predict by changing t_{n+1}.

Therefore the final autoregressive model, combined with the interval scaling, to predict x_n is given by:

$$\boldsymbol{x}_{n+1} = \boldsymbol{a}_0 + t_1^* A_1 \boldsymbol{x}_1 + t_2^* A_2 \boldsymbol{x}_2 + \ldots + t *_n A_n \boldsymbol{x}_n.$$

3.5   Classification Layer

The predicted feature vector $\boldsymbol{x}_{n+1}$ is then classified using the standard dense, fully-connected layer within deep learning. For binary prognosis problems, such as progression/non-progression, the sigmoid activation function may be used, whereas, in multiclass prognosis, the softmax activation is used. When the sigmoid activation is used, the classification layer becomes equivalent to logistic regression.

4   Experiments

We demonstrate the proposed method on the Age-Related Eye Disease Study (AREDS) dataset, the most extensive study in AMD and which included a control epidemiological group [2].

As our proposed model is comprised of two novel concepts, we will consider each of them individually and then combined, to assess the contribution of each.

The first model uses a single time point to predict the outcome at the next time point and does not take into account the time between the observation and the prediction; this is similar to previously used methods such as [3,5,36]; however, we do not include additional fields, stereo images, or genetic data, as this information is not routinely available. The second model uses a single time point with the proposed interval scaling method to account for the uneven time intervals. The third model uses vector autoregression (VAR) with multiple time points but does not use the interval scaling. Finally, the proposed model uses both VAR and interval scaling.

4.1   Dataset

AMD is a degenerative condition of the retina and a leading cause of vision loss worldwide, with a predicted prevalence of 288 million by 2040 [35]. A meta-analysis of 2007–2009 data estimated that there are 71,000 new cases of late AMD per year in the UK [22].

AMD can be divided into three primary stages, early, intermediate and advanced [2]. Early and intermediate AMD are characterized by small- to medium-sized drusen; these are lipid deposits that form under the retina and are visible on color retinal photographs. Advanced AMD is sub-divided into atrophic and neovascular AMD; it is possible to have both forms. Neovascular AMD (nAMD) is treatable with courses of repeated intravitreal injections of an antibody against vascular endothelial growth factor, a protein that upregulates the production of new vessels. The maintenance or restoration of vision in nAMD depends significantly on early identification and intervention [35]. Atrophic AMD is currently untreatable.

Images were taken from AREDS, a longitudinal prospective study into AMD. Patients with early/intermediate AMD, between the ages of 55 and 80, were recruited and followed up to 12 years from baseline. Images were taken at time points with varying intervals. From the AREDS dataset, we extracted 4903 eyes, which had three visits displaying early/intermediate AMD and a fourth visit with either early/intermediate or advanced AMD; from these images, we aimed to predict the diagnosis at visit four using the first three images. We split the images into training (60%), validation (20%), and testing (20%) subsets; this is shown in Table 1. Example patient profiles are shown in Fig. 2.

4.2   Computation

Analyses were carried out on a Linux machine running Ubuntu 18.04, with 32GB of memory and a Titan X 12 GB GPU. Preprocessing and model development were carried out with Python 3.7 [33], with Keras 2.2.4 [10] and Tensorflow [1] used for deep learning. Statistical analysis of the results was conducted in R 2.4.4 [24], with the pROC [25], ReportROC [13], dca [28] and PredictABEL [20] packages..

In model development, we used the Adam optimizer [19] with an initial learning rate of 0.0001, which was reduced to one-fifth if performance did not improve

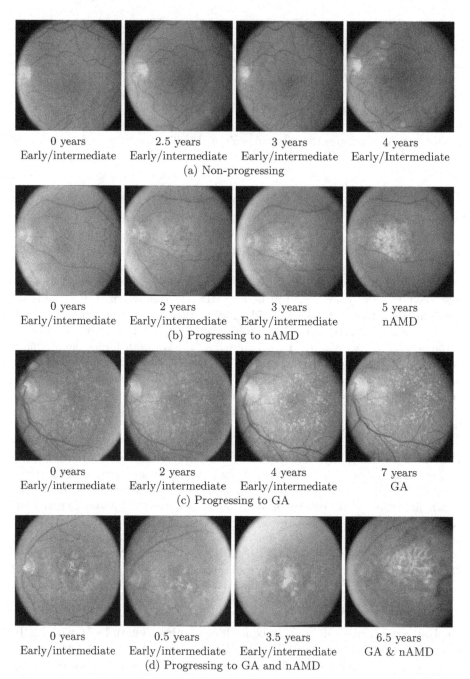

| 0 years | 2.5 years | 3 years | 4 years |
| Early/intermediate | Early/intermediate | Early/intermediate | Early/Intermediate |

(a) Non-progressing

| 0 years | 2 years | 3 years | 5 years |
| Early/intermediate | Early/intermediate | Early/intermediate | nAMD |

(b) Progressing to nAMD

| 0 years | 2 years | 4 years | 7 years |
| Early/intermediate | Early/intermediate | Early/intermediate | GA |

(c) Progressing to GA

| 0 years | 0.5 years | 3.5 years | 6.5 years |
| Early/intermediate | Early/intermediate | Early/intermediate | GA & nAMD |

(d) Progressing to GA and nAMD

Fig. 2. 35° color retinal images from 4 cases in the AREDS dataset illustrating different progression histories: (a) non-progressing patient, (b) patient progressing to neovascular AMD (nAMD) at time point four, (c) patient progressing to geographic atrophy (GA) at time point four, (d) patient progressing to both GA and nAMD at time point four.

after five epochs. Binary cross-entropy was used as the loss function. Early stopping was utilized with a patience of 10 epochs, and the best model was chosen according to the validation Youden's index [37] to prevent overfitting.

Table 1. The data were split into training, validation, and testing.

	Training	Validation	Testing
Progressing	272	91	90
Non-progressing	2670	890	890

4.3 Preprocessing

We first cropped the images to the boundary. An automated algorithm calculated the difference between the image pixel values and the black background; this difference map was then used to calculate the bounding box, and the image was cropped to this bounding box. All images were rescaled to 256×256 pixels, using the Lanczos resampling algorithm [31], to reduce computational requirements. Images of right eyes were flipped horizontally, such that the optic disc is located on the left-hand side of each image. Finally, pixel values were rescaled to between 0 and 1.

4.4 Metrics

We assess model performance in four areas: overall performance (Brier score [8]), discrimination (area under the receiver operating characteristic (AUC) [16], sensitivity, and specificity), reclassification (net reclassification index (NRI) [21]), and clinical usefulness (decision curve analysis (DCA) [34]).

The overall performance of the model is evaluated using the Brier score, which is defined as the mean square difference between the predicted probability and the observed outcome, with lower values indicating better performance. As the data is highly imbalanced, with only around 10% of patients progressing to AMD, we report the area under the receiver operating characteristic (AUC), as well as the optimum sensitivity and specificity, based on Youden's index [37]. Confidence intervals are reported to show whether differences in results are statistically significant. We used De Long's method [12] to calculate the 95% confidence intervals and De Long's test to test for a statistically significant difference between AUCs. For sensitivity and specificity, we calculated bootstrapped 95% confidence intervals with 2000 samples. Youden's index was used in model selection and is not reported; however, the index can be easily calculated by $J = sensitivity + specificity - 1$. The NRI is a measure of how well a new model reclassifies subjects, with positive values indicating that the new model classifies subjects better than the previous model. DCA displays graphically how

Table 2. Performance metrics in the testing set. Brier score. Area Under the Receiver Operating Characteristic Curve (AUC) with 95% confidence intervals constructed by De Long's method. Sensitivity and specificity with 95% confidence intervals constructed by bootstrapping with 2000 samples. The proposed method achieves an improved Brier score, statistically significantly higher AUC and sensitivity, and a non-significant difference in sensitivity. Bold indicates statistically significant results, at the 95% level.

Method	Brier	AUC	Sensitivity	Specificity
Single time point without scaling	0.121	0.873 (0.838, 0.907)	0.911 (0.852, 0.970)	0.764 (0.736, 0.792)
Single time point with scaling	0.114	0.881 (0.856, 0.906)	0.956 (0.913, 0.998)	0.816 (0.790, 0.841)
VAR without scaling	0.124	0.889 (0.862, 0.915)	0.911 (0.852, 0.970)	0.792 (0.765, 0.819)
VAR with scaling	0.050	**0.966 (0.947, 0.984)**	0.878 (0.810, 0.945)	**0.930 (0.914, 0.947)**

well diagnostic and prognostic models perform by calculating the net benefit at various thresholds and showing at what threshold a model ceases to have a net benefit.

4.5   Results

The proposed method, using VAR with interval scaling, attains a Brier score of 0.050, an AUC of 0.966 (95% CI: 0.947, 0.984), a sensitivity of 0.878 (0.810, 0.945), and a specificity of 0.930 (0.914, 0.947). The single time point without scaling method attained a Brier score of 0.121, an AUC of 0.873 (0.838, 0.907), a sensitivity of 0.911 (0.852, 0.970), and a specificity of 0.764 (0.736, 0.792). The single time point without scaling method attained a Brier score of 0.114, an AUC of 0.881 (0.856, 0.906), a sensitivity of 0.956 (0.913, 0.998), and a specificity of 0.816 (0.790, 0.841). The single time point without scaling method attained a Brier score of 0.124, an AUC of 0.889 (0.862, 0.915), a sensitivity of 0.911 (0.852, 0.970), and a specificity of 0.792 (0.765, 0.819).

The results show that the proposed method achieves a statistically significant improvement in AUC and specificity over the other methods, with a non-significant difference in sensitivity. De Long's test for difference in AUCs gave a p-value of <0.0001 against the second-highest AUC, indicating that the improvement in AUC is highly significant. These results are presented in Table 2, with the ROC curve displayed graphically in Fig. 3.

Compared to the proposed VAR with the scaling method, the net reclassification improvements were 0.146 (0.053, 0.239) net improvement over the single time point without scaling, 0.334 (0.225, 0.443) over the single time point with scaling, and 0.176 (0.065, 0.287) over the VAR without scaling method, showing

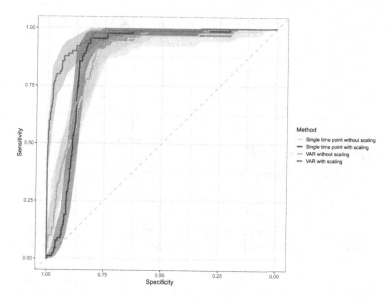

Fig. 3. Receiver operating characteristic curve of the proposed model using VAR with interval scaling and comparison models.

a significant improvement in classification over all other models. The decision curve, shown in Fig. 4, also shows that the VAR with scaling method provides an increased net benefit over the other methods.

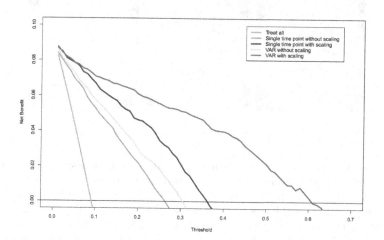

Fig. 4. Decision curve showing the net benefit of the treat all approach, the single time point, and the proposed method. Models with a higher threshold before hitting zero net benefit are considered to more clinically useful. This shows graphically that our model has a greater clinical usefulness over the single time point method.

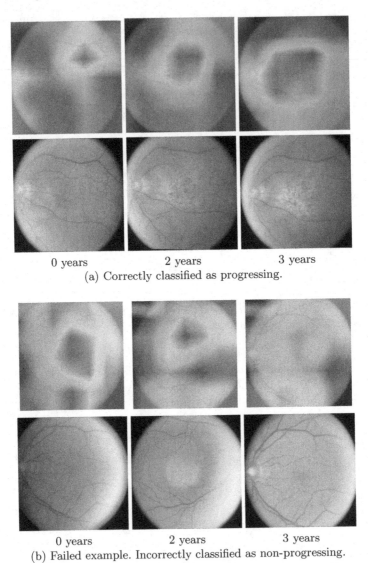

0 years   2 years   3 years
(a) Correctly classified as progressing.

0 years   2 years   3 years
(b) Failed example. Incorrectly classified as non-progressing.

Fig. 5. Two example class activation maps from the testing dataset. Areas in red are considered more important by the network during classification. (a) The centre of the image show a high concentration of drusen, which is expected to be important. At 0 years, the algorithm highlights an incorrect area, as drusen is not easily observed at this time point. At 2 years, drusen is more visible and the algorithm begins to correctly identify the expected features. At 3 years, drusen is highly visible and the algorithm correctly identifies the area of interest. (b) A failed example where the algorithm incorrectly classified the image as non-progressing. The class activation map shows no particular region of interest in the final image. The disease is difficult to see and the second image appears to be from a different eye to the other two, suggesting a mistake within the dataset, which confuses the network and leads to an incorrect prognosis. (Color figure online)

4.6   Class Activation Maps

Class activation maps [38] visualize how the network reaches the decision. Class activation maps showed that areas with high levels of drusen are considered important, which is expected. The class activation maps suggest that the model correctly identifies features in the majority of images and is working as intended. In earlier images, the class activation maps are not as accurate as those of later images; this suggests that the model becomes more confident of the outcome as the disease progresses, which is also true of human graders. Examples of class activation maps are shown in Fig. 5; red shows areas of high importance and blue shows areas of low importance.

5   Conclusions

We have proposed a novel method, using VAR, to predict the future course of diseases, applicable to diseases with multiple longitudinal 2D images with uneven time intervals between images. The proposed method allows implicit feature extraction to be trained simultaneously alongside a traditional statistical model, namely an autoregressive model. Our method extracts feature vectors from previous images and aims to predict the feature vector of some future disease using an autoregressive model, which is then classified.

Experiments on a dataset of age-related macular degeneration showed good performance, with particularly high specificity, which is desirable in a screening setting. Class activation maps display that the algorithm is identifying the expected features of an image as being relevant to the final prediction. The method is easily generalizable to many diseases and modalities and requires only the disease outcome as ground truth during training. Future models may include clinical and genetic variables to improve the prognostic performance further.

References

1. Abadi, M., et al.: Tensorflow: Large-scale machine learning on heterogeneous systems (2015). Software available from tensorflow.org
2. Age-Related Eye Disease Study Research Group: The Age-Related Eye Disease Study (AREDS): design implications. AREDS report no. 1. Control. Clin. Trials **20**(6), 573–600 (1999). https://doi.org/10.1016/s0197-2456(99)00031-8
3. Arcadu, F., Benmansour, F., Maunz, A., Willis, J., Haskova, Z., Prunotto, M.: Deep learning algorithm predicts diabetic retinopathy progression in individual patients. NPJ Digit. Med. **2**, 92 (2019). https://doi.org/10.1038/s41746-019-0172-3
4. Arenja, N., et al.: Prognostic value of novel imaging parameters derived from standard cardiovascular magnetic resonance in high risk patients with systemic light chain amyloidosis. J. Cardiovasc. Magn. Reson. **21**(1), 53 (2019). https://doi.org/10.1186/s12968-019-0564-1
5. Babenko, B., et al.: Predicting progression of age-related macular degeneration from fundus images using deep learning (2019). arXiv preprint arXiv:1904.05478

6. Banerjee, I., et al.: A deep-learning approach for prognosis of age-related macular degeneration disease using SD-OCT imaging biomarkers (2019). arXiv preprint arXiv:1902.10700

7. Bridge, J., Harding, S., Zheng, Y.: Development and validation of a novel prognostic model for predicting amd progression using longitudinal fundus images. BMJ Open Ophthalmol. **5**(1) (2020). https://doi.org/10.1136/bmjophth-2020-000569

8. Brier, G.W.: Verification of forecasts expressed in terms of probability. Monthly Weather Rev. **78**(1), 1–3 (1950)

9. Choi, H., Jin, K.H.: Predicting cognitive decline with deep learning of brain metabolism and amyloid imaging. Behav. Brain Res. **344**, 103–109 (2018). https://doi.org/10.1016/j.bbr.2018.02.017

10. Chollet, F., et al.: Keras (2015). https://keras.io

11. Cox, D.R.: Regression models and life-tables. J. Roy. Stat. Soc. Ser. B (Methodol.) **34**(2), 187–220 (1972). www.jstor.org/stable/2985181

12. DeLong, E.R., DeLong, D.M., Clarke-Pearson, D.L.: Comparing the areas under two or more correlated receiver operating characteristic curves: a nonparametric approach. Biometrics **44**(3), 837–845 (1988)

13. Du, Z., Hao, Y.: reportROC: An Easy Way to Report ROC Analysis (2019). https://CRAN.R-project.org/package=reportROC, R package version 3.4

14. Grassmann, F., et al.: A deep learning algorithm for prediction of age-related eye disease study severity scale for age-related macular degeneration from color fundus photography. Ophthalmology **125**(9), 1410–1420 (2018). https://doi.org/10.1016/j.ophtha.2018.02.037

15. Gregor, K., Danihelka, I., Mnih, A., Blundell, C., Wierstra, D.: Deep autoregressive networks (2013). arXiv preprint arXiv:1310.8499

16. Harrell, F.E., Califf, R.M., Pryor, D.B., Lee, K.L., Rosati, R.A.: Evaluating the yield of medical tests. Jama **247**(18), 2543–2546 (1982). https://jamanetwork.com/journals/jama/article-abstract/372568

17. Hilario, A., et al.: A prognostic model based on preoperative MRI predicts overall survival in patients with diffuse gliomas. Am. J. Neuroradiol. **35**(6), 1096 (2014). https://doi.org/10.3174/ajnr.A3837

18. Keane, P.A., Patel, P.J., Liakopoulos, S., Heussen, F.M., Sadda, S.R., Tufail, A.: Evaluation of age-related macular degeneration with optical coherence tomography. Surv. Ophthalmol **57**(5), 389–414 (2012). https://doi.org/10.1016/j.survophthal.2012.01.006

19. Kingma, D.P., Ba, J.: Adam: a method for stochastic optimization (2014). arXiv preprint arXiv:1412.6980

20. Kundu, S., Aulchenko, Y.S., Janssens, A.C.J.: PredictABEL: Assessment of Risk Prediction Models (2014). https://CRAN.R-project.org/package=PredictABEL, R package version 1.2-2

21. Leening, M.J., Vedder, M.M., Witteman, J.C., Pencina, M.J., Steyerberg, E.W.: Net reclassification improvement: computation, interpretation, and controversies: a literature review and clinician's guide. Ann. Internal Med. **160**(2), 122–131 (2014). https://doi.org/10.7326/m13-1522

22. Owen, C.G., Jarrar, Z., Wormald, R., Cook, D.G., Fletcher, A.E., Rudnicka, A.R.: The estimated prevalence and incidence of late stage age related macular degeneration in the uk. Brit. J. Ophthalmol. **96**(5), 752 (2012). https://doi.org/10.1136/bjophthalmol-2011-301109

23. Priestley, M.: Non-linear and Non-stationary Time Series Analysis. Academic Press, Cambridge (1988)

24. R Core Team: R: A language and environment for statistical computing. R Foundation for Statistical Computing (2019). https://www.R-project.org/

25. Robin, X., et al.: pROC: an open-source package for R and S+ to analyze and compare ROC curves. BMC Bioinf. **12**(1), 77 (2011). https://doi.org/10.1186/1471-2105-12-77

26. Russakovsky, O., et al.: Imagenet large scale visual recognition challenge. Int. J. Comput. Vision **115**(3), 211–252 (2015)

27. de Sisternes, L., Simon, N., Tibshirani, R., Leng, T., Rubin, D.L.: Quantitative SD-OCT imaging biomarkers as indicators of age-related macular degeneration progression. Invest. Ophthalmol. Vision Sci. **55**(11), 7093–103 (2014). https://doi.org/10.1167/iovs.14-14918

28. Sjoberg, D.D.: dca: Decision Curve Analysis (2020). R package version 0.1.0.9000

29. Steyerberg, E.W., et al.: Prognosis research strategy (PROGRESS) 3: prognostic model research. PLoS Med. **10**(2), e1001381–e1001381 (2013)

30. Szegedy, C., Vanhoucke, V., Ioffe, S., Shlens, J., Wojna, Z.: Rethinking the inception architecture for computer vision. In: Proceedings of the IEEE Conference on Computer Vision and Pattern Recognition, pp. 2818–2826

31. Turkowski, K.: Filters for common resampling tasks (1990)

32. Uria, B., Côté, M.A., Gregor, K., Murray, I., Larochelle, H.: Neural autoregressive distribution estimation. J. Mach. Learn. Res. **17**(1), 7184–7220 (2016)

33. Van Rossum, G., Drake Jr, F.L.: Python tutorial. Centrum voor Wiskunde en Informatica Amsterdam (1995)

34. Vickers, A.J., Elkin, E.B.: Decision curve analysis: a novel method for evaluating prediction models. Med. Decis. Mak. **26**(6), 565–574 (2006). https://doi.org/10.1177/0272989x06295361

35. Williams, B.M., Burgess, P.I., Zheng, Y.: Chapter 13 - Drusen and macular degeneration, pp. 245–272. Academic Press, Cambridge (2019). https://doi.org/10.1016/B978-0-08-102816-2.00013-7

36. Yan, Q., et al.: Deep-learning-based prediction of late age-related macular degeneration progression. medRxiv, p. 19006171 (2019). https://doi.org/10.1101/19006171

37. Youden, W.J.: Index for rating diagnostic tests. Cancer **3**(1), 32–35 (1950)

38. Zhou, B., Khosla, A., Lapedriza, A., Oliva, A., Torralba, A.: Learning deep features for discriminative localization. In: Proceedings of the IEEE Conference on Computer Vision and Pattern Recognition, pp. 2921–2929 (2016)

Radiomics-Led Monitoring of Non-small Cell Lung Cancer Patients During Radiotherapy

Roushanak Rahmat[1,2,4,5,6](✉), David Harris-Birtill[2], David Finn[3], Yang Feng[4], Dean Montgomery[4], William H. Nailon[4,5], and Stephen McLaughlin[6]

[1] Computational Imaging Group, The Institute of Cancer Research, London, UK
roushanak.rahmat@icr.ac.uk
[2] School of Computer Science, University of St Andrews, St Andrews, UK
[3] Department of Radiation Oncology, Western General Hospital, Edinburgh, UK
[4] Department of Oncology Physics, Western General Hospital, Edinburgh, UK
[5] School of Engineering, The University of Edinburgh, Edinburgh, UK
[6] School of Engineering and Physical Sciences, Heriot-Watt University, Edinburgh, UK

Abstract. Co-locating the gross tumour volume (GTV) on cone-beam computed tomography (CBCT) of non small cell lung cancer (NSCLC) patients receiving radiotherapy (RT) is difficult because of the lack of image contrast between the tumour and surrounding tissue. This paper presents a new image analysis approach, based on second-order statistics obtained from gray level co-occurrence matrices (GLCM) combined with level sets, for assisting clinicians in identifying the GTV on CBCT images. To demonstrate the potential of the approach planning CT images from 50 NSCLC patients were rigidly registered with CBCT images from fractions 1 and 10. Image texture analysis was combined with two level set methodologies and used to automatically identify the GTV on the registered CBCT images. The Dice correlation coefficients ($\mu \pm \sigma$) calculated between the clinician-defined and image analysis defined GTV on the planning CT and the CBCT for three different parameterisations of the model were: 0.69 ± 0.19, 0.63 ± 0.17, 0.86 ± 0.13 on fraction 1 CBCT images and 0.70 ± 0.17, 0.62 ± 0.15, 0.86 ± 0.12 on fraction 10 CBCT images. This preliminary data suggests that the image analysis approach presented may have potential for clinicians in identifying the GTV in low contrast CBCT images of NSCLC patients. Additional validation and further work, particularly in overcoming the lack of gold standard reference images, are required to progress this approach.

Keywords: Image segmentation · Level set · Radiomics · Radiotherapy · Lung cancer

© Springer Nature Switzerland AG 2021
B. W. Papież et al. (Eds.): MIUA 2021, LNCS 12722, pp. 532–546, 2021.
https://doi.org/10.1007/978-3-030-80432-9_39

1   Introduction

In 2014 in the United Kingdom (UK) there were 46,400 new cases of lung cancer and 35,900 deaths were directly attributable to this disease [1]. Treatment options include radiotherapy (RT) in which a high dose of radiation is delivered to diseased tissue whilst the dose to healthy tissue is minimised [2]. In RT planning, delineation of the volume of interest is typically based on a visual assessment of CT. The accuracy of the volume is dependent upon the ability to visualise the tumour and identify potential areas of tumour involvement based on tumour biology.

At time-of-treatment CBCT images are used for patient alignment by identifying the position of landmarks such as bone [3]. However, establishing the location of the GTV on CBCT images is extremely challenging because of the poor soft tissue contrast. Furthermore, it is currently not feasible to estimate changes in the GTV on CBCT that are due to disease progression or in response to radiotherapy. This limits the use of CBCT in NSCLC for adapting, or personalising treatment, where the planned dose to healthy tissue and the planned dose to the tumour are preserved.

The use of semi or fully automatic image segmentation approaches can help to increase accuracy and reduce the time spent defining radiation target volumes and organs at risk (OAR) of damage from radiation. However, segmenting tumours, and OARs, on CT and CBCT images is challenging and the aim of this work was to develop a radiomics-based approach for CBCT image analysis for (1) automatically identifying the GTV on follow-up CBCT images and (2) detecting significant changes in shape and size indicative of disease progression or response to radiotherapy. In relation to this a bespoke image segmentation framework was developed that combined image texture analysis with level set analysis, which is shown in Fig. 1. The framework has the potential to lead to improvements in radiation targeting when the morphology of the tumour has changed, which could be practice changing. At present changes in tumour morphology are analysed after specific fractions of RT throughout the course of treatment [4]. Here this was done at fraction 10 (#10), that is the 10th day of treatment, when a volumetric CBCT image set was acquired and algorithm applied. The central hypothesis proposed here was that the texture of a cancerous region changes in response to radiation, which can be detected by texture/radiomics analysis. To ensure that the radiomics analysis was performed only on the tumour volume a novel level set technique proposed in [5], which combines two or more different level set methods in parallel, was used.

In this paper, Sect. 2 provides an overview of the proposed methodology and the framework for monitoring NSCLC cancer. Further on, this section provides a focused literature review covering the level set and texture analysis methods used. Section 3 provides details of the dataset, including contouring information, and the registration procedure used. Finally, Sect. 4 presents the results obtained following application of the framework to a large cohort of patients and ground truth data, which was available for one of the patients.

2   Overview of the Combined Texture and Level Set Model

The proposed framework for monitoring lung cancer during RT is shown in Fig. 1. The process begins with the registration of the CBCT acquired at fraction 1, (#1) to the planning CT image on which the GTV is defined. This provides a set of parameters for starting the combined texture and level set segmentation on CBCT #1. A Dice similarity coefficient, typically greater than 90%, is then used to assess the similarity between the GTV on the planning CT and the CBCT. Such a high value is used because the CBCT #1 images are acquired before RT and therefore before any change in shape as a result of radiation. A similar approach was used for the segmentation of the GTV on CBCT #10 and on the planning-CT images, however, because this CBCT scan takes place at the 10th day of treatment less similarity is expected between the planning CT and CBCT images and as a result the Dice coefficient value was reduced to 70%. With these settings if the similarity is far more than the chosen threshold RT should be continued but if far less, it is an indication that there has been a significant change in the GTV and repositioning, or further investigation, is required.

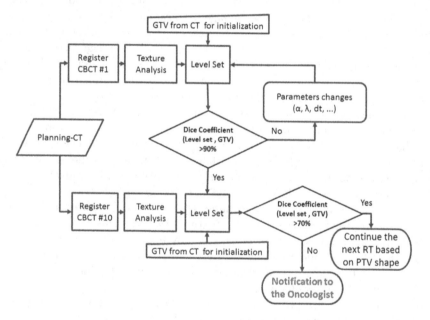

Fig. 1. The proposed framework for monitoring lung cancer during RT using a combined texture and level set model, which can be used for detecting large deviations in the GTV on CBCT images.

There are many different level set techniques each with propagation characteristics that may be suited to particular applications and there are several review articles covering the use of these techniques in different imaging applications. These include region based image segmentation in [6], medical imaging in [7,8], inverse problems and optimal design in [9], piecewise constant applications in [10], deformable models in [11,12]. There are also comprehensive reviews covering more general level set applications in [13–16]. What is common in all of these articles is that each approach is robust to a specific application but not for all types of images, which is a problem in complex medical images.

In general, a level set defines an evolving boundary as the zero level set of a higher dimension of function ϕ, where the height (z axis) corresponds to the minimum distance from each point in a rectangular coordinate (image plane) from the contour C. To start this evolving function an initial value (initialization) is required, which can be done automatically or semi-automatically in two or more phases depending on how many different batches of segmentation are expected in an image. The two-phase level set method segments the image into two regions. Wherever three or four-phase level set methods exist, they can divide an image into three or four categories respectively by applying two separate level set functions at the same time. Re-initialization is repeated during evolution to prevent the occurrence of sharp corners by calculating new ϕ values depending on the specified speed function.

In the1990 s Chan-Vese extended the Osher-Sethian model by applying an energy minimization term which allows automatic detection of interior contours [17]. This is performed using the piecewise constant and piecewise smooth optimal approximations proposed by Mumford-Shah [18]. In [17], further modifications resulted in a two-phase level set method without edges that could segment the image into two regions. This model is based on the mean intensity values in each region of the level set (inside or outside in two-phase) in an image. In 2000, Chan and Vese developed their model further to deal with vector-valued images, which performed robustly in the presence of noise [19]. This model is widely used in colour imaging and video imaging for detecting motion of objects and texture images. In 2002, they presented a multi-phase level set method that uses the log numbers of the level set function to separate n phases by using piecewise constant [20]. Each level set Multi-Phase Chan-Vese function is initialized separately but the same level set function is applied for both initialisations.

The Li level set model is another popular and robust model, which has developed different forms based on gradient flow for solving the inhomogeneity in intensity as an edge based level set model [21–23]. This model uses the energy minimization technique, similar to the snake model, by reducing the fitting energy in image segmentation.

2.1   Parallel Level Sets in Vector-Valued Image Model

We recently proposed a new level set model in [5] which had the least amount of dependency to the parameter setting. This model is most useful for the type of image where there is a lack the ground truth information. This model takes advantage of the presence of existing models by combining them in vector-valued imaging or multi-phase. Also, level set in segmentation of the tumour infiltration boundary showed robust performance in construction of radiotherapy target volumes [24].

Here texture analysis was used as a pre-segmentation aid to before applying the main segmentation model based on the level set. It has been previously reported that the texture of healthy tissues is different from cancerous tissues [25] and the texture may also change within these regions during the course of radiotherapy. There are many different ways of calculating texture features such as [26–29]. In this paper, second-order statistics were applied using the GLCM approach. First defined by Haralick in 1979 [26], this statistical method is used to calculate second-order texture features and commonly 14 features are calculated. A modified GLCM approach proposed by Randen [30], which uses histogram-equalising, was also used in this work.

3   Tests Conducted on NSCLC Cohort

Images from 50 lung cancer patients treated with radiotherapy between 2010 and 2011 at the Edinburgh Cancer Centre, Edinburgh, UK were used in this study. Planning CT images were available for all patients and CBCT images acquired at #1 and #10. All CT images were acquired with 512×512 pixels and a pixel size of $0.977\,\mathrm{mm} \times 0.977\,\mathrm{mm}$. All CBCT images were acquired with 384×384 pixels and a pixel size of $1.172\,\mathrm{mm} \times 1.172\,\mathrm{mm}$. The number of CT and CBCT image slices containing the tumour volume varied between different patients. A radiation oncologist evaluated all images and the outlined the GTV, clinical tumour volume (CTV), planning tumour volume (PTV) and organs at risk (OAR).

Special attention was given to CBCT #10, which is the point where the patient has received approximately 30% of their treatment and where an indication of response to treatment would be extremely valuable. In Fig. 2, the planning-CT and relevant CBCT #1 and #10 for one of the patients in the dataset are shown. In Fig. 2 the GTV is shown in red and on CBCT images the estimated region of GTV is circled in red highlighting the difference in tumour volume between these two images being acquired.

3.1   Registration of CT and CBCT

Registration is a necessary step in medical imaging to allow clinicians to compare images from different time intervals and different modalities. From an image processing point of view, it is very challenging to find the transformation between different images to align them properly. In RT, radiographers capture CBCT images and record the change in the position of a patient compared to the planning-CT images. In other words, radiographers register CBCT to CT when preparing the patient for treatment. Finding this translation is vital for transferring the tumour shape defined by the oncologist on CT to CBCT before starting any treatment by making use of DICOM data.

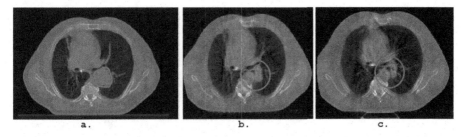

Fig. 2. Mapping of the GTV over the course of treatment: a. one slice of the planning-CT with the GTV contour shown in red, b. CBCT #1 acquired 13 days after the planning CT and before any radiotherapy is given and c. the CBCT #10, 26 days after the initial planning CT (Color figure online).

The CBCT images were registered to CT images by a radiographer resulting in a transformation matrix, which was considered as the clinical gold standard transform. In this registration three transformations take place. When CBCT images are transferred to CT coordinates they need to be transferred to the centre of the CBCT and then transferred based on image, patient position and patient orientation. Next, the radiology translation file, attached to each CBCT, can be used to find the changes between CBCT and CT to bring the image into the origin domain/space of CT. The last translation is based on embedded information in the CT images which defines the current position of the patient compared to their original position. DICOM data embedded in CBCT images contains the values of rotation and translation of the patient body for comparison to CT images.

As shown in Fig. 3, three transformation matrices, T_1, T_2 and T_3 (all of them 4×4), are the key elements required to calculate the mapping between the reference and target images. T_1 and T_2 can be calculated using Eq. 1 on DICOM information from the reference and target images respectively. T_3 is based on the information from the reference coordinate system (RCS) which are the DICOM data embedded in CBCT measured and recorded by the radiographer.

$$T_1 = \begin{bmatrix} X_x \Delta i & Y_x \Delta j & 0 & S_x \\ X_y \Delta i & Y_y \Delta j & 0 & S_y \\ X_z \Delta i & Y_z \Delta j & 0 & S_z \\ 0 & 0 & 0 & 1 \end{bmatrix} \qquad (1)$$

where S_{xyz} are the three values of the Image Position Patient (DICOM tag), which indicates the location from the origin of the RCS in mm. X_{xyz} are the first three values of the Image Orientation Patient (DICOM tag) and Y_{xyz} are the last three values of the Image Orientation Patient (DICOM tag). Δi and Δj are the pixel size, which can be extracted from Pixel Spacing (DICOM tag). After reading T_3 from an RT registration file generated by the radiographers, the transformation matrix T, which represents the gold standard in rigid registration, can be calculated using Eq. 2.

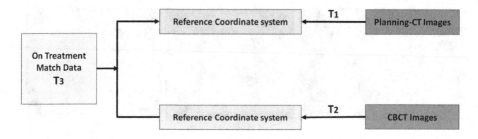

Fig. 3. Registration of CBCT images to planning-CT images is performed using the RT information determined by radiographers at time-of-treatment. The relationship between the reference image (planning-CT), the target image (CBCT) and their RCS which are embedded in the image meta information are all used to find the transformation matrix of image registration. Here T_1 is the transformation matrix from the planning-CT image to the RCS, T_2 is the transformation matrix from CBCT image to the RCS and T_3 is the transformation matrix from the RCS of the target image to the RCS of the reference image.

$$T = T_1^{-1} T_3 T_2 \qquad (2)$$

From the initial planning CT scan to radiotherapy at each fraction patients are positioned in exactly the same way. Throughout treatment patients are aligned to reference tattoos made on the skin surface, which make lateral adjustments straightforward through couch shifts. No rotational adjustment was made for the cohort of patients used here.

4   Results and Discussion

Radiomics provides a nice range of 1st order, 2nd order statistical features (descriptors of the relationships between image voxels (e.g. gray-level co-occurrence matrix (GLCM), run length matrix (RLM), size zone matrix (SZM), and neighborhood gray tone difference matrix (NGTDM) derived textures, textures extracted from filtered images), size and shape based–features, image intensity histogram feature and fractal features [31–33]. We have selected the 2nd order Haralick features in 3D lung volumes to calculate feature maps in voxel-based [34,35].

To identify the best texture features the GLCM approach was applied in four symmetric directions in a 5 × 5 region. Feature values were equalised to enable comparison and CBCT images were converted to unit 8 format. Of the fourteen texture features, several were found to be highly correlated, which would result in a low-quality input for level set convergence. The best texture feature that could be combined with Chan-Vese method was found to be the sum variance, which had the best distributed intensity levels among all texture images. Figure 4 illustrates all features followed by their histogram and the equalised histogram image of CBCT #1 and CBCT #10 respectively. The red contour refers to the

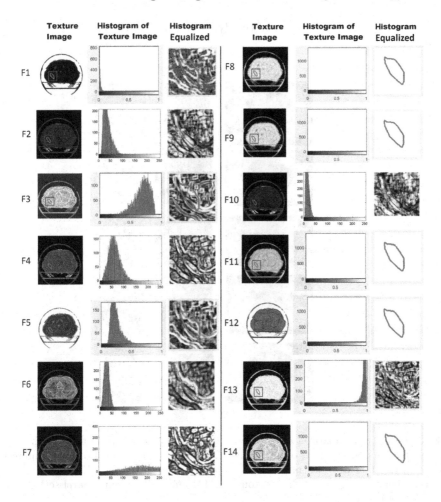

Fig. 4. Haralick texture results for one of the dataset patient's CBCT #1, F1: angular second moment, F2: contrast, F3: correlation, F4: variance, F5: inverse difference moment, F6: sum average, F7: sum variance, F8: sum Entropy, F9: entropy, F10: difference variance, F11: difference entropy, F12: information measures of correlation, F13: information measures of correlation and F14: maximal correlation coefficient.

GTV on CT. Only F7, sum variance, has the original smooth histogram and can highlight the tumour more accurately than other features for all slices of CBCT #1 and CBCT #10.

Besides demonstrating the strength of the sum variance feature, Fig. 4 also shows that many of the features are not suitable for this application. For example, features such as F12 to F14 may disrupt the performance of other features when combined with them. The main reason for this is that these texture features have a narrow range of intensity levels as shown in F1, F3, F5, F8, F9, F11, F12, F13 and F14. On the other hand F7, the sum variance, has sufficient variation in intensity and preserves more of the dynamic range of the change in intensity as shown in Fig. 4.

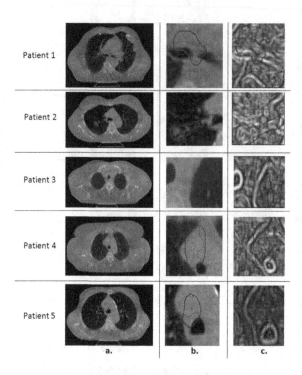

Fig. 5. Level set comparison on registered CBCT images and their textured images for five patients, a. Level set on registered CBCT image, b. cropped region of interest in each patient of previous part and c. cropped region of interest when level set on textured images is applied (blue contour is GTV and red contour is level set) (Color figure online).

Figure 5 shows examples of the CBCT images and the contours produced by Chan-Vese level set with and without being combined with textured features. Columns b show the results obtained when level set analysis was applied on the CBCT images and column c when it was applied on the sum variance of the same image. Of the fourteen features used, the image produced by the sum variance feature consistently produced the best results as shown in Fig. 5. Patient 3 has a small volume of tumour on the left lung which the level set alone was not able to detect, however, by applying the level set approach to the sum variance image the tumour was detected up. In general texture images are likely to offer a better start for the level set since the tumour cells are changing during treatment.

4.1   Assessing the Clinical Performance in the Absence of a Ground Truth

One of the challenges of this work was the lack of ground truth data on the true location of the GTV on the CBCT images. In other words, how do we know which method for automatically identifying the GTV is the best? To investigate this further and to investigate the impact of different parameterisations of the model on the final GTV contour different combinations were tested and the results presented to an experienced a radiation oncologist. The different combinations tested were:

1. Sum variance + Parallel level set (Model 1)
2. Sum variance + Chan-Vese level set (Model 2)
3. Sum variance + Li level set (Model 3)

The success of the proposed models, shown in Fig. 6, depends on three factors: initialisation of the GTV, the texture image and the choice of parameters. The level set is very much related to its parameter tuning, therefore choosing appropriate parameters was important for convergence.

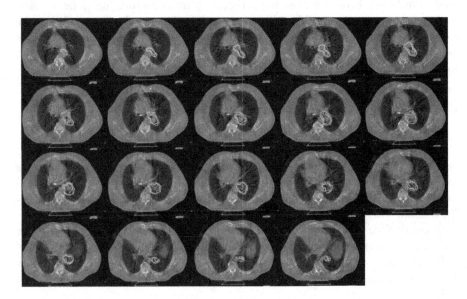

Fig. 6. Three best selected segmentations by the oncologist. Blue contour is GTV, pink refers to Model 1, red is Model 2 and yellow is Model 3 (Color figure online).

Figure 7 shows the Dice coefficient comparison between the GTV on the planning CT and the three selected contours presented to the oncologist for all fifty patients on CBCT #10. Model 2 has the highest Dice coefficient for all patients which highlights the inability of this method to move far from the

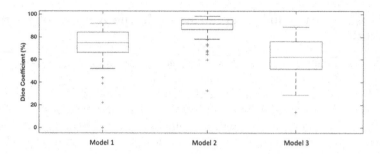

Fig. 7. The strength of proposed models on CBCT #10 by comparing the Dice coefficient for all fifty patients compared to the relevant GTV. All models were selected by the oncologist. In general, the performance of Model 2 is better than Model 1 and also Model 3.

initial point. Model 3 has the best performance with generally more than 50% similarity to the original GTV.

The volume, as well as position of the tumour, can affect the accuracy of the proposed model. Smaller tumours and more superior tumours in the chest cavity are less affected by breathing, therefore from an imaging point of view the images are less distorted by breathing motion. The oncologist noted that the proposed model can be wrong when the Dice coefficient is small since the GTV itself might not be accurate. To evaluate this further the approach was tested in non-medical images.

4.2 Accessing Performance on Non-medical Data with a Ground Truth

The performance of the proposed models was tested on non-medical images for the purpose of establishing accuracy and overall performance. The non-medical images were chosen from the level set segmentation literature which have ground truth data. The results are shown in Fig. 8 as Cha-Vese, Li and Parallel (combined Chan-Vese and Li) level set models.

The presence of ground truth on the non-medical dataset provides a better evaluation on a more comprehensive and quantitative demonstration. The results in Fig. 9 show the ability of the models (Model 1, Model 2 and Model 3) in segmentation and the improvement of the level set methods when combined with texture feature.

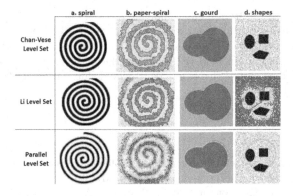

Fig. 8. Non-medical images: a. spiral, b. paper-spiral, c. gourd and d. shapes, used for the segmentation with different level set methods: Chan-Vese [17], Li [21] and parallel level set [5].

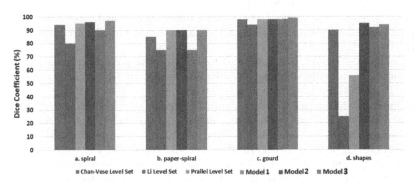

Fig. 9. Testing proposed texture+level set models (Model 1, Model 2 and Model 3) compared to the level set models without combining them with texture (Chan-Vese [17], Li [21] and parallel level set [5]), on four non-medical test images with ground truth: a. gourd, b. spiral, c. shapes and d. gradient-ball.

5   Conclusion

Assessing lung cancer patients during radiotherapy is very challenging because of the lack of contrast on CBCT. With further validation the proposed approach for automatically analysing lung tumours on CBCT images would help clinicians better estimate the size and shape of tumours during a course of radiotherapy. However, there are multiple factors that must be considered when combining texture and a level set model. First is the matter of choosing the appropriate method from each technique that can be practically combined as well as perform satisfactorily on lung CBCT images. The second challenging issue in the proposed model was initialisation, which the level set can be very dependent on. The third and most challenging issue was parameter setting for the level set which can influence its success or failure.

Acknowledgements. The authors would like to thank Dr Allan Price for the clinical validation along many fruitful discussions. Also, all members of Oncology Physics and Radiography Department at the Edinburgh Cancer Centre. We would like to thank EPSRC impact acceleration fund (EP/K503940/1) for helping support this project. RR was supported as part of the James-Watt Scholarship during her PhD research at the Heriot-Watt University.

References

1. https://www.cancerresearchuk.org/health-professional/cancer-statistics/ statistics-by-cancer-type/lung-cancer . Accessed July 3 2016
2. Grills, I.S., Yan, D., Martinez, A.A., Vicini, F.A., Wong, J.W., Kestin, L.L.: Potential for reduced toxicity and dose escalation in the treatment of inoperable non-small-cell lung cancer: a comparison of intensity-modulated radiation therapy (IMRT), 3D conformal radiation, and elective nodal irradiation. Int. J. Radiat. Oncol. Biol. Phys. **57**(3), 875–890 (2003)
3. Wen, Q., et al.: The value of CBCT-based tumor density and volume variations in prediction of early response to chemoradiation therapy in advanced NSCLC. Sci. Rep. **7**(1), 1–10 (2017)
4. Grégoire, V., Mackie, T.R.: State of the art on dose prescription, reporting and recording in intensity-modulated radiation therapy (ICRU report no. 83). Cancer/Radiothérapie **15**(6–7), 555–559 (2011)
5. Rahmat, R., Nailon, W.H., Price, A., Harris-Birtill, D., McLaughlin, S.: New level set model in follow up radiotherapy image analysis. In: Valdés Hernández, M., González-Castro, V. (eds.) MIUA 2017. CCIS, vol. 723, pp. 273–284. Springer, Cham (2017). https://doi.org/10.1007/978-3-319-60964-5_24
6. Jiang, Y., Wang, M., Xu, H.: A survey for region-based level set image segmentation. In: 2012 11th International Symposium on Distributed Computing and Applications to Business, Engineering & Science, pp. 413–416. IEEE (2012)
7. Suri, J.S., Liu, K.: Level set regularizers for shape recovery in medical images. In: Proceedings 14th IEEE Symposium on Computer-Based Medical Systems. CBMS 2001, pp. 369–374. IEEE (2001)
8. Angelini, E., Jin, Y., Laine, A.: State of the art of level set methods in segmentation and registration of medical imaging modalities. In: Suri, J.S., Wilson, D.L., Laxminarayan, S. (eds.) Handbook of Biomedical Image Analysis. Springer, Boston, USA (2005). https://doi.org/10.1007/0-306-48608-3_2
9. Burger, M., Osher, S.J.: A survey on level set methods for inverse problems and optimal design. Eur. J. Appl. Math. **16**(2), 263 (2005)
10. Tai, X.-C., Chan, T.F.: A survey on multiple level set methods with applications for identifying piecewise constant functions. Int. J. Numer. Anal. Model **1**(1), 25–47 (2004)
11. Montagnat, J., Delingette, H., Ayache, N.: A review of deformable surfaces: topology, geometry and deformation. Image Vis. Comput. **19**(14), 1023–1040 (2001)
12. Suri, J.S., Liu, K., Singh, S., Laxminarayan, S.N., Zeng, X., Reden, L.: Shape recovery algorithms using level sets in 2-D/3-D medical imagery: a state-of-the-art review. IEEE Trans. Inf. Technol. Biomed. **6**(1), 8–28 (2002)
13. Rahmat, R., Harris-Birtill, D.: Comparison of level set models in image segmentation. IET Image Process. **12**(12), 2212–2221 (2018)

14. Cremers, D., Rousson, M., Deriche, R.: A review of statistical approaches to level set segmentation: integrating color, texture, motion and shape. Int. J. Comput. Vis. **72**(2), 195–215 (2007)
15. Bhaidasna, Z.C., Mehta, S.: A review on level set method for image segmentation. Int. J. Comput. Appl. 63(11) (2013)
16. Vineetha, G., Darshan, G.: Level set method for image segmentation: a survey. IOSR J. Comput. Eng **8**(6), 74–78 (2013)
17. Chan, T., Vese, L.: An active contour model without edges. In: Nielsen, M., Johansen, P., Olsen, O.F., Weickert, J. (eds.) Scale-Space 1999. LNCS, vol. 1682, pp. 141–151. Springer, Heidelberg (1999). https://doi.org/10.1007/3-540-48236-9_13
18. Mumford, D.B., Shah, J.: Optimal approximations by piecewise smooth functions and associated variational problems. Commun. Pure Appl. Math. (1989)
19. Chan, T.F., Sandberg, B.Y., Vese, L.A.: Active contours without edges for vector-valued images. J. Vis. Commun. Image Represent. **11**(2), 130–141 (2000)
20. Vese, L.A., Chan, T.F.: A multiphase level set framework for image segmentation using the mumford and shah model. Int. J. Comput. Vis. **50**(3), 271–293 (2002)
21. Li, C., Kao, C.-Y., Gore, J.C., Ding, Z.: Minimization of region-scalable fitting energy for image segmentation. IEEE Trans. Image Process. **17**(10), 1940–1949 (2008)
22. Li, C., Kao, C.-Y., Gore, J.C., Ding, Z.: Implicit active contours driven by local binary fitting energy. In: 2007 IEEE Conference on Computer Vision and Pattern Recognition, pp. 1–7. IEEE (2007)
23. Li, C., Huang, R., Ding, Z., Gatenby, J.C., Metaxas, D.N., Gore, J.C.: A level set method for image segmentation in the presence of intensity inhomogeneities with application to MRI. IEEE Trans. Image process. **20**(7), 2007–2016 (2011)
24. Rahmat, R., Brochu, F., Li, C., Sinha, R., Price, S.J., Jena, R.: Semi-automated construction of patient individualised clinical target volumes for radiotherapy treatment of glioblastoma utilising diffusion tensor decomposition maps. Brit. J. Radiol. **93**(1108), 20190441 (2020)
25. Orlhac, F., Soussan, M., Chouahnia, K., Martinod, E., Buvat, I.: 18F-FDG PET-derived textural indices reflect tissue-specific uptake pattern in non-small cell lung cancer. PLoS One **10**(2), e0145063 (2015)
26. Haralick, R.M.: Statistical and structural approaches to texture. Proc. IEEE **67**(5), 786–804 (1979)
27. Reyes-Aldasoro, C.C., Bhalerao, A.: The Bhattacharyya space for feature selection and its application to texture segmentation. Pattern Recogn. **39**(5), 812–826 (2006)
28. Ojala, T., Pietikainen, M., Maenpaa, T.: Multiresolution gray-scale and rotation invariant texture classification with local binary patterns. IEEE Trans. Pattern Anal. Mach. Intell. **24**(7), 971–987 (2002)
29. Malpica, N., Ortuño, J.E., Santos, A.: A multichannel watershed-based algorithm for supervised texture segmentation. Pattern Recogn. Lett. **24**(9–10), 1545–1554 (2003)
30. Randen, T., Husoy, J.H.: Filtering for texture classification: a comparative study. IEEE Trans. Pattern Anal. Mach. Intell. **21**(4), 291–310 (1999)
31. Moons, K.G.M., et al.: Transparent reporting of a multivariable prediction model for individual prognosis or diagnosis (TRIPOD): explanation and elaboration. Ann. Intern. Med. **162**(1), W1–W73 (2015)
32. Vallières, M., Zwanenburg, A., Badic, B., Le Rest, C.C., Visvikis, D., Hatt, M.: Responsible radiomics research for faster clinical translation (2018)

33. Mahon, R.N., Ghita, M., Hugo, G.D., Weiss, E.: Combat harmonization for radiomic features in independent phantom and lung cancer patient computed tomography datasets. Phys. Med. Biol. **65**(1), 015010 (2020)
34. Zwanenburg, A., et al.: The image biomarker standardization initiative: standardized quantitative radiomics for high-throughput image-based phenotyping. Radiology **295**(2), 328–338 (2020)
35. Traverso, A., Wee, L., Dekker, A., Gillies, R.: Repeatability and reproducibility of radiomic features: a systematic review. Int. J. Radiat. Oncol. Biol. Phys. **102**(4), 1143–1158 (2018)

Deep Learning Classification of Cardiomegaly Using Combined Imaging and Non-imaging ICU Data

Declan Grant[1], Bartłomiej W. Papież[2], Guy Parsons[3], Lionel Tarassenko[1], and Adam Mahdi[1(✉)]

[1] Department of Engineering Science, University of Oxford, Oxford, UK
declan.grant@exeter.ox.ac.uk, {lionel.tarassenko,adam.mahdi}@eng.ox.ac.uk
[2] Big Data Institute, Li Ka Shing Centre for Health Information and Discovery,
University of Oxford, Oxford, UK
bartlomiej.papiez@bdi.ox.ac.uk
[3] Kadoorie Centre and Intensive Care Registrar, Thames Valley Deanery,
NIHR Academic Clinical Fellow at Oxford University, Oxford, UK
guy.parsons@nhs.net

Abstract. In this paper, we investigate the classification of cardiomegaly using multimodal data, combining imaging data from chest radiography with routinely collected Intensive Care Unit (ICU) data comprising vital sign values, laboratory measurements, and admission metadata. In practice a clinician would assess for the presence of cardiomegaly using a synthesis of multiple sources of data, however, prior machine learning approaches to this task have focused on chest radiographs only. We show that non-imaging ICU data can be used for cardiomegaly classification and propose a novel multimodal network trained simultaneously on both chest radiographs and ICU data. We compare the predictive power of both single-mode approaches with the joint network. We use a subset of data from the publicly available MIMIC-CXR and MIMIC-IV datasets, which contain both chest radiographs and non-imaging ICU data for the same patients. The approach from non-imaging ICU data alone achieves an AUC of 0.684 and the standard chest radiography approach an AUC of 0.840. Our joint model achieves an AUC of 0.880. We conclude that non-imaging ICU data have predictive value for cardiomegaly, and that combining chest radiographs with non-imaging ICU data has the potential to improve model performance for the same subset of patients, with further work required to demonstrate a significant improvement.

Keywords: Deep learning · Chest X-ray · Cardiomegaly · Multimodal approach

1 Introduction

Cardiomegaly is an abnormal enlargement of the heart usually indicating an underlying pathology warranting further investigation. In clinical practice, car-

© Springer Nature Switzerland AG 2021
B. W. Papież et al. (Eds.): MIUA 2021, LNCS 12722, pp. 547–558, 2021.
https://doi.org/10.1007/978-3-030-80432-9_40

diomegaly can be detected visually by examining the size of the heart on a postero-anterior chest radiograph. A standard parameter for diagnosis is the cardiothoracic ratio (CTR), which is the ratio of the cardiac width (maximum horizontal cardiac diameter) to the thoracic width (maximum horizontal distance between the inner edges of the ribs). Automated detection of cardiomegaly originally used edge detection to measure the CTR [10] but more recently there has been growing interest into deep learning approaches to the problem. This has been accelerated by the availability of large, publicly available clinical imaging datasets [12,24]. Deep learning methods focus on using convolutional neural networks to either assign a binary cardiomegaly diagnosis based on the input image [19], or to use U-Net networks to segment the heart and lungs from the image and estimate the CTR [18,23].

Until now, classification methods have been almost exclusively applied to imaging datasets (primarily radiographs), which occasionally provide some demographic data such as gender and age [12]. However, this is in contrast to how clinicians would diagnose cardiomegaly. A clinician may use a chest radiograph alone to diagnose the condition but would also consider other types of medical data, including demographics, results from a number of blood tests, other imaging information and vital sign data to assess severity and underlying pathology.

Previous approaches, which introduced multimodal network architectures, have combined medical images with basic demographics to predict the outcome of endovascular treatment from clinical metadata and imaging [20], or to classify skin lesions from dermoscopic images and patient age and sex [7]. Limited amount of work has been done in relation to cardiomegaly classification from combining imaging and non-imaging data. An example is the method proposed by [1], in which the neural network integrates the X-ray image with the limited non-imaging data (patient age, gender and acquisition type) in the classification process.

In this work, we aim to improve prediction of cardiomegaly by proposing a novel model capable of combining imaging with non-imaging Intensive Care Unit (ICU) data. To the best of our knowledge, this is the first study that combines chest radiographs and extensive non-imaging information collected during the ICU stay including vital sign values, laboratory measurements, and demographics. We use a combination of the MIMIC-IV and MIMIC-CXR, which constitutes one of the first publicly available datasets of X-rays and cardiomegaly labels with additional modalities of vital sign values, laboratory measurements, and patient metadata. We assess the relative predictive power of each modality, and then compare our joint framework with the approaches that utilize imaging or non-imaging data alone.

2   Data and Methods

2.1   Datasets

We used the publicly available MIMIC-CXR [12] and MIMIC-IV [11] datasets. Both of these databases used the same patient IDs, stays, admissions and the dates and times are consistent for each patient.

MIMIC-CXR Database. MIMIC-CXR is a large publicly available database of patient chest radiographs collected from the Beth Israel Deaconess Medical Center (BIDMC) emergency department between 2011 and 2017. It contains 227,835 X-ray studies for 64,588 patients. Each study may contain multiple images from different view positions and in total there are 377,110 radiographs. Every study also has an associated free-text radiology report, written at the time of the study. Here, we used MIMIC-CXR in JPG format provided with MIMIC-CXR-JPG [13] as it additionally contains structured labels derived from these reports.

MIMIC-IV Database. MIMIC-IV contains data from hospital stays for patients who were admitted to the BIDMC between 2008 and 2019. MIMIC-IV is separated into five modules: core (patient stay information), hospital (laboratories and microbiology), ICU data (ICU stays and events), emergency department and CXR (lookup tables to allow linking to MIMIC-CXR).

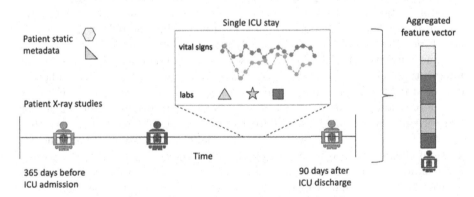

Fig. 1. Aggregating feature vectors from multimodal ICU data (see Sect. 2.3). We built a consistent, high-quality dataset, which combines imaging and non-imaging data including patient static metadata (e.g. age, gender, ethnicity, length of stay) as well as laboratory values (e.g. glucose, magnesium, hemoglobin) and vital sign values (e.g. heart rate, respiration rate, oxygen saturation).

2.2   Preprocessing

Cardiomegaly was detected using posterior-anterior radiographs, to prevent the artificial enlargement of the cardiac silhouette that can occur with antero-posterior and other radiograph views. We therefore filtered the image datasets to only include posterior-anterior views. MIMIC-CXR-JPG provides four types of cardiomegaly labels (positive, negative, uncertain, no mention) derived using two natural language processing tools NegBio [15] and CheXpert [9]. We removed any images where these tools disagreed with each other over the label assigned, to prevent introducing error. We note that we cannot conclude that a report with no mention of cardiomegaly indicates the absence of cardiomegaly. There-fore, we created a subset of images, which only have either a positive or negative label for cardiomegaly. These images were then re-sized to 244 × 244 pixels and underwent normalisation of the mean and standard deviation. We also performed standard data augmentation consisting of a random rotation of up to 10 degrees as well as random horizontal and vertical flips to improve robustness.

We grouped vital sign values (e.g. heart rate, respiration rate, oxygen sat-uration) and laboratory values (e.g. glucose, magnesium, hemoglobin) for each ICU admission, before averaging them across the stay to produce a single set of summary features. We then added patient metadata (e.g. age, gender, ethnicity, length of stay) to each feature vector.

2.3   Combining Imaging and Non-imaging Data

There is a unique patient ID across MIMIC-CXR-JPG and MIMIC-IV. We linked ICU stays to radiographic studies, which took place up to 365 d before the patient entered the ICU and upto 90 days after leaving the ICU. Cardiomegaly is a chronic condition that usually slowly progresses if the underlying pathology remains untreated, and it does not resolve once present without considerable intervention. This period was therefore identified as providing a reasonable win-dow of stability in the condition for assessment in our modelling [5]. For each unique ICU stay, we first collected the studies from our images subset which took place within the specified window for that same patient (See Fig. 1). We then checked for consistency between the image labels. We assumed cardiomegaly does not change from positive to negative within the chosen window. If a patient has conflicting (both positive and negative) image labels within this window, we cannot be confident which is the correct label to assign to the ICU stay, so we remove ICU stays surrounded by conflicting cardiomegaly labels. If the labels show consistency, we link the ICU stay and the image study which took place closest to ICU admission. While the labels generated from free-text reports [9,15] can be inconsistent, the presented data curation process led us to create a high-quality dataset (both imaging and non-imaging), which was later used to train, validate, and test the models presented in the next section.

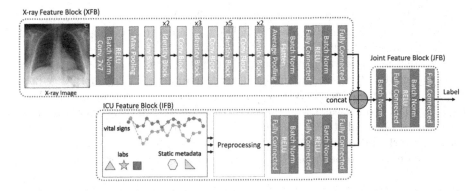

Fig. 2. Our approach for combining X-ray imaging and non-imaging ICU data.

2.4 Models

Image Classification. We implemented a ResNet [8] architecture for classifying cardiomegaly presence in X-ray images. ResNet architectures have been evaluated on similar radiograph databases and shown to achieve state-of-the-art results for this task. For example, ResNet-32 achieved 0.84 cardiomegaly accuracy [2] on the CheXpert database [9], which also uses labels automatically derived from the reports. All networks were pre-trained on ImageNet [6]. We used a binary cross entropy loss function, Adam optimizer [14] and cyclical learning rates [22]. Each image-based model was trained in two stages. First, we froze the convolutional layers of the ResNet and trained just the final fully connected layers. Then we unfroze all of the layers and continued training. Before each stage we conducted an initial learning rate test and determined optimal boundaries for the learning rate to cycle in during training [22].

ICU Data Classification. We implemented XGBoost [4] to classify cardiomegaly from ICU data consisting of vital sign values, laboratory measurements and demographics. XGBoost has been shown to perform well at similar classification tasks in clinical machine learning [3,16,17]. We use a binary cross entropy loss function.

Multimodal Classification. We propose a multimodal network for the classification of cardiomegaly (see Fig. 2), which combines simultaneously the imaging data (chest radiographs) and non-imaging ICU data (vital sign values, laboratory values, static patient metadata including demographics). We use a ResNet-50 architecture in the X-ray Feature Extraction Block (XFB) to extract relevant features from the images. The ICU data is fed into the ICU Feature Block (IFB), which contains a neural network consisting of three fully connected layers. To join the learned image and ICU features, we concatenate the outputs of the XFB (32 nodes) and IFB (16 nodes) using a fully connected layer. After concatenation, we add two more fully connected layers to produce a label. Previous approaches have used similar network architectures to combine medical imaging with small amounts of patient metadata [1,7,20], however, here we extend on this to include all routinely collected ICU data including vital sign values and laboratory measurements.

The joint network was trained similarly to the ResNet model, using a binary cross entropy loss function, Adam optimizer [14] and cyclical learning rates [22]. We also pre-trained the XFB and IFB separately before concatenating them together. The model was then trained in two stages. First, we froze the early layers and trained only the fully connected layers of the JFB. Then we unfroze the entire network and continued training. Optimal bounds for the cyclical learning rate were determined before each stage from a learning rate test.

Table 1. Patient characteristics.

Variable	Cardiomegaly	All patients
ICU Stays	1,795	2,571
Male/Female	756/904	1077/1327
Age [years], mean (SD)	66.0 (14.9)	64.9 (15.6)
Ethnicity		
White, N (%)	1,084 (65.3)	1,601 (66.9)
Asian, N (%)	58 (3.5)	93 (3.9)
Black, N (%)	309 (18.6)	390 (16.3)
Hispanic/Latino, N (%)	94 (5.7)	127 (5.3)
Other/Mixed/Unknown, N (%)	78 (10.2)	183 (7.6)
LOS [days], median (IQR)	2.1 (2.5)	2.0 (2.4)

3   Results

The original MIMIC-CXR contains 227,835 studies (377,110 radiographs) for 64,588 patients including 53,565 X-ray studies, with positive or negative cardiomegaly labels, for 22,914 patients. After applying the inclusion and exclusion criteria described in Sect. 2.2, the dataset used in this work contains 2,571 ICU stays and 2,404 patients. A brief overview of the patient characteristics including Sex, Age, Ethnicity and length of stay (LOS) is given in Table 1.

We divided our curated dataset (see Sect. 2.2) into 5-folds for cross-validation. No patient appears in more than one fold and every fold has the same ratio of positive to negative labels. Each model was trained on the same 5-folds but only made use of the relevant modalities: ResNet (Images), XGBoost (ICU data) and the joint network (Images and non-imaging ICU data). The results of classification averaged across 5-folds are summarised in Table 2 using Accuracy, F1-score and AUC (area under the receiver operating characteristic curve).

XGBoost has a number of hyperparameters which we optimised through a grid search within the ranges defined in Table 3. The results of 5-fold cross validation for image-based ResNet-50 approach showed stronger predictive value (AUC 0.840) than the non-imaging ICU data-based XGBoost model (AUC 0.684) with the multimodal approach showed the strongest predictive skill (AUC 0.880). For comparison we showed the results for a model using demographics/metadata only (AUC 0.647) and laboratory and vital sign values only (AUC 0.671).

Table 2. Results for the three classes of models considered in this work on the combination of demographics/metadata (D), laboratory and vital sign values (LV) and imaging data (I). The multimodal network, which merges imaging and non-imaging data, has the strongest predictive power showing the overall moderate improvement over ResNet-50. The values show the mean Accuracy, F1-score and AUC (the area under the ROC curve) for the 5-fold cross validation with the corresponding standard deviation.

Model	Data type	Accuracy	F1-score	AUC
Multimodal	D+LV+I	0.837 (0.012)	0.886 (0.009)	0.880 (0.011)
ResNet-50	I	0.797 (0.034)	0.857 (0.028)	0.840 (0.033)
XGBoost	D+LV	0.700 (0.084)	0.771 (0.029)	0.684 (0.046)
XGBoost	LV	0.694 (0.020)	0.769 (0.017)	0.671 (0.036)
XGBoost	D	0.642 (0.030)	0.712 (0.031)	0.647 (0.021)

Table 3. The hyperparameters for each XGBoost model were optimised within the ranges shown below.

Hyperparameters	Range
Learning rate	[0.0001, 0.1]
Max tree depth	[2, 8]
Gamma	[0, 2]
Colsample	[0, 1]
Subsample	[0, 1]

Examples of radiographs with correct and incorrect classification are given in Fig. 3. Finally, the heat maps from the image-based ResNet-50 model for true positive and true negative instances of cardiomegaly are shown in Fig. 4.

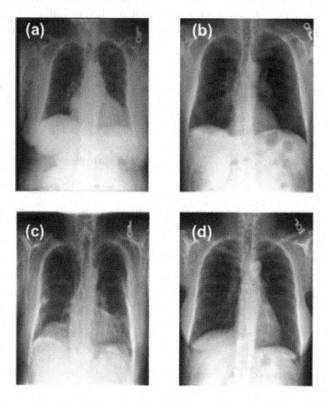

Fig. 3. Examples of X-ray images showing correct and incorrect classification by our proposed joint model: (a) true positive, (b) false negative (clinical review of a radiograph suggests cardiomegaly, based on pattern of cardiac silhouette), (c) false positive (clinical review of a radiograph suggests anatomical distortion by other pathology), (d) true negative.

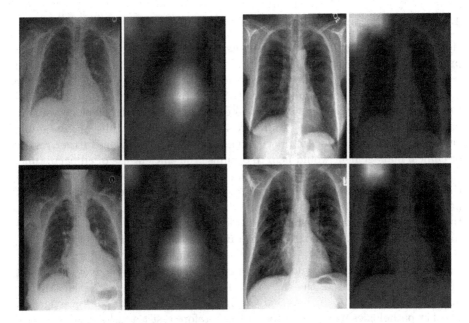

Fig. 4. Examples of heat maps from the image-based ResNet-50 model for true positive (left two columns) and true negative (right two columns) instances of cardiomegaly. The heat maps were created using a Grad-Cam [21] approach and show which areas of the X-ray image were important during prediction.

4   Discussion

4.1   Principal Findings

In this work, we proposed a multimodal network which is able to use both images and ICU data to determine the presence of cardiomegaly. Further it provides a proof-of-concept for a multi-modal deep network approach for detecting other cardiopulmonary diseases, in general, from a combination of images, vital sign values, laboratory measurements and other patient metadata. We used two recently published datasets to uniquely combine various modalities for the same patients. This allowed us to build a multimodal approach with the aim to imitate the thought process of a clinician. Clinicians tend to use multiple sources of data to draw their conclusions, unlike most machine learning approaches which have relied exclusively on chest radiographs. Finally, as cardiomegaly usually indicates an underlying pathology that warrants further investigation and clinical management, and can convey significant ill health if unrecognised, the early detection of cardiomegaly is beneficial. The automatic detection of cardiomegaly on radiographs taken for the investigation of other pathologies using these tools could therefore assist as a form of preventative healthcare.

4.2 Strengths and Weaknesses of the Study

Advantages of Using Automated Systems for X-ray Labelling. Current clinical practice is for the requesting clinician or responsible radiologist to review and report the chest radiograph performed. Radiologist availability and clinician time is often a limited resource and tools with the potential to optimise the application of this limited resource have the potential to increase the efficiency of care pathways. A tool which can accurately label chest radiographs as containing either no elements of concern or with relevant labels warranting further review or clinical management would help to optimise the efficient use of clinician time.

Further, when a clinician/radiologist reviews a chest radiograph a potential cognitive bias can occur whereby once a pathology is identified secondary pathologies are not as readily recognised. Automated tools are not subject to this bias and so the use of such tools could help prevent the missed detection of further pathologies on radiographs with multiple elements of concern.

Finally, as cardiomegaly usually indicates an underlying pathology that warrants further investigation and clinical management, and can convey significant ill health if unrecognised, the early detection of cardiomegaly is beneficial. The automatic detection of cardiomegaly on radiographs taken for the investigation of other pathologies using these tools can therefore assist as a form of preventative healthcare.

Imitating an ICU Doctor. We use a unique, recently published dataset that combines various modalities for the same patients. This allowed us to build a multimodal approach with the aim to imitate the thought process of a clinician. Clinicians tend to use multiple sources of data to draw their conclusions, unlike most machine learning approaches which have relied exclusively on chest radiographs.

Automatic Labelling Accuracy. The labels for our data are automatically generated from free-text reports [9,15]. It is well known that these types of automatic procedures result in noisy data, which may affect the estimation of the performance of the models [13].

Time Between X-rays and ICU Data. Although we choose the closest X-ray to each ICU admission the period of time in between is not a constant value. Some X-rays and associated ICU stays will take place closer together in time than other X-rays and ICU stays. This introduces inconsistency into our dataset.

Label Bias. The cardiomegaly labels originate from free-text reports derived from X-ray images. This may bias the estimated predictive power in favour of images over signals.

Time Series Data Averaged. As part of the preprocessing we averaged the ICU time-dependent data (e.g. vital signs, laboratory measurement) into summary vectors over the whole stay in order to construct our joint network.

Acknowledgements. GP is supported by an NIHR fellowship. AM and LT are supported by the NIHR Oxford Biomedical Research Centre (BRC). The views expressed are those of the author(s) and not necessarily those of the NHS, the NIHR or the Department of Health. BWP acknowledges Rutherford Fund at Health Data Research UK (HDR UK) and Nuffield Department of Population Health (NDPH) Senior Research Fellowship.

References

1. Baltruschat, I.M., Nickisch, H., Grass, M., Knopp, T., Saalbach, A.: Comparison of deep learning approaches for multi-label chest X-ray classification. Sci. Rep. **9**(1), 1–10 (2019). https://doi.org/10.1038/s41598-019-42294-8

2. Bressem, K.K., Adams, L.C., Erxleben, C., Hamm, B., Niehues, S.M., Vahldiek, J.L.: Comparing different deep learning architectures for classification of chest radiographs. Sci. Rep. **10**(1), 13590 (2020). https://doi.org/10.1038/s41598-020-70479-z

3. Chang, W., et al.: A machine-learning-based prediction method for hypertension outcomes based on medical data. Diagnostics **9**(4), 178 (2019). https://doi.org/10.3390/diagnostics9040178

4. Chen, T., Guestrin, C.: XGBoost: a scalable tree boosting system. In: Proceedings of the 22nd ACM SIGKDD International Conference on Knowledge Discovery and Data Mining. ACM, New York, NY, USA

5. Davies, M.J.: The cardiomyopathies: an overview. Heart **83**(4), 469 (2000). https://doi.org/10.1136/heart.83.4.469

6. Deng, J., Dong, W., Socher, R., Li, L.J., Li, K., Fei-Fei, L.: ImageNet: a large-scale hierarchical image database. In: Institute of Electrical and Electronics Engineers, pp. 248–255. (IEEE) (2010). https://doi.org/10.1109/cvpr.2009.5206848

7. Gessert, N., Nielsen, M., Shaikh, M., Werner, R., Schlaefer, A.: Skin lesion classification using ensembles of multi-resolution Efficient Nets with metadata. MethodsX **7**, 100864 (2020). https://doi.org/10.1016/j.mex.2020.100864

8. He, K., Zhang, X., Ren, S., Sun, J.: Deep residual learning for image recognition. In: Proceedings of the IEEE Computer Society Conference on Computer Vision and Pattern Recognition. vol. 2016, pp. 770–778. IEEE Computer Society (2016). https://doi.org/10.1109/CVPR.2016.90

9. Irvin, J., et al.: CheXpert: a large chest radiograph dataset with uncertainty labels and expert comparison. In: 33rd AAAI Conference on Artificial Intelligence, AAAI 2019, 31st Innovative Applications of Artificial Intelligence Conference, IAAI 2019 and the 9th AAAI Symposium on Educational Advances in Artificial Intelligence, EAAI 2019. vol. 33, pp. 590–597. AAAI Press (2019). https://doi.org/10.1609/aaai.v33i01.3301590, www.aaai.org

10. Ishida, T., Katsuragawa, S., Chida, K., MacMahon, H., Doi, K.: Computer-aided diagnosis for detection of cardiomegaly in digital chest radiographs. In: Medical Imaging 2005: Image Processing. vol. 5747, p. 914. SPIE (2005). https://doi.org/10.1117/12.595755

11. Johnson, A., Bulgarelli, L., Pollard, T., Horng, S., Celi, L.A., Mark, R.: MIMIC-IV v0.4. Technical Report (2020). https://physionet.org/content/mimiciv/0.4/

12. Johnson, A.E.W., et al.: MIMIC-CXR, a de-identified publicly available database of chest radiographs with free-text reports. Sci. Data **6**(1), 1–8 (2019). https://doi.org/10.1038/s41597-019-0322-0

13. Johnson, A.E.W., et al.: MIMIC-CXR-JPG, a large publicly available database of labeled chest radiographs. arXiv (2019)

14. Kingma, D.P., Ba, J.L.: Adam: a method for stochastic optimization. In: 3rd International Conference on Learning Representations, ICLR 2015 - Conference Track Proceedings (2015)

15. Peng, Y., Wang, X., Lu, L., Bagheri, M., Summers, R., Lu, Z.: NegBio: a high-performance tool for negation and uncertainty detection in radiology reports (2017)

16. Pimentel, M., Mahdi, A., Redfern, O., Santos, M., Tarassenko, L.: Uncertainty-aware model for reliable prediction of sepsis in the ICU. In: 2019 Computing in Cardiology Conference (CinC). vol. 45 (2019). https://doi.org/10.22489/cinc.2019.209

17. Pimentel, M.A.F., et al.: Detecting deteriorating patients in hospital: development and validation of a novel scoring system. Am. J. Respir. Crit. Care Med. (2021). https://doi.org/10.1164/rccm.202007-2700oc

18. Que, Q., et al.: CardioXNet: automated detection for cardiomegaly based on deep learning. In: Proceedings of the Annual International Conference of the IEEE Engineering in Medicine and Biology Society, EMBS. vol. 2018, pp. 612–615. Institute of Electrical and Electronics Engineers Inc. (2018). https://doi.org/10.1109/EMBC.2018.8512374

19. Rajpurkar, P., et al.: CheXNet: radiologist-level pneumonia detection on chest X-rays with deep learning. arXiv (2017). http://arxiv.org/abs/1711.05225

20. Samak, Z.A., Clatworthy, P., Mirmehdi, M.: Prediction of thrombectomy functional outcomes using multimodal data. In: Papież, B.W., Namburete, A.I.L., Yaqub, M., Noble, J.A. (eds.) MIUA 2020. CCIS, vol. 1248, pp. 267–279. Springer, Cham (2020). https://doi.org/10.1007/978-3-030-52791-4_21

21. Selvaraju, R.R., Cogswell, M., Das, A., Vedantam, R., Parikh, D., Batra, D.: Grad-cam: Visual explanations from deep networks via gradient-based localization. In: Proceedings of the IEEE international conference on computer vision. pp. 618–626 (2017)

22. Smith, L.N.: Cyclical learning rates for training neural networks. In: Proceedings - 2017 IEEE Winter Conference on Applications of Computer Vision, WACV 2017, pp. 464–472. Institute of Electrical and Electronics Engineers Inc. (2017). https://doi.org/10.1109/WACV.2017.58

23. Sogancioglu, E., Murphy, K., Calli, E., Scholten, E.T., Schalekamp, S., Van Ginneken, B.: Cardiomegaly detection on chest radiographs: segmentation versus classification. IEEE Access 8, 94631–94642 (2020). https://doi.org/10.1109/ACCESS.2020.2995567

24. Wang, X., Peng, Y., Lu, L., Lu, Z., Bagheri, M., Summers, R.M.: ChestX-ray8: hospital-scale chest X-ray database and benchmarks on weakly-supervised classification and localization of common thorax diseases. In: Proceedings of the 30th IEEE Conference on Computer Vision and Pattern Recognition, CVPR 2017. vol. 2017, pp. 3462–3471. Institute of Electrical and Electronics Engineers Inc. (2017). https://doi.org/10.1109/CVPR.2017.369

Author Index

Printed in the United States
by Baker & Taylor Publisher Services